Topics in Applied Physics
Volume 85

Springer-Verlag Berlin Heidelberg GmbH

Physics and Astronomy ONLINE LIBRARY

http://www.springer.de/phys/

Topics in Applied Physics

Topics in Applied Physics is a well-established series of review books, each of which presents a comprehensive survey of a selected topic within the broad area of applied physics. Edited and written by leading research scientists in the field concerned, each volume contains review contributions covering the various aspects of the topic. Together these provide an overview of the state of the art in the respective field, extending from an introduction to the subject right up to the frontiers of contemporary research.
Topics in Applied Physics is addressed to all scientists at universities and in industry who wish to obtain an overview and to keep abreast of advances in applied physics. The series also provides easy but comprehensive access to the fields for newcomers starting research.
Contributions are specially commissioned. The Managing Editors are open to any suggestions for topics coming from the community of applied physicists no matter what the field and encourage prospective editors to approach them with ideas.
See also: http://www.springer.de/phys/series/TAP

Managing Editors

Dr. Claus E. Ascheron

Springer-Verlag Heidelberg
Topics in Applied Physics
Tiergartenstr. 17
69121 Heidelberg
Germany
ascheron@springer.de

Dr. Hans J. Kölsch

Springer-Verlag Heidelberg
Topics in Applied Physics
Tiergartenstr. 17
69121 Heidelberg
Germany
koelsch@springer.de

Assistant Editor

Dr. Werner Skolaut

Springer-Verlag Heidelberg
Topics in Applied Physics
Tiergartenstr. 17
69121 Heidelberg
Germany
skolaut@springer.de

K. Wandelt S. Thurgate (Eds.)

Solid–Liquid Interfaces

Macroscopic Phenomena – Microscopic Understanding

With 228 Figures

Springer

Prof. Dr. Klaus Wandelt
Universität Bonn
Institut für Physikalische Chemie
Wegelerstr. 12
53115 Bonn
Germany
k.wandelt@uni-bonn.de

Prof. Dr. Stephe Thurgate
Murdoch University
School of PSET
South Sreet
Murdoch, 6150 WA
Australia
thurgate@central.murdoch.edu.au

Library of Congress Cataloging-in-Publication Data.

Solid-liquid interfaces: macroscopic understanding / K. Wandelt, S. Thurgate (eds.). p.cm -(Topics in applied physics, ISSN 0303-4216; v.85) Includes bibliographical references and index.
ISBN 978-3-662-30781-6 ISBN 978-3-540-44817-4 (eBook)
DOI 10.1007/978-3-540-44817-4
1. Solid-liquid interfaces. I. Wandelt, K. (Klaus), 1944- II. Thurgate S. (Stephe), 1952- III. Series. QD509.S65 S5527 2003 530.4'17-dc21 2002022722

Physics and Astronomy Classification Scheme (PACS):
68, 68.08, 68.37, 68.43, 68.49, 73, 75, 75.70, 75.80, 78.30

ISSN print edition: 0303-4216
ISSN electronic edition: 1437-0859
ISBN 978-3-662-30781-6

http://www.springer.de

Originally published by Springer-Verlag Berlin Heidelberg New York in 2003
Softcover reprint of the hardcover 1st edition 2003

Typesetting: DA-TEX Gerd Blumenstein, Leipzig
Cover design: *design & production* GmbH, Heidelberg

Printed on acid-free paper SPIN: 11017974 57/3111/mf - 5 4 3 2

Preface

Processes at interfaces play a key role in our world in general, and in many modern technologies in particular. Phenomena like adsorption, adhesion, crystallisation, evaporation, dissolution, corrosion, film growth, heterogeneous catalysis etc. are of fundamental importance in areas like materials science, solid state physics, physical chemistry, electrochemistry, mineralogy etc. Interfaces exist between any two phases which may be solid, liquid or gaseous. An understanding of interfacial processes is based on a detailed knowledge of the properties of the respective interface. In particular *Surface Science* has been extremely successful in the advancement of our understanding of properties and processes at free surfaces under UltraHigh Vacuum (UHV) conditions. This success has its origin in the development of a plethora of highly sensitive analytical methods which are applicable under UHV conditions like, for instance, all spectroscopies based on electron, ion or atom beams. As a result we know by now that the properties at the surface differ from those in the bulk of the same material in almost every respect. In this sense surfaces have even been specified as "a new state of matter", and it is obvious that their properties need to be know in detail before physical and chemical interactions and processes with and at surfaces like, e.g. in tribology, adsorption, corrosion, catalysis etc. can be understood and optimized.

Inasmuch as the ultrahigh vacuum condition was a fundamental prerequisite for this progress in *Surface Science*, it also implies a serious limitation, because most of the investigations in UHV have "only" model character. In reality the processes mentioned above take place at interfaces between more or less dense phases. It is rather unlikely that the results obtained under UHV conditions straightforwardly apply also to interfaces between such phases.

Some of the largest-scale industrial processes such as flotation, metal recovery and refinement, galvanization, electrolysis, passivation etc. are based on phenomena at solid–liquid interfaces. Macroscopically these phenomena are well described. Their characterization on the atomic and molecular level, however, constitutes a new dimension of complexity and, thus, scientific challenge, because the concealment of the interface between two condensed phases excludes the application of most of the established *Surface Science* techniques.

There are two possible ways to investigate the properties of solid–liquid interfaces, namely the in-situ and the ex-situ approach. The in-situ approach seeks for spectroscopic and microscopic information directly from the solid–liquid interface and, hence, requires experimental probes which permeate either one (or both) of the two condensed phases, mostly the liquid one. It is only since recently that some in-situ techniques are available which fulfil this requirement. These new methods are primarily based on photon beams and scanning probe techniques and start to give insight in the properties and processes at solid–liquid interfaces with the same precision as it is current standard in UHV-based surface science. The number of these techniques, however, is still limited, and in particular in-situ information about the concentration and spatial distribution of chemical species is difficult to obtain. Hence, there is still a need to rely on the large variety of UHV-based surface techniques, which implies an emersion of the solid from the liquid phase and its transfer into a UHV-chamber. Such ex-situ investigations of the solid surface, however, are afflicted by the uncertainty that the properties of the surface may have changed upon transfer. It is therefore advisable to complement ex-situ investigations by measurements with at least one method which works in both ambients. Of course, only if this latter method yields the same results in the liquid and in UHV confidence into the information given by the other ex-situ techniques is more justified.

The combined application of in-situ and ex-situ experimental methods has, indeed, led to a tremendous progress in the microscopic and spectroscopic characterization of solid–liquid interfaces. The present book gives an account of state-of-the-art investigations of properties and processes at selected examples of solid–liquid interfaces which are technologically relevant. Thirteen chapters are arranged into the four parts: "Adsorption in Processing Technologies", "Anion Adsorption", "Electrochemical Surface Modification" and "Interface Reactions".

Flotation is probably the technological process with the highest throughput (by mass) in which control of the chemistry, physics and engineering at solid–liquid interfaces determines the economic profit. This control aims at the design of hydrophobic solid–liquid interfaces for bubble attachment only on the surface of the valuable phase, i.e. sulphide particles. This chemical design requires detailed surface information in order to optimize the surface treatment and modification procedure, namely surface oxidation and the addition of "collector molecules". The first contribution by *Smart* and co-workers gives a very complete review of the oxidation and collector adsorption at sulphide mineral surfaces using ex-situ scanning and imaging surface analytical techniques. The complementary chapter by *Buckley, Hope* and *Woods* concentrates on the adsorption of organic reagents in the processing of sulphide minerals, and the next one by *Mayer* and co-workers specifically addresses the use of synchrotron radiation to the investigation of chemical binding, molecular orientation and oxidation of organic adsorbates on sulphide surfaces.

Another large scale and economically important process is the recovery of gold from its ores. New technologies have been developed over the past three decades which enable a profitable recovery of gold from previously uneconomic waste. These new technologies rely on dissolution and adsorption processes from complex slurries of crushed ores. While they have been optimised to a high level of efficiency, a microscopic understanding is still lacking. *Poinen* and *Thurgate* present a detailed in-situ study on the adsorption of the aurocyanide ion complex $(Au(CN)_2)^-$ on activated carbon using Scanning Tunnelling Microscopy (STM). This process is actually used to extract gold from low-grade oxide ores and is based on the fact that the gold complex binds very strongly to carbon, even though the concentration of many other ions and molecules in the solution is higher by several orders of magnitude.

An understanding of all electrochemical reactions at or with solid surfaces requires detailed information on the inter-facial structure and on its variation as a function of the applied potential and the adsorption of ions and molecules. In recent years substantial progress has been made in the elucidation of the so-called specific adsorption of anions on electrode surfaces, the most fundamental and ubiquitous electrochemical process. *Broekmann* et al. present state-of-the-art in-situ STM studies on the structure and morphology of a Cu(111) electrode in dilute sulphuric acid solution using a newly designed electrochemical STM. This new instrument enables a direct correlation of cyclovoltametric measurements and high-resolution STM images monitored in three different detection modes, namely potentiostatically, potentiodynamically and quasi-spectroscopically. The whole set of surface structures is presented and discussed covering the full range of the cyclic voltammogram from the cathodic regime of hydrogen evolution over the sulphate adsorption and desorption to the anodic copper dissolution and redeposition. A very interesting sulphate-induced restructuring of the Cu(111) surface is found and adsorbed hydronium superstructures are discovered. The interest in the detailed properties of Cu surfaces in contact with electrolytes has strongly increased since electrochemical Cu deposition is being used in modern chip production.

Stuhlmann et al. describe similar STM studies combined with ex-situ electron- and ion-spectroscopic measurements on the interaction of chloride ions with a Cu(111) electrode. The combined in-situ and ex-situ results not only provide a complete picture of the chloride adsorption and desorption but also shine light on a chloride-induced enhancement of the kinetics of the hydrogen evolution reaction and, thereby, lay the basis for a simple model for the simulation of the corresponding voltametric behavior.

As important as information about the atomic and molecular structure at the interfaces are data about their electronic and vibrational properties. For this purpose, a series of modern in-situ optical spectroscopies is used, such as InfraRed (IR) and Fourier Transform InfraRed (FTIR) spectroscopy as well as Raman and Surface-Enhanced Raman Spectroscopy (SERS). These vibrational spectroscopies are powerful tools to identify adsorbates and intermediates and to study their interaction with the substrate. Even more suitable are nonlinear optical spectroscopies of second order because they exhibit an inherent interface sensitivity and yield a radiation which is easily separable in space, frequency and time. *Pettinger, Bilger* and *Lipkowski* give a detailed description of the recently developed method of Interference Second-Harmonic Generation Anisotropy (ISHGA) and demonstrates its capabilities with results on halide ion adsorption on Au(111) electrodes.

One important outcome of *Surface Science* is certainly the controlled growth and preparation of low-dimensional structures such as ultrathin films and clusters under UHV conditions. As mentioned before, however, electrodeposition and electrochemical structure formation is becoming an attractive alternative; electrochemical deposition and processing can be performed under cleanliness conditions equivalent to those in ultrahigh vacuum. And the fact that these electrochemical processes are conducted close to equilibrium promises an even better control. The part "Electrochemical Surface Modification" includes three representative examples.

Schindler demonstrates that ultrathin technologically important magnetic films of Co and Fe in the thickness range of a few atomic monolayers can be prepared by electrodeposition from an aqueous electrolyte. Using in-situ measurements of the Magneto-Optical Kerr Effect (MOKE) and in-situ surface x-ray diffraction it is shown that these films have the same high magnetic and structural quality as Co and Fe films grown by physical vapor deposition in UHV.

The contribution by *Dretschkow* and *Wandlowski* deals with the in-situ structural characterisation of organic adsorbate layers on the molecular scale using in-situ STM in combination with conventional electrochemical methods. Such organic films play an important role as corrosion inhibitors and as surfactants in plating processes. The relatively simple control of their chemical properties and, thus, functionality makes them, among others, important electronic and optical materials, additives in electrocatalysis, and biological and chemical sensors. The selected examples presented in this chapter, once again, demonstrate the unique capability of STM to study molecular structures and structural transitions also in solution.

2D and 3D arrangements of nanometer particles have become very interesting research objects, both theoretically and experimentally, since techniques are available to produce them. *Xia, Nagle* and *Schuster* describe a procedure how small Au islands can be prepared within a Cu monolayer on a Au(111) electrode by merely controlling the electrochemical potential.

The final part is devoted to the detection and modification of chemical reactions at interfaces. *Wittstock* describes how Scanning ElectroChemical Microscopy (SECM) may contribute to a better understanding of macroscopic phenomena at heterogeneous electrode surfaces by enabling a microscopic view on reactivities at the solid–liquid interface. Using so-called ultramicroelectrodes it is possible to map out local electrochemical reactivities, to induce localised electrochemical surface modifications, or to investigate heterogeneous and homogeneous reaction kinetics with micrometer spatial resolution.

Electrochemical processes are normally driven by the potential difference at the solid–liquid interface. This potential difference influences parameters like adsorption rate, surface concentration and adsorbate structure of educts, intermediates and products on the electrode surface. These parameters may be influenced by the injection of "hot" electrons into the solid–liquid interface. "Hot" electrons may, thus, stimulate interfacial reactions at potentials different from the equilibrium potential or may overcome a kinetic hindrance of a reaction due to an overpotential. Hence, "hot" electrons may modify and reveal reaction paths and kinetics. A practical application of "hot" electron injection in solid–liquid interfaces may lead to the design of electrochemical redox sensors for the detection of oxidizable or reducible species in solutions. *Diesing, Kritzler* and *Otto* describe the physics of "hot" electron production at Ag electrodes and address a few examples of their influence on electrode reactions.

The final contribution by *O'Connor* provides a short introduction into low energy negative recoil spectroscopy, an ex-situ technique which is particularly sensitive to adsorbed species like hydrogen and oxygen, common constituents in electrochemical double layers and passivation products. This spectroscopy is a very valuable variation of Low-Energy Ion Scattering (LEIS) described in the contribution by *Stuhlmann* et al.

Solid–liquid interfaces will become an increasingly important area of research due to both their scientific challenge and undoubtedly growing technological relevance. The present book describes a number of techniques and experimental approaches and their application to a few selected examples of solid–liquid interfaces. On the one hand it is meant to give an overview over the state of the art in some areas, on the other hand it aims at stimulating an expansion of this type of *Interface Science.*

Bonn, Murdoch
August 2002

Klaus Wandelt
Steve Thurgate

Contents

Part I. Adsorption in Processing Technologies

Part I

Adsorption in Processing Technologies

Surface Analytical Studies of Oxidation and Collector Adsorption in Sulfide Mineral Flotation

Roger S. C. Smart, John Amarantidis, William M. Skinner,
Clive A. Prestidge, Lori La Vanier, and Stephen R. Grano

Ian Wark Research Institute, The Commonwealth ARC Special Research Centre
in Particle and Material, University of South Australia
The Levels Campus, Mawson Lakes, South Australia 5095, Australia
roger.smart@unisa.edu.au

Abstract. The physical and chemical forms of sulfide mineral surfaces are reviewed. The initial surfaces and oxidation products have been studied by Scanning Auger Microscopy (SAM), X-ray Photoelectron Spectroscopy (XPS), Scanning Tunneling Microscopy (STM), Atomic Force Microscopy (AFM), Scanning Electron Microscopy (SEM) and Time-of-Flight Secondary Ion Mass Spectrometry (ToF-SIMS). Changes to surface speciation as a function of time, pH, Eh and collector adsorption, related to mineral flotation, have been followed with these techniques. Oxidation products are formed in different processes, namely: metal-deficient sulfides, polysulfides and sulfur; oxidized fine sulfide particles; colloidal hydroxide particles and flocs; continuous surface layers (e.g. hydroxide, oxyhydroxide, oxide species) of varying depth; sulfate and carbonate species; isolated, patchwise and face-specific oxide, hydroxide and hydroxycarbonate development. The actions of collector molecules (e.g. xanthates, dithiophosphinates) have been identified in several modes, namely: adsorption to specific surface sites; colloidal precipitation from solution; detachment of small sulfide particles from larger particle surfaces; detachment of small oxide/hydroxide particles; removal of adsorbed and amorphous oxidized surface layers; inhibition of oxidation; disaggregation of larger particles; and patchwise or face-specific coverage. The different modes of oxidation and collector action are exemplified using case studies from the literature and recent research.

1 Introduction

The process of selective separation of valuable sulfide minerals from gangue (e.g. other non-valuable sulfide minerals, oxides, silicates, etc.) by flotation probably has the highest throughput (by mass) of any process in which control of the chemistry, physics and engineering at the solid/liquid interface decides the economic outcome. Worldwide, this industry has been estimated to generate more than US\$100 bill. p.a. with throughput in excess of 10^9 tonnes p.a.

Flotation separation is heavily dependent on processing steps involving control of surface oxidation products on the mineral surfaces and the addition

K. Wandelt, S. Thurgate (Eds.): Solid–Liquid Interfaces, Topics Appl. Phys. **85**, 3–60 (2003)

of collector molecules designed to induce a hydrophobic surface for bubble-particle attachment. Separation is then achieved by flotation of the valuable mineral into the froth collected from the top of the flotation cell. A highly schematic representation of the process is shown in Fig. 1. Bubbles usually much larger than the ground particles are generated in the flotation cell, rising through the pulp (or slurry) to collect hydrophobic particles into the froth concentrate. Hence, the primary objective is to achieve a hydrophobic solid/liquid interface only on the valuable sulfide phase to be concentrated in this step.

The complex nature of this process is well-established both from plant operation and laboratory research (e.g. [1,2,3,4]). To describe the full set of complexities would fill, and indeed has filled, several books (e.g. [5,6,7,8,9,10]). In this review we will focus on the use of scanning and imaging surface analytical techniques applied to the study of sulfide surfaces in flotation. In this context, it is useful to identify some of these complexities that will be relevant to later discussion. They include:

- the extent of liberation of individual mineral phases by grinding or, conversely, the extent of remaining composite particles;
- chemical alteration of the surface layers of the sulfide minerals induced by oxidation reactions in the pulp solution;
- the presence of a wide range of particle sizes in the ground sulfide ores;
- galvanic interactions between different sulfide minerals to produce different reaction products on the mineral surfaces;
- interaction between particles in the form of aggregates and flocs;
- the presence of colloidal precipitates arising from dissolution of the sulfide minerals and grinding media;
- the mechanism of adsorption of reagents to specific surface sites;
- competitive adsorption between oxidation products, conditioning reagents and collector reagents.

The chemically altered layers on the surfaces of metal sulfides, particularly as oxides and hydroxides, are known to interfere with the recovery and selectivity of minerals separation by flotation (e.g. [11,12,13,14,15,16]). As part of the complete description of flotation chemistry, it is clearly necessary to understand the chemical and physical forms of these layers and the changes that occur in them during conditioning and flotation. The range of information available from the effective use of surface sensitive techniques has greatly assisted this understanding. The techniques have included Scanning Auger Microscopy (SAM), X-ray Photoelectron Spectroscopy (XPS), Scanning Tunneling Microscopy (STM), Atomic Force Microscopy (AFM), Analytical Scanning Electron Microscopy (SEM) and the recent application of Time-of-Flight Secondary Ion Mass Spectrometry (ToF-SIMS) ([4,17]). This review will consider information arising from conventional use of these techniques applied to surface characterization, reactions and the role of surface layers. We will concentrate on two specific aspects of the surface layers,

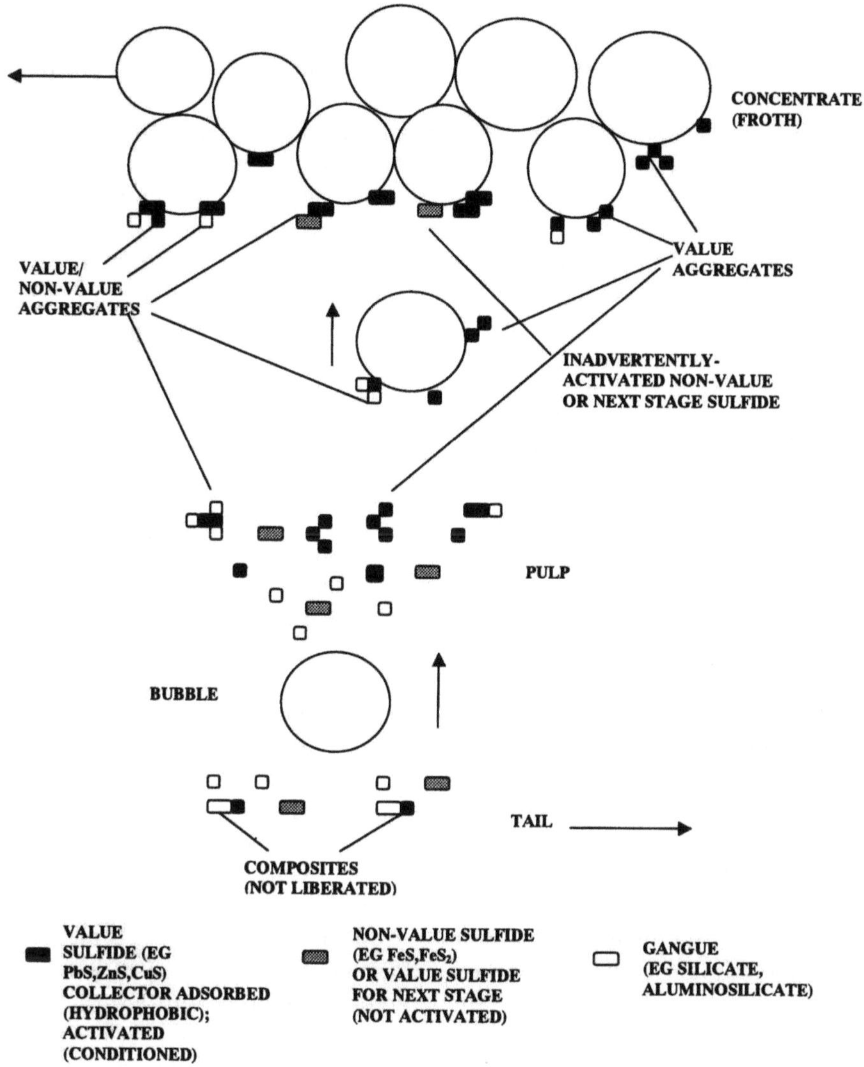

Fig. 1. Flotation process (schematic)

namely surface oxidation and the effects of collector addition on these layers, with applications drawn from both research on single minerals/mineral mixtures and industrial plant practice. The complementary chapters will specifically address the application of monolayer-sensitive synchrotron radiation XPS to mineral surface and adsorption of organic reagents in processing of sulfide minerals.

1.1 Techniques

Several surface sensitive techniques, capable of analyzing the first few atomic layers of the mineral surface, have now been used for more than ten years in a variety of studies related to the mechanisms of oxidation and adsorption in sulfide mineral flotation. The significance of these techniques is that they provide not only a compositional analysis of the surface but also information on chemical states (e.g. oxidation, bonding) and spatial distribution of adsorbed species on individual particles and complex mixtures of minerals as a function of depth through the surface layers. It is recognized that, since they are ex situ techniques that operate in ultra high vacuum, validation of the relationship between the measured surface compositions or chemical states and those prevailing in the original pulp solution in the flotation cell or circuit, is required. Sampling methodologies have been developed and tested over extensive (i.e. more than 1000 samples) sets of correlated surface analysis/flotation response testing in major projects in Australia (e.g. [18,19]) and Canada (e.g. [20]. A summary of this work in our laboratory to 1995 has been produced for the Australian Mineral Industries Research Association [21]. Surface chemical changes corresponding to changes in process conditions have been clearly demonstrated in this work.

In one sampling methodology [1] the sample is taken directly from the circuit or cell, dissolved oxygen removed from the solution before freezing, with the thawed slurry introduced to the spectrometer without air exposure of the mineral surfaces. The majority of the solution phase is removed (or exchanged) by decantation before evaporation in the fore-vacuum of the spectrometer. In a second approach [22] the sample is size-separated, washed and dried under vacuum before storage under an inert gas or vacuum. The storage vials contain silica-gel to remove residual moisture and are transported in frozen condition. The mineral phases for examination by LIMS or ToF-SIMS are hand-selected under an optical microscope in air.

Single mineral studies of surface oxidation and adsorption of collector molecules have used a variety of sample preparation procedures for the mineral surfaces. These have included fracture under high vacuum, surface abrasion under inert gas followed by transfer to the spectrometer, washing by water or prepared pH/Eh solutions, and solid powders pressed into conductive tape or soft metal substrates.

The characteristics and operating conditions of SAM, XPS, SEM and ToF-SIMS techniques have been fully described in previous publications (e.g. [1,17,22]). STM and AFM studies of mineral surfaces and their oxidation have also been described in a review [23] and several papers (e.g. [24,25]). These techniques have provided important complementary information to that from SAM and SEM with resolution down to atomic scale.

1.2 Surface Oxidation

It is well established that all metal sulfide minerals exhibit oxide and hydroxide species on their surface after exposure to air or aqueous solution. They have been observed in studies of pyrite (e.g. [26,27]), pyrrhotite (e.g. [28,29,30,31]), chalcopyrite (e.g. [1,12,32]), galena (e.g. [33,34]), pentlandite (e.g. [35,36]), cobaltite (e.g. [37]) and sphalerite (e.g. [37,38,39,40]). However, the mechanisms of surface oxidation are substantially different between different sulfides and are influenced by variables such as conditioning time, pH, Eh and the gas atmosphere above the sample.

Evidence from the combination of techniques has now shown that the mechanisms of oxidation are considerably more complex than those represented by simple reactions such as:

$$MS + xH_2O + \frac{1}{2}xO_2 \rightarrow M_{1-x}S + xM(OH)_2 \,. \quad (1)$$

The chemical nature of the $M_{1-x}S$ product, the spatial distribution of the oxidation products, dissolved and reprecipitated species, other higher oxidation products (e.g. sulfate) and interactions with other dissolved species (e.g. CO_2) all complicate the real situation. As examples from the case studies below will show, hydroxide products are not formed uniformly over the surface of the sulfide mineral, the chemical form of the metal-deficient or sulfur-rich surface is highly variable, and the presence of small, oxidized particles and particle interactions must be considered.

The mechanisms of surface oxidation and the consequent physical and chemical forms of oxidation products on the surface, derived from studies using surface analytical techniques, can be summarized as:

- metal-deficient (sulfur-rich), oxide surfaces, polysulfides and elemental sulfur;
- oxidized fine particles attached to larger sulfide particle surfaces;
- colloidal metal hydroxide particles and flocs;
- continuous surface layers (e.g. oxide/hydroxide) of varying depth;
- formation of sulfate and carbonate species;
- non-uniform spatial distribution with different oxidation rates, e.g. isolated, patchwise oxidation sites, face specificity.

Each of these oxidation mechanisms will now be discussed using evidence from published literature and research undertaken in our own group. In reviewing the research literature, it rapidly becomes apparent that a myriad of different oxidation and adsorption conditions have been used. Oxidation in air and oxygen at different temperatures and times, in solution at different pH and Eh, for different times, with different added species (dissolved and colloidal) all severely complicate comparisons. Similarly, for collector adsorption studies, different preconditioning (e.g. oxidation), molecular structures, concentrations and conditioning times are found in most papers. We have attempted to place the studies in some systematic order and to distinguish different regimes based on particular factors where possible. The remaining research issues, identified at the end of this review, note that the dependence of surface chemistry and flotation on a number of these factors is still distinctly unclear.

Where studies are available, we have ordered results on different minerals in the sequence: pyrite; pyrrhotite; chalcopyrite; pentlandite; galena; sphalerite; and other sulfide minerals.

2 Case Studies of Oxidation Mechanisms

2.1 Metal-Deficient Sulfides, Polysulfides and Elemental Sulfur

It is now well established that iron-containing sulfide minerals (e.g. pyrite, pyrrhotite, chalcopyrite, pentlandite) essentially follow a reaction mechanism similar to that in (1) above in that iron hydroxide products and an underlying metal-deficient or sulfur-rich sulfide surface are formed. The seminal work of *Buckley, Woods* and their colleagues [26], using a combination of XPS and electrochemical techniques, has clearly demonstrated this mechanism in single mineral studies. In their work oxidation of abraded pyrite surfaces exposed to air for a few minutes produced a high binding energy doublet component of the S2*p* spectrum in addition to ferric oxide/hydroxide reaction products. The sulfur product was attributed to an iron-deficient $Fe_{1-x}S_2$ surface layer with the later proposition that polysulfide-like species S_n^{2-} are formed. Specifically, *Mycroft* et al. [41] have correlated XPS and Raman spectra of electrochemically oxidized pyrite surfaces with polysulfide model compounds but only at $E_h > 600$ mV, pH 5. Recently, monolayer-sensitive Synchrotron Radiation XPS (SRXPS) has been used [42] to demonstrate that in situ fractured pyrite surfaces have two surface chemical states, corresponding to monosulfide S^{2-} and the surface atom of the first disulfide S_2^{2-} layer, as well as the less reactive sulfur atoms of S_2^{2-} groups beneath the surface layer in bulk coordination. Air oxidation of pyrite begins with the oxidation of S^{2-} sites, resulting from surface fracture of disulfide ions, and a structural model of the oxidation mechanism is proposed. Confirmation of the reactants in (1) has also resulted from the work of *Nesbitt* and *Muir* [43]. Fresh pyrite fracture surfaces exposed to water vapour for 7 h display no change to their Fe2*p* or S2*p* XPS spectra although oxygen adsorption occurs as H_2O, OH^- and O^{2-}. The absence of oxygen inhibits the full reaction so that metal-deficient (i.e. polysulfide-like) and oxy-sulfur species are not observed in S2*p* spectra. Exposure of the same surface to air changed the proportion of oxygen species dramatically with OH^- and O^{2-} rapidly increasing accompanied by development of a broad Fe(III) peak in the Fe2*p* spectrum due to oxyhydroxide surface species. In their experiments, no sulfate peak developed in the S2*p* spectrum during 24 h exposure to air but it is well-recognized that different pyrite samples (both synthetic and natural) can display widely varying kinetics in their oxidation (e.g. [26,44]).

Marcasite (FeS_2) surfaces, representing the dimorph of pyrite formed at low temperatures often as an alteration product of pyrrhotite oxidation, have been analyzed using XPS and AES as vacuum-fractured surfaces and after reaction in oxygenated acidic (pH3) solution. Similar S^{2-} and S_2^{2-} species to those observed with vacuum-fractured pyrite surfaces were found but, for marcasite, a small additional S2p peak near 163.5 eV was observed and interpreted to be polysulfide species. After reaction in oxygenated pH 3 solution, this polysulfide increases at the expense of disulfide. A new iron species appears in the Fe2$p_{3/2}$ spectrum near 709 eV representing 10–15 % of the total iron. This iron species was assigned to Fe(II) associated with either OH^- or Cl^- but may also be due to Fe(III)-S [43]. In studies of the dissolution kinetics under these conditions, analyses of aqueous sulfur speciation revealed fluctuations in sulfur content for oxidation states lower than sulfate. Correlated XPS analysis suggested that these fluctuations may result from periodic release of polysulfide to solution after accumulation on the reactive marcasite surface [45].

Another pyrite analogue (arsenopyrite) examined in a similar series of experiments [46] revealed that arsenic, sulfur and possibly iron exist in multiple oxidation states in the near surface of the unoxidized mineral. XPS finds sulfur as monosulfide S^{2-}, disulfide S_2^{2-} and polysulfide S_n^{2-} and, although As^{-1} predominates, $\sim$15 %As° is also observed. Iron appears to be present as Se(II) bonded to both arsenic and sulfur but some Fe(III) may be present also bonded to As-S. Air exposure reveals development of Fe(III) oxyhydroxide species and the development of As^{5+}, As^{3+} and As^{1+} with minor polysulfides and possibly thiosulfate. The nature of the oxidized species produced and their rates of formation are strikingly similar between reactions in air and with air-saturated distilled water suggesting that the mechanisms are likely to be closely similar in both media. Auger depth profiles of oxidized arsenopyrite demonstrate that arsenic diffuses from the interior of the mineral to the surface during oxidation where it is the most readily oxidized species producing large amounts of As^{3+} and As^{5+}. This mechanism promotes rapid, selective leaching of arsenites and arsenates with the well-recognized risks to water quality and biota.

Collectorless flotation of pyrite in alkaline solution, correlated to electrochemical oxidation, can be explained by the production of a hydrophobic sulfur-rich surface together with hydrophilic iron hydroxide species [47]. After grinding, the surface becomes substantially covered by the hydrophilic species and no significant flotation is observed without addition of collector. Collectorless flotation can, however, be easily obtained after complexing the iron with EDTA in solution indicating that the underlying hydrophobic sulfur-rich layer is responsible for pyrite flotation under these conditions. It is noted that elemental sulfur was not evident at pyrite surfaces exposed to air or neutral to alkaline solutions [26,27]. Thin layers of elemental sulfur were, however, observed on pyrite surfaces exposed to aerated, dilute sodium sulfide solutions [26,48]. Prolonged exposure of the pyrite to solution resulted in the appearance of sulfate, together with the ferric oxide/hydroxide, in the surface layers (see below).

Similarly, pyrrhotite surfaces oxidized in air or aqueous solution form iron oxide/hydroxide and iron-deficient sulfide surfaces [28,29]. *Knipe* et al. [49] also confirmed that, like pyrite, there is no evidence of oxidation products on pyrrhotite surfaces exposed to water (D_2O) alone. In mildly acidic solution, a soluble iron product is formed leaving a surface sulfur species with a S2p binding energy only 0.2 eV less than that for elemental sulfur. In contrast, alkaline solutions resulted in ferric hydroxide, sulfate species and iron-deficient sulfide. As with pyrite, electrochemical oxidation to Eh $>$ 400 mV produced both elemental sulfur and sulfate species.

Further evidence for the mechanism of formation of the sulfur-rich surface has come from combined XPS and XRD studies of pyrrhotite surfaces exposed to air, water and deoxygenated acid solution [31]. After acid reaction, the surface partly restructures to a crystalline, defective tetragonal Fe_2S_3 product in which linear chains of S_n atoms have a S–S distance similar to elemental sulfur but the high binding energy doublet S2p associated with the sulfur-rich surface is still 0.2 eV less than that of S_8. The shift to high binding energy proceeds systematically from 1.0 to 1.8 eV as the reaction progresses. Combined XPS and SAM studies of air-oxidized pyrrhotite surfaces [30,50] add further evidence to the oxidation mechanism with the observation of monosulfide through disulfide to polysulfide species. After 50 hours of air oxidation, three compositional zones are found with the outermost iron and oxygen-rich layer $<$ 10 Å thick and a sulfur-rich layer beneath displaying a continuous, gradual decrease in S/Fe ratio through to that of the unaltered pyrrhotite. These studies have also demonstrated the presence of $\approx$ 30%Fe(III) in the natural pyrrhotite surface freshly fractured under high vacuum.

Recent studies of the role of surface sulfur species in pyrrhotite dissolution in acid conditions have also thrown some light on oxidation mechanisms [51]. Synthetic, hexagonal pyrrhotite ($Fe_{1-x}S$, $x \approx 0.1$), dissolving in deoxygenated acidic conditions exhibits an induction period before rapid dissolution occurs. The length of the induction period is controlled by the amount of surface oxidation products on the initial surface, acid strength, and solution temperature. During the induction period, best described as a period of inhibited dissolution, there is slow release of iron but little or no production of H_2S. Correlated XPS and dissolution kinetic analysis has identified four stages of dissolution, namely: immediate dissolution of the outermost layer of oxidized iron oxyhydroxide and oxy-sulfur species; inhibited oxidative dissolution of the sulfur-rich (i.e. polysulfide-like), metal-deficient underlayer with kinetics limited by iron diffusion through this layer; sudden, rapid, acid-consuming non-oxidative reaction of monosulfide species with production of HS^- and H_2S and inhibited dissolution due to reoxidation by oxidizing solution species (e.g. Fe^{3+}) producing polysulfide, elemental sulfur and oxy-sulfur species. Fluctuations in sulfur content of the solution during dissolution, similar to those found with marcasite, were also measured in the pyrrhotite dissolution kinetics but, in this case, the sudden releases were correlated with non-oxidative production of HS^- and H_2S. A reductive mechanism in which surface states associated with S atoms accumulate electrons in a metastable condition before sudden reduction of S_n^{2-} to S^{2-} species and non-oxidative protonation has been proposed with XPS evidence for this electron accummulation.

Chalcopyrite also oxidizes with the formation of a ferric oxide/hydroxide overlayer and an iron-deficient sulfur-rich, copper-rich underlying sulfide in air or alkaline solution [32,33,52]. It is not yet clear whether a specific copper sulfide phase is formed in the reacted sulfide surface: CuS [52] and CuS_2 [32,33] have both been suggested. The collectorless flotation of chalcopyrite after air exposure or solution oxidation has been directly correlated with the surface composition determined by XPS [12]. Removal of iron hydroxide species during conditioning in alkaline solution, to leave the hydrophobic sulfur-rich sulfide surface, showed strong flotation. Conversely, oxidized chalcopyrite surfaces reduced in situ become copper deficient and were unfloatable. Combined XPS and electrochemical studies of chalcopyrite oxidation [53] have revealed different kinds of passive films depending on the pH of the solution and the applied potential. In strongly acidic solution at relatively low anodic potentials, the passive film is sulfur-rich with stoichiometry CuS_2. Although the structure of this surface phase has not been determined at higher anodic potentials in acid solution, elemental sulfur is formed. In weakly alkaline solution, the passive film consists of the metastable CuS_2 phase plus Fe(III) oxyhydroxides for lower potentials and elemental sulfur plus Cu(II) and Fe(III) oxyhydroxides at higher potentials. Their work suggests that in weakly acidic or strongly alkaline electrolytes, the passivating effect is less evident in electrochemical kinetics and XPS analysis.

The iron-containing pentlandite $(Fe, Ni)_9S_8$ mineral has also been studied with similar results using XPS, SAM, Mössbauer and spectral reflectance measurements. The initial vacuum-fractured surface of natural pentlandite exhibits two doublets in the S2p spectrum near 161.4 and 162.2 eV interpreted as sulfur in a four-coordinate environment and five-coordinate environment respectively with intensities in agreement with the theoretical ratio predicted from the pentlandite structure [54]. Oxidation of the pentlandite surfaces gives $\approx$10Å thickness layer of iron oxide/hydroxide, with later appearance of nickel oxides/hydroxides and sulfate species, and the formation of a sulfur-rich, nickel-rich subsurface layer [35,36,54]. In dilute acid, the oxide/hydroxide layer is largely soluble whereas, in alkaline media, the products were similar to those from air oxidation but the reaction rate was faster. The resulting subsurface is believed to restructure from pentlandite to violarite $FeNi_2S_4$ [35].

The surface chemistry of millerite (NiS) and niccolite (NiAs) are also of significant interest because of their relationship with the surface chemistry of pentlandite and their structural similarity to pyrrhotite. In the absence of iron in the structure, millerite oxidizes in air and aqueous solution to form hydroxynickel and oxysulfur nickel species whilst the sulfur is also oxidized to polysulfide-like species despite the nickel oxidation state remaining Ni(II) [55]. These oxidation products appear to either form islands or very thin ($\approx$ 1 nm thick) layers at the millerite surface in both air and solution apparently because the sulfate and hydroxy nickel species are soluble in solution giving the dynamically equilibrated surface layer. As with arsenopyrite, the arsenic in niccolite is more reactive than nickel with XPS showing the development of small As^{1+} and As^{3+} peaks in the As3d spectrum. The O1 *s* spectrum reveals the formation of hydroxide but also adsorbed H_2O and possibly atomic oxygen produced by the association of O_2 at the NiAs surface. The oxidation mechanism suggested by the authors involves initiation of arsenic oxidation by reduction of adsorbed atomic oxygen radicals followed by hydration to produce hydroxide surface species. In air, an apparently passivating thin oxidized overlayer ($\approx$ 1 nm thick) is formed but, in aqueous solution, a thicker oxidized overlayer ($\approx$ 12 nm thick) is produced after 7 days reaction. These overlayers contain the same secondary products as those found at the air-oxidized surface.

Galena surface oxidation proceeds by considerably different and more complex mechanisms. In air, initial stages of the oxidation show (XPS and STM) the PbS surface becoming enriched in lead oxide/hydroxide and carbonate species [24,33]. The spatial distribution of those products is described later (Fig. 7). However, no metal-deficient sulfide S2*p* high binding energy shift was observed in these conventional XPS studies, and sulfate species were only observed after extended exposure to air.

More recently, *Kartio* et al. [56] have demonstrated, using synchrotron radiation tuned close to the XPS signal, that the information obtained may be limited by surface sensitivity. Optimal XPS surface sensitivity is only attained with kinetic energies of the exciting XPS electrons in the region 30–100 eV where the escape depth is typically a fraction of a nanometer. For pyrite (100) surfaces cleaved in ultra high vacuum, two new types of sites corresponding to surface states of FeS_2 and FeS type in addition to bulk FeS_2 states, were found [57]. In contrast, the (100) surface of a cleaved galena crystal does not give any evidence for additional states in the Pb4f or S2*p* spectra [56].

In neutral or alkaline aqueous solution, the mechanism of oxidation apparently involves the congruent dissolution of lead and sulfide ions, their oxidation/hydrolysis in solution or near the surface with the formation of lead oxide/hydroxide, lead hydroxy carbonate and sulfur oxy species (e.g.

$S_2O_3^{2-}$, SO_4^2). Dissolution occurs from spatially separated pits on the surface (see Fig. 8 below) but XPS also finds lead hydroxy and carbonate surface species. The dissolution products may be subsequently readsorbed or precipitated on the galena surface at higher solution concentration. In particular, at pH 9.2, bulk lead hydroxide was established to be the major product of galena oxidation at 400 mV SHE from both XPS and ATR/FTIR studies [58]. These authors also found that wet polishing galena under distilled water could produce either lead- or sulfur-rich surface layers depending on the semiconducting nature (i.e. p- or n-type) of the sample, a result which has some support from other studies reviewed in the contribution by *Buckley, Hope* and *Woods*. After cathodic polarization at −500 mV SHE, the surface appeared to be lead-rich regardless of the origin of the galena.

Under acidic conditions metal-deficient sulfide, polysulfide and elemental sulfur phases have been detected by XPS [33], SAM [59] and STM [60]. The oxidation mechanism is apparently incongruent:

$$n(\mathrm{PbS}) \rightarrow \mathrm{Pb}_{n-x}S_n + x\mathrm{Pb}^{2+} + 2xe^- \tag{2}$$

with the metastable, sulfur-rich surface presumably responsible for the strong collectorless flotation under acidic conditions. *Higgins* and *Hamers* [61] have established, using electrochemical STM, that dissolution of galena is crystallographically anisotropic with the step edges becoming aligned principally along the [110] directions. Their results suggest that the surface dissolution kinetics under acidic conditions (pH 2.7) are most likely controlled by desorption of sulfide species, particularly HS^-. In their observations, dissolution appears to occur by selective removal of step edge species although, under neutral pH conditions, no preferential step edge orientation is observed. The combined XPS and AFM studies of galena oxidation in acetate buffer (pH 4.9) by *Wittstock* et al. [60] produced dramatic imaging of elemental sulfur protusions 10–200 nm after initial roughening of the galena surface. These protusions are separated by several hundred nm and appear to result from a process of diffusion in the aqueous phase. They are only retained for XPS examination by cooling the sample before the beginning of the evacuation of the sample in the spectrometer entry chamber. XPS shows the formation of elemental sulfur starting at potentials more anodic than 160 mV SHE. AFM imaging first detects the protusions at +236 mV SHE. The authors therefore propose that the process causing surface roughening is dissolution of PbS to Pb(II) ions and HS^- ions while the deposition reaction is the electrochemical oxidation of HS^- ions to elemental sulfur. It appears likely that sulfur formation starts at impurity locations leading to different rates and sizes of protusion development.

Hydrodynamic conditions can also extensively influence the form of the galena mineral surfaces. Shear conditions and dilution during cyclosizing have been shown [62] to enhance the incongruent dissolution of galena removing the lead hydroxide species and exposing metal-deficient sulfide or polysulfide surfaces with high measured contact angles and flotabilities.

There has been some controversy surrounding the nature of the surface species formed during mineral oxidation and, in particular, which species (i.e. metal-deficient sulfides, polysulfides or elemental sulfur) are responsible for the collectorless flotation of galena [63]. Evidence has been provided that elemental sulfur can be extracted with cyclohexane from floated wet-ground pyrrhotite [64] and recently multilayer quantities of sulfur were apparently extracted from dry-ground galena and chalcopyrite [65]. It is certainly present on galena surfaces as pH 4.9 after anodic oxidation above 200 mV SHE [60]. *Buckley* and *Riley* [66] have, however, shown that whilst ethanol does extract a sulfur species from galena, sulfur is not present in elemental form on the freshly ground mineral surface as evidenced by sequentially cooled samples in XPS spectra. Cyclohexane did not extract the same sulfur species. It remains possible that the elemental sulfur extracted by ethanol may have been transformed from polysulfide species on the mineral surface during extraction or that the solvent may have chemically altered the surface. It is also not simple to distinguish elemental S_8 from polysulfide species in UV-visible spectra.

Other aspects of this controversy with relevant references are summarized in the discussion from the Workshop on Flotation-Related Surface Chemistry of Sulfide Minerals [67].

The surface oxidation of sphalerite has been less systematically studied than those of other sulfide minerals but the pattern of reaction appears to be similar. After three weeks in air or conditioning in alkaline solution (1 h), there are no significant changes in the Zn Auger peak ($L_3M_{4,5}M_{4,5}$) or S2p peaks and very little evidence for oxidized products. It has been suggested that sphalerite oxidizes considerably more slowly than the other sulfide minerals under these conditions [38].

In contrast, under acid leaching conditions, dissolution of zinc occurs with the formation of a metal-deficient sulfide layer. Under oxidative leaching conditions, the metal-deficient sulfide is again observed in thicker layers and, at high acid concentrations, elemental sulfur ($\approx$10 %) was also formed protecting the sphalerite from further leaching [40]. With addition of 0.1 M Fe(III) under these conditions, thick layers of elemental sulfur developed.

Important information on oxidation (and collector adsorption) mechanisms can also be derived from the now extensive studies of copper activation of sphalerite in flotation research recently comprehensively reviewed by *Finkelstein* [68]. The mechanism of ion exchange of Cu(II) for zinc ions, followed by reduction of Cu(II) to Cu(I) with corresponding oxidation of adjacent S sites, has recently been described structurally and chemically on the basis of combined XPS, SIMS, XAFS and solution studies [69]. Under conditions of high pH (> 7.5) and high nominal surface coverage of the sphalerite by Cu(II), $Cu(OH)_2$ colloidal particles are observed on the sphalerite surface using SIMS and XPS [69,70]. Under other conditions, SIMS has shown that adsorption of the Cu is essentially uniform and not related to low coordination sites on the sphalerite surface. Depth profiles under these conditions with activation time of 15 min show that the copper is largely in the first few atomic layers but longer periods of activation at pH < 7 and high nominal surface coverage can result in exchange and redox reaction to depths exceeding 40 monolayers. XAFS analysis reveals that the Cu has an oxidation state $< +1$ and occupies a distorted trigonal planar geometry coordinated to 3S atoms in both surface and bulk sites. In the surface sites, the 1:1 replacement of Zn^{2+} in this configuration by Cu(II), confirmed as 1:1 by both solution and XPS analysis [68,69,71], is followed by reduction in situ to Cu(I) and oxidation of the three neighbouring S atoms to an oxidation state of ≈ -1.5. On bulk absorption into the sphalerite lattice, the distorted trigonal planar configuration is achieved through the breakage of a formerly tetrahedral Zn–S bond. The three remaining coordinated S atoms still have an oxidation state of ≈ 1.5 but the disbonded S atom has an oxidation state near -0.5. These observations fully explain the high binding energy S2p components of the XPS spectra [68,69] without invoking polysulfide, metal-deficient sulfide or elemental sulfur rearrangements in which S–S bonding is implied. The issue of possible photoreduction of Cu(II) to Cu(I), possibly compromising this mechanism, has been fully examined by *Skinner* et al. [72]. They showed that the primary influence of X-ray irradiation in conventional XPS experiments was in dehydration of $Cu(OH)_2$ overlayers with associated reduction in the concentration of surface oxygen. Only after dehydration to Cu(0) was photoreduction to Cu(I) found and this process generally took more than 1 h in the spectrometer. Since no signals for Cu(II) are found under the adsorption conditions relating to the mechanism described within the first 2–5 min of XPS measurement of the surface, it can safely be concluded that the Cu(I) results from the reduction/oxidation mechanism not X-ray-irradiation. The presence of the oxidized S species resulting from this mechanism has also been shown to impart hydrophobicity through increased contact angles and flotation recovery as well as hydrophobic aggregation [73,74] in a similar manner to that observed with pyrite and chalcopyrite.

In general, given the different rates of reaction and congruency/incongruency differences in dissolution between the different sulfides, the mechanisms of formation of metal-deficient sulfide are quite similar. In neutral or alkaline solution, oxide/hydroxide and hydroxy carbonate species are formed. Fe(III) oxidized species remain on the surface, Pb(II) and Zn(II) species initially dissolve and then reprecipitate. The underlying mineral is left metal-deficient producing, for iron-containing sulfides, polysulfide S_n^{2-} species but, for PbS, soluble sulfur-oxy species. After prolonged periods in solution, all surfaces show the formation of sulfate ions. In acid conditions, the removal of the metal oxide/hydroxide species (at different pH for each metal), exposes a metal-deficient sulfide surface with hydrophobic character for all of the sulfides studied to date. Elemental sulfur is only found after severe acid leaching, in highly oxidative (i.e. Eh $>$ 400 mV) solutions, or after the addition of sodium sulfide in conditioning. The evidence that the metal-deficient sulfide surface corresponds to polysulfide species is now quite strong [31,41]. At least some forms of these polysulfides can clearly impart hydrophobicity (e.g. [62]) and mineral floatability.

Recently, the evidence for assignment of the high binding energy components of S2p XPS spectra to metal-deficient, polysulfide defect sites and elemental sulfur has been reviewed [75]. It is apparent that several different model structures have been attributed to these components after fracture, polishing, oxidation or reaction sometimes to the same surface prepared under nominally similar conditions. From XPS evidence alone, the binding energy ranges of S2$p_{3/2}$ components of these model structures overlap so that assignment cannot be reliably based on this criterion alone. The central issues in the assignment relate to:

- whether S–S bonding exists (as in disulfide ions, polysulfide ions and elemental sulfur);
- whether the lattice structure of the original mineral surface is retained (with vacancies, coordination defects or substitution) or is restructured to a new surface mineral phase (confirmed by diffraction or X-Ray Absorption Fine Structure (XAFS) coordination evidence);
- whether elemental sulfur species are present (lost to vacuum at room temperature) requiring sample cooling below 200 K for retention;
- whether altered electronic states associated with substituted atoms and defect (coordination) sites can be confirmed (by XAFS or valence band SRXPS) without invoking S–S bonding or lattice restructuring.

On the basis of present evidence, the following conclusions concerning the reliability of assignment of the high BE S2p components were drawn [75]. Assignment of high BE components of S2$p_{3/2}$ spectra at 163.6–164.0 eV to elemental sulfur requires confirmation by evaporative loss at 295 K and/or S–S distances (XAFS, XRD) or vibrations (IR, Raman). Assignment of high BE components at 162.0–163.6 eV to polysulfides S_n^{2-} requires confirmation of S–S bonding by IR, Raman or XAFS. Assignment of restructuring in surface layers requires confirmation by diffraction (XRD, ED) or XAFS for altered lattice structure, M–S and/or S–S bond coordination and distances. Assignment of electronic defect sites as vacancies, lower coordination sites, or substituted sites (e.g. activation, impurities) requires confirmation by XAFS and/or VBXPS (SR).

The combination of evidence from XAFS, diffraction, vibrational spectroscopy and possibly valence band SRXPS with XPS evidence is likely to become of increasing importance in further studies of the reactions of metal sulfide surfaces.

2.2 Oxidized Fine Sulfide Particles

The presence of oxidized fine particles (i.e. $< 2\,\mu m$) attached to larger sulfide mineral surfaces has been established in the SAM/XPS work of *Smart* [1] with

La Vanier. The removal of fine particles by successive ultrasonication/decantation, from ground pyrite surfaces was shown to systematically reduce the surface oxygen concentration (as hydroxide O1 s at 531.5 eV binding energy) for pyrite from $\approx$ 40 at.% to $\approx$ 22 at.%. Ion etching this surface to 25 nm depth without removal of the attached fine particles reduced the oxygen concentration to $\approx$ 28 % but, after fine particle removal, ion etching effectively removed all of the hydroxide from the surface. Chalcopyrite surfaces behaved similarly but were less oxidized.

The physical nature of the oxidized layer formed initially on these surfaces can be seen in Fig. 2 where a chalcopyrite sample, ground initially in distilled water and allowed to condition for 1.5 h, was reground at that time and examined immediately using high-resolution field emission SEM without coating at 5 kV.

The oxidized surface layer and fine oxidized particles are clearly visible on the conditioned surface at the lower right of the micrograph with the fracture face, and newly fractured fine particles on this fresh face, at upper left of the micrograph.

The surface layer, at high magnification in other micrographs, appears to be relatively thin, i.e. of the order of 10 nm (Fig. 3). The secondary electron image clearly illustrates a chemical and physical difference between the oxidized, conditioned surface and the fractured face exposed to the solution for only a few minutes. Similar images are obtained from fractured pyrite surfaces prepared using the same methodology. In general, SEM and SAM images from pyrite and chalcopyrite surfaces examined in the early stages of oxidation show two general features, namely oxidized fine particles ap-

Fig. 2. Field emission scanning electron micrograph of a ground chalcopyrite surface conditioned in water for 1.5 h. then reground immediately before imaging. The fresh fracture face on the left-hand side can be contrasted with the oxidized face on the right

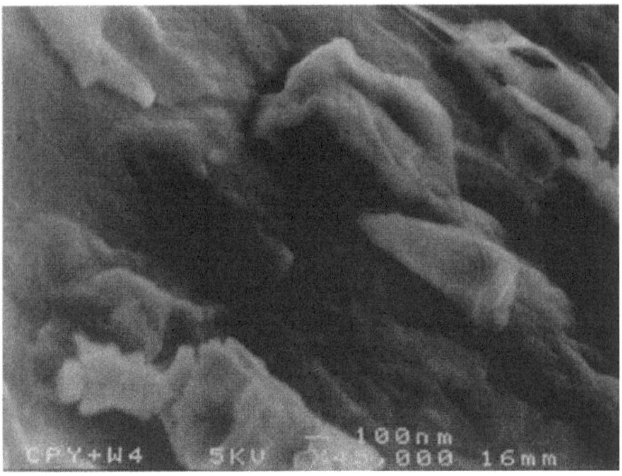

Fig. 3. Oxidized surface of sample as in Fig. 2 at higher magnification

parently attached to larger particle surfaces and continuous oxidized layers formed on both large and fine particle surfaces. At higher magnification, as in Fig. 3, it is apparent that fine particles expose a wide variety of different facets, morphologies and sizes (from dimensions less than 50 nm to several microns).

The attached oxidized fine sulfide particles are not easily removed by sedimentation/decantation techniques [1]. Successive ultrasonication/decantation (eight cycles) reduces the surface concentration of attached fine particles but the shearing action of the cyclosizer produces larger particles almost completely free of adhering fine particles [76]. Additionally, cyclosizing resulted in considerably reduced surface oxygen concentrations, greater exposure of sulfide minerals and thinner layers of oxidation products. The oxidized fine particles are clearly strongly adsorbed to the oxidized surfaces of larger particles. These hydrophilic fines may, on the one hand, reduce the hydrophobicity of larger particles and flotation selectivity. On the other hand, the attached particles may also be collected to the concentrate thereby affecting flotation grade. Size-by-size assays are affected by the removal of these fine particles from coarser particle surfaces during cyclosizing, particularly in assays of the finest fractions [76].

2.3 Colloidal Metal Hydroxide Particles and Flocs

The SAM results reported by *Smart* [1] have shown that $Fe(OH)_3$ flocs form on the surface of pyrite and chalcopyrite particles conditioned for relatively long periods (> 14 d) in nitrogen-purged pH 9 solution. These flocs comprise clumps of loose aggregates, with dimensions 1–3 µm, consisting of smaller spheroidal particles each with approximate diameter 0.1–0.5 µm. They are not observed on similar surfaces conditioned for shorter periods (i.e. several hours) but have been seen in plant samples following fine grinding in which accelerated dissolution of pyrrhotite and iron grinding media has occurred [18,19]. They appear to have been precipitated from saturated ferric hydroxide solution as colloidal $Fe(OH)_3$ particles. Mechanical agitation or ultrasonication/decantation easily removes these flocs from the sulfide surfaces showing that they are only weakly bonded to the oxidized sulfide minerals.

Colloidal iron oxide particles prepared synthetically in solution as small spheroids can be attached to galena surfaces, following classical particle adsorption isotherms, to full monolayer coverage at pH values below 6 [59]. In the pH range 6–10 electrostatic repulsion between iron oxide particles and unoxidized galena surfaces opposes adsorption. Adsorption in this pH range is due to attractive interactions or hydrogen bonding between iron oxide and oxidized regions on the galena surface. The mechanism for the interaction of hydrolysed iron (III) species with galena surfaces also appears to rely largely on electrostatic interactions [77]. The highest adsorption densities occur at pH 2–4 and pH 10–11 with values significantly higher than those in the intermediate pH range. Iron (III) adsorption densities, determined from EDTA extraction, have been shown to give excellent agreement with surface atomic concentrations determined from XPS analysis. The adsorption is dramatically influenced by iron (III) concentration, pH and the extent of galena oxidation. At pH values between 2 and 4, the galena surface is weakly negatively charged whereas the iron hydroxide species, e.g. $Fe(OH)_2^+$, are positively charged. Under alkaline conditions, oxidized galena surfaces with lead hydroxy groups (isoelectric point near pH 10) may interact with negatively charged iron species, e.g. $Fe(OH)_4^-$.

The effect of surface oxidation products as metal hydroxides and sulfuroxy species, either adsorbed in thin layers or precipitated from solution as colloidal particles, on flotation of sphalerite, chalcopyrite and galena has been directly demonstrated [16,78]. *Clarke* et al. [16] have shown that interaction between these oxidation products and the mineral is generally weak, e.g. electrostatic and/or hydrophobic in nature. The products can be removed from the surface by a variety of methods including mechanical (sonication or attrition with quartz) or chemical (pH change or complexation). The improvement in flotation recovery can be directly correlated with the amount of oxidation products removed by these methods. XPS studies have confirmed the "cleaning" actions by re-exposure of the underlying, reacted sulfide surfaces [125]. These methods were also selective in removing oxidation products in mixed mineral systems.

In situ STM images of galena surfaces in the presence of 10^{-3} M Pb^{2+} ions at pH 7 have also shown the development of elongated, oval colloidal projections with dimensions of $\approx 50\,nm \times 20\,nm$ [25]. The spatial distribution on the surface displays a directionality apparently corresponding to the [110] lattice directions of the galena surface (Fig. 4). XPS analysis has confirmed that these species are predominantly lead hydroxide presumably formed from

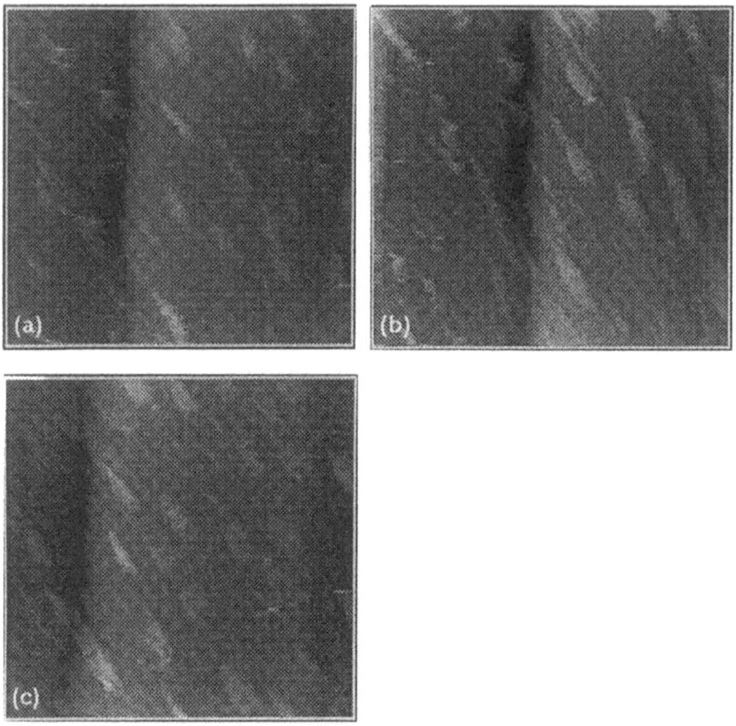

Fig. 4. Top-view STM images of galena treated with 10^{-3} M lead (II) ions at pH 7, as a function of time (xy scale = 500 nm): (**a**) 10 min (z scale = 11.2 nm), (**b**) 30 min (z scale = 13.7 nm) and (**c**) 50 min (z scale = 15.4 nm). Constant current mode (0.2–0.25 nA), bias $\sim$ 0.35 V

the hydrolysis of lead ions followed by surface attachment. It is not yet clear whether the mechanism of surface attachment involves the formation of lead hydroxide colloids in solution and their precipitation on to the galena surface or adsorption of $Pb^{2+}/Pb(OH)_2$ molecular species at specific sites on the galena surface before in situ growth. However, the formation of these patchy surface layers shows that the galena surface is heterogeneous and that its overall hydrophobicity and flotation response will be controlled not only by the surface chemistry but by the surface arrangement of hydrophilic and hydrophobic patches (see below).

2.4 Continuous Oxide/Hydroxide Surface Layers

The presence of continuous layers of metal hydroxide products, particularly iron hydroxides, can be directly inferred from the case studies in the previous sections where longer periods of conditioning in air-saturated solutions, electrochemical oxidation and in situ dissolution/precipitation has occurred. SAM evidence for these continuous layers after long conditioning periods has also been previously published [1]. Spot analyses on the sulfide mineral surfaces after long periods of conditioning at pH 9 has shown that both pyrite and chalcopyrite surfaces are oxidized at all points measured, although the depth of oxidation in SAM depth profiles is highly variable, i.e. < 10 nm to > 300nm (Fig. 5c).

The presence of continuous oxidized layers on chalcocite surfaces is supported by the work of *Mielczarski* using a combination of XPS and FTIR studies [79,80,125]. The extent of surface coverage of n- and p-type chalcopyrite after exposure to oxygen or air-saturated solution has been shown to increase systematically with time [81]. After 30 min of conditioning, XPS oxygen depth profiles showed complete coverage of the chalcopyrite surface to depths averaging 6–10 nm.

The formation of a continuous oxidized surface layer is a function of oxidation environment (e.g. air, solution), time, pH, Eh and the defect properties of the sulfide mineral. The presence of the considerable number and variety of fine particles, with dimensions from < 0.1 µm to 10 µm, providing substantial coverage of larger particle surfaces (as above) helps to explain the wide variation in spot analyses from XPS and SAM surface compositions. It is apparent

Fig. 5. SAM images from particles at successively higher magnification, i.e. white bar = (**a**) 100 µm, (**b**) 10 µm, (**c**) 5 µm. Note the five marked points (i.e. 1–5) for Auger analysis in (**c**) [1]. Oxygen depth profiles gave approximately 5, 30, >300, 80, 10 nm respectively for each point

that even 200 μm areas of analysis can contain widely varying proportions of oxidized fine particles, flocs, and precipitated particles (Fig. 5). Nevertheless, in many cases, it is also established that sulfide surfaces can be effectively continuously covered by oxidation products, as amorphous, adsorbed layers between the particles and flocs.

2.5 Sulfate and Carbonate Species

The formation of sulfur-oxy species, particularly sulfate, as a function of time of oxidation in air or solution, has been identified in the preceding section. Specifically, sulfate species have been observed for pyrite species in alkaline solution and in air (14 days) [26]. Pyrrhotite surfaces form sulfate only slowly after more than 10 days exposure to air [28] but more rapidly in air-saturated alkaline solution (i.e. 5 min) [29]. Grinding pyrrhotite in air increases the rate of formation of sulfate species on the surface (i.e. < 10 min) [31]. Chalcopyrite exposed to air for extended periods (i.e. 10 days) shows weak sulfate peaks in the XPS S2p spectra [32,33,82]. Oxidizing pentlandite surfaces in air, steam, alkaline solution or with oxidizing agents produced a surface layer which included sulfate species with the iron oxides and hydroxides [35] although recent studies [36] of the surface composition of pentlandite under flotation-related conditions did not find that sulfate species were present unless relatively high Eh conditions in alkaline media were used.

The initial oxidation of galena in air, studied by combined XPS and STM techniques, showed only the formation of carbonate, hydroxycarbonate and hydroxide species without sulfate formation [24]. Exposure of galena to air for extended periods (i.e. 4 days) resulted in sulfate formation and the binding energy of this species is more consistent with a lead hydroxy sulfate than bulk lead sulfate [33,83]. Lead hydroxide and lead carbonate still constitute the major part of the oxidized layer even after 10 months exposure to air. The mechanism for the dissolution of lead and sulfur from galena, via oxidation in solution near the surface, requires the formation of sulfur-oxy species which are initially soluble but may readsorb or precipitate as the oxidation/dissolution proceeds [84].

The presence of carbonate in the oxidized surface products is not always acknowledged. It has been explicitly found on oxidized pyrrhotite surfaces [28], galena surfaces [24,33] and in sulfide ore samples [1] but it is likely that it is present in surface layers of all sulfide minerals after exposure to alkaline air-saturated solution for extended periods of time.

The role of sulfate and carbonate species in directing hydrophobicity of surfaces, passivation of oxidation reactions and insoluble precipitate formation (e.g. $CaSO_4$, $CaCO_3$) [14] has not been fully recognized relative to the role of hydroxides, oxyhydroxides and oxides. It is important to note that in some of these sulfide mineral systems, these species are not removed from the surface by repeated solution exchange. There is evidence, for instance, that sulfate species play a direct role in controlling pyrrhotite dissolution during ore grinding [51].

2.6 Patchwise and Face Specific Oxidation

Face specificity of the oxidation has been discussed in relation to the depth of the oxide layers, derived from SAM point analyses (Fig. 5c), at different places in the same chalcopyrite mineral surface [1]. Further evidence of the distribution of the oxidation in the same pyrite:chalcopyrite system can be seen in Fig. 6 where lateral distribution maps for Fe, Cu, S and O are compared before and after ion etching. On the initial surface (Fig. 6a) there is a very clear association between Fe and O with the notable exception of the much less oxidized region of dark contrast containing Point 1 from Fig. 5c. It is also noted that there is a distinct line of oxidation on the chalcopyrite surface at the site of the linear defect containing Point 4 from Fig. 5c and some association with Fe along this defect. Other regions of oxidation in the O map on the chalcopyrite surface can be directly associated with $Fe(OH)_3$ flocs. Conversely, there is also a strong association between Cu and S with the notable addition of the high-sulfur region containing Point 1 from Fig. 5c on the pyrite surface. The contrast in these Auger maps depends on selection of the discrimination levels between high and low surface concentrations. Nevertheless, they clearly illustrate the much higher level of surface oxidation of the pyrite compared with the chalcopyrite surface and the exceptional nature of the small region of dark contrast on the pyrite surface.

After ion etching (Fig. 6b) interpretation of the images is more difficult but it is clear that most of the oxidation products have been removed with the exception of large floc aggregates and regions shadowed from the ion beam

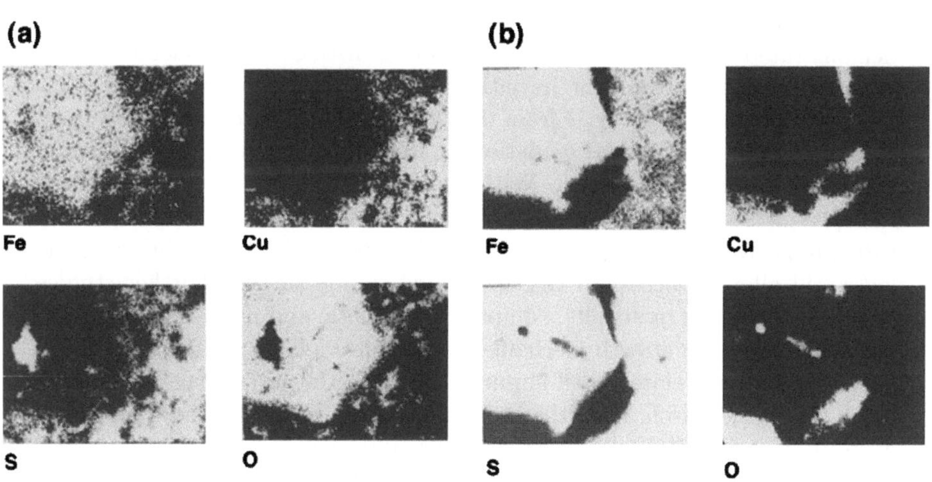

Fig. 6. SAM elemental distribution maps from the region in Fig. 5c for (**a**) Fe, Cu, S, O on the initial surface and (**b**) Fe, Cu, S, O after ≈ 300 nm ion etch

(i.e. close to the edges of the pyrite overlap with the chalcopyrite surface). Similarly, signals for S have been exposed in both the pyrite and chalcopyrite surfaces in all regions except those where oxidation remains. The Cu map distinguishes the regions shadowed from the ion beam because the Cu concentration on the initial surface is higher than that even on the chalcopyrite surface after ion etching. This result may indicate some preferential sputtering of Cu by the ion beam since the surface concentrations are lower than those expected for the clean chalcopyrite surface.

These and other analyses from different spots on the same particle surface of either pyrite or chalcopyrite strongly suggest that different depths of oxidation are found where fracture has exposed different facets of the mineral structure. High index faces may be expected to oxidize faster, or deeper, in the same time period, whilst analyses like that from Point 1 from Fig. 5c above suggest that some faces or regions of the surface are relatively unreactive. The ability of xanthate ions to displace oxidation products and adsorb must, at least in part, be determined by these differences in oxidation depths.

The development of isolated, patchwise oxidation in air and solution has been very well illustrated by STM studies of galena surfaces. *Eggleston* and *Hochella* [85,86,87] have imaged (001) surfaces of galena at atomic scale after exposure to water for 1 min. Apparent vacancies at the sulfur sites are correlated with oxidation in their model of this process. The oxidized regions do not initiate randomly but, once oxidation has begun at a site, these regions tend to nucleate and grow without initiation of new sites. The boundaries of the oxidized regions tend to lie along the [110] directions apparently due to S atoms across this front having only one nearest neighbor oxidized sulfur whereas an unoxidized S across a [100] boundary would have two nearest neighbor oxidized sulfurs. As with crystallization processes, [100] fronts move fast and disappear leaving the slow-moving [110] dominant.

At lower magnification, the process of galena oxidation in air has also demonstrated random sites of oxidation and growth on (001) galena surfaces with no clear preference for initiation at step edges or corners [24]. This process is illustrated in Fig. 7 from that work and correlated with XPS spectra showing that the initial oxidation products are peroxide, hydroxide and carbonate species successively. With time up to 270 min in air, the oxidation products grow from surface features with lateral dimensions < 0.6 nm through to overlapped regions > 9 nm diameter with "holes" in the overlayer still allowing access to the underlying sulfide surface. Further studies of galena oxidation in air [88], comparing synthetic and natural galena samples, confirmed the growth mechanism on natural galena with the oxidation initiation sites correlated with impurity atoms in the surface layer. The very much slower oxidation of synthetic galena did occur preferentially on edges, dislocations and lattice defect sites on the (001) faces of the galena crystal. The XPS spectra in this case show predominantly lead hydroxide and sulfate with a smaller contribution from carbonate in the oxidation products.

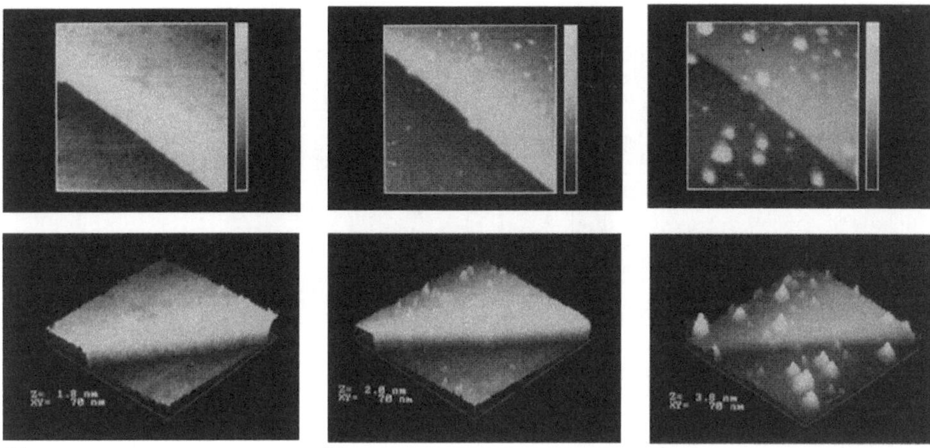

(a) (b) (c)

Fig. 7. STM images from a 70 nm × 70 nm area of: (**a**) freshly cleaned galena surface; (**b**) the same surface after 70 min standing in air; (**c**) after 270 min in air. The upper row are grey-scale images; the lower row are 3-D (rotated) images with the vertical scales 1.8, 2.0 and 3.8 nm, respectively. Constant current mode (0.2–0.25 nA), bias ≈ 0.35 V

In solution, STM (and AFM) imaging showed the development of subnanometer pits with increasing reaction time in air-purged water at pH 7 (Fig. 8) [25]. The boundaries of the pits lie in the (100) and (010) directions in the galena surface with depths corresponding to unit cell dimensions of galena (i.e. 0.3, 0.6 nm). The process occurring in solution is congruent dissolution, confirmed by XPS spectra showing unaltered Pb4f and S2p signals.

The x, y dimensions of the pits and their rates of formation depend strongly on the pH and purging gas (i.e. O_2, air, N_2) used, e.g. Fig. 9. Dissolution rates, determined directly from STM images of monolayers removed, decrease with increasing pH in agreement with the reported dissolution studies on galena [34,89]. For all pH values studied, the growth of the dissolution pits is significantly greater in the x- and y-directions than in the z-direction suggesting that the edges are more active towards dissolution than the faces [25]. At the relatively low surface-area-to-volume ratios used in the STM studies, there is no evidence for the growth of surface oxidation products similar to those observed in air or to adsorption/precipitation of lead hydroxide colloids from solution. Increasing the lead ion concentration to 10^{-3} M in solution resulted in surface product formation with a distinct [110] directionality (Fig. 4) [25].

The pH dependency of galena dissolution is consistent with the mechanism proposed by *Hsieh* and *Huang* [89] where surface protonation occurs initially as:

$$PbS + 2H^+ \rightarrow PbSH_2^{2+} \tag{3}$$

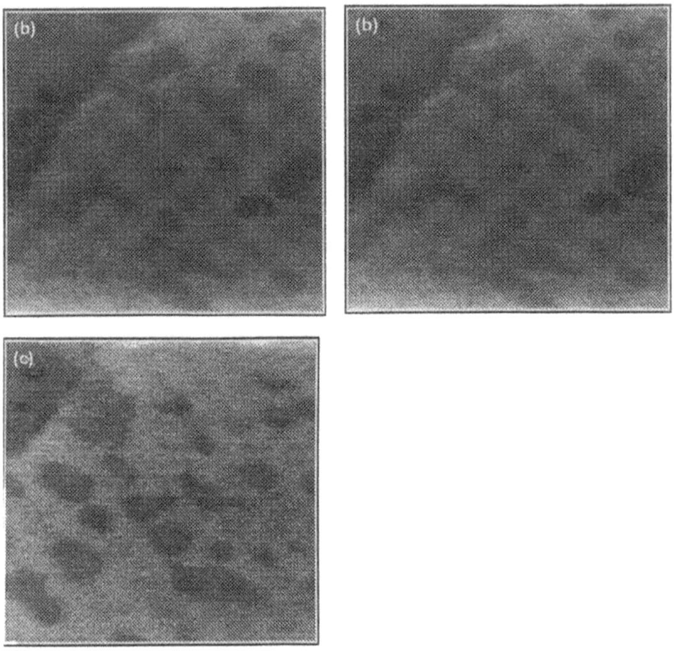

Fig. 8. STM top-view images of galena in air-purged water at pH 7 (xy scale $= 500\,\mathrm{nm}$): (**a**) 20 min (z scale $= 1.7\,\mathrm{nm}$), (**b**) 40 min (z scale $= 2.1\,\mathrm{nm}$), and (**c**) 60 min (z scale $= 1.8\,\mathrm{nm}$). Conditions as for Fig. 4

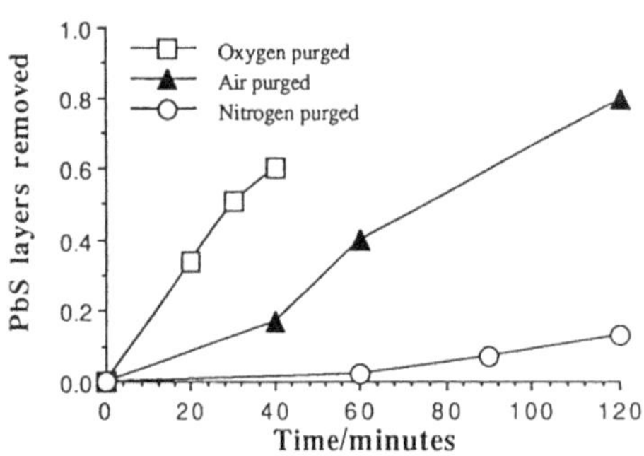

Fig. 9. The number of equivalent lead sulfide monolayers removed from the galena surface as a function of time in air-purged water, at different pH values

followed by oxygen adsorption:

$$PbSH_2^{2+} + 2O_2 \rightarrow PbSH_2^{2} + (O_2)_2 \tag{4}$$

and the dissolution step:

$$PbSH_2^{2+}(O_2)_2 \rightarrow Pb^{2+} + SO_4^{2-} + 2H^+ \,. \tag{5}$$

It should be noted that under strongly acidic conditions or mechanical shearing action (e.g. cyclosizer), there is XPS evidence for incongruent dissolution with the formation of a metal-deficient sulfide surface as discussed in the previous section.

2.7 Actions of Collector Molecules

Having established the modes of oxidation in sulfide mineral conditioning, it is now appropriate to consider the effects of adsorption of selective collector molecules on the metal sulfide surfaces and the oxidation products, in their various forms, found on these surfaces.

In the context of complex sulfide ores, from previous work [1,13,20,90,91], a number of general observations of the effects of conditioning and collector reagents in operating flotation circuits have already been made. For instance, considerable reduction in hydroxide concentration on the sulfide minerals in concentrates compared with these concentrations in conditioned feed and tails samples, is a general observation from XPS studies [1].

Removal of hydroxide layers by steam cleaning [1], dithiophosphate (DTP) addition [14], grinding and reducing environments [13] and the addition of complexing agents for metal hydroxide species [13] have all been observed. The surface layers of concentrate samples, however, still have a relatively complex composition even when quite high grades of the concentrates are obtained. In order to understand the processes producing changes in the chemical and physical nature of the sulfide minerals during flotation, we will describe the actions of collector molecule adsorption in several different modes, with case studies from literature and our recent research, namely:

- adsorption to specific surface sites;
- colloidal precipitation of metal-collector species from solution;
- detachment of oxidized fine sulfide particles from larger particle surfaces;
- detachment of colloidal metal oxide/hydroxide particles and flocs;
- removal of adsorbed, oxidized surface layers;
- inhibition of oxidation;
- aggregation and disaggregation of particles; and
- patchwise or face-specific coverage.

In relation to these several modes, the role of the collector in sulfide ore flotation has been reviewed previously [2,3]. Interference by metal hydroxides, in addition to collector adsorption to create hydrophobic surfaces, has been a generic theme in these reviews. *Shannon* and *Trahar* [2] note that the function of added collector may, in some cases, be primarily to counteract the hydrophilic effects of metal hydroxides (by participating in side reactions with oxidation products) rather than to directly increase the hydrophobic character of the floating minerals. It has been shown [36] that the coverage of xanthate required on freshly abraded galena surfaces to give flotation response may constitute only a small fraction of a monolayer, a result which is in accord with plant practice where xanthate concentrations used usually correspond to $< 10\%$ of monolayer coverage of the selected sulfide mineral.

3 Actions of Collectors: Case Studies

The selective flotation of sulfide minerals is normally carried out using thiol-based collectors including xanthates (alkyl dithiocarbonates) (X), dialkyl dithiophosphates and dithiophosphinates (DTP) and dialkyl dithiocarbamates.

3.1 Adsorption to Specific Surface Sites

Early theories of collector adsorption have been refined to the currently accepted mixed potential model in which reaction between the collector and a specific surface site takes place by an anodic process with dissolved oxygen reduced at other surface sites to remove the electrons donated by the oxidation process

$$X^- \rightarrow X_{ads} + e^- , \tag{6}$$

$$MS + 2X^- \rightarrow MX_2 + S + 2e^- , \tag{7}$$

$$O_2 + 4H^+ + 4e^- \rightarrow 2H_2O . \tag{8}$$

The mechanism has been reviewed by *Richardson* [3] including the necessity for the presence of oxygen in the adsorption process. The S entity may, as previously discussed, be in the form of a metal-deficient sulfide or polysulfide species rather than elemental sulfur.

With xanthate adsorption, oxidation to dixanthogen at the surface or in solution may provide a mechanism additional to the charge transfer chemisorption of the xanthate ion, i.e.

$$2X^- \rightarrow X_2 + 2e^- . \tag{9}$$

The expectation is that any of these four species (i.e. X_{ads}, MX_2, S, X_2) constitute entities contributing to the hydrophobicity of the surface.

Direct observations of xanthate adsorption and the molecular form of the collector molecules have been provided by the extensive XPS, synchrotron and FTIR studies of *Buckley, Mielcarzski, Suoninen* and their colleagues using idealized metal and sulfide surfaces. The specific interaction of xanthate with copper atoms and ions has been widely studied in the work of *Mielczarski* et al. and *Kartio* et al. on elemental copper substrates [56,92]. With conventional XPS, adsorption to monolayer coverage is observed but, with the very high surface sensitivity of synchrotron radiation, an additional

S2 *p* doublet is found attributed to different binding of the CS_2 group between the first chemisorbed monolayer and subsequent layers deposited over the first layer in pH 9.2 with 6×10^{-5} M ethyl xanthate for 1 min. Further consideration of experimental factors in XPS studies of xanthate adsorption on metal and mineral surfaces [93] has confirmed this interpretation and explained the dominating role of preadsorbed contaminants in alternative interpretations [94]. Orientation in the first monolayer also explains the relative intensities of C1 *s*, S2 *p* and O1 *s* spectra with the thiol groups at the copper surface and the methyl groups furthest from the interface.

Xanthate adsorption on Cu(II) activated pyrite shows that the oxidation state of copper is changed to Cu(I) before xanthate adsorption [1,95]. Both copper xanthate and iron hydroxy xanthate species are inferred from the spectra for air-saturated neutral to alkaline (i.e. pH 10) 10^{-4} M xanthate solutions. Dixanthogen is only formed and adsorbed on the pyrite surface at Eh values above 400 mV. The work has recently been reviewed [96].

On copper-activated sphalerite surfaces [37] with low copper (II) additions and high affinity adsorption behavior, copper (I) ethyl xanthate is the predominant surface species. The rate and extent of ethyl xanthate adsorption are, however, decreased by extended conditioning periods apparently due to penetration of copper ions into the zinc sulfide lattice confirmed by XPS depth profiling. Time dependence of xanthate adsorption is then related to subsequent back diffusion to the solid–aqueous solution interface. At high copper additions, both dixanthogen and copper (I) ethyl xanthate are detected on the sphalerite surface but this appears to be a precipitation mechanism rather than surface reaction (see next section).

The presence of adsorbed xanthate on freshly fractured galena surfaces has been confirmed from both S2 *p* spectra and the more surface-sensitive X-ray induced Auger spectra (i.e. S LMM and Pb NOO) signals. This work [97] has correlated xanthate coverage (using voltammetry) with XPS spectra and flotation recovery showing that only a fraction of the monolayer is adsorbed at maximum recovery. Submonolayer, perpendicularly oriented adsorbed lead ethyl xanthate was confirmed in combined XPS, FTIR and controlled potential studies [24]. There is some contention in the different literature [6,98] that pre-oxidized galena surfaces are either necessary for xanthate adsorption or facilitate the adsorption of xanthates. According to *Buckley* and *Woods* [97], xanthate is adsorbed immediately on freshly exposed galena surfaces following immersion in air-saturated solutions at pH 9.2. The coverage varies from 0.15 after 10 min to 0.37 after 1 h in 10^{-4} M solution. Surface oxidation must therefore be concurrent with adsorption if this is necessary for uptake of the xanthate. The results from STM studies of surface oxidation of galena in solution, with congruent dissolution forming (100)-based pits rather than oxidation products, are also not consistent with the requirement for preoxidation of the surface [25]. Precipitation of lead xanthate species may however result from this oxidation/dissolution reaction either in situ or in solution.

The influence of grinding on single and mixed mineral systems with collector adsorption has been studied by Cases' group using a combination of FTIR and XPS [99,100,101]. For pyrite, the results were essentially similar to those discussed above with dixanthogen and iron hydroxyxanthate as the main species. On galena, they found monocoordinate lead xanthate, non-stoichiometric and stoichiometric lead xanthate and amyl carbonate disulfide species. Dixanthogen is formed in a second adsorption stage at higher surface coverage corresponding to complete flotation and a sharp decrease in zeta potential.

The influence of galvanic interactions in mixed mineral systems on the form of the adsorbed xanthate species has been investigated by *Bozkurt* et al. [102]. They used ATR/FTIR combined with open circuit potential measurements to show that, in the pentlandite/pyrrhotite system, dixanthogen was the main adsorption product on both minerals. Dixanthogen adsorption on pentlandite was increased and on pyrrhotite decreased for these minerals in direct contact or sharing the same solution. The open circuit potential measurement showed that pentlandite, which initially had a higher mixed potential than pyrrhotite, reversed this order in the presence of xanthate leading to oxidation by anodic reaction.

The detection of collector molecules on mineral surfaces from operating flotation plants has also been confirmed by XPS [1,90]. More recently, a series of studies using ToF-SIMS by *Chryssoulis* and his group [103,104,105] and by *Brinen* and co-workers [106,107] have confirmed that this technique can distinguish fragment ions from the collectors and spatially map their distribution on single mineral grains. Sphalerite particles lost to scavenger tails have been shown [105] to have less amylxanthate and isobutyl xanthate compared to sphalerite recovered to the zinc final concentrate. Xanthates were also detected on the surface of sphalerite particles into a copper/lead final concentrate. The face specificity of adsorption of DTP collectors will be discussed in the final section.

3.2 Precipitation from Solution

There are now many examples of studies in the literature in which uptake of the collector molecules on the sulfide mineral surface occurs through the formation of colloidal metal-xanthate or metal-hydroxyxanthate species in solution. In some cases, it is likely that these colloidal particles form close to the surface as a result of dissolution of metal ions or reaction with the oxidation product sites on the surface. Hence, precipitation from solution may imply several different processes resulting in adsorbed precipitates.

For instance, *Plescia* [108] used a combination of XRD and SEM to show that lead ethyl xanthate nucleates and crystallizes on small isolated areas of a galena surface in contact with 10^{-2} M ethyl xanthate solution. The kinetics of the formation of the crystalline, mainly stoichiometric PbX_2, show that there is widespread nucleation in the first few minutes but the initial crystallites tend to dissolve. After 5 min the crystalline form becomes stable and the kinetics of growth can be related to xanthate concentration, temperature and the oxidation state of regions on the galena surface. The presence of pre-oxidized regions or parallel oxidation/xanthate adsorption may be necessary for the formation of precipitates of lead xanthate [98].

In situ STM images of a freshly cleaved galena crystal in contact with an air equilibrated 10^{-3} M ethyl xanthate solution are shown in Fig. 10. Colloidal particles of lead ethylxanthate (as confirmed by XPS and FTIR) form at the surface and correspond to multilayer surface coverage:

$$n\mathrm{PbS} + 2x\mathrm{EX}^- \rightarrow \mathrm{Pb}_{n-x}\mathrm{S}_\mathrm{n}x\mathrm{Pb(EX)}_2 + 2x\mathrm{e}\,. \qquad (10)$$

In situ STM studies of ethyl xanthate treated pre-oxidized galena surfaces have shown the removal of oxidized lead species and the formation of colloidal lead ethylxanthate particles [25,109]

$$\mathrm{PbS}, y\mathrm{Pb(OH)}_2 + 2y\mathrm{EX}^- \rightarrow \mathrm{PbS} \times y\mathrm{Pb(EX)}_2 + 2y\mathrm{OH}^-\,. \qquad (11)$$

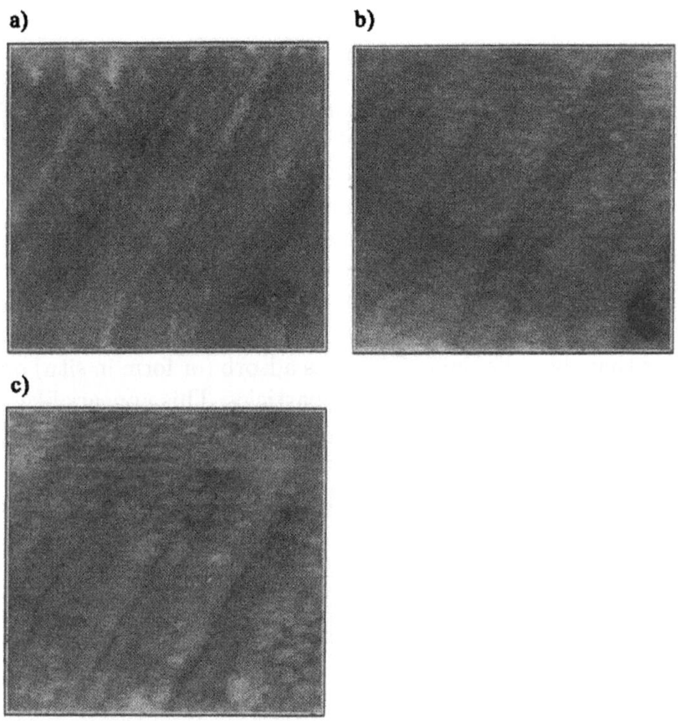

Fig. 10. In situ STM images of freshly cleaved galena surface in the presence of ethyl xanthate (10^{-4} M) at pH 9: (**a**) 20 min, (**b**) 40 min and (**c**) 60 min recorded at constant current 0.25 nA with tip bias 0.35 V ($x, y = 500$ nm, $z = 6.0$ nm)

XPS, FTIR and UV visible studies have shown that copper (I) ethyl xanthate and diethyl dixanthogen can be detected on the sphalerite surface as a result of precipitation [37]. The reaction mechanism proposed is:

$$Cu^{2+} + 2EX^{-} \rightarrow Cu(EX)_2 \rightarrow CuEX + 1/2(EX)_2 \tag{12}$$

A similar mechanism leads to the inadvertent activation of sphalerite during galena flotation by the adsorption and reaction of lead ions and their reaction with xanthate. Lead concentrates obtained towards the end of the lead rougher flotation circuit showed, in XPS analysis, increased Pb/Zn atomic ratios over bulk values. Single mineral samples of sphalerite immersed in the flotation pulp also showed significant amounts of lead on the surface and, when contacted, with added xanthate, FTIR spectra confirmed the presence of lead xanthate and dixanthogen [110].

It is also possible that iron hydroxy xanthate species are formed both in solution and on the surface of pyrite contributing to its flotation behavior [84]. A dissolution/complexation/adsorption model has been developed to explain zeta potential changes during the interaction of xanthate at concentrations from 10^{-4} to 10^{-3} M. The model suggests that, in addition to dixanthogen, these iron hydroxy xanthate complexes contribute to the flotation of pyrite as adsorbed or precipitated species.

Hence, the previously discredited reaction mechanism for sulfide mineral flotation based on the formation of insoluble metal xanthates [3,111,112] should not be discounted. The *Taggart* model was criticized on the grounds that stoichiometry would require much higher collector concentration than those effective in practice (i.e. like EDTA complexation). However, the presence of precipitated, colloidal metal-collector species for lead, copper and iron with relatively low surface coverage can be demonstrated and correlated with flotation response. In some cases, e.g. galena, pyrite at Eh < 400 mV, this mechanism may be the dominant influence on flotation response. Selectivity in flotation requires that the colloidal precipitates adsorb (or form in situ) on sulfide particles rather than non-sulfide gangue particles. This appears likely but has not yet been directly studied.

3.3 Detachment of Oxidized Fine Sulfide Particles

It is generally agreed that, if sufficient hydrophilic, oxidized material is present on the mineral surfaces, this will overcome any natural or self-induced floatability as well as modifying the collector-induced floatability of the sulfide particles [2]. Collector molecules are suggested to have the dual action of removing oxidized products from surfaces and providing a hydrophobic surface for bubble attachment and flotation. As demonstrated above, the forms of oxidized products on the sulfide mineral surfaces include fine oxidized sulfide particles together with colloidal metal hydroxide particles and oxidized surface layers [1]. The surface cleaning action of xanthate has been recently reviewed by *Senior* and *Trahar* [78] in studies of the flotation of chalcopyrite in the presence of metal hydroxides. The "cleaning" mechanism does not appear to be simple dissolution of the oxidized metals as can be achieved with other complexing agents, e.g. EDTA, because the amount of xanthate needed to restore floatability is stoichiometrically orders of magnitude below that required for the complete conversion of the metal hydroxides to dissolved metal xanthate species. This fact is also well established in plant practice where EDTA additions have been found to be effective in giving increased recovery but are prohibitively expensive.

Indications of the cleaning action in removing oxidized surface layers were noted previously [1]. Figure 11 compares the Cu2*p* signals from a ground chalcopyrite sample before and after addition of 5×10^{-5} M butyl xanthate. The initial high binding energy shoulder on the Cu2*p* peaks, attributed to charged hydroxide species, have been removed by the action of the xanthate.

Recent research in our group has followed the oxidation behavior of pyrite as a function of conditioning time at pH 9 under oxygen-purged conditions and the actions of xanthate adsorption.

The morphology and composition of the pyrite particle surfaces, including adsorbed oxidized particulates and overlayers, have been obtained from a SAM study whilst EDTA extraction has determined the amounts of oxidized iron species in the solution phase and on the mineral particle surfaces separately. The behavior of oxidized fine sulfide particles, with substantial hydroxide surface coverage, will clearly be similar to that of metal hydroxide precipitates and particles since their surface chemistry is closely similar. Some examples from this systematic study are selected for discussion.

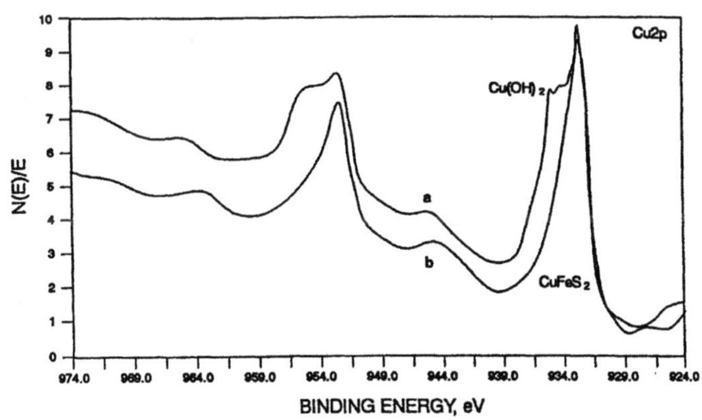

Fig. 11. XPS Cu2*p* spectra from ground chalcopyrite surfaces derived from (**a**) distilled water (*upper curve*), (**b**) pH 9 pulp with 5×10^{-5} M butyl xanthate addition (*lower curve*). Note the removal of the high binding energy copper hydroxide species after addition of xanthate collector. Some iron hydroxide remains, however, in Fe2*p* spectra

Ground natural pyrite, oxidized for 150 min at pH 9 under oxygen purged conditions reveals surface features consisting of oxidized fine particles attached to larger particle surfaces together with continuous oxidized layers formed on both fine and large particle surfaces. The SAM secondary electron image and oxygen map presented in Figs. 12a,b show clearly the surface morphology of an oxidized pyrite particle. The oxygen map shows that the colloidal precipitates at point 2 are particularly rich in oxygen as are the cleavage edges at point 3. It is also evident that oxidized products cover most of the particle surface as exemplified by point 1, an apparently featureless fracture face, free of precipitate particles.

Spot SAM analyses from points 1 and 2 are presented in Table 1 in comparison with the surface assay at the start of conditioning. Compared with the freshly ground pyrite surfaces, dramatic increases in oxygen exposure are evident both on precipitates and on fracture faces with a commensurate decrease in the sulfur signal. This, together with surface iron enrichment, confirms that iron oxide/hydroxide is present both as fine adsorbed precipitates and as a continuous overlayer on the pyrite surface.

Solution speciation calculations [77] indicate that $Fe(OH)_3$ and $Fe(OH)_4^-$ are the predominant species present under these conditions. At pH 9, $Fe(OH)_3$ colloidal species are negatively charged (i.e. $\approx$ pH 6.7) as are pyrite surfaces [84] so it is unlikely that $Fe(OH)_3$ precipitates would adsorb by electrostatic interaction. This is supported by the results summarized in Fig. 13 for iron-EDTA extraction analysis as a function of oxidation time at pH 9.

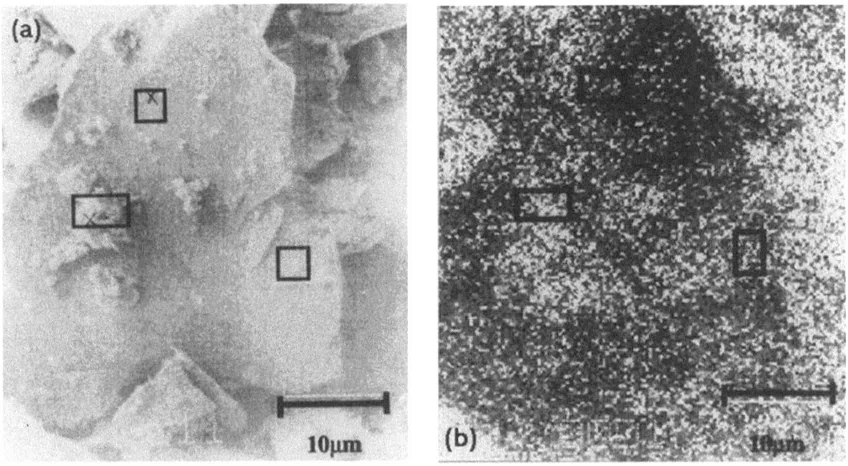

Fig. 12. (**a**) SAM secondary electron micrograph of a pyrite particle oxidized in solution for 150 min at pH 9. Analysis points indicated are point 1 *(top)* – featureless surface region, point 2 *(centre)* – colloidal precipitate and point 3 *(right)* – cleavage edges. (**b**) SAM oxygen map of the same pyrite particle as in (**a**)

Table 1. Typical SAM spot analyses for pH 9 conditioned pyrite at start ($t = 0$ min) and end ($t = 150$ min) of conditioning

Analysis	Atomic concentration (%)			
	C	O	S	Fe
pH 9, No KEX $t = 0$ min Start of conditioning	33	11	41	15
End of conditioning $t = 150$ min	24	28	19	28
Spot 1:Fracture face	20	40	11	29
Spot 2: Colloidal ppt.				

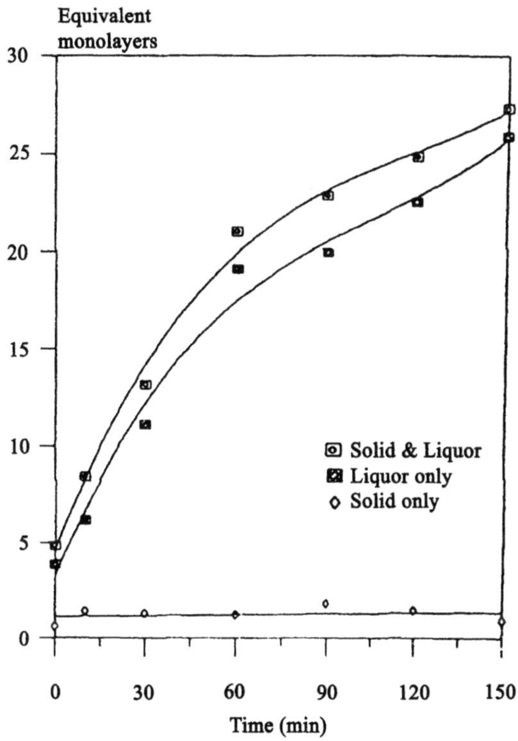

Fig. 13. EDTA extractable iron in equivalent monolayers as a function of conditioning time at pH 9 for solid+liquor, and liquid and solid separately

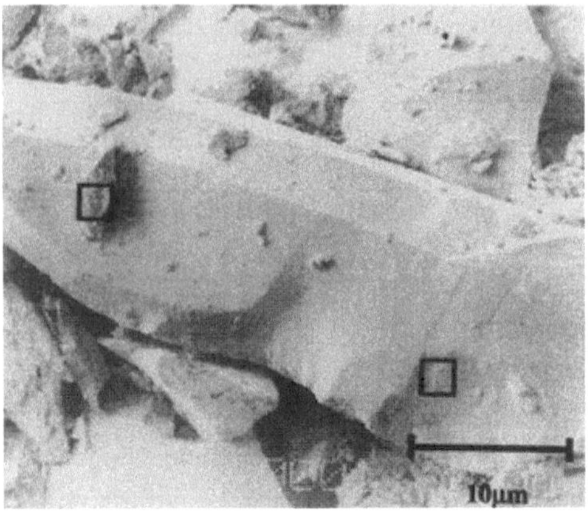

Fig. 14. SAM secondary electron micrograph of a pyrite particle oxidized for 150 min at pH 9 in the presence of 10^{-3} M KEX. Spot 1 (*right*), Spot 2 (*left*) as marked

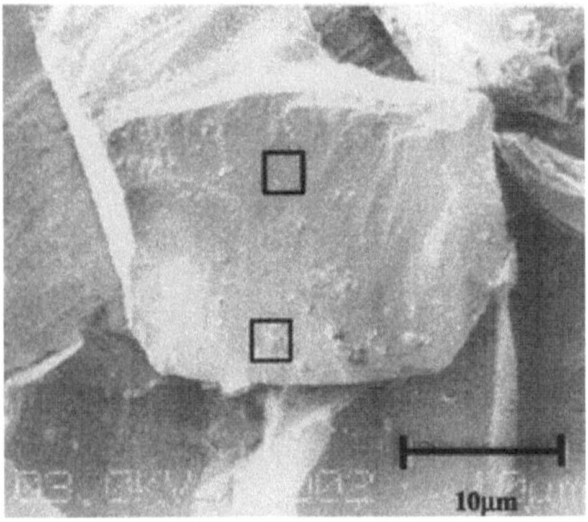

Fig. 15. SAM secondary electron micrograph of a pyrite particle oxidized for 150 min at pH 9 in the presence of 10^{-5} M KEX. Spot 1 (*upper*), spot 2 (*lower*) as marked

It is clear that at pH 9 the majority of EDTA extractable iron is present in solution and not on the surface.

The mechanism of attachment of the oxidized fine particles and colloidal iron hydroxides must therefore rely on the in situ formation or hydrogen bonding between the hydroxyl groups on each surface.

Figures 14, 15 display SAM secondary electron images of a typical pyrite particle surfaces after pH 9 oxidation for 150 min in the presence of 10^{-3} M and 10^{-5}M potassium ethyl xanthate (KEX) respectively. In the presence of 10^{-3}M KEX, there appears to be no evidence of the colloidal precipitates observed before xanthate was present. Spot analyses as indicated in Fig. 14 yield

Table 2. Typical SAM spot analyses for pyrite conditioned at pH 9 in the absence and presence of 10^{-5} M and 10^{-3} M KEX

Auger Analysis	Atomic Concentration (%)			
	C	O	S	Fe
pH 9, No KEX				
t = 0 min	33	11	41	15
t = 150 min	24	28	19	28
Spot 1: Fracture face				
Spot 2: Colloidal ppt	20	40	11	29
pH 9, 10^{-3}M KEX				
$t = 150$ min	24	8	24	42
Spot 1: Fracture face	19	10	25	46
Spot 2: Fine particles				
pH 9, 10^{-3}M KEX				
$t = 150$ min	30	30	38	31
Spot 1: Fracture face	31	31	35	34
Spot 2: Fine particles				

the surface compositional information summarized in Table 2, which shows a complete absence of detectable oxygen together with a sulfur-enriched surface. Again considering Eh/pH stability data [84], the most probable iron-xanthate complex at pH 9 and high Eh is the $Fe(OH)_2EX$ species. It is also likely that, due to the highly oxidizing conditions (> 300 mV), the formation of dixanthogen (EX_2) may have occurred. The fact that an oxygen signal was not detectable on the particle surfaces suggests that any adsorbed oxidized iron hydroxyxanthate complexes are either in volatile form (i.e. desorbed in the ultra-high vacuum) or they have not formed. $Fe(OH)_2EX$ is not known to volatilize easily. The level of EDTA extractable iron, summarized in Fig. 16, at 10^{-3} M KEX is extremely low from either solution or surface phase, supporting the proposition that a surface barrier or layer on the mineral surface has severely limited oxidation. The species most likely to be present on the surface is EX_2 which is volatile in vacuum. Pyrite surfaces exposed to such high xanthate concentrations would certainly be rendered hydrophobic, increasing their recovery to concentrates, but this concentration is well above that used in flotation. This oxidation inhibition mechanism may, however, explain different literature results and interpretation between 10^{-5} M and 10^{-3} M xanthate additions.

At a concentration of 10^{-5} MKEX, particle surface morphology and composition is distinctly different. Colloidal precipitates (Fig. 15, point 2) are still observed and the surface compositional results (Table 2), by virtue of significant oxygen exposure, suggest the presence of oxidized iron species. It

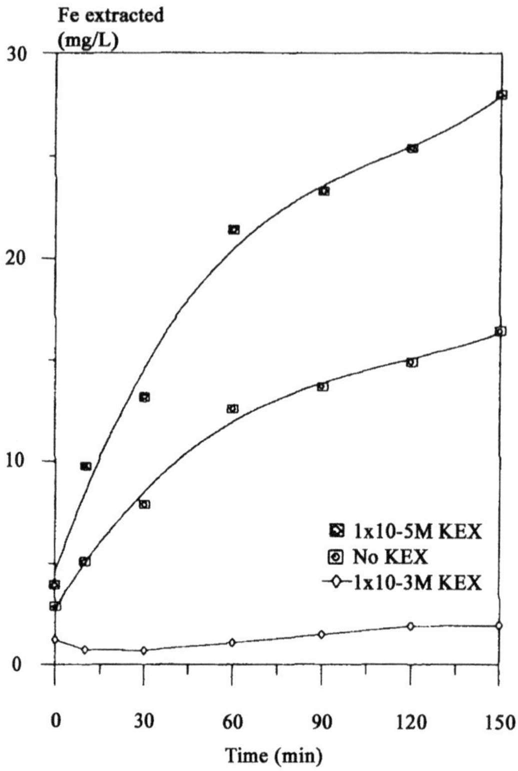

Fig. 16. EDTA extractable iron from pyrite as a function of conditioning time at pH 9 in the presence of No KEX, 10^{-5} M KEX and 10^{-3} M KEX

is evident therefore that some surface oxidation still takes place on pyrite at lower xanthate concentration.

EDTA extractable iron results confirm that the effects of xanthate at 10^{-3} M and 10^{-5} M KEX concentrations are indeed very different. The level of extractable iron at 10^{-5} M KEX is almost an order of magnitude greater than that of the 10^{-3} M case and double the levels when no xanthate is present. Correlation with SAM results indicate that the passivating layer has not formed and that the dominant effect of xanthate at 10^{-5} M concentration is to partially clean the pyrite surface of hydroxides as they form, ensuring available iron for further oxidation and dissolution, and increased iron in solution. This mechanism is consistent with the dramatic increase in EDTA-complexable iron.

The effects of xanthate on the oxidation behavior of pyrite at pH 9 are highly dependent on xanthate concentration. At low concentrations (10^{-5} M) the primary xanthate action is that of a cleaning and dispersal agent, displacing oxidized iron species from the pyrite surface, allowing further oxidation and maintaining a hydrophilic pyrite surface. At much higher xanthate concentrations (10^{-3} M), the cleaned pyrite surfaces appear to be almost immediately passivated by an overlayer likely to consist of dixanthogen which, due to the high Eh conditions and high initial concentration of xanthate, may form more rapidly and segregate at particle surfaces due to its hydrophobicity. Dixanthogen is highly volatile and would not be detected in the surface by SAM under ambient ultra-high vacuum conditions.

Other examples [14] have also demonstrated dramatic reductions in surface oxygen concentration using XPS analysis. A pyrite:chalcopyrite mineral mixture ground in an iron mill gave oxygen percentages without xanthate of 41 at.% on the initial surface and 32% after ion etching to 2.5 nm depth. Addition of isobutyl xanthate at 0.1 kg/tonne (i.e. $\approx 10^{-5}$ M) for 5 min to the same samples reduced the initial oxygen concentration to 21% with increased exposure of Cu(1.5×), S(2.8×) and reduced Fe exposure (0.75×) consistent with the mechanism of surface cleaning and dispersal of iron hydroxides.

A similar action is found with the collector dicresyldithiophosphate [113]. Table 3 shows that the surface oxygen percentage is dramatically reduced by addition of 7.5×10^{-5} M at pH 4 (maximum adsorption). Removing the fine particles from the ground pyrite sample, by three successive ultrasonication/decantation steps with solution replacement, shows that the majority of the material removed by the collector addition is in the form of fine particles rather than oxidized surface layers. Table 3 shows a level of surface oxygen concentration before fines removal similar to that after collector addition.

This surface concentration is affected by addition of the dithiophosphate collector to the deslimed pyrite sample. After collector addition, both sulfate and ferric hydroxide signals are reduced in the XPS spectra and the contamination by calcium, present on the surface before collector addition, is also removed. There has clearly been a process in which oxidized fine particles are detached from the pyrite surface by the collector addition similar to that in mechanical removal of these fine particles.

Table 3. Survey of the AC% of elements for undeslimed pyrite particles conditioned at pH 4, N_2 purged

Element	Initial	Etch	Initial	Etch
	([DCDTP]=0 M)		([DCDTP] = 7.5×10^{-3} M)	
Carbon (C)	28	7.8	26	8.3
Oxygen (O)	28	24	11	5.2
Iron (Fe)	10	27	21	39
Sulfur (S)	27	33	40	47
Calcium (Ca)	3.6	4.6	0.45	< 0.05
Phosphorous (P)	-	-	0.28	0

Table 4. Survey of the AC% of elements for deslimed pyrite particles conditioned at pH 4, N_2 purged

Element	Initial	Etch	Initial	Etc
	([DCDTP]=0 M)		([DCDTP] = 7.5×10^{-3} M)	
Carbon (C)	33	8.3	45	18
Oxygen (O)	13	2.4	14	4.0
Iron (Fe)	20	49	11	40
Sulfur (S)	33	39	28	36
Phosphorous (P)	-	-	0.47	0

3.4 Detachment of Colloidal Oxide/Hydroxide Particles

Results and discussion presented in the previous section. are clearly relevant to this action of collector adsorption. In addition, ethyl xanthate has been specifically shown to detach colloidal iron oxide particles [59] and freshly precipitated layers of ferric hydroxide [77] from galena particle surfaces. There is still some controversy on the xanthate concentrations necessary to achieve effective removal of attached particles and surface layers. In some cases, concentrations well in excess of those in flotation practice have been required to observe removal of surface oxygen [77] whereas substantial reductions of oxygen concentrations have also been observed at concentrations of the order of 10^{-5} M xanthate [14,113]. The difference between these two concentrations may, however, lie in separation of the dispersing action, removing detached fine particles, from the complexation action dissolving oxidized surface layers. For instance, removal of $Cu(OH)_2$ precipitates, giving rise to the high binding energy component of the XPS Cu2p signals and the satellite structure for Cu(II) above 940 eV (see Fig. 11), is achieved using 10^{-4} M ethyl xanthate solution at pH 6–10 [95]. In contrast, iron hydroxides are only wholly removed (i.e. quantitatively) from galena surfaces using ethyl xanthate concentrations similar to those required for EDTA complexation (i.e. $> 10^{-3}$ M) [77]. Separate XPS investigations [98,114] have confirmed this displacement reaction. Xanthate ions exchange with hydroxide species on the preoxidized galena surface to form soluble OH species [98]. There is a significant reduction in the surface oxygen concentration according to a reaction like (11).

The displacement of oxidized products by the xanthate chemisorption reaction (11) also gives increased particle contact angle and floatability [114].

3.5 Removal of Oxidized Surface Layers

Much of the discussion in the previous two sections has been concerned not only with collector displacement of oxidized fine sulfide particles and metal hydroxide colloidal precipitates but also with the actions of collectors on amorphous, oxidized surface layers. Some additional comments are, however, relevant.

Mielcarzski and *Minni* [79,125] have proposed four stages in the mechanism of dithiophosphate interaction with chalcocite based on correlated XPS and FTIR measurements:

1. In the first minute, preadsorbed hydrocarbon contaminants begin to be removed with corresponding adsorption of the collector on to the oxidized (i.e. oxide/hydroxide/carbonate) surface layer;
2. Between 1 and 5 min, further removal of hydrocarbons and some oxide/hydroxide/carbonate groups continues with increasing orientation of the adsorbed molecules.
3. At adsorption times of 5–15 min, desorption of the oxidized products is completed and a relatively well-ordered, oriented layer of collector molecules is formed;
4. Continuation of the adsorption process results in the formation of a layer of precipitated copper diethyldithiophosphate on top of the close-packed monolayer.

Mielczarski [115] has also observed the formation of a monolayer of iron xanthate, followed by multilayer coverage including the formation of diethyl dixanthogen and the simultaneous removal of oxidation products from marcasite surfaces.

Removal of all iron (III) species and oxidized surface layers with EDTA has been directly demonstrated by XPS [77]. However, it must be pointed out that the action of xanthate in removing oxidized fine particles and surface layers is subject to the extent of prior oxidation of the mineral surface, the length of time of conditioning after xanthate addition and the subsequent length of time in the pulp solution before flotation. The presence of these oxidation products in a pyrite:chalcopyrite mixture, stored in deoxygenated solution at pH 8.5 for long periods (i.e. up to 100 days) has been demonstrated in SAM analyses with or without the addition of isobutyl xanthate (10^{-5} M) to the ground product [1]. XPS analyses using the small spot (i.e. 200 µm) XPS analyzer of the PHI 5000LS system, has added some further information to the SAM analyses.

Table 5 shows the surface elemental concentrations of six different particles with widely varying percentages of all elements examined. The sample from which the six particles were selected in this case was that with the added isobutyl xanthate although similar results are obtained from the sample without xanthate after long storage times. The wide variation in composition between the particles is difficult to explain without the evidence from SAM (e.g. Figs. 2,3,5). In general, however, there is an increase in the concentration of oxygen and a corresponding reduction in exposure of copper and sulfur signals particularly compared to the samples with xanthate addition discussed previously. The variability of the carbon percentage (i.e. 15–44%) suggests that different mineral particles have considerably different surface coverages of adsorbed xanthate and, in general, it appears that more xanthate is adsorbed on surfaces with lower levels of surface oxidation as represented

Table 5. XPS surface concentrations (at.%) from individual particles in the sample with added isobutyl xanthate after longer storage

Particle	C1 *s*	O1 *s*	Fe2 *p*	S2 *p*	Cu2 *p*3/2
1	44	32	8.7	12	3.1
2	27	48	50	19	2.0
3	36	40	6.5	13	5.0
4	20	50	16	12	2.6
5	15	62	5.3	16	1.8
6	20	58	8.2	11	2.9
Average	27	48	8.3	14	2.9

in the oxygen atomic concentration or, conversely, where xanthate adsorption occurs the oxygen atomic concentration is reduced through the cleaning mechanism. The analyses also suggest that, after the longer times of storage, the iron concentration is increased at the expense of the copper concentration.

The C1 *s* spectra from both the area-average and individual particle surfaces, in addition to hydrocarbon contamination signals near 284.8 and 286.5 eV, also show a small contribution apparently due to carbonate species near 288.5 eV. The area average S2 *p* signal (Fig. 17) shows curve fits for a sulfide doublet ($2p_{3/2}$ near 160.8 eV), a higher binding energy doublet ($2p_{3/2}$ near 162.8 eV) and a sulfate doublet ($2p_{3/2}$ near 168.6 eV). Hence, at least three different species contribute to the S atomic concentration in Table 5. The $2p_{3/2}$ sulfide peak at 160.8 eV corresponds closely to that from chalcopyrite fracture surfaces at 150 K [32]. The $2p_{3/2}$ peak at 162.8 eV can be attributed to pyrite (i.e. 162.6 eV), adsorbed xanthate (i.e. 162.5 eV) [95], or to a high binding energy-shifted metal-deficient sulfide species from oxidation of chalcopyrite [32]. The O1 *s* spectra in all cases, i.e. area average and individual particles, were very broad (529.5–533 eV) suggesting overlap

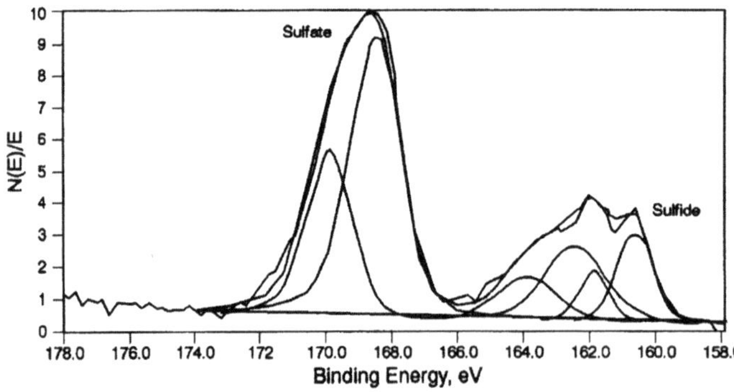

Fig. 17. XPS area-average S2 *p* spectra three curve fitted doublets for S2 $p_{1/2}$ (2:1 intensities, separation 1.2 eV) components corresponding to sulfide, adsorbed xanthate (or metal-deficient sulfide) and sulfate (see text)

of contributions from several different species, i.e. O^{2-}, OH^-, carbonate, sulfate, etc.

The multispecies nature of other signals can also be seen in Fig. 18 where the signals from Fe2p, S2p and Cu2p are compared between area average (AA) and a single particle (SP). The Fe2p spectra (Fig. 18a) clearly show that individual particles (SP) can give signals for iron sulfide species (≈ 707 eV) whereas the overall exposure (AA) of these sulfide species, compared to the

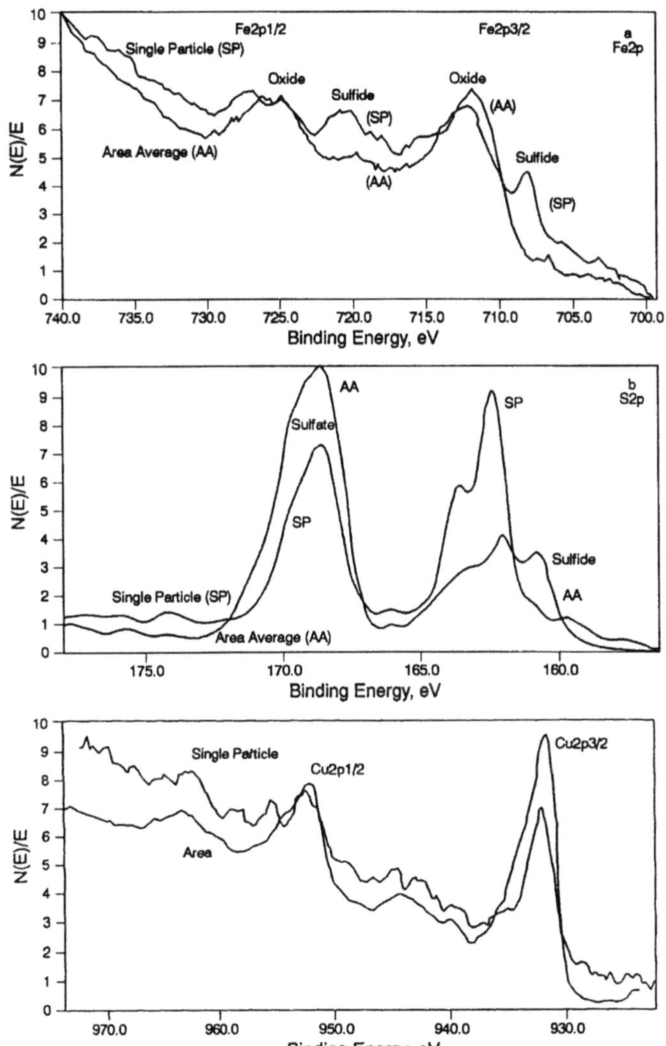

Fig. 18. XPS spectra comparing area-average (AA) and single-particle (SP) surface analysis for (**a**) Fe2p, (**b**) S2p and (**c**) Cu2p regions

predominant ferric hydroxide species, is very low. Correspondingly, the S2p signal comparison (Fig. 18a) shows a predominant doublet with $2p_{3/2}$ near 162.8 eV (i.e. pyrite, adsorbed xanthate or metal-deficient sulfide) on the single particle with a reduced proportion of sulfate relative to the area average. The Cu2p spectra (Fig. 18c) show a higher proportion of Cu(II), seen in the shake-up satellite intensity near 944 eV, when compared with the predominantly Cu(I) species normally found on surfaces conditioned for shorter times [1]. The comparison between the area-average and single particle spectra also shows distinct high binding energy shoulders on both the Cu2$p_{3/2}$ and Cu2$p_{1/2}$ signals attributable to charge-shifted hydroxide, sulfate or oxide species. The intensities of these shoulders are clearly higher in the area average spectrum than the individual particle spectrum suggesting that the more extensive xanthate adsorption on the individual particle has displaced some of this material. This result is in accord with that reported in previous work (e.g. Fig. 20) [1].

In summary, the samples after longer storage show wide variation of surface compositions of individual particles, some particles exhibiting relatively high levels of adsorbed xanthate and correspondingly lower levels of surface oxidation whilst other particles had significant increases in surface oxidation. Area-average spectra show an overall increase in surface oxidation compared with samples conditioned for shorter times and a corresponding reduction in the exposure of copper signals although all of the individual particles examined gave distinct copper surface concentration. Sulfate and carbonate species were also detected in area-average and individual-particle spectra. Some individual particles exposed iron as sulfides but the overall level of exposure in the area-average spectra was very low, i.e. predominately as ferric hydroxide species.

The presence of the considerable number and variety of attached fine particles with dimensions from $< 0.1\,\mu m$ to tens of micrometers, providing substantial coverage of larger (i.e. 100–200 µm) particle surfaces as in Fig. 5, helps to explain the wide variation in XPS surface compositions in Table 5. It is now apparent that 200 µm diameter areas of analysis can contain widely varying proportions of oxidized flocs, fine particles, surface layers and adsorbed species (e.g. sulfate, carbonate).

3.6 Inhibition of Oxidation

Our SAM results described in Sect. 3.3 have demonstrated that, at high xanthate concentrations (i.e. 10^{-3} M), the formation of a surface layer, likely to consist of the oily, hydrophobic dixanthogen species, can inhibit oxidation of pyrite surfaces which would otherwise produce iron hydroxide oxidation products on the surface and in solution at lower xanthate concentrations. These results call into question the interpretation and relevance of mechanisms derived from studies at higher collector concentrations where the collector has been added in the grinding or early in the conditioning stage.

3.7 Aggregation and Disaggregation

The aggregation or disaggregation action of collector molecules has been difficult to quantify even in conditioned feeds to flotation. The action of aggregation caused by bubble attachment during flotation also makes it very difficult to follow in the separation between concentrate and tails. There are very few references in the literature to studies of the aggregate (rather than particle) size and composition likely to be most successfully separated in flotation.

The question of whether the addition of collectors causes aggregation or disaggregation of larger (i.e. $> 5\,\mu m$) particles appears to depend heavily on the conditions under which the collector is added. In high-intensity conditioning (i.e. > 850 rpm), shear flocculation can occur [116]. In normal conditioning and flotation, xanthates have been suggested [9] to act in a similar manner to classical dispersing agents, e.g. hexametaphosphates. Their effectiveness in dispersing the particles clearly depends on such factors as: selective adsorption; surface charge on the minerals and colloids (i.e. pH relative to i.e.p values); ionic strength of the solution (i.e. dissolved species); and surface coverage. These factors are highly variable between different flotation pulps and in different laboratory experiments.

Fine/coarse particle interactions and aggregation have been studied for the copper activation of sphalerite and the effect of xanthate addition using correlated microflotation and on-line particle size distribution analysis coupled with surface analysis. Significant aggregation of $< 20\,\mu m$ fine particles to 20–30 μm aggregates is already observed during initial conditioning at pH 5.5 and 8.5 during the first 20 min before any reagent addition. Addition of Cu(II) at pH 5.5 produces some development of larger aggregates 100–120 μm but flotation response of the fine particles remains poor. At pH 8.5 there is no evidence of the larger aggregates on copper addition and flotation recovery of the fine particles is similarly poor. Addition of xanthate at pH 5.5 dramatically increases the extent of aggregation $> 100\,\mu m$ and flotation recovery ($> 95\%$) whereas, at pH 8.5, only weak larger aggregates appear to form and the recovery does not increase to the same extent (i.e. 60–80%). In the presence of coarser particles (38–75 μm) at pH 5.5, with Cu(II) addition alone, a very high percentage of the fine particles were recovered and a fine/coarse particle aggregation (i.e. "piggy-backing") mechanism was confirmed by on-line particle size analysis and optical microscopy. At the higher pH 8.5, the interaction between fine and coarse sphalerite is slower and much less complete with correspondingly poorer flotation of both the fine and coarse fractions. Surface analysis confirms that this is due to partial coverage of the surface by colloidal hydroxides and overall hydrophilicity inhibiting strong hydrophobic interactions between the particles [73].

A direct comparison of the dispersing action of ethyl xanthate with anionic dispersants (e.g. polyphosphates) [59] has been made for iron oxide/hydroxide on quartz surfaces. Full dispersion is only achieved at xanthate concentrations greatly in excess of those generally used in flotation practice. Nevertheless, the results described above in which colloidal oxidized fine particles are detached from galena and other sulfide surfaces does suggest that the collector may have a role, albeit a minor one, in disaggregation under normal flotation conditions. This action of collectors deserves more systematic study than it has yet received.

In several industrial research projects [18,19,117] and in experience from plant practice, changes in the particle size distribution can be directly related to the addition of collector, its concentration, point of addition (e.g. during grinding, before or after cyclone classification) and length of conditioning time.

3.8 Patch-wise and Face-specific Collector Interaction

The observation of patch-wise attachment of colloidal lead xanthate from in situ studies of ethylxanthate-treated galena surfaces [25] has been previously described (Fig. 9). The formation and attachment of precipitated species can clearly give rise to relatively low surface coverages in isolated regions or patches.

The early microradiographic analysis of *Plaksin* et al. [118] first proved that the xanthate distribution on the galena surface was not uniform and that an increase in xanthate uptake did not necessarily result in increased flotation recovery. More recently, *Brinen* and co-workers [106,107] have used ToF-SIMS, in near static SIMS conditions, to detect and map the distribution of dithiophosphinate on galena crystal surfaces. Non-uniform adsorption was directly observed from 5×10^{-6} M solution conditioning for 5 min at pH 6–8. The concentrate from flotation (+75–150 µm fraction) was examined. The studies provided evidence to suggest that adsorption of the collector occurred in oxygen-rich areas and that there is some face-specificity on single galena particles. The uneven distribution of dithiophosphinates is attributed to mineral surface hetereogeneity arising from defects, impurities, lattice imperfections and small variations in stoichiometry but the correlation with oxygen-rich areas strongly supports the mechanism of parallel oxidation/adsorption or adsorption on to preoxidized areas of the galena surfaces discussed in previous sections. ToF-SIMS has also been used by *Chryssoulis* and colleagues [20,105] to image a mixture of amyldithiophosphate and amylxanthate on laboratory-treated mineral surfaces. The distribution of these reagents was again distinctly non-uniform.

To date, no direct imaging and mapping of the distribution of collectors on sulfide surfaces other than galena have been published. It remains to be seen whether the DTP and xanthate adsorption systems on galena surfaces, in which colloidal precipitates are known to form, is a specific example of non-uniform coverage or whether this phenomenon is general. Correlation of xanthate and other collector distributions with oxidized and unoxidized regions of sulfide surfaces in, for instance, pyrite and copper-activated sphalerite flotation systems is an obvious extension of the highly sensitive ToF-SIMS technique.

3.9 Implications for Flotation Practice

It may be thought that these examples of mechanisms of surface oxidation of sulfide minerals and the actions of collector adsorption in modifying this surface chemistry now provide a respectable body of work on which to base strategies for improved flotation practice. However, these mechanisms are much more complex, diverse and interactive than we might have envisaged even ten years ago. Our understanding of the factors directly influencing flotation in any particular set of conditions is still rudimentary. More questions are generated by each attempt to systematically analyse the effect of one particular variable. It remains true [2] that the "flotation operation is usually knife-edged, control difficult and results variable".

Nevertheless, there are some examples in which the surface analytical studies, combined with flotation metallurgy and solution chemistry, have directly contributed to improvements in flotation recovery and grade.

For example naturally floatable iron sulfides have been identified and separately removed in copper flotation at Mount Isa [13]. This study also identified the selective removal of ferric hydroxides and carbonates by collector addition. Characterization of flotation products from a lead-zinc-copper concentrator in south eastern Missouri using XPS [91] verified the action of surface modifiers used in conditioning and correlated their surface concentrations and oxidation states with flotation response. The study noted that lead in the bulk galena-chalcopyrite scavenger concentrate was significantly more oxidized than the lead in the bulk galena-chalcopyrite rougher concentrate. The effects of fine grinding on flotation performance have been surveyed in a correlated XPS-flotation study [119] and specific application to the zinc regrind at Cominco Alaska's Red Dog mine has been reported [120]. The action of an extended period of aerated conditioning before copper activation and collector conditioning in increasing sphalerite flotation at the Murchison Zinc (Australia) concentrator [15] was explained, using XPS, by the removal of zinc hydroxides from sphalerite surface and the concomitant appearance of a metal-deficient sphalerite surface. XPS also demonstrated an increase in the oxidation state of pyrite after aerated preconditioning. The presence of excessive surface oxidation in copper reflotation at Western Mining Corporation's Olympic Dam operation [121], identified by XPS analysis, led to improved operation of Lasta filters. A low flotation rate of galena in lead roughers at the Hilton Concentrator of Mt. Isa Mines was analyzed [122,123] using XPS. The presence of precipitated species and their removal by a change of conditioning reagents (i.e. lime to soda ash) and collector reagent (i.e. ethyl xanthate to dicresyldithiophosphate, a collector stable in the presence of sulfite species over a wide pH range) has been used to address this problem.

A variety of industrial research projects, not normally reported in the literature, have used and are now using these techniques in relatively routine studies [18,19,20,103,104,105,117].

4 Remaining Research Issues

In several parts of the discussion in this review, it will become apparent that there are unresolved questions relating to both oxidation mechanisms and collector adsorption mechanisms. In some cases, argument continues concerning the nature of surface species, in other cases there are conflicting research results on nominally similar systems and, in yet other cases, alternative explanations for observed phenomena do not have direct evidence to substantiate one or other hypothesis. We will summarize some of these issues in this section.

4.1 Oxidation Mechanisms

Surface species confined to literally the top atomic layer of the sulfide surfaces have only recently been revealed by synchrotron XPS studies of sulfide surfaces before oxidation. The application of this highly surface-sensitive technique to the definition of these species in the top layer, sub-surface and bulk of the sulfide during oxidation will considerably assist our understanding of the oxidation mechanisms.

The sites of oxidative attack on the different mineral sulfide surfaces are not yet well defined. Evidence for oxidation at steps, edges and corners on pure minerals have been found but impurity sites appear to dominate oxidative attack on natural mineral surfaces. Chemical and structural information at the atomic level is still needed to define the mechanisms of oxidation in air and in different solution conditions. This is likely to come from the scanned probe microscopies (STM and AFM) particularly as their compositional capabilities are extended.

The question of the relevance of ex situ surface analytical results to real flotation conditions requires further correlation of these techniques with in situ studies from FTIR, Raman spectroscopy and the scanned probe microscopies. The body of evidence from XPS, SAM, SEM and ToF-SIMS now available provides impressive correlations with chemical changes expected in oxidation, collector adsorption and flotation processing but independent verification will continue to be required in closely correlated studies.

Recent developmental work within our institute [124], comprising correlated XPS and static ToF-SIMS measurements of oxidation of galena surfaces, exemplify such studies. A freshly fractured PbS surface reacted in air-equilibrated pH 8 aqueous solution for increasing periods of time has shown a systematic variation of S_n/S secondary ion yield ratios with increasing atomic percent S2p contribution from oxidized sulfur species. Figure 19

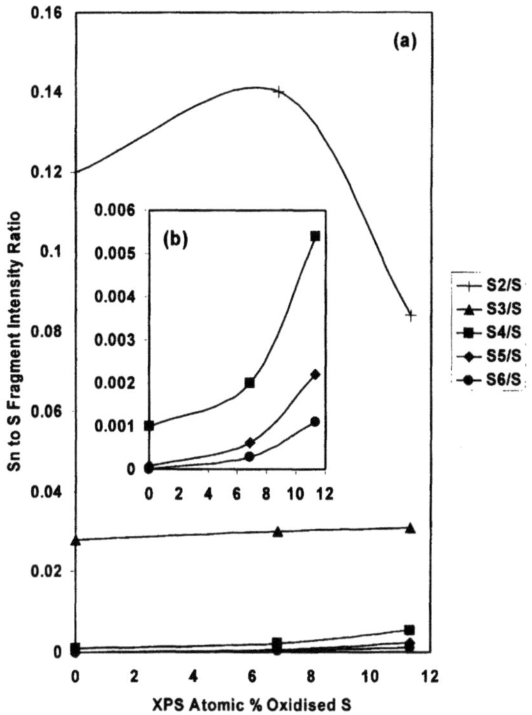

Fig. 19. (**a**) Atomic percent S2p contribution from oxidized sulfur species vs. S_n/S secondary ion yield ratio for sequentially oxidized galena. (**b**) Expanded S_4/S, S_5/S, S_6/S region

summarizes these results. In particular, the S_2/S ratio initially increases and then decreases with oxidation time, the S_3/S ratio increases slightly and the S_4/S, S_5/S and S_6/S ratios increase systematically with oxidation. This trend does not appear to correlate with total sulfur exposure (coverage) alone as measured by low take-off angle XPS. The results therefore strongly suggest increasing polymerization of S–S species at the expense of lower-order oligomers with increasing oxidation. These results are in accord with the XPS assignment of the oxidized sulfur species with polysulfides of increasing chain length. ToF-SIMS imaging of fractured and oxidized galena surfaces, at the micron scale, show a uniform coverage of oxidation products but preferential adsorption of monovalent ions (Na, K) at step edges and corners (Fig. 20). Previous STM studies [25,88] have suggested that impurities and defects in the galena surface may be prefered over high-energy edges as sites for oxidation. The role of species such as Na and K, undetected by XPS, SEM, STM, etc., on oxidation and reaction mechanisms is unclear.

In collectorless flotation, the species responsible for hydrophobicity and bubble-particle attachment in flotation still generates intense discussion at meetings concerned with the surface chemistry of flotation of sulfide minerals.

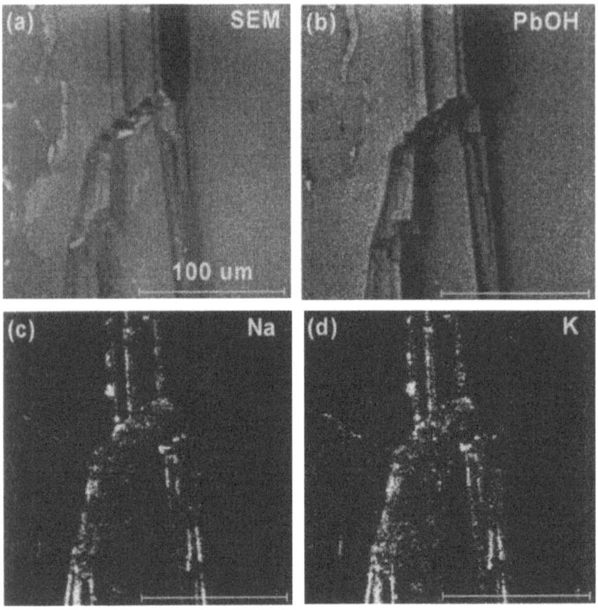

Fig. 20. (**a**) Ion induced secondary electron image; (**b**) $PbOH^+$; (**c**) Na^+ and (**d**) K^+ ToF-SIMS secondary ion images of a PbS fracture surface oxidized for 30 min at pH 8 under air equilibrated aqueous solution

The contenders, metal-deficient sulfide surfaces, polysulfides and elemental sulfur, each have strong support groups. The review has discussed specific examples of conditions where one or another of these species has been reliably shown to dominate the surface layers. In many other cases, this distinction is much less obvious and highly technique-dependent. For instance, extraction of the same surfaces with ethanol or cyclohexane produces elemental sulfur with the first solvent but not with the second solvent. In cases where elemental sulfur has been extracted from the surface into the solution phase, surface analytical techniques suggest that the elemental sulfur was not present in the surface layers before extraction (i.e. metal-deficient sulfide or polysulfide species only were found). Hence, the surface species may be altered by interaction with the solvent. Studies of the species present on mineral surfaces concentrated in the flotation process, using both ex situ and in situ techniques, may help to resolve conditions under which one or another of these species is responsible for flotation.

The chemical forms and distribution of non-colloidal oxidized species on surfaces are still significantly uncertain. In some cases, high surface concentrations of sulfate cannot be associated with stoichiometric amounts of cations, relatively thick surface layers of apparently soluble species are found and they are not removed by agitation or ultrasonication, and the thickness of these layers varies from a few nm to hundreds of nm over the same particle surface. The development of these surface layers over time in different solution conditions will require study over many years.

Differences in oxidation between particles of different size have been noted in many research papers. These differences may be confined to rates of oxidation related to surface area and defect site concentrations but it is also possible that other mechanisms (e.g. surface free energy differences, impurity segregation, fracture faceting) may be directing these processes. More systematic study of particle size effects in oxidation of specific minerals is still required.

4.2 Collector Adsorption and Reaction Mechanisms

Perhaps the most obvious issue requiring further research lies in studies of the collector species that directly contribute to flotation. This will require comparison of concentrate with tails samples during flotation separation. The majority of research to date has focused on the conditioned mineral before flotation separation. Very few studies comparing species between concentrate and tails have been reported.

Conflicting results from different techniques and different adsorption conditions have been reported on the association of adsorbed collector with oxidized or non-oxidized regions of sulfide mineral surfaces. This question has not yet been systematically studied despite the variety of mechanisms postulating hydroxide displacement, metal or sulfide site attachment and colloid displacement. More direct evidence correlating in situ scanned probe microscopy studies with XPS, SAM and ToF-SIMS would increase our confidence in these mechanisms. A recent XPS/ToF-SIMS study of collector adsorption on oxidized galena surfaces [124] has imaged, at high isopropyl xanthate collector concentration, the formation of droplet-like regions depleted of Na, K and Ca (Fig. 21b) while imaging of the isopropyl xanthate molecular ion, IPX^- (Fig. 21c) shows uniform coverage of the collector. These features, which appear to cluster near cleavage edges, are suggested to be due to dixanthogen formation and subsequent volatilization in ultra-high vacuum (no sample cooling was used). The use of the collector molecular ion and the sensitivity of the ToF-SIMS technique is also demonstrated by the detection and discrimination of isobutyl xanthate impurity in the isopropyl product (Fig. 21d). In future SIMS investigations, particularly where real pulp and stream samples are studied and where combinations of reagents are used, it will be essential to monitor molecular parent ions of organic reagents in order to be assured of measuring the actual distribution of the species of interest.

A related question concerns the role of colloidal metal-collector species. Are these formed by replacement of adsorbed OH species (or colloids) by the collector anion during adsorption or does the metal-collector species form first in solution and then displace metal hydroxide colloids? Do metal-hydroxy-collector species form on surfaces? The presence of metal-collector colloids on

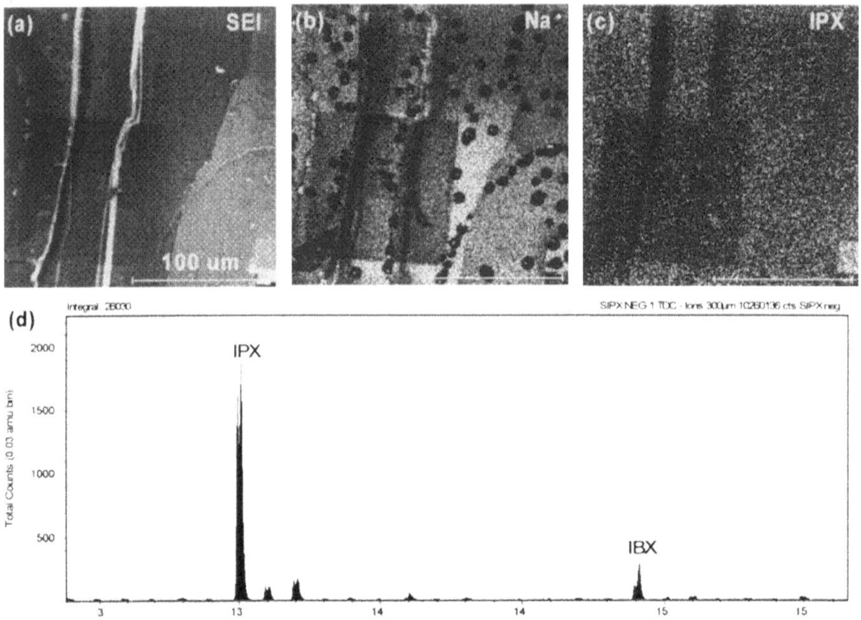

Fig. 21. (**a**) Ion induced secondary electron image; (**b**) Na^+ ion; (**c**) iso-propyl xanthate, IPX^- molecular ion images and (**d**) selected range ToF-SIMS mass spectrum of a PbS fracture surface oxidized for 30 min at pH 8 under air-equilibrated aqueous solution and subsequently exposed to 10^{-3} M sodium iso-propyl xanthate. The square region to bottom left is the raster pattern left by a 100 μm × 100 μm sputter

surfaces has been clearly established but the mechanism of their formation or attachment requires clarification.

The lateral distribution of collector species on sulfide mineral surfaces has now been shown by a variety of techniques to be non-uniform but it is not yet clear whether this non-uniformity is determined by faceting, oxidation, colloid growth and adsorption and/or defects.

There are some examples in the literature of major differences in behavior of the mineral surfaces between high ($> 10^{-3}$ M) and low (e.g. 10^{-5} M) collector concentrations. Differences in the structure and reactivity of the adsorbed layer and the sulfide surface are implied by these results but there are very few reports in the literature. This question is potentially of major importance because mechanisms derived at one concentration may not be applicable to another concentration (i.e. flotation conditions).

There are also conflicting reports in the literature regarding the tendency of collector molecules to cause aggregation or disaggregation of particles under different conditions of concentration, shear and pH. This aspect of flotation related to collector adsorption would benefit from systematic study but reliable observations of particle size distribution changes are difficult to achieve.

There is no shortage of interesting and useful research problems in the application of surface analytical techniques to the understanding of mechanisms in the flotation separation of minerals to occupy us into the future.

Acknowledgements

Much of the work reported here has involved extensive collaboration with other members of the Ian Wark Research Institute, namely Professor John Ralston, Dr. Daniel Fornasiero, Mr. Angus Netting, Dr. Pawittar Arora, Mr. Darren Simpson, Ms Angela Lange (Thiel). Collaboration with Dr. John Frew, Mr. Kevin Davey and Mr. Ross Glen at CSIRO Division of Minerals is also gratefully acknowledged. The assistance of the Australian Mineral Industries Research Association (Dr. Jim May, Mr. Bruce Fraser, Mr. Phillip Campbell, Mr. David Stribley) in providing support for more than eight years of research under industrial contracts has been of paramount importance. In particular, valuable discussions with industrial sponsors has provided much of the focus for this work, particularly: Dr. Bill Johnson (MIM); Mr. Geoff Richmond (Aberfoyle); Dr. Ian Clark, Dr. Greg Lane (Western Mining Corporation); Dr. Mike Fairweather (Cominco). The Australian Research Council has also provided funding for parallel fundamental studies in a Senior Research Fellowship (R. St.C. S.) and Research Fellowship (C.A.P.).

The first draft of this contribution was critically reviewed by Prof. Alan Buckley. His comments and questions have been most useful in refocusing many issues and are gratefully acknowledged.

References

1. R. St. C. Smart: Surface layers in base metal sulfide flotation, Min. Eng. **4**, 891–909 (1991)
2. L. K. Shannon, W. J. Trahar: The role of collector in sulfide ore flotation, in P. Somasundaran (Ed.): *Advances in Mineral Processing* (Soc. Mining. Eng., Colorado 1986) pp. 408–425
3. P. E. Richardson: Surface chemistry of sulfide flotation, in D. J. Vaughan, R. A. D. Pattrick (Eds): *Mineral Surfaces* (Chapman and Hall, New York 1995) pp. 261–302
4. R. St. C. Smart: Minerals, ceramics and glasses, in J. C. Revière, S. Myhra (Eds.): *Handbook of Surface and Interface Analysis* (Dekker, New York 1998) pp. 543–606
5. M. C. Fuerstenau (Ed.): *Flotation*, A. M. Gaudin Memorial, Vols. 1, 2. (Amer. Inst. Min. Met. Pet. Eng., New York 1976)
   M. C. Fuerstenau, J. D. Miller, M. C. Kuhn: *Chemistry of Flotation* (Amer. Inst. Min. Met. Pet. Eng., New York 1985) pp. 28–29
6. J. Leja: *Surface Chemistry of Froth Flotation* (Plenum, New York 1982)

7. P. Somasundaran (Ed.): *Advances in Minerals Processing* (Soc. Mining Engineers, Colorado 1986)
8. K. S. E. Forssberg (Ed.): *Flotation of Sulfide Minerals* (Elsevier, Amsterdam 1986)
9. J. S. Laskowski, J. Ralston (Eds.): Colloid Chemistry in Mineral Processing, Dev. Min. Process. 12 (Elsevier, Amsterdam 1992)
10. D. J. Vaughan, R. A. D. Pattrick (Eds): *Mineral Surfaces* (Chapman and Hall, New York 1995)
11. P. J. Guy, W. J. Trahar: The effects of oxidation and mineral interaction on sulfide flotation, in E. C. Forssberg (Ed.): *Flotation of Sulfide Minerals* (Elsevier, Amsterdam 1985) pp. 28–46
12. J. B. Zachwieja, J. J. McCarron, G. W. Walker, A. N. Buckley: Correlation between the surface composition and collectorless flotation of chalcopyrite, J. Coll. Int. Sci. **132**, 462–468 (1989)
13. S. Grano, J. Ralston, R. St. C. Smart: Influence of electrochemical environment on the flotation behavior of Mt. Isa copper and lead-zinc ore, Int. J. Min. Proc. **30**, 69–97 (1990)
14. P. Clarke, D. Fornasiero, J. Ralston, R. St. C. Smart: Min. Eng. **8**, 1347 (1995)
15. Z. Kristall, S. R. Grano, K. Reynolds, R. St. C. Smart, J. Ralston: *An investigation of sphalerite flotation in the Murchison Zinc concentrator*, Proc. Fifth Mill Operators' Conference (Aust. Inst. Min. Metall. Publ. 1994) pp. 171–180
16. P. Clarke, D. Fornasiero, J. Ralston, R. St. C. Smart: A study of the removal of oxidation products from sulfide mineral surfaces, Mineral. Eng. **8**, 1347–1357 (1995)
17. J. O'Connor, B. A. Sexton, R. St. C. Smart (Eds): *Surface Analysis Methods in Materials Science*, Springer Ser. Surf. Sci **23**, (Springer, Berlin, Heidelberg 1992)
18. J. Ralston, R. St. C. Smart (Eds.): *Interaction of Iron Sulfide Minerals and their Influence on Sulfide Mineral Flotation*, Australian Minerals Industry Research Association, Project P260 (AMIRA, Melbourne 1992)
19. T. J. Napier-Munn (Ed.): *The Methods and Benefits of Fine Grinding Ores*, Australian Minerals Industry Research Association, Project P336 (AMIRA, Melbourne 1994)
20. K. G. Stowe, S. L. Chryssoulis, J. Y. Kim: Mapping of composition of mineral surfaces by ToF-SIMS, Min. Eng. **8**, 421–430 (1995)
21. R. St. C. Smart, W. M. Skinner, A. K. O. Netting, D. Simpson, P. Bandini, C. Greet (Eds.): *Protocols for Surface Aanalysis in Minerals Processing*, Australian Minerals Industry Research Association (AMIRA, Melbourne 1996)
22. S. L. Chryssoulis: Mineral surface characterisation by TOF-LIMS and ToF-SIMS, AMTEL Report no. 21 (UWO, London 1994)
23. M. F. Hochella Jr.: Mineral surfaces: their characterisation and their chemical, physical and reactive natures, in D. J. Vaughan, R. A. D. Pattrick (Eds.): *Mineral Surfaces* (Chapman and Hall, New York 1995) pp. 17–60
24. K. Laajalehto, R. St. C. Smart, J. Ralston, E. Suoninen: STM and XPS investigation of reaction of galena in air, Appl. Surf. Sci. **64**, 29–39 (1993)
25. B. S. Kim, R. A. Hayes, C. A. Prestidge, J. Ralston, R. St. C. Smart: Scanning tunneling microscopy studies of galena: the mechanisms of oxidation in aqueous solution, Langmuir, **11**, 2554–2562 (1995)
26. A. N. Buckley, R. Woods: The surface oxidation of pyrite, Appl. Surf. Sci. **27**, 347–452 (1987)

27. A. N. Buckley, I. C. Hamilton, R. Woods: Investigation of the surface oxidation of sulfide minerals by linear potential sweep voltammetry and X-ray photoelectron spectroscopy, in K. S. E. Forssberg (Ed.): *Flotation of Sulfide Minerals* (Elsevier, Amsterdam 1985) pp. 41–60
28. A. N. Buckley, R. Woods: X-ray photoelectron spectroscopy of oxidized pyrrhotite surfaces. I. Exposure to air, Appl. Surf. Sci. **2/23**, 280–287 (1985)
29. A. N. Buckley, R. Woods: X-ray photoelectron spectroscopy of oxidized pyrrhotite surfaces. II. Exposure to aqueous solutions, Appl. Surf. Sci. **20**, 472–480 (1985)
30. A. R. Pratt, I. J. Muir, H. W. Nesbitt: X-Ray photoelectron and Auger electron spectroscopic studies of pyrrhotite and mechanism of air oxidation, Geochim. Cosmochim. Acta. **58**, 827–841 (1994)
31. C. F. Jones, S. LeCount, R. St. C. Smart, T. White: Compositional and structural alteration of pyrrhotite surfaces in solution: XPS and XRD studies, App. Surf. Sci. **55**, 65–85 (1992)
32. A. N. Buckley, R. Woods: An X-Ray photoelectron spectroscopic study of the oxidation of chalcopyrite, Aust. J. Chem. **37**, 2403–13 (1984)
33. A. N. Buckley, R. Woods: An X-ray photoelectron spectroscopic study of the oxidation of galena, Appl. Surf. Sci. **17**, 401–414 (1984)
34. D. Fornasiero, F. Li, J. Ralston, R. St. C. Smart: Oxidation of galena surfaces, I. X-ray photoelectron spectroscopic and dissolution kinetics studies, J. Coll. Int. Sci. **164**, 333–344 (1994)
35. S. Richardson, D. J. Vaughan: Surface alteration of pentlandite and spectroscopic evidence for secondary violarite formation, Min. Mag. **53**, 213–22 (1989)
36. A. N. Buckley, R. Woods: Surface composition of pentlandite under flotation-related conditions, Surf. Int. Anal. **17**, 675–680 (1991)
37. A. N. Buckley: The surface oxidation of cobaltite, Aust. J. Chem. **40**, 231–9 (1987)
38. A. N. Buckley, R. Woods, H. J. Wouterlood: Surface characterization of natural sphalerites under processing conditions by X-ray photoelectron spectroscopy, in P. E. Richardson, R. Woods (Eds): Proc. Int. Symp. Electrochemistry in Mineral and Metal Processing II (Electrochem. Soc., Pennington 1988) pp. 211–233
39. A. N. Buckley, R. Woods, H. J. Wouterlood: An XPS investigation of the surface of natural sphalerites under flotation-related conditions, Int. J. Min. Proc. **26**, 29–49 (1989)
40. A. N. Buckley, H. J. Wouterlood, R. Woods: The surface composition of natural sphalerites under oxidative leaching conditions, Hydrometall. **22**, 39–56 (1989)
41. J. R. Mycroft, G. M. Bancroft, N. S. McIntyre, J. W. Lorimer, I. R. Hill: Detection of sulfur and polysulfide on electrochemically oxidized pyrite surfaces by XPS and Raman spectroscopy, J. Electroanal. Chem. **292**, 139–152 (1990)
42. A. G. Schaufuß, H. W. Nesbitt, I. Kartio, K. Laajalehto, G. M. Bancroft, R. Szargan: Reactivity of surface chemical states of fractured pyrite, Surf. Sci. **411**, 321–328 (1998)
43. H. W. Nesbitt, I. J. Muir: XPS study of a pristine pyrite surface reacted with water vapour and air, Geochim. Cosmochim Acta. **58**, 4667–4679 (1994)

44. G. M. Bancroft, M. M. Hyland: Spectroscopic studies of adsorption/reduction reactions of aqueous metal complexes on sulphide surfaces, in M. F. Hochella, A. F. White (Eds.): *Mineral-Water Interface Geochemistry*, Rev. Mineral. **23**, 511–558 (1990)
45. N. J. Rinker, H. W. Nesbitt, A. R. Pratt: Marcasite oxidation in low temperature acidic (pH 3.0) solutions: mechanisms and rate laws, Amer. Mineral. **82**, 900–912 (1999)
46. H. W. Nesbitt, I. J. Muir, A. R. Pratt: Oxidation of arsenopyrite by air and air-saturated distilled water, and implications for mechanism of oxidation, Geochim. Cosmochim. Acta **59**, 1773–1786 (1995)
47. E. Ahlberg, K. S. E. Forssberg, X. Wang (Eds.): The surface oxidation of pyrite in alkaline solution, J. Appl. Electrochem. **20**, 1033–1039 (1990)
48. J. J. McCarron, G. W. Walker, A. N. Buckley: An X-ray photoelectron spectroscopic investigation of chalcopyrite and pyrite surfaces after conditioning in sodium sulfide solutions, Int. J. Min. Proc. **30**, 1–16 (1990)
49. S. W. Knipe, J. R. Mycroft, A. R. Pratt, H. W. Nesbitt, G. M. Bancroft: XPS study of water adsorption on iron sulfide minerals, Geochim. Cosmochim. Acta **59**, 1079–1090 (1995)
50. J. R. Mycroft, H. W. Nesbitt, A. R. Pratt: X-ray photoelectron and Auger electron spectroscopy of air-oxidized pyrrhotite: distribution of oxidized species with depth, Geochim. Cosmochim. Acta **59**, 721–733 (1995)
51. J. E. Thomas, C. F. Jones, W. M. Skinner, R. St. C. Smart: The role of sulfur species in the inhibition of pyrrhotite dissolution in acid conditions, Geochim. Cosmochim. Acta **62**, 1555–1565 (1998)
52. G. H. Luttrell, R. H. Yoon: Surface studies of the collectorless flotation of chalcopyrite, Coll. Surf. **4**, 271–281 (1984)
53. Q. Yin, G. H. Kelsall, D. J. Vaughan, K. E. R. England: Atmospheric and electrochemical oxidation of the surface of chalcopyrite ($CuFeS_2$), Geochim. Cosmochim. Acta. **59**, 1091–1100 (1995)
54. D. L. Legrand, G. M. Bancroft, H. W. Nesbitt: Surface characterisation of pentlandite $(Fe,Ni)_9S_8$ by XPS, Int. J. Mineral. Proc. **51**, 217–228 (1997)
55. D. L. Legrand, H. W. Nesbitt, G. M. Bancroft: XPS survey of a pristine millerite surface and the effect of air and water oxidation, Amer. Mineral. **83**, 1256–1265 (1998)
56. I. Kartio, K. Laajalehto, E. Suoninen: Application of electron spectroscopy to characterization of mineral surfaces in flotation studies, Coll. Surf. **93**, 149–158 (1994)
57. S. Karthe, R. Szargan, E. Suoninen: Oxidation of pyrite surfaces: a photoelectron spectroscopic study, Appl. Surf. Sci., **72**, 157–170 (1993)
58. I. V. Chernyskova, S. I. Andreev: Spectroscopic study of galena surface oxidation in aqueous solution. I. XPS and ATR/FTIR spectroscopy, Appl. Surf. Sci. **108**, 225–236 (1997)
59. P. Bandini, C. A. Prestidge, J. Ralston, R. St. C. Smart: Min. Eng. **14**, 487 (2001)
60. G. Wittstock, I. Kartio, D. Hirsch, S. Kunze, R. Szargan: Oxidation of galena in acetate buffer investigated by AFM and photoelectron spectroscopy, Langmuir **12**, 5709–5721 (1996)
61. S. R. Higgins, R. J. Hamers: Chemical dissolution of the galena (001) surface observed using electrochemical STM, Geochim. Cosmochim. Acta. **60**, 3067–3077 (1996)

62. C. A. Prestidge, J. Ralston: Contact angle studies of galena particles, J. Colloid Interface Sci. **172**, 302–310 (1995)
63. A. N. Buckley: The application of X-ray photoelectron spectroscopy to flotation research, Surf. A. **93**, 159–172 (1994)
64. G. W. Heyes, W. J. Trahar: The flotation of pyrite and pyrrhotite in the absence of conventional collectors, in P. E. Richardson, S. Srinivasan, R. C. Woods (Eds): Proc. Int. Symp. Electrochemistry in Metals Processing (Electrochem. Soc., Pennington 1984) pp. 219–232
65. S. Kelebek, G. W. Smith: Collectorless flotation of galena and chalcopyrite: correlation between flotation rate and the amount of extracted sulfur, Min. Metall. Process. **6**, 123–129 (1989)
66. A. N. Buckley, K. W. Riley: The self-induced floatability of sulfide minerals: examination of recent evidence for elemental sulfur as being the hydrophobic entity, Surf. Interf. Anal. **17**, 655–659 (1991)
67. G. W. Walker, P. E. Richardson, A. N. Buckley: Workshop on the flotation-related surface chemistry of sulfide minerals, Int. J. Miner. Process. **25**, 153–158 (1989)
68. N. P. Finkelstein: The activation of sulfide minerals for flotation: a review, Int. J. Mineral. Proc. **52**, 81–120 (1997)
69. A. R. Gerson, A. G. Lange, K. E. Prince, R. St. C. Smart: The mechanism of copper activation of sphalerite, Appl. Surf. Sci. **137**, 207–223 (1999)
70. C. A. Prestidge, W. M. Skinner, J. Ralston, R. St. C. Smart: Copper (II) activation and cyanide deactivation of zinc sulfide under mildly alkaline conditions, Appl. Surf. Sci. **108**, 333–344 (1997)
71. I. J. Kartio, C. I. Basilio, R. H. Yoon: An XPS study of sphalerite activation by copper, Langmuir **14**, 5274–5278 (1998)
72. W. M. Skinner, C. A. Prestidge, R. St. C. Smart: Irradiation effects during XPS studies of Cu(II) activation of zinc sulfide, Surf. Interf. Anal. **24**, 620–626 (1996)
73. A. G. Lange, W. M. Skinner, R. St. C. Smart: Fine:coarse particle interactions and aggregation in sphalerite flotation, Mineral. Eng. **10**, 681–693 (1997)
74. T. V. Subrahmanyan, C. A. Prestidge, J. Ralston: Constant angle and surface analysis studies of sphalerite particles, Mineral. Eng. **9**, 727–741 (1996)
75. R. St. C. Smart, W. M. Skinner, A. R. Gerson: XPS of sulfide mineral surfaces: metal-deficient polysulfides, defects and elemental sulfur, Surf. Interf. Anal. **28**, 101 (1999)
76. C. Greet, R. St. C. Smart: The effect of size separation by cyclosizing and sedimentation/decantation on mineral surfaces, Miner. Eng. **10**, 995–1011 (1995)
77. C. A. Prestidge, W. Skinner, J. Ralston, R. St. C. Smart: The interaction of iron (III) species with galena surfaces, Coll. Surf. A **105**, 325–339 (1995)
78. G. D. Senior, W. J. Trahar: The influence of metal hydroxide and collector on the flotation of chalcopyrite, Int. J. Min. Proc. **33**, 321–341 (1992)
79. J. Mielcarzski, E. Minni: Adsorption of diethyldithiophosphate on cuprous sulfide, Surf. Int. Anal. **6**(5) 222–234 (1984)
80. P. Clarke: The interaction of metal ions and their hydrolysis products with sulfide mineral surfaces, Ph.D. thesis, University of South Australia, Adelaide (1995)
81. P. Eadington: Study of the oxidation layers on surfaces of chalcopyrite by use of Auger electron spectroscopy, Trans. Inst. Min. Metall. **77**, C186–189 (1968)

82. D. Brion: Etude par spectroscopie de photoelectrons de la degradation superficielle de FeS_2, ZnS et PbS à l'air et dans l'eau, Appl. Surf. Sci. **5**, 133–152 (1980)
83. S. Evans, E. Raftery: Electron spectroscopic studies of galena and its oxidation by microwave-generated oxygen species and by air, J. Chem. Soc. Faraday Trans. **78**, 3545–3560 (1982)
84. D. Fornasiero, J. Ralston: Iron hydroxide complexes and their influence on the interaction between ethyl xanthate and pyrite, J. Coll. Int. Sci. **151**(1) 225–235 (1992)
85. C. M. Eggleston, M. F. Hochella Jr.: Scanning tunneling microscopy of sulfide surfaces, Geochim. Cosmochim. Acta. **54**, 1511–1517 (1990)
86. C. M. Eggleston, M. F. Hochella Jr.: Scanning tunneling microscopy of galena (100) surface oxidation and sorption of aqueous gold, Science **254**, 983–986 (1991)
87. C. M. Eggleston, M. F. Hochella Jr.: Tunneling spectroscopy applied to PbS(001) surfaces: fresh surfaces, oxidation and sorption of aqueous Au, Amer. Mineral. **78**, 877–883 (1992)
88. B. S. Kim, R. A. Hayes, C. A. Prestidge, J. Ralston, R. St. C. Smart: Scanning tunneling microscopy studies of galena: the mechanism of oxidation in air, Appl. Surf. Sci. **78**, 385–397 (1994)
89. Y. H. Hsieh, C. P. Huang: The dissolution of PbS in dilute aqueous solutions, J. Coll. Interf. Sci. **131**, 537–549 (1989)
90. R. K. Clifford, K. C. Purdy, J. D. Miller: Characterisation of sulfide mineral surfaces in froth flotation systems using electron spectroscopy for chemical analysis, Amer. Inst. Chem. Eng. Symp. Ser. **71**, 138–147 (1975)
91. K. L. Purdy, R. K. Clifford: ESCA characterisation of flotation products from a lead-zinc-copper concentrator on the new lead belt of southeast Missouri, Proc. Amer. Inst. Min. Met. Pet. Eng., (New York 1975) pp. 30
92. J. Mielczarski, F. Werfel, E. Suoninen: XPS studies of interaction of xanthate with copper surfaces, App. Surf. Sci. **17**, 160–174 (1983)
93. K.-S. Johansson, J. Juhanoja, K. Laajalehto, E. Suoninen, J. Mielcarzski: XPS studies of xanthate adsorption on metals and sulfides, Surf. Int. Anal. **9**, 501–505 (1986)
94. K. C. Pillai, V. Y. Young, J. O. M. Bockris: XPS studies of xanthate adsorption on galena surfaces, Appl. Surf. Sci. **16**, 322–334 (1983)
95. R. Szargan, S. Karthe, E. Suoninen: XPS studies of xanthate adsorption on pyrite, Appl. Surf. Sci. **55**, 227–232 (1992)
96. E. Suoninen, K. Laajalehto: Structure of thiol collector layers on sulfide surfaces, Proc. XVIII Int. Min. Proc. Congr. (Sydney, Aust.) (Aust. Inst. Min. Met, Melbourne 1993) pp. 625–629 (1993)
97. A. N. Buckley, R. Woods: Adsorption of ethyl xanthate on freshly exposed galena surfaces, Coll. Surf. **53**, 33–45 (1991)
98. P. W. Page, L. B. Hazell: X-Ray photoelectron spectroscopy (XPS) studies of potassium amyl xanthate (K AX) adsorption on precipitated PbS related to galena flotation, Int. J. Min. Proc. **25**, 87–100 (1989)
99. J. M. Cases, M. Kongolo, P. de Donato, L. Michot, R. Erre: Interaction between finely ground galena and pyrite with potassium amylxanthate in relation to flotation, 1. Influence of alkaline grinding, Int. J. Min. Proc. **28**, 313–337 (1990)

100. J. M. Cases, M. Kongolo, P. de Donato, L. Michot, R. Erre: Interaction between finely ground galena and pyrite with potassium amylxanthate in relation to flotation, 2. Influence of grinding media at natural pH, Int. J. Min. Proc. **30**, 35–68 (1990)
101. M. Kongolo, J. M. Cases, P. de Donato, L. Michot, R. Erre: Interaction of finely ground galena and potassium amylxanthate in relation to flotation, 3. Influence of acid and neutral grinding, Int. J. Min. Proc. **30**, 195–215 (1990)
102. V. Bozkurt, Z. Xu, J. A. Finch: Pentlandite/pyrrhotite interaction and xanthate adsorption, Int. J. Mineral. Proc. **52**, 203–214 (1998)
103. S. L. Chryssoulis, K. G. Stowe, F. Reich: Characterisation of composition of mineral surfaces by laser-probe microanalysis, Trans. Inst. Min. Met. C. **101**, C1–C6 (1992)
104. S. L. Chryssoulis, J. Kim, K. G. Stowe: LIMS study of variables affecting sphalerite flotation, *Proc. 26th Canad. Min. Proc. Conf.* (Ottawa 1994)
105. S. Chryssoulis, K. Stowe, E. Niehuis, H. G. Cramer, C. Bendel, J. Kim: Detection of collectors on mineral grains by ToF-SIMS, Trans. Inst. Min. Met., **8** C 141 (1995)
106. J. S. Brinen, F. Reich: Static SIMS imaging of the adsorption of Diisobutyl dithiophosphinate on galena surfaces, Surf. Interface Anal. **18**, 448–452 (1992)
107. J. S. Brinen, S. Greenhouse, D. R. Nagaraj, J. Lee: SIMS and SIMS imaging studies of adsorbed dialkyl dithiophosphinates on PbS crystal surfaces, Int. J. Min. Proc. **38**, 93–109 (1993)
108. P. Plescia: Study of galena/potassium ethyl xanthate system by X-ray diffractometry and scanning electron microscopy, App. Surf. Sci. **72**, 249–257 (1993)
109. J. Ralston: The chemistry of galena flotation: principles and practice, Min. Eng. **7**, 715–735 (1994)
110. C. I. Basilio, I. J. Kartio, R.-H. Yoon: Lead activation of sphalerite during galena flotation, Mineral. Eng. **9**, 869–879 (1996)
111. A. F. Taggart: Flotation, in *Handbook of Mineral Dressing* (Wiley, New York 1945) Sect. 12
112. A. F. Taggart, G. R. M. del Guidici, D. A. Ziehl: The case for chemical theory of flotation, Trans. AIME **112**, 348–381 (1934)
113. K. van der Stelt, W. Skinner, S. Grano: A study of the interaction of di-cresyl, dithiophosphate with galena and pyrite using micro flotation, z potential measurements and X-ray photoelectron spectroscopy, Ian Wark Research Institute Report (1993)
114. C. A. Prestidge, J. Ralston: Contact angle studies of ethyl xanthate coated galena particles, J. Colloid Interface Sci. **184**, 512 (1996)
115. J. Mielcarzski: In situ ATR-IR spectroscopic study of xanthate adsorption on marcasite, Coll. Surf. **17**, 251–271 (1986)
116. L. J. Warren: Shear flocculation. in J. Laskowski, J. Ralston (Eds.): *Colloid Chemistry in Mineral Processing* (Elsevier, Amsterdam 1992) pp. 309–330
117. G. Jameson, J. Ralston (Eds.): *Investigation of High Energy Conditioning as a Pretreatment for Fine Particle Flotation*, Australian Minerals Industry Research Association, Project P397 (AMIRA, Melbourne 1995)
118. I. N. Plaskin, S. P. Zaitseva, G. A. Myasnikova, L. P. Starchik, V. I. Turnikova, G. N. Khazinskiya, R. S. Shaefeyev: Trans. Inst. Min. Met. **67**, 1–24 (1957)
119. J. A. Frew, R. St. C. Smart , E. V. Manlapig: Effects of fine grinding on flotation performance: generic statements, Proc. Fifth Mill Operators' Conference (Aust. Inst. Min. Metall, Carlton 1994) pp. 245–250

120. J. A. Frew, K. J. Davey, R. M. Glen, R. St. C. Smart: Effects of fine grinding on flotation performance: Zinc regrind at Cominco Alaska's Red Dog Mine, Proc. Fifth Mill Operators' Conference (Aust. Inst. Min. Metall., Carlton 1994) pp. 287–288
121. R. St. C. Smart, B. Judd: Improved Lasta filter and copper reflotation performance through surface analysis surveys at WMC's Olympic Dam Operation, Proc. Fifth Mill Operators' Conference (Aust. Inst. Min. Metall., Carlton 1994) pp. 1–4
122. S. R. Grano, D. W. Lauder, N. W. Johnson, S. Sobieraj, R. St. C. Smart, J. Ralston: Surface analysis as a tool for problem solving: a case study of the Hilton concentrator at Mt. Isa Mines Ltd., Proc. Symp. Polymetallic Sulfides Iberian Pyrite Belt (Evora, Portugal), (Portug. Min. Ind. Assoc., Lisbon 1993), Sect. 3–13, pp. 1–15
123. S. R. Grano, P. L. Wong, W. Skinner, N. W. Johnson, J. Ralston: Detection and control of calcium sulfate precipitation in the Hilton concentrator of Mt. Isa Mines, Ltd., Australia: Proc. XIX Int. J. Min. Proc. Congress **3**, Ch 29 171–179 (1996)
124. J. E. Thomas, R. St. C. Smart, W. M. Skinner: Kinetic factors for oxidative and non-oxidative dissolution of iron sulfides, Min. Eng. **13** 1149 (2000)
125. J. Mielcarzski: XPS study of ethyl xanthate adsorption on oxidized surface of cuprous sulfide, J. Coll. Int. Sci. **120**, 201–209 (1987)

Metals from Sulfide Minerals: The Role of Adsorption of Organic Reagents in Processing Technologies

Alan N. Buckley[1], Gregory A. Hope[2], and Ronald Woods[2]

[1] School of Chemistry, University of New South Wales
Sydney, 2052, Australia
a.buckley@unsw.edu.au

[2] Institute of Physical Chemistry, Griffith University
Kessels Rd, QLD 4111, Brisbane, Australia
g.hope@sct.gu.edu.au

Abstract. The extraction of base metals from sulfide mineral ores is of major industrial and economic importance, and as a consequence considerable effort has been, and continues to be, expended on research to maximize the extraction efficiency. Several unit operations in the extraction process involve interactions at the solid–liquid interface, and these interactions can be modified to achieve enhanced processing efficiency by the adsorption of organic reagents at the surface of the solid phase. In this chapter, the different ways in which organic reagents can interact with a solid surface are first described generally and then more specifically in relation to froth flotation and electrowinning, two of the most important unit operations in base metal production. In addition to a detailed treatment of the relevant adsorption mechanisms, the principal surface analytical techniques that have been used to elucidate these mechanisms are described. For the study of collector adsorption in flotation, both in situ and ex situ techniques are covered, with spectroelectrochemical studies and UV-visible spectroscopy included in the former, conventional (anode-generated) X-ray photoelectron and secondary ion mass spectroscopies representing the latter, and with FTIR and Raman scattering spectroscopies applicable in both situations. For the study of adsorption of reagents to influence deposit morphology in electrowinning, electrochemical techniques and surface enhanced Raman scattering spectroscopy are emphasized. It is shown that the combination of in situ and ex situ electrochemical and spectroscopic techniques represents a powerful approach for investigating the adsorption of organic reagents in the recovery of metals from sulfide ores.

Metal sulfides constitute the dominant ore minerals for many important metals, including copper, lead, molybdenum, nickel, silver, and zinc. Gold and platinum metals are often present as minor components of sulfide ores and are recovered as by-products of the extraction of a base metal, or as the major product. The winning of metals from sulfide ores involves a number of unit processes. Following mining, the valuable component minerals are usually concentrated and separated by froth flotation. The resulting concentrates are generally converted to metal pyrometallurgically and then electrorefined, or

K. Wandelt, S. Thurgate (Eds.): Solid–Liquid Interfaces, Topics Appl. Phys. **85**, 61–96 (2003)

treated hydrometallurgically to produce an aqueous solution from which the metal may be recovered by electrowinning. The adsorption of organic species at the solid–solution interface is a consequential process in a number of the constituent operations. In this chapter we consider the most important areas, those of flotation, and of electrowinning and refining.

1 Adsorption at the Solid–Solution Interface

In order for adsorption at a solid–solution interface to take place, solvent molecules, which may themselves be strongly adsorbed, must be displaced from the surface to make way for the adsorbing species. Also, the adsorbate must proceed through an electrified double layer in which a negative or a positive charge on the solid is balanced by a surface excess of ions in the solution phase at the interface. There will be a potential at which the surface charge is zero. The charge on the solid is positive or negative depending on whether the potential across the solid–solution interface is above or below the point of zero charge, respectively.

Ions can accumulate at the interface between a solid and an electrolyte solution in excess of those required to balance the charge on the surface at the operating potential. This phenomenon is termed specific adsorption, contact adsorption, or superequivalent adsorption. In the flotation of oxide minerals, the specific adsorption of long alkyl chain anions or cations at the solid–solution interface is utilized to render such non-conducting minerals hydrophobic and hence floatable. The free energy of adsorption of these species largely arises from the electrostatic attraction between the charged head group of the adsorbate and the charge on the mineral surface, but also includes a contribution from association between the hydrocarbon chains [1]. Molecular organic species can also accumulate at the solid–solution interface and this phenomenon is applied in metal deposition processes such as those described in Sect. 4.

Adsorption can occur by charge transfer between a conducting solid, such as a metal or a metal sulfide, and a solution species which, as a result, becomes covalently bonded to atoms in the surface of the solid phase. The adsorbate can be an intermediate in an overall electrocatalytic reaction, or can be a stable chemisorbed product. In many cases, chemisorption occurs as a precursor to the development of a bulk phase on the electrode surface. For isolated particles, the charge passed in the chemisorption process must be balanced by the occurrence of a simultaneous electrochemical process in what is termed a "mixed potential" system [2]. This situation occurs in the interaction of thiol collectors with sulfide mineral surfaces (Sect. 3). The cathodic process in this case is usually the reduction of oxygen.

There is a degree of overlap between specific adsorption and chemisorption. It is now accepted [3] that there can be a partial charge transfer in the adsorption of ions at the solid–solution interface and hence that the integral charge exhibited in solution is not necessarily retained at the interface. This results in adsorbed ions being partially covalently bonded to the surface.

Chemisorption of solution species can also occur without charge transfer, with donor bonds being formed between lone pairs of electrons in one or more atoms of the adsorbate and atoms in the surface. Co-adsorption of organic and inorganic ions can also occur to form a surface complex. It will be shown in Sect. 6 that this is the situation with the adsorption of thiourea on copper surfaces in electrowinning and electrorefining systems.

In addition, aromatic compounds may adsorb through π-bonding; in this case, adsorption occurs through overlap between the electrons in the solid surface and π-bonding orbitals in the organic species.

Knowledge of the mode and structure of adsorbed organic species has been significantly extended in recent years as a result of investigation of Self-Assembled Monolayers (SAMs) [4]. SAMs have received attention because of the significance of these layers in a wide range of applications (adhesion, lubrication, microelectronics, chemical sensors, etc.), and because their investigation provides a means of understanding ordered organic layers that are analogues of biological systems such as lipid bilayers. The focus of attention in SAMs has been on the structure and orientation of the adsorbate, but the findings have also proved valuable in understanding systems of metallurgical interest.

2 Flotation

The flotation process involves crushing and wet grinding an ore to liberate separate grains of the various valuable minerals and gangue components, pulping the ore particles with additional water, and then selectively rendering hydrophobic the surface of the mineral of interest by interaction with an organic collector species. A stream of air bubbles is passed through the pulp; the bubbles attach to, and levitate, the hydrophobic particles and collect them in a froth layer that flows over the weir of the flotation cell. The chemical composition of the remaining pulp can then be altered to render, in turn, further valuable minerals hydrophobic, and hence floatable.

3 Interaction of Thiol Collectors with Metal and Sulfide Mineral Surfaces

The key chemical step in the flotation process is the adsorption of the organic collector on the mineral surface. The adsorption process has been investigated by electrochemical methods and a range of complementary spectroscopic techniques.

3.1 Electrochemical Approaches

3.1.1 Voltammetry

Voltammetric studies of thiol collectors on a range of metal and metal sulfide surfaces have revealed the appearance of a prewave on the positive-going scan at a potential below the reversible value for the formation of the metal thiolate. Systems in which such underpotential deposition has been investigated include ethyl xanthate on galena (PbS) [5,6,7], chalcocite (Cu_2S) [8,9,10,11,12,13,14] and silver sulfide (Ag_2S) [15], and ethyl xanthate on the corresponding metals, lead [16], copper [13,14,17,18] and silver [15,18,19]. Voltammetric prewaves have also been observed for methyl and butyl xanthates on galena [6], diethyl dithiophosphate on chalcocite [20,21,22] and isopropyl and isobutyl xanthates on silver [23]. In each case, the charge associated with the prewave corresponds to that expected for the charge-transfer adsorption of a monolayer of the collector molecule.

Voltammetric prewaves have been observed for many systems. In the 1940s, *Brdicka* [24,25] showed that a polarographic prewave is observed when the product of the electrochemical reaction is adsorbed on the surface of the electrode, and a post-wave when the electroactive species itself is adsorbed. The shift in the half-wave potential results from the free energy of adsorption. The deposition of the first monolayer of one metal on another at underpotentials to the reversible value for the bulk phase is a general phenomenon, having been characterized for many systems [26]. The underpotential deposition of metals is discussed in detail in other chapters in this book.

Figure 1 shows a comparison of the voltammograms recorded in the presence of ethyl xanthate for copper, lead and silver with those for the corresponding metal sulfides, PbS, Cu_2S and Ag_2S. The arrows mark the reversible potentials for the formation of the bulk metal xanthates.

For galena, the reversible potential shown on the figure corresponds to that for the process:

$$PbS + 2C_2H_5OCS_2^- \rightarrow Pb(C_2H_5OCS_2)_2 + S + 2e^- . \tag{1}$$

The difference in the reversible potentials reflects the difference in the activity of lead in the metal and in the sulfide. It can be seen that the prewaves for lead and its sulfide occur at similar underpotentials, supporting the conclusion that xanthate is adsorbed on metal sites in the galena surface. The initial process forming lead xanthate is considered to be the removal of lead atoms from the sulfide lattice with the concomitant formation of either elemental sulfur (reaction(1)) or a metal-deficient lead sulfide:

$$PbS + 2y(C_2H_5OCS_2^-) \rightarrow yPb(C_2H_5OCS_2)_2Pb_{1-y}S + 2ye^- . \tag{2}$$

Reaction (2) will be favoured at low rates with reaction (1) occurring when the composition of the sulfidic product phase has a sulfur content above the stable, or metastable, stoichiometry range. Thus, the continuous increase in potential on a voltammogram could result in rapidly proceeding through the conditions under which reaction (2) occurs into the domain of reaction (1).

Copper and chalcocite display analogous behaviour to that of lead and galena. The reversible potential for the chalcocite/xanthate system corresponds to the reaction:

$$Cu_2S + 0.07C_2H_5OCS_2^- \rightarrow Cu_{1.93}S + 0.07Cu(C_2H_5OCS_2) + 0.07e^- . \quad (3)$$

In contrast, the voltammetric features displayed by silver sulfide in Fig. 1 occur at similar potentials to those for silver. This is because the silver sulfide used in obtaining the voltammogram in Fig. 1 was in the β-form which contains excess silver [15]. The silver activity in such a silver-rich material is greater than that in stoichiometric silver sulfide and is close to that of the metal when the silver/sulfur ratio reaches 2.002:1, the upper limit of the homogeneity range of β-$Ag_{2+\delta}S$ [27]. Thus, the reaction forming bulk silver xanthate on the sulfide is represented by:

$$Ag_{2+\delta}S + xC_2H_5OCS_2^- \rightarrow Ag_{2+\delta-x}S + xAg(C_2H_5OCS_2) + xe^- . \quad (4)$$

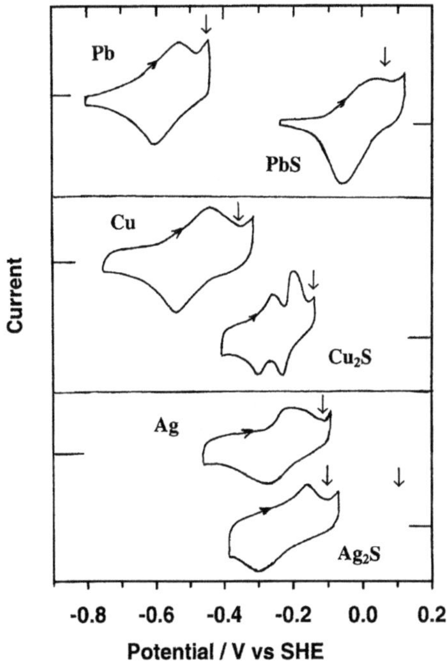

Fig. 1. Comparison of voltammograms for metals and their sulfides in the underpotential region. Potential scans at 20 mV/s in pH 9.2 solution containing 10^{-3} mol/dm^3 ethyl xanthate. The vertical arrows mark the calculated reversible potentials for the formation of the metal xanthates; the solid arrow for Ag_2S is that for the formation of sulfur and the dashed arrow for reaction (4)

As with the other two systems, the prewave occurs at similar underpotentials for silver and silver sulfide. The correspondence between the voltammetric behaviour of the metal and its sulfide suggests that the metal electrode can be applied as a model for characterizing the adsorption of thiol collectors.

There is no clearly discernible prewave in the voltammogram for ethyl xanthate oxidation at a gold electrode [5]. The multilayer product in this case is diethyl dixanthogen formed by the anodic process:

$$2C_2H_5OCS_2^- \rightarrow (C_2H_5OCS_2)_2 + 2e^- . \qquad (5)$$

The dithiolate is formed on gold because the potential at which reaction (5) occurs is less than that for the formation of gold ethyl xanthate [28]. The electrochemical kinetics showed [5], however, that chemisorbed xanthate is an intermediate in reaction (5), and that it is present in a high fractional coverage.

A prewave has been observed in the anodic oxidation of ethyl xanthate on silver-gold alloys [29]. The charge associated with monolayer coverage was found to be proportional to the silver content, and hence xanthate adsorbs only on silver sites in the alloy surface in the prewave region.

The large surface area created in the comminution stage preceding flotation, together with the limited supply of oxygen, normally means that most sulfide mineral particles are largely unoxidized before interaction with the collector occurs. Nevertheless, at least some surface oxidation would be typical, especially in the case of highly reactive sulfides such as the nickel minerals. Therefore, it is pertinent to study the interaction of such oxidized surfaces with collector molecules. Few voltammetric investigations have been reported on the interaction of pre-oxidized sulfide mineral electrode surfaces with thiol collectors. It is generally considered that a voltammogram after pre-oxidation will be the same as one for an unoxidized surface, but commencing at a more positive potential. Studies on galena [30] showed, however, that pre-oxidation of this mineral resulted in the potential dependence of the interaction of ethyl xanthate with galena differing from that for a fresh mineral surface, and becoming similar to that for lead. This was explained in terms of the loss of sulfur as thiosulfate during pre-oxidation through the reaction:

$$2PbS + 5H_2O \rightarrow 2PbO + S_2O_3^{2-} + 10H^+ + 8e^- . \qquad (6)$$

The lead oxide can subsequently interact by ion exchange with ethyl xanthate:

$$PbO + 2C_2H_5OCS_2^- + 2H^+ \rightarrow Pb(C_2H_5OCS_2)_2 + H_2O . \qquad (7)$$

Since there is no sulfur associated with this product on the surface, this lead xanthate gives rise to a similar voltammogram to that for the lead/ethyl xanthate system.

3.1.2 Isotherms

Equilibrium coverages in the prewave potential region have been determined as a function of potential for the chalcocite/ethyl xanthate [14], copper/ethyl xanthate [14], chalcocite/diethyl dithiophosphate [22], silver/ethyl xanthate [19], silver-gold alloy/ethyl xanthate [28], galena/ethyl xanthate [31,32] and lead/ethyl xanthate [16] systems. In each case, the data were found to comply with the Frumkin adsorption isotherm [33]:

$$[\theta/(1-\theta)]\exp g\theta = K[\mathrm{X}]\exp(\gamma \mathrm{FE/RT})\,, \tag{8}$$

where θ is the fractional surface coverage, g and K are constants, [X] is the concentration in solution of the thiol collector X, γ is the electrosorption valency, and the other terms have their usual meaning. The Frumkin isotherm is a phenomenological equation based on macroscopic considerations but has been explained in terms of molecular model theories. The term $g\theta$ arises from variation of the free energy of adsorption with coverage, through either heterogeneity of surface sites or interactions between adsorbed molecules. The electrosorption valency, γ, reflects changes in the double layer associated with the adsorption process, in addition to the number of electrons involved in the charge transfer reaction.

As pointed out in the previous section, ethyl xanthate was shown [29] to chemisorb on silver sites in the surface of silver–gold alloys. The coverage data as a function of potential was found to fit the Frumkin isotherm previously established for silver [19] when an additional term was included in the isotherm to account for the variation of activity of silver with alloy composition.

The fact that coverage data in flotation-related systems are delineated by the Frumkin isotherm implies that the attachment of the collector to the sulfide mineral corresponds to that of chemisorption, with collector species being bonded to metal atoms still retained in the surface of the mineral. It also indicates that the chemisorbed species is not characterized by a single Gibbs free energy, but that the free energy varies with coverage.

A further implication of fitting to an isotherm is that thermodynamic information becomes available to characterize chemisorption, and hence chemisorbed species can be incorporated in E_{uh}–pH diagrams describing the interaction of thiol collectors with mineral surfaces. Thus, the stability domains of all thiol species formed at the interface can be included [33]. Since the chemisorbed thiol is formed at lower potentials than the metal thiol compound, it occupies a stability domain in a potential region lower than that for the bulk species. This is exemplified in the E_h–pH diagram for the chalcocite/ethyl xanthate system shown in Fig. 2. It can be seen that the region of chemisorption extends nearly 0.3 V below the stability zone of copper(I)ethyl xanthate. At higher E_h values, chalcocite is oxidized to copper sulfides of progressively lower copper content [14]. Chemisorbed xanthate will not co-exist with these sulfides when CuO is formed since the oxide is expected to overlay the surface

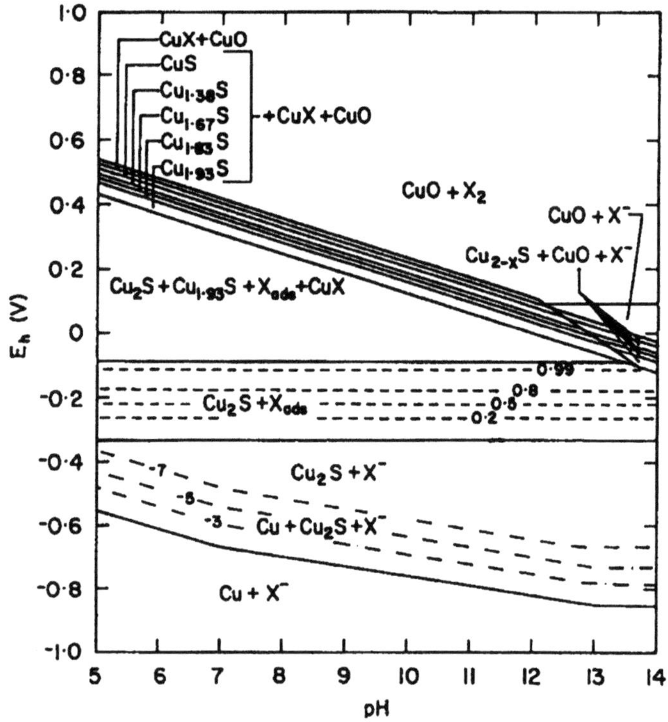

Fig. 2. E_{h}–pH diagram for the chalcocite/water/ethyl xanthate system for an initial xanthate concentration of 10^{-3} mol/dm^3. Sulfur-oxygen anions not considered. Species designated $Cu_{2-x}S$ cover the same range of stoichiometries identified in the acid region. Fractional coverage of chemisorbed xanthate as indicated; copper coverage marked in log mol [14]

of the mineral. This explains the observation of an upper, as well as a lower, potential limit of flotation [42].

The rate of adsorption of ethyl xanthate on galena was found [5] to correspond to Elovich kinetics. The Elovich rate equation is based on similar assumptions to that of the Frumkin adsorption isotherm, and has been found to apply to chemisorption on solids from the gas phase and also to charge transfer adsorption of oxygen and organic species on platinum.

3.2 Potential Dependence of Hydrophobicity and Flotability

The adsorption of organic collector species can be characterized by the degree of surface hydrophobicity produced. *Wark* and co-workers [34] refined methods for the determination of the three-phase contact angle at the gas–solid–solution interface and the application of contact angle measurements to the prediction of flotation response of minerals when treated with collector solutions that simulate commercial practice. The contact angle is a thermodynamic quantity since it is related by the Young equation [34] to the interfacial tensions between the three phases.

Later workers [33] extended this approach by measuring contact angle as a function of the potential across the solid–solution interface. This allowed the determination of how effective the initial, chemisorbed layer is in establishing hydrophobicity. Since chemisorption is the thermodynamically favoured process in the interaction of xanthates and other thiols with sulfide minerals, chemisorbed xanthate is expected to be the major species formed in the presence of xanthate concentrations relevant to flotation.

Contact angle measurements have demonstrated that the chemisorbed ethyl xanthate layer gives rise to the maximum angle characteristic of the alkyl group of this collector [34] before the monolayer is established on galena [6], silver [19] and silver–gold alloys [29]. A similar result was observed for the galena/butyl xanthate [6] and chalcocite/diethyl dithiophosphate [35] systems.

Flotation recovery has also been determined as a function of potential, and this approach is clearly the most appropriate measure of the efficacy of a flotation collector, since the purpose of the flotation process is the efficient recovery of minerals. The control of potential of mineral particles has been established by the use of a particulate bed electrode, or by the controlled addition of suitable redox reagents. The former approach has been applied to gold in the presence of methyl, ethyl, propyl and butyl xanthates [36], and to the galena/ethyl xanthate [6], chalcocite/ethyl xanthate [9,10,37], pyrite (FeS_2)/ethyl xanthate [6,37], chalcopyrite ($CuFeS_2$)/ethyl xanthate [37] and bornite (Cu_5FeS_4)/ethyl xanthate [37,38] systems. Control of potential through the addition of redox reagents has been employed for the galena/ethyl xanthate [39,40,41], chalcocite/ethyl xanthate [11,42], and chalcocite/diethyl dithiophosphate [35] systems.

The potential dependence of galena flotation in the presence of ethyl xanthate was found [39] to depend on the nature of the grinding medium. With mild steel media, the onset of flotation correlated with the anodic voltammetric wave assigned to chemisorption of xanthate on galena [5,6,7]. With stainless steel or ceramic media, flotation occurred at much lower potentials and correlated with the voltammetric current observed after pre-oxidation [30] (see Sect. 3.1). It was concluded that oxidation of galena by dissolved air took place under the latter conditions, but was inhibited by cathodic protection by the mild steel in the former situation. The potential dependence of flotation in the galena/xanthate system was shown [40] to be the same as that previously found [39] for grinding with mild steel media. Thus, it would appear that flotation in the plant situation follows that observed for galena ground under reducing conditions.

Prior oxidation of other sulfides does not appear to change the potential dependence in the same way as for galena. It was shown, however, that in the flotation of chalcocite, a higher thiol collector concentration is required after oxidation to achieve similar recoveries as for the unoxidized mineral [20,21,43]. This results from interaction of collector with the oxidation products either forming a weakly bound species because it is formed by ion exchange with a thick oxide layer, or occurring away from the surface if the products are soluble.

Correlations have been established between the potential dependence of flotation recovery and the corresponding chemisorption isotherm for the galena/ethyl xanthate [44], chalcocite/ethyl xanthate [14] and chalcocite/diethyl dithiophosphate [35] systems. The results are shown in Fig. 3, and it can be seen that, in each case, significant flotation occurs at potentials at which the coverage is low, 50% recovery corresponds to a fractional coverage of < 0.2 and 90% recovery to about half-coverage.

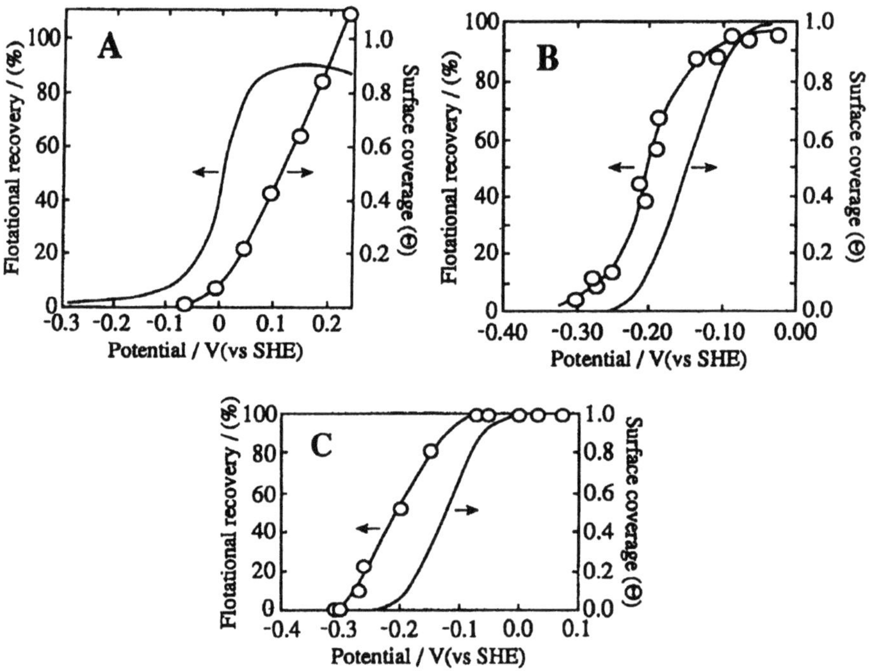

Fig. 3. Comparison of flotation recovery and fractional coverage for (**A**) galena with 2.3×10^{-5} mol/dm^3 ethyl xanthate [44], (**B**) chalcocite with 10^{-5} mol/dm^3 ethyl xanthate [14] and (**C**) chalcocite with 10^{-5} mol/dm^3 diethyl dithiophosphate [35]

3.3 Spectroelectrochemical Studies

Complementing electrochemical investigations with studies using a range of spectroscopies has provided valuable information on the nature of the surface species formed by the interaction of flotation collectors with metal and metal sulfide surfaces.

3.3.1 UV-Vis Spectroscopy

Combining electrochemical techniques with UV-Vis spectroscopy of the solution phase circulated through high surface area electrodes has been applied to confirm the identity of the products of various reactions. This approach was applied to demonstrate that ethyl xanthate is abstracted from solution as the potential of a chalcocite particulate bed electrode is increased through the prewave region, and that it is returned to solution when the direction of the potential scan is reversed [9,10]. It has also been applied to the silver/ethyl xanthate [19] and copper/ethyl xanthate [45] systems, and has indicated that chemisorption involves attachment of xanthate to the surface without change in the integrity of the collector molecule.

The application of UV-Vis spectroscopy is particularly useful in situations where it is difficult to identify thiol oxidation on voltammograms because there is a significant background current arising from oxidation of the substrate. This spectroscopic technique allowed the potential dependence of the adsorption of ethyl xanthate on bornite to be determined [37,38]. It was demonstrated that xanthate adsorbs on bornite at underpotentials and that the surface species formed induces flotation. A similar result was obtained for ethyl xanthate on chalcopyrite [37]. Whereas voltammetry only revealed the formation of dixanthogen on chalcopyrite, UV-Vis spectroscopy showed that xanthate was abstracted from solution at potentials lower than that at which dixanthogen began to be produced. The potential at which the xanthate concentration began to decrease correlated with the potential at which the flotation of a chalcopyrite particle bed was observed to commence. From an estimate of the reversible potential for the formation of copper xanthate on chalcopyrite, and assuming that xanthate chemisorption begins at a similar underpotential to that for chalcocite, it was concluded [33] that the onset of flotation correlates with the potential at which xanthate chemisorbs.

3.3.2 FTIR Spectroscopy

Infrared Spectroscopy provides a valuable means of detecting the presence of organic species at surfaces, identifying the chemical nature of surface compounds, and determining the structure and orientation of the adsorbed entity. Water is a strong absorber of infrared radiation and this poses problems for carrying out FTIR spectroscopy in situ. Such studies have, however, been achieved with metal sulfides under potential control using a specially designed cell in which mineral particles are placed on a carbon paste electrode and the electrode is moved to be adjacent to the cell window during the recording of spectra [46]. The window was an attenuated total reflection element composed of a material such as germanium which is transparent to

IR radiation. Using this approach it was shown [46] that both chalcocite and galena in the presence of ethyl xanthate gave rise to FTIR spectra in the prewave potential region similar to those of the metal xanthates. Diethyl dixanthogen was the only product identified on pyrite and it was formed at the potential at which previous authors had shown pyrite to commence floating [37]. Dixanthogen was found to be the initial product of anodic oxidation of ethyl xanthate on chalcopyrite, with copper(I)xanthate also being formed as the potential was increased. No surface xanthate species were, however, detected in the potential region in which this mineral had been shown [37] to begin to float. Chalcopyrite is known to display floatability in the absence of collectors [47] as a result of surface oxidation, and it was considered [46] that this phenomenon was responsible for the initial flotation previously observed. No flotation had, however, been observed in the absence of collector under the conditions employed [37] and, as pointed out above, some xanthate is abstracted from solution at the flotation potential edge. Hence, it would appear that the xanthate coverage required to induce the flotation of chalcopyrite is below the limit of detection of the FTIR technique that was employed.

FTIR spectroscopy has been applied ex situ by many workers to identify organic species adsorbed on metal and metal sulfide surfaces. To obtain useful information, it is necessary to take care to ensure that negligible changes occur during transfer of the electrode from the electrochemical cell to the spectrometer. This approach has been adopted to determine the potential dependence of adsorption in systems in which voltammetry does not yield unequivocal information. Adsorption can then be correlated with flotation recovery [48] in an analogous manner to that discussed in Sect. 3.2.

Ex situ FTIR has been employed to distinguish chemisorbed xanthate on gold in the presence of the major product, dixanthogen [28]. Figure 4 shows FTIR spectra of a gold surface after treatment in $5 \times 10^{-4}\,\mathrm{mol/dm^3}$ ethyl xanthate solution at a range of potentials together with FTIR spectra of ethyl xanthate and diethyl dixanthogen. It can be seen that a band appears in the spectrum from the gold electrode at $1204\,\mathrm{cm^{-1}}$ at potentials below that at which dixanthogen develops. This was assigned to chemisorbed xanthate. The presence of chemisorbed xanthate confirmed the conclusion reached from previous analysis of electrode kinetics (Sect. 3.1). Previous studies on gold particles [36] showed that flotation commences at $\approx 0.02\,\mathrm{V}$ below the reversible potential for the formation of dixanthogen and hence indicated that the initiation of the flotation of gold results from the chemisorption process.

The intensity of the $1204\,\mathrm{cm^{-1}}$ band in the FTIR spectrum for chemisorbed xanthate (Fig. 4) is enhanced in comparison with the other bands which are barely discernible when only chemisorbed xanthate is present. This behaviour has been interpreted [49] as resulting from orientation effects. Similar behaviour is observed with silver surfaces [19]; this can be seen from the comparison shown in Fig. 5 of the FTIR spectra at different potentials with the spectrum for silver xanthate. As with gold, the band at $1210\,\mathrm{cm^{-1}}$ is

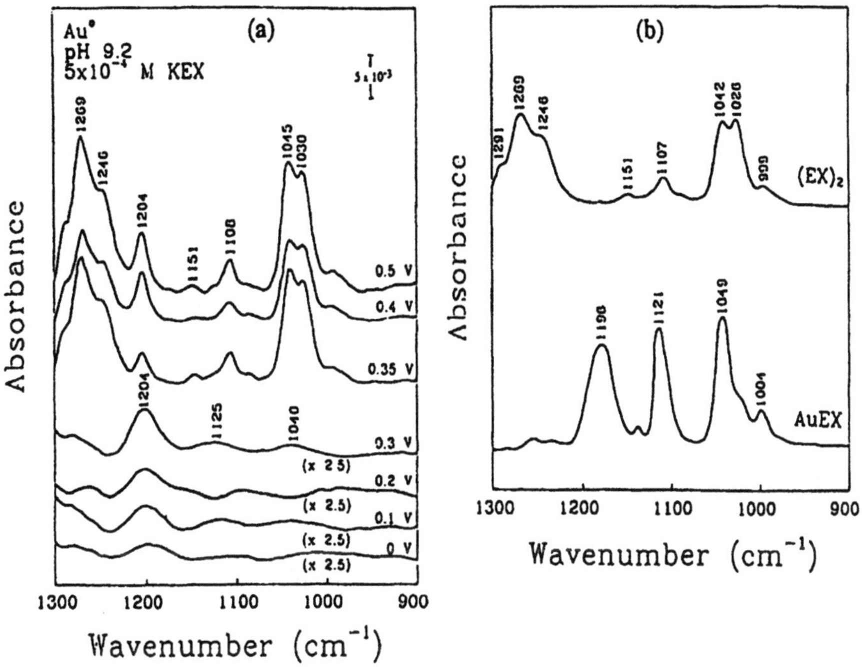

Fig. 4. FTIR spectra of (**a**) a gold electrode held at different potentials in pH 9.2 solution containing 5×10^{-4} mol/dm^3 ethyl xanthate; (**b**) diethyl dixanthogen and gold ethyl xanthate

shifted slightly from that in the bulk compound, and is enhanced relative to that arising from other vibrations. Also, the appearance of this band correlated with the potential at which a finite contact angle was observed on a silver surface.

3.3.3 Raman Scattering Spectroscopy

Raman spectra provide information which is complementary to that from infrared; both spectra result from transitions between the vibrational energy levels of molecules, but different selection rules apply. Raman spectroscopy has the advantage that it can be carried out in aqueous media and with glass cells.

Raman scattering is enhanced by a factor of 10^4–10^6 when adsorption occurs on suitably roughened copper, silver and gold surfaces, and this allows characterisation of surface species at the submonolayer level. Surface Enhanced Raman Scattering (SERS) spectroscopy has been applied to confirm the identity of the chemisorbed ethyl xanthate layers on silver [50,51], copper [51] and gold [52] surfaces.

For all three metals, SERS spectra displaying all the bands expected from vibrations giving rise to Raman bands from the corresponding bulk metal

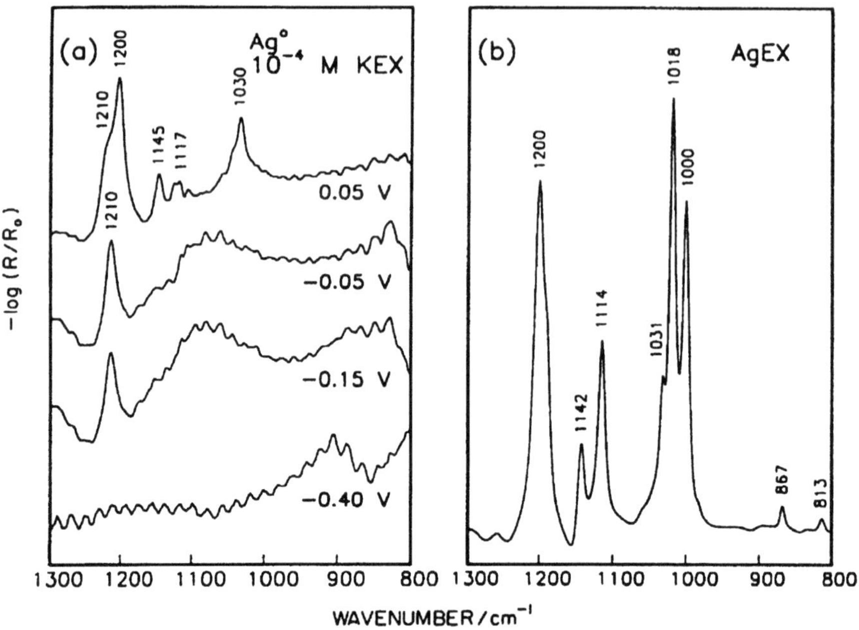

Fig. 5. FTIR spectra of (**a**) a silver electrode held at different potentials in pH 9.2 solution containing 10^{-4} mol/dm^3 ethyl xanthate and (**b**) silver ethyl xanthate [19]

xanthate were observed. For silver [50], the identity of the surface species was confirmed as chemisorbed xanthate from its electron spectra (Sect. 4).

SERS spectra from a silver electrode, after holding at a potential just below that corresponding to monolayer coverage in 5×10^{-4} mol/dm^3 ethyl xanthate solution, and recorded in situ and after emersion, are shown in Fig. 6. The figure also presents the Raman spectrum for silver xanthate. The difference between the two spectra is largely due to absorption of the scattered radiation by water at wavenumbers $> 1800\,\mathrm{cm}^{-1}$. Corresponding spectra for gold are shown in Fig. 7.

It can be seen from both these figures that each band in the bulk compound is present in the SERS spectrum. A similar result was found for ethyl xanthate on copper [51]. Thus, these Raman investigations confirm the integrity of the xanthate molecule when chemisorbed on copper, silver and gold surfaces. A comparison of Figs. 6 and 7 with Figs. 5 and 6, respectively, shows that SERS spectroscopy provides more convincing identification of chemisorbed species than FTIR for these systems.

SERS spectroscopy has also been applied to the study of the adsorption of *O*-isopropyl-*N*-ethylthionocarbamate (IPETC) with copper surfaces [53]. This collector molecule can form a coordination complex with copper(I) chloride, CuCl(IPETC)$_2$, as well as a copper(I) compound. Thus, in chloride media, IPETC can either chemisorb on copper, or form a surface complex similar

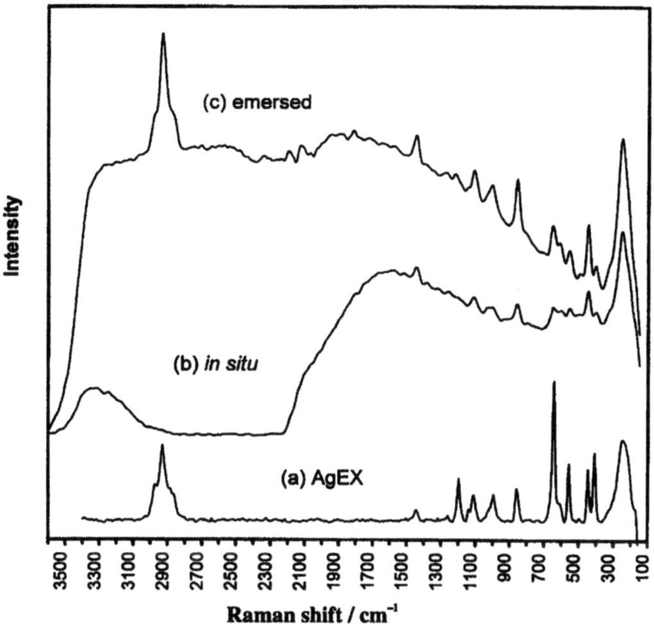

Fig. 6. SERS spectra of silver electrode at $-0.09\,V$ vs SHE in pH 9.2 solution containing 5×10^{-4} mol/dm^3 ethyl xanthate recorded (**b**) in situ, (**c**) after emersion. Curve (**a**) is the Raman spectrum of solid silver ethyl xanthate [51]

in form to that of the thiourea complexes found to be important in controlling the electrodeposition of copper (Sect. 5). The SERS spectra revealed that in both chloride and sulfate media, IPETC chemisorbs; it adsorbs through a charge transfer process in which the sulfur atom in the organic species becomes bonded to a copper atom in the metal surface and the hydrogen atom is displaced from the nitrogen to form a hydrogen ion in solution. The reversible potential for the formation of bulk copper(I) IPETC was derived from measurements of the rest potential of a copper electrode in IPETC solutions. This indicated that, as with xanthate, IPETC adsorption commences at an underpotential.

3.4 Electron Spectroscopy

X-ray Photoelectron Spectroscopy (XPS), which is usually understood to include X-ray excited Auger electron spectroscopy, provides both elemental and chemical state information concerning the outermost few nanometres of the specimen under investigation. Thus, an organic species adsorbed on a mineral or metal surface can be characterized by this technique.

The precise analysis depth for XPS depends, inter alia, on the kinetic energy of the Auger or photoelectron, and on the electron take-off angle. In

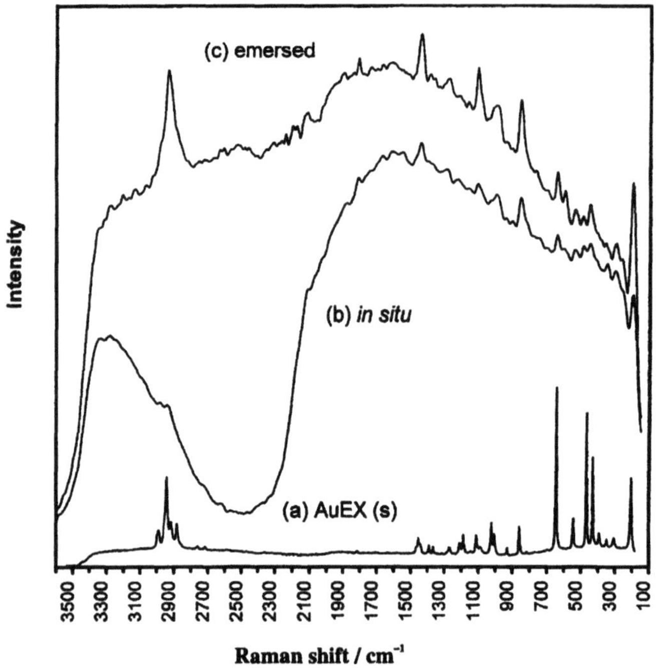

Fig. 7. SERS spectra of gold electrode at 0.10 V vs SHE in pH 9.2 solution containing 5×10^{-4} mol/dm^3 ethyl xanthate recorded (**b**) in situ, (**c**) after emersion. Curve (**a**) is the FT-Raman spectrum of solid gold ethyl xanthate [52]

conventional XPS, Mg or Al K_α radiation is mostly used, normally monochromatized in the latter case. For either X-ray source, S $2p$ photoelectrons, for example, would have kinetic energies greater than 1000 eV and the analysis depth would be of the order of 6 nm for an electron take-off angle of 45°. Sulfur atoms in an adsorbed monolayer can be distinguished from sulfur atoms in a mineral substrate with Mg or Al K_α radiation, provided the binding energies are different, because the intensity of emitted electrons decreases exponentially with depth. Nevertheless, the separation of the signal from the adsorbed layer from that of the substrate can be enhanced by ejecting electrons from the specimen with lower energy X-rays, and/or using a lower take-off angle. Yttrium or zirconium M_ζ radiation (132 or 151 eV) is sometimes useful, but X-ray satellites are more intrusive than with Mg K_α radiation.

The variation of electron escape depth with kinetic energy exhibits a minimum near 50 eV [54], so that the maximum surface sensitivity is obtained when the electrons have a kinetic energy close to this value. Thus, if information about the surface chemical state of sulfur atoms is being sought from the precise binding energies of S $2p$ photoelectrons, maximum surface sensitivity would be obtained if the S $2p$ electrons were ejected by $\approx$ 215 eV X-rays, since S $2p$ binding energies are in the range 160–170 eV. At an electron take-off angle of 45°, the analysis depth at maximum surface sensitivity is approximately three atomic layers. Reduction of the take-off angle enhances the surface sensitivity, by a factor of 2.7 if the angle relative to the specimen surface is reduced from 45° to 15°, however this approach is only feasible if the specimen surface is quite smooth.

Currently, an adequate flux of X-rays of tunable energy can only be obtained from a synchrotron beam line incorporating optical components suitable for X-rays in the range of interest (typically 100 to 1000 eV for surface analysis). Synchrotron radiation XPS (SR-XPS) and its application to sulfide mineral flotation is considered in detail in the chapter by Szargan et al.

Electron spectroscopy is a very informative surface analysis technique, however because it is ultra-high vacuum based, it suffers from the disadvantage of being ex situ. Ideally, in research directed towards improving the efficiency of separation by flotation, surface analysis of the mineral particles should be carried out in situ, i.e., while the surface under investigation is in contact with the aqueous flotation medium. For an ex situ analysis to be applicable, it is necessary to establish that the chemical species present at the solid–vacuum interface are essentially the same as those that determine the contact angle at the solid–liquid–air interface. This requirement is beyond ensuring that no gross chemical changes occur. In particular, precautions must be taken to retain at the surface, high vapour-pressure species such as dixanthogen or even elemental sulfur, and to prevent any alteration to the chemical composition resulting from the surface no longer being at the electrochemical potential of interest. Not surprisingly, there have been numerous reports of anaerobic cell chambers designed to minimize any alteration to an electrode surface arising from transfer from an electrochemical cell to a vacuum-based spectrometer [54].

The generally good correlation between surface compositions deduced by an ex situ analytical method such as XPS, and observed flotation behaviour, has given researchers confidence that the information provided by these ex situ techniques is a reliable representation of the surface chemical composition of minerals in aqueous flotation media.

3.4.1 Interaction of Thiol Collectors with Unoxidized Sulfide Mineral Surfaces

Conventional XPS was applied [15,22,44,55,56,57] to characterize xanthate layers adsorbed on unoxidized galena surfaces. In these investigations, galena was cleaved under xanthate solution to ensure that no oxidation occurred prior to interaction between the mineral surface and the collector. The results of these studies established that monolayer adsorption of thiol collectors on sulfide minerals can be differentiated from multilayer adsorption of the corresponding thiolate by means of XPS. The basis for the differentiation is that any substrate core electron energy shifts arising from collector chemisorption are usually too small to be detectable by conventional XPS, whereas the corresponding metal core electron energy shifts for multilayer metal-collector compound formation are clearly discernible. This situation was first observed for the ethyl and amyl xanthate/galena systems [56]. The S 2*p* spectrum, and to a lesser extent the C 1*s* and O 1*s* spectra, indicated submonolayer adsorption of xanthate while the Pb 4*f* spectra determined before and after immersion remained indistinguishable. Similar behaviour was observed for the *n*-butyl xanthate/galena system when electron spectra were determined with an electron take-off angle of 15° to enhance surface sensitivity [57].

The XPS study of the galena/butyl xanthate system included investigations of surfaces under potential control. At underpotentials to lead xanthate formation, the S 2*p* spectrum displayed an additional doublet shifted by approximately 1.5 eV from that of the substrate and no shifted Pb 4*f* component. After the potential was held above the reversible value of the $PbS/[Pb(C_4H_9OCS_2)_2 + S]$ couple, two additional doublets were evident in the S 2*p* spectrum, shifted by 1.35 and 1.8 eV (Fig. 8) as well as a Pb 4*f* component shifted by ≈ 1 eV. The two additional S 2*p* doublets were assigned to monolayer xanthate and lead xanthate, respectively.

A negligible difference in the Cu(LMM) Auger spectrum for chalcocite before and after treatment with diethyl dithiophosphate under chemisorption conditions (Fig. 9) was also interpreted as being consistent with submonolayer adsorption, and confirmation that deposition of molecular copper dithiophosphate had not occurred [22]. In this case, chemisorption of the collector was revealed not only by an additional S 2*p* component, but also by the presence of a P 2*p* peak (Fig. 9). In contrast with the galena/xanthate system, the

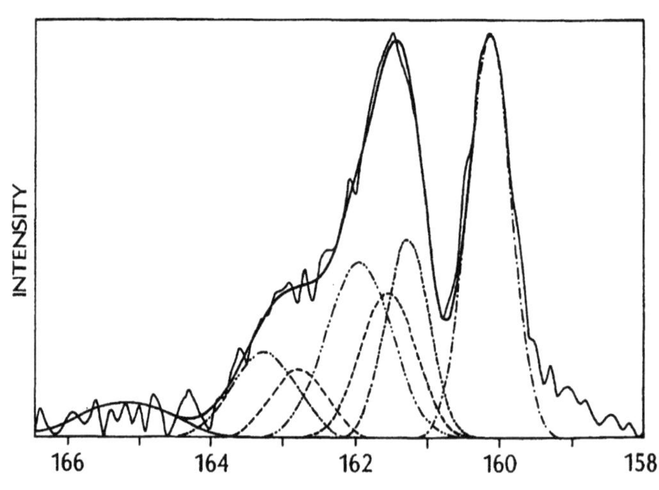

Fig. 8. The fitted S 2*p* photoelectron spectrum from abraded galena surfaces maintained for 5 min in 0.05 mol/dm^3 sodium tetraborate (pH 9.2) containing 10^{-3} mol/dm^3 *n*–butyl xanthate at 0.2 V (SHE) [57]

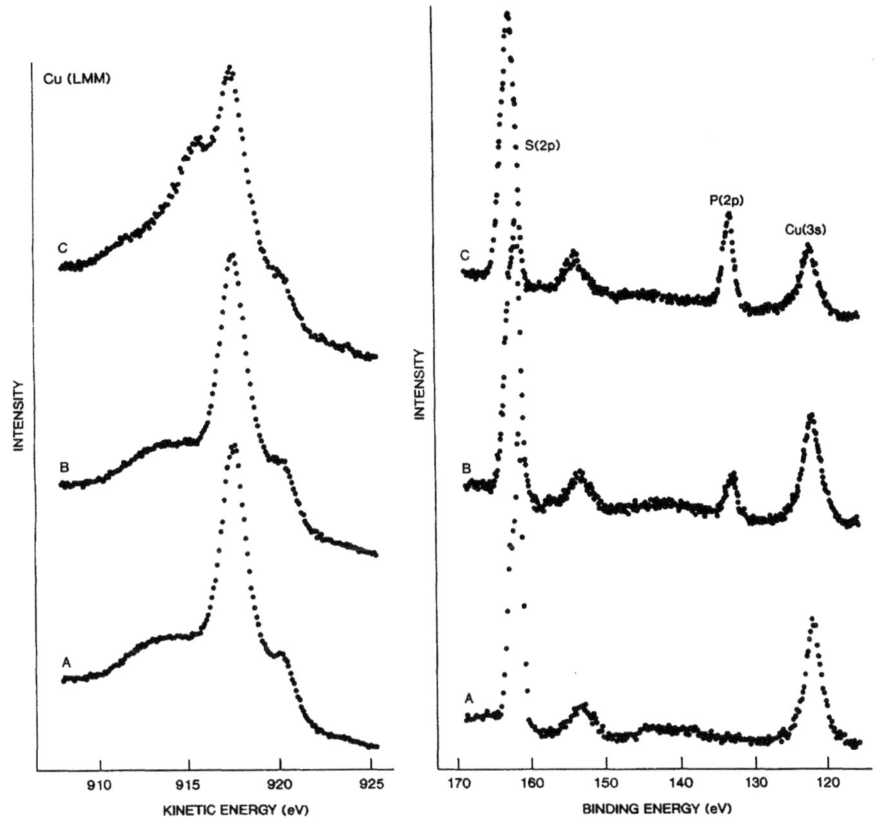

Fig. 9. Cu(LMM) X-ray excited Auger electron spectra and the corresponding S $2p$/P $2p$/Cu $3s$ photoelectron spectra for chalcocite surfaces before (**A**) and after adsorption of a monolayer (**B**) and multilayer (**C**) of dithiophosphate at pH 9 [22]

S $2p$ binding energy for the monolayer was indistinguishable from that for copper diethyl dithiophosphate. The difference probably arises from the stoichiometry of the surface species being the same in the monolayer and the bulk compound for copper, whereas chemisorbed xanthate on galena involves one thiol radical per lead atom compared with two in lead xanthate.

The interpretation of the lack of a shifted component in the Cu(LMM) Auger spectrum in terms of chemisorption differed from that given by previous workers. For example, *Mielczarski* et al. [58] concluded that the absence of any significant change in the Cu(LMM) Auger spectrum for chalcocite before and after interaction with ethyl xanthate within the potential range where voltammetry indicates chemisorption to have taken place was evidence for adsorption of xanthate *not* having occurred. Their conclusion was based on the assumption that since molecular copper xanthate exhibits Auger peaks at energies shifted ≈ 1 eV from those of the copper sulfide substrate, then any adsorption should result in a discernible shift in the Auger electron energies. The new component they observed in the S 2*p* spectrum from chalcocite in this potential range was attributed to a xanthate decomposition product.

Contini et al. [59] also observed no change in the Cu(LMM) spectrum from chalcocite after immersion in 10^{-5} mol/dm^3 5-methyl-2-mercaptobenzoxazole, whereas the S 2*p* and N 1*s* spectra provided clear evidence for adsorption. While they accepted that these observations could be explained in terms of submonolayer chemisorption, they did not exclude the possibility that a change in the Cu(LMM) spectrum was not evident because of inadequate sensitivity.

Further work was carried out [15] to confirm the absence of a discernible shift in the Auger spectra of the metal component of a surface when monolayer thiol adsorption occurs. The surfaces selected for this study were silver and sulfidized silver, and the adsorbate was ethyl xanthate. Similar behaviour was observed with silver and silver sulfide electrodes maintained at potentials below the reversible value for molecular silver xanthate formation. No discernible changes in the Ag(MNN) spectra were evident, notwithstanding the observation of a S 2*p* component consistent with the chemisorption of xanthate. Ag(MNN) electrons provide greater surface sensitivity than Cu(LMM) or Pb 4*f* in conventional XPS because the kinetic energy of the silver Auger electrons is approximately 350 eV compared with more than 900 eV for Cu(LMM) electrons or more than 1 keV for Pb 4*f* electrons. In this regard, it should be noted that the absence of a shifted Pb 4*f* component, following the adsorption of xanthate on galena, has been confirmed by electron spectra excited by synchrotron radiation of an energy (215 eV) optimized to maximize the Pb 4*f* and S 2*p* surface sensitivity [60].

Notwithstanding the generally satisfactory situation concerning the reliability of vacuum-based analyses, some in situ and ex situ infrared spectroscopic work reported in 1995 disputed the applicability of electron spectroscopic analyses in flotation-related research. In that infrared investigation, the existence of a monolayer or submonolayer of thiol radicals chemisorbed at potentials below the reversible value for the formation of the bulk metal thiolate was challenged [61,62]. It was argued that the underpotential deposition involved the adsorption of thiol decomposition products rather than thiol radicals. Because electron spectroscopic data had been interpreted as providing strong support for the non-dissociative chemisorption of thiol collector radicals, this questioning of the thermodynamically favoured collector interaction mechanism was tantamount to a questioning of the reliability of vacuum-based analyses of surfaces modified in aqueous media. However, as pointed out in Sect. 3.3, the validity of such ex situ analyses has since been confirmed by in situ and ex situ SERS investigations of the silver/xanthate system, together with complementary electron spectroscopy (*vide supra*). A silver surface is an ideal one for investigating the monolayer chemisorbed from xanthate solution, because not only is an electrochemically roughened silver surface a suitable substrate for SERS spectroscopy, but also the X-ray excited Ag(MNN) electron spectrum is able to provide surface-sensitive ex situ chemical information. In situ and ex situ SERS spectra can be determined before and after XPS analysis of such a monolayer-covered silver electrode.

In the light of suggestions that the absence of a discernible change in the Pb $4f$ spectrum could be due either to the relatively low surface sensitivity of conventional XPS for low binding energy electrons, or to masking of the substrate Pb atoms by the adsorbed xanthate, rather than chemisorption leading to a negligible shift in substrate core electron binding energies, it is pertinent to consider whether such mechanisms preclude the use of XPS to distinguish monolayer and multilayer adsorption. Even if the absence of a discernible change in the substrate spectra were due to inadequate sensitivity, then the observation of negligible change would still be indicative of monolayer or submonolayer adsorption. Thus, XPS is a useful adjunct in the preparation of adsorbed monolayers for ex situ analysis when maintenance of potential control at the end of the adsorption process can be a problem. Indeed, in the preparation of xanthate adsorbed on electrochemically roughened silver substrates for SERS spectroscopic analysis, considerable care had to be exercised in order to limit adsorption to a monolayer [50]. For successful monolayer deposition from $10^{-4}\,\mathrm{mol/dm^3}$ xanthate in pH 9 borate solution, the roughened silver electrode surface was held at $-0.1\,\mathrm{V}$ in borate solution that had been deoxygenated with a stream of pure nitrogen until the current passing was negligible. This procedure was carried out in order to remove any residual oxygen adsorbed on the silver. Sufficient xanthate in a small volume of deoxygenated borate solution was then added to result in a xanthate concentration of $10^{-4}\,\mathrm{mol/dm^3}$ in the cell. The potential was subsequently increased to $-0.05\,\mathrm{V}$ and maintained at this value for 5 min. At the end of the deposition period, the electrode was washed free of xanthate solution without significant exposure to oxygen immediately after the circuit was opened. Removal of the aqueous phase from the electrode was facilitated by the hydrophobicity of the surface, and subsequent brief exposure of the dry surface to air was not expected to result in significant alteration. If instead of this procedure, the roughened silver surface was treated at $-0.05\,\mathrm{V}$ after direct immersion of the electrode into deoxygenated $10^{-4}\,\mathrm{mol/dm^3}$ xanthate solution, XPS indicated the formation of multilayer silver xanthate. It was presumed that multilayer adsorption occurred because of the presence of oxygen adsorbed within the roughened surface of the silver electrode. A significant time is required to remove all the oxygen from a roughened silver surface at $-0.05\,\mathrm{V}$.

The formation of thiolate multilayers at the surface of sulfide minerals is not expected to be the major reaction under flotation conditions because chemisorption is the thermodynamically favoured process in the interaction between the sulfides and thiol collectors, and collector concentrations are relatively low in practice. Nevertheless, given that air is normally the flotation gas, the E_h conditions in a flotation cell are such that some multilayer formation is possible. Thiolate formation at the surface of a sulfide mineral, unlike the situation with a metal, should be accompanied by the formation of a sulfur-rich surface. For example, the formation of lead xanthate at the surface of galena involves the removal of lead atoms from the sulfide lattice to leave either elemental sulfur (reaction (1)), or a lead-deficient sulfide lattice (reaction (2)).

The S $2p$ spectra for butyl xanthate on galena (Fig. 8) show no evidence for any sulfur-rich species. This was explained by relaxation of the surface stoichiometry within a time scale comparable with that of the XPS measurement [63,64]. The relaxation process results in the surface stoichiometry becoming close to that of the bulk after the treated mineral has been removed from the aqueous phase, but before the specimen can be evacuated and the photoelectron spectra obtained. This study was carried out with galena from Broken Hill which is known to be n-type and lead-rich. It would be of interest to carry out similar studies on p-type galena which has a stoichiometry containing excess sulfur rather than lead. It has been confirmed in a conventional and SR-XPS study [65] that relaxation of surface composition is much slower with galena that is sulfur-excess than with mineral containing excess lead.

As noted in the preceding sections, the interaction of thiols with some minerals and metals results in a dithiolate, such as dixanthogen, as the multilayer product (reaction (4)). Because of the typically high vapour pressure of a dithiolate (diethyl dixanthogen has a melting point of 32 °C and other dialkyl dixanthogens are liquid at normal temperature and pressure), considerable care is required to ensure that it is retained at the surface for detection by XPS. Methods for cooling specimens to retain volatile species, without concomitantly condensing water or other molecules on top of the surface to be examined, have been described by *Buckley* and *Woods* [66] and *Kartio* et al. [67]. The latter group used their technique to detect dixanthogen produced on pyrite held for 15 min at 450 mV in 6×10^{-5} mol/dm^3 ethyl xanthate solution at pH 4.0. When the specimen was subsequently allowed to warm to ambient temperature in the spectrometer, the S 2p peaks corresponding to dixanthogen were no longer evident. Residual peaks of low intensity were observed that could have arisen from chemisorbed xanthate [68]. Their specimen cooling technique has since been modified to eliminate partial evacuation during the dry inert gas flushing stage [69].

3.4.2 Adsorption of Collectors at Pre-oxidized Sulfide Mineral Surfaces

As pointed out in Sect. 3.1, it is expected that flotation involves sulfide surfaces that are largely unoxidized. In some situations, however, such as with partially oxidized ore or with extensive recycling of floats and tails, flotation will be carried out on sulfide surfaces that have undergone oxidation.

It is apparent from electron spectra that the initial oxidation of sulfide minerals results in a metal-deficient sulfide surface layer rather than elemental sulfur or sulfur–oxygen species [33,70]. Although that conclusion has gained only limited acceptance, there is general agreement that the metal atoms at the surface react to form a hydrated oxide or hydroxide. Whether or not these metal–oxygen species remain at the surface under the turbulent conditions of a flotation cell depends largely on the precise value of the typically alkaline pH of the aqueous medium. For example, it has been shown by XPS that the initial iron hydroxy species formed at the surface of chalcopyrite exposed to air is removed from the surface in an oxygen-free aqueous medium of pH 8–11 [71]. If metal–oxygen species are removed from the surface of the particles, then any collector-mineral interaction would involve a metal–deficient sulfide surface. On the other hand, if the metal–oxygen species remain at the surface, most probably in patches, then the thiol molecules could interact either with these patches of hydroxide, or with unoxidized regions of the surface. There is substantial evidence to suggest that the presence of the collector facilitates removal of the hydroxide from the surface.

In addition to information derived from inadvertently oxidized surfaces, there have been several XPS studies of the interaction of thiol collectors with deliberately oxidized sulfide minerals. In one such study [72], surfaces of oxidized chalcocite were characterized by XPS before and after reaction with pH 9.5 ethyl xanthate solution for various periods. Oxidation was effected by exposure to air in a pH 9.5 aqueous medium for 20 min, and electron spectra indicated cupric oxide, hydroxide and carbonate as the principal surface oxidation products. No sulfur oxidation products were detected. Brief treatment with xanthate solution lowered both the O $1s$ intensity near 531.5 eV and the Cu $2p$ intensity from oxidation products markedly, and resulted in the appearance of a S $2p_{3/2}$ component near 162.3 eV due to adsorbed xanthate. The spectra indicated that the oxidation products were removed from the chalcocite surface at the beginning of the xanthate adsorption. Increasing the treatment time in xanthate solution essentially eliminated all oxidized species from the surface to leave an adsorbed ethyl xanthate layer. The observed variation in the intensities of the 284.9 eV and 286.5 eV xanthate C $1s$ peaks relative to the Cu $2p$ and S $2p$ intensities was consistent with the initial formation of a well-oriented monolayer of adsorbed xanthate and its subsequent covering by a multilayer of randomly oriented cuprous ethyl xanthate. Cu(LMM) Auger spectra were not recorded in the study, so that electron spectroscopic evidence to corroborate the formation of a chemisorbed monolayer was not available. *Mielczarski* et al. [58] did include these Auger spectra in a later investigation of the chalcocite/xanthate system, however as already discussed, it was presumed that xanthate chemisorption would give rise to a discernible Cu(LMM) shift, and it was concluded that rather than being chemisorbed in the underpotential region, xanthate was functioning as a source of sulfur for the formation of a sulfur-rich surface.

In another study, galena surfaces were exposed to air for predetermined periods prior to treatment with a concentrated xanthate solution at pH $\approx$ 6 [73]. From the electron spectra, less oxidized lead was evident at the surface after treatment with the collector than before, and this observation was interpreted in terms of lead xanthate peeling off the surface following reaction between the collector and the oxidation products. It was acknowledged that chemisorbed xanthate could also have been present if the Pb $4f$ binding energies for that species were similar to those for galena (as had been proposed [56]). A high binding energy component of low intensity in the corresponding S $2s$ spectrum was assigned to adsorbed xanthate. The O $1s$ intensity (from species other than physisorbed water) remained high after the treatment in collector solution.

XPS and other techniques were used to characterize oxidized chalcopyrite surfaces before and after treatment with butyl ethoxycarbonyl thiourea (BECTU) [74]. Ground chalcopyrite was conditioned for 20 min in an aqueous medium purged with either oxygen or nitrogen gas; at pH 9.5, the measured potentials were 210 mV and 70 mV on the SHE scale, respectively. In each case, the mineral was subsequently allowed to reach equilibrium in BECTU solution. The thiourea N $1s$ spectra indicated that more collector was adsorbed following preconditioning in oxygen than in nitrogen. The O $1s$ and Fe $2p$ spectra revealed that treatment with the collector resulted in a substantial reduction in the concentration of oxidized iron, whereas, apart from an increase in intensity (arising from the removal of iron hydroxy species), the S $2p$ and Cu $2p$ spectra were similar before and after BECTU treatment. It was concluded that adsorption of BECTU removes surface oxidation products from the mineral surface, particularly iron hydroxide/oxides.

It is known that sulfide minerals exhibit floatability in the absence of collectors [75] and that this phenomenon is inhibited by the presence of metal hydroxides on the mineral surface derived from mineral oxidation, other ore components, or grinding media [76]. The inhibition can be eliminated by the addition of small amounts of collector [76] and it was suggested [33] that this results from rejection of hydrophilic species from a surface rendered hydrophobic by collector adsorption. The XPS results outlined above support this hypothesis.

3.5 Static Secondary Ion Mass Spectrometry

Secondary Ion Mass Spectrometry (SIMS) is more surface sensitive, and potentially allows a much greater differentiation of surface species, than XPS. However, it is an inherently destructive technique because information is obtained from molecular fragments sputtered from the surface of the material being analysed. The primary ion dose for static SIMS is usually considerably less than 10^{13} ions/ cm^2, so that the lifetime of the surface monolayer is greater than the analysis time. Nevertheless, substrate and even adjacent monolayer damage, including primary ion implantation, is inevitable since the primary ion energy used is commonly several keV and can be as high as 25 keV. Thus, given that some chemical rearrangement would be caused by one or more of the ion ejection and collection processes, not all the ions detected would be expected to reflect the unaltered chemical composition of the surface. It is also more difficult to extract quantitative information from secondary ion mass spectra than from photoelectron spectra as secondary ion emission for a given species depends markedly on the chemical environment of that species. Nonetheless, it is possible to obtain some quantitative information from the ion spectra obtained with modern Time-of-Flight (ToF) spectrometers [77].

There have been relatively few reports of surface characterization of sulfide minerals by static SIMS, either before or after adsorption of thiol collectors. Most of the work to date in this area has been carried out by Brinen and co-workers [74,78,79]. Their initial application [73] involved the imaging of diisobutyl dithiophosphate (DTP) adsorbed at pH 9 on a laboratory flotation concentrate of 75–150 μm galena particles. Imaging was carried out by ToF-SIMS with a 25 keV ^{69}Ga ion beam of spot size 200 nm. The strongest peak in the negative ion spectrum was the parent ion of the collector at m/z 209, while the strongest peak in the positive ion spectrum was the ^{208}Pb peak at m/z 208. Areas of crystal faces giving rise to intense collector parent ions in the negative ion spectra were associated with weak $^{208}Pb^+$ peaks, and *vice versa*, consistent with patches of adsorbed collector. In later work [79], static SIMS spectra were determined using 4 keV Xe ions. These spectra confirmed the presence of a 1:1 lead dithiophosphate species at the surface of galena treated with the collector at pH 6. A series of peaks at m/z 416, 417, 418 and 419, with an intensity pattern similar to that for the 206, 207, 208 and 209 Pb isotope peaks, was observed in the positive ion spectrum. The former peaks were assigned to a Pb-DTP-H species. The abundance of the Pb-DTP-H^+ ions was interpreted as strong evidence in support of adsorption of the collector to lead atoms in the galena surface.

ToF-SIMS with a ^{69}Ga primary ion beam was also used to investigate the nature and distribution of butyl ethoxycarbonyl thiourea (BECTU) on chalcopyrite [74]. The most intense peaks in the positive ion spectrum from the chalcopyrite before treatment with collector were at m/z 56 and 63, due to the most abundant isotopes of iron and copper respectively. The corresponding spectrum from a chalcopyrite surface conditioned with BECTU (molecular mass 204) revealed peaks at m/z 203 and 205 due to BECTU $\pm$ H ions, and relatively intense peaks at m/z 267 and 269 corresponding to a copper BECTU species with ^{63}Cu and ^{65}Cu isotopes, respectively. Apparently these species contain an additional H in an analogous manner to that discussed above for the lead DTP system. Peaks at m/z 471 and 473, which would indicate the presence of a $Cu(BECTU)_2$ species, or at m/z 259 and 261, which would indicate the presence of Fe-BECTU, were not observed.

This was strong evidence for interaction of the collector with copper rather than iron atoms in the mineral surface.

These initial investigations illustrate the potential use of ToF-SIMS in the study of organic reagents adsorbed on mineral sulfides and their corresponding metals. The technique is currently being used for mineral processing research in several laboratories around the world.

4 Electrowinning and Electrorefining

Metal deposition processes involve the initial nucleation of metal clusters which grow and develop into a new phase. A theoretical framework has been established to describe the formation of nuclei and the rate of growth of individual and multiple nuclei as well as the overall kinetics of the metal deposition process [3]. On the macroscopic scale, metal layers can form in a range of textures and with different degrees of surface roughness, and the current density is an important parameter in determining the physical nature of the deposit [80]. At high current densities, where mass transport becomes an important factor in the deposition kinetics, dendrites and powdery deposits are formed. The initial nucleation step is dependent on inter alia the presence of adsorbed surface active species on the electrode surface.

In electrowinning and refining, the aim is to produce a deposit that can be readily converted into a saleable product that meets customer purity specifications. Silver is refined at current densities in the region of dendritic growth in Thum and Moebius cells [81] and this allows the product to be recovered directly from the cell without removing the cathodes. With nickel, copper and zinc, a compact product is required and the current density is selected at a value that gives rise to a dense, microcrystalline electrodeposit. The exchange current density for the Ni/Ni^{2+} system is relatively low and hence deposition occurs at significant overpotentials which facilitate nucleation and give smooth deposits. Metals such as zinc and copper that have low plating overpotentials tend to form coarse, rough deposits. This is due to the slow nucleation rates at low potentials which compels the new metal to grow on few nuclei. To overcome this problem, various additives are included in the solution that adsorb on the metal surface, preferentially at the most active growth sites. Ideally, these additives should not become incorporated into the deposit to any significant extent, since this would diminish product quality.

Control of electrolyte composition, particularly that of additive levels, is very important in maintaining optimum plant efficiency and economy, which in present practice relies on empirical observations. Electrochemical techniques have been developed for the continuous monitoring of additive performance in the electrowinning of zinc and lead [82,83] and in the winning and refining of copper [83,84,85,86]. These techniques have not, however, been widely applied by metallurgical industries.

The mode of action of additives in electrodeposition is an active area for research and development. Thiourea is a widely used addition reagent in copper refineries, and is applied in combination with protein colloid (glue) to avoid nodular growth which could lead to short circuits between anodes and cathodes. It is considered [87,88,89] that thiourea effectively refines the deposit grain size by adsorption on active nucleation sites at the cathode during plating. Chloride ion is also an addition reagent in all copper refineries and is added to precipitate silver as its chloride, and this is recovered from the anode slimes. Chloride also increases cathode polarization and hence improves deposit morphology. Clearly, the extent of polarization resulting from the inclusion of addition reagents in electrorefining cells must be relatively small, or the efficiency of separation from impurity elements in the anode will be diminished.

5 Adsorption of Thiourea on Copper Cathodes

The mechanism of the deposition of copper from sulfate electrolytes involves the formation of Cu^+ as an intermediate with the first step, the reduction of Cu^{2+} to Cu^+, being rate determining [90]. The presence of Cu^+ ions at the electrode surface during deposition of copper, and the reverse process of metal dissolution, can be demonstrated using a Rotating Ring Disc Electrode (RRDE). In this technique, any solution species formed at the rotating disc are swept past the ring where they can be detected from their current/potential response.

5.1 Electrochemical Techniques

Figure 10 shows the RRDE response from a stainless steel disc with a platinum ring in a synthetic copper electrorefining electrolyte, viz., $45\,g/dm^3$ $CuSO_4$, $2\,mol/dm^3 H_2SO_4$ solution [91]. Stainless steel is the material used

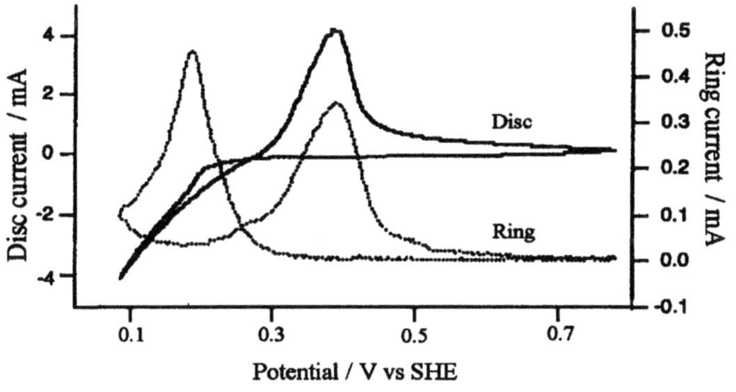

Fig. 10. RRDE response in synthetic copper electrorefining electrolyte [91]

for the cathodic blanks in the commercial ISA [92] and KIDD [93,94] copper processes. The potential of the disc was scanned while the ring potential was set at 1.28 V vs SHE, where Cu^+ ions are oxidized to Cu^{2+}. It can be seen that the onset of the cathodic current on the disc resulting from copper deposition during the negative going scan is accompanied by an anodic current on the ring indicating the intermediate formation of Cu^+ during the initial stage. On the subsequent positive going scan, the copper stripping peak on the disc is also associated with the detection of Cu^+ at the ring.

Figure 11 shows the corresponding RRDE response with the same electrolyte but containing 0.5 ppm thiourea. There is no evidence for an anodic process and hence, in contrast to what is observed with thiol collectors (Sect. 3), thiourea is not adsorbed on copper by an oxidative charge transfer chemisorption process.

The current signal from the disc in the copper deposition region is less than that in thiourea-free solution due to the action of thiourea in blocking the initial stage of the deposition reaction. The ring current in Fig. 11 indicates that thiourea does not change the mechanism of copper deposition through a Cu^+ intermediate. It also shows that the presence of thiourea has little influence on Cu^+ formed on the negative going scan. The ring response on the positive going scan shows an additional feature when thiourea is present; a new Cu^+ peak appears at $\sim$ 0.28 V vs SHE. In some cases, particularly at high thiourea concentrations, several peaks were reported [91]. These peaks were thought to be related to copper(I)-thiourea complexes which are released from the disc surface. The RRDE investigations established that thiourea is adsorbed on the copper surface when the metal is being deposited and identified the potential region in which adsorption occurs. Such electrochemical methods provide a means of studying electrodeposition systems

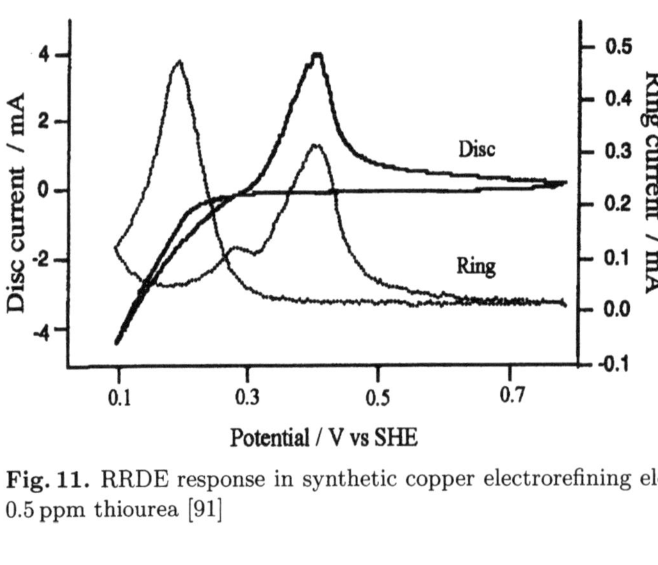

Fig. 11. RRDE response in synthetic copper electrorefining electrolyte containing 0.5 ppm thiourea [91]

under the same conditions as those that pertain in practice, and determining the influence of additives on various plating parameters. They do not, however, provide direct identification of the structure and bonding of surface species formed; for this purpose SERS spectroscopy has been applied.

5.2 Raman Scattering Spectroscopy

SERS has been applied by a number of authors [91,94,95,96,97,98,99,100,101] to elucidate the interaction of thiourea with copper and silver surfaces in sulfate-, acid- and halide-containing solutions. It was shown that there is a decrease in the C–S bond character of thiourea on adsorption, and a corresponding increase in the C–S bond character. This is illustrated in Fig. 12. An FT-Raman spectrum of aqueous 1 mol/dm^3 thiourea is shown in curve (a) and a typical in situ SERS spectrum of a copper electrode in thiourea solution (polarized at −0.2 V vs SHE in 2 mol/dm^3 H_2SO_4 containing 5 ppm thiourea) presented in curve (b).

The spectrum for the thiourea solution (curve (a)) shows vibrational modes for both C–S (743 cm^{-1}) and C–S (480 cm^{-1}) and this was attributed [100] to the presence of resonance structures. It can be seen that the C=S band at 743 cm^{-1}, which is the most intense band in curve (a), is absent in

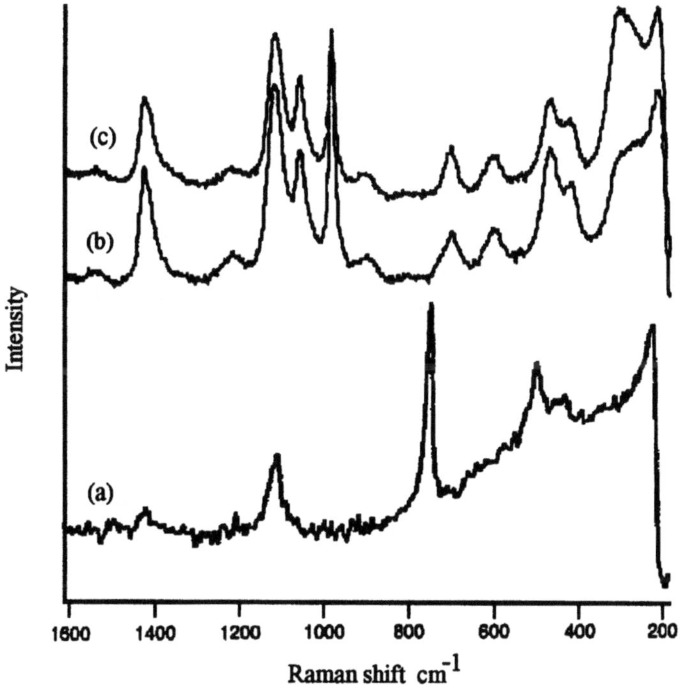

Fig. 12. (**a**) Raman spectrum of 1 mol/dm^3 aqueous thiourea; (**b**) SERS spectra from copper in 10 ppm thiourea /2 mol/dm^3 H_2SO_4; (**c**) as (**b**) +60 ppm chloride [91]

the SERS spectrum, curve (b). The C–S band is still present and a band at $283\,cm^{-1}$ is apparent that was assigned to the Cu–S stretch. These finding are consistent with adsorption of the organic molecule on the copper electrode occurring by coordination with the sulfur atom.

In addition to other bands in the SERS spectrum in curve (b) that can be assigned to vibrations of the thiourea molecule, *Hope* and co-workers [91,100,101] noted that there are also bands at 217, 605, 982 and $1055\,cm^{-1}$ and attributed these bands to the presence of adsorbed SO_4^{2-} and HSO_4^- derived from the electrolyte. SERS spectra for the interaction of SO_4^{2-} and HSO_4 ions with a copper electrode have been observed in sulfuric acid solutions without thiourea [102]. The adsorption of sulfate and thiourea on copper was shown [91,100,101], however, to be a co-operative rather than a competitive process, since the intensity of the sulfate bands as well as those of thiourea, increased with increase in thiourea concentration. Furthermore, the sulfate band at $982\,cm^{-1}$ is significantly shifted from the corresponding band in thiourea-free solution ($970\,cm^{-1}$). The co-adsorption of thiourea and sulfate suggested that there were further interesting characteristics of the adsorption of thiourea at the copper surface in sulfuric acid solution than had been reported previously.

Complex formation has been reported [103,104] between copper(I), thiourea (tu) and anions such as sulfate and the crystal structure of the previously unknown $[Cu_4(tu)_7](SO_4)_2 \cdot H_2O$ and the vibrational spectra of a range of copper (I) thiourea complexes were determined [104]. The copper atoms were found to lie in a tetrahedral arrangement forming $[Cu_4(tu)_7]^{4+}$ clusters interlinked by sulfate ions, which strongly interact with thiourea ligands through hydrogen bonds. The bond lengths about the thiourea ligands indicated a decrease in the carbon–sulfur double-bond character, consistent with the co-ordination of thiourea with copper being through the sulfur atoms. The Raman bands in the sulfate complexes were found to appear at similar wavenumbers to those of the SERS bands for the species adsorbed on copper. The stretching vibration for Cu–S for these complexes was found to show a strong dependence on the copper coordination environment, and the Raman shift observed in the SERS indicated that the coordination in the Cu/thiourea species have is relatively low. This is as expected for a surface analogue of a copper(I) thiourea sulfate complex.

The adsorption of thiourea was found [100] to display a dependence on the applied potential and solution thiourea concentration. This dependence has implications for electrodeposition processes where control of the electrode potential is a major requirement for optimum process efficiency.

In situ SERS spectroscopy was applied [101] to investigate the influence of chloride ion on the co-adsorption of thiourea and sulfate ions in sulfuric acid solution at the concentration levels used in the electrorefining of copper. It was found that chloride was co-adsorbed, as evidenced by the appearance of a sharp band at $300\,cm^{-1}$ (see curve (c) of Fig. 12), and that adsorption of the halide was favoured at low negative potentials. The presence of the chloride in solution at low concentrations was also shown to result in enhancement of the adsorption of thiourea and sulfate. Furthermore, the adsorption of chloride at the copper electrode altered the molecular structure of the interface due to an interaction with co-adsorbed thiourea and sulfate species at the electrode surface. The chloride apparently interacts with the nitrogen-containing groups of the thiourea molecule adsorbed at the electrode surface as evidenced by changes in band shape and intensity for the NH_2 torsion and C–N stretching vibrational modes. In this manner, the adsorption of chloride results in a rearrangement or alteration of the molecular structure at the electrode surface that is observed by changes in signal intensity for the various vibrational bands associated with the adsorption of thiourea, sulfate and chloride at the electrode surface.

Glue was found [91] not to be SERS active at the concentrations used in practice by the electrowinning and refining industry. SERS spectra could, however, be recorded for higher glue concentrations and the results provide useful information on the interaction of this additive with a copper surface. Figure 13 presents the normal Raman spectrum of the dry solid glue and the SERS spectrum from a copper electrode in a sulfuric acid electrolyte containing 50 ppm glue.

It was concluded that a glue layer was deposited on the copper surface and that this exhibited a Raman spectrum similar to that of the solid compound. Three additional features are apparent, however, in the SERS spec-

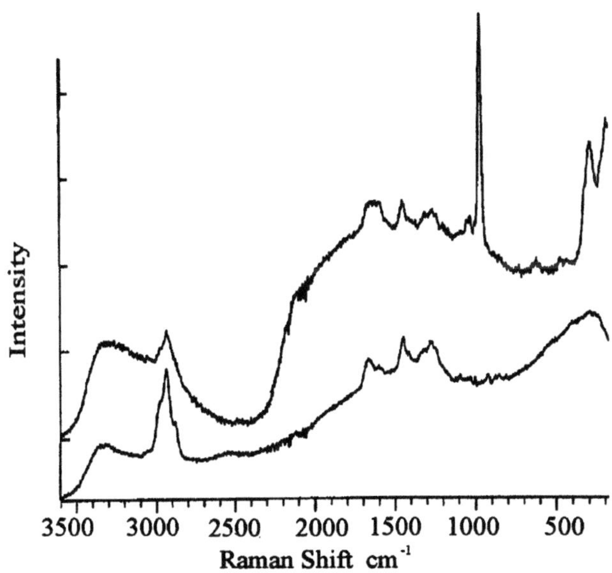

Fig. 13. Raman spectrum of dry solid glue *(lower)* and SERS spectrum from a copper electrode in 50 ppm glue /2 mol/dm^3H_2SO_4 [91]

trum; a strong SO_4^{2-} band appears at $982\,cm^{-1}$, a Cu–N band occurs at $269\,cm^{-1}$ and the N–H band at $1594\,cm^{-1}$ of increased intensity. It would appear that there are additional protons bonded to the nitrogen atoms of the gelatine molecule, and that sulfate anions balance the additional charge. This observation of the nature of the glue–electrode interaction is supported by other results [105] which have demonstrated potential dependence and the displacement of sulfate anions by the "harder" chloride species under particular conditions.

6 Concluding Remarks

The combination of electrochemical and spectroscopic techniques provides a powerful approach to elucidating the adsorption of organic reagents on surfaces that are of importance to the effective operation of various processes in the recovery of metals from sulfidic ores. Electrochemical methods provide a means of determining the conditions under which adsorption takes place and the kinetics of adsorption. Vibrational and electron spectroscopies identify the nature of the adsorbed species and the mode of bonding to the surface. The application of these techniques has shown that adsorption of organic reagents in metallurgical systems takes place by a number of different mechanisms.

Charge transfer chemisorption occurs in the interaction of thiol flotation collectors with metal sulfide surfaces and is the thermodynamically favoured process. The formation of metal/thiol compounds and dithiolates can also occur and play a role in rendering sulfide minerals hydrophobic and hence floatable. The chemisorbed thiol is attached to the surface through metal–sulfur covalent bonds. The chemisorbed species can be distinguished from metal compound formation by FTIR and electron spectroscopies, while SERS demonstrates that the thiol retains its molecular integrity on chemisorption.

Surface complexes are the important species formed by thiourea additives that are used to control the morphology of copper electrodeposits in electrowinning and electrorefining. The complexes are formed by co-operative adsorption of electrolyte anions with the thiourea. Sulfate and bisulfate are the major anions in these systems. Chloride ion is normally present in low concentrations and when adsorbed alters the structure of the thiourea complex through bonding with the nitrogen atom in the thiourea molecule.

References

1. D. W. Fuerstenau, R. H. Urbana: in P. Somasundaran, B. J. Moudgil (Eds.): *Reagents in Mineral Technology* (Marcel Dekker, New York 1988) p. 1
2. R. Woods: in M. H. Jones, J. T. Woodcock (Eds.): *Principles of Mineral Flotation* (Aus. Institute Mining Metallurgy, Melbourne 1984) p. 91
3. J. O'M. Bockris, S. U. M. Khan: *Surface Electrochemistry* (Plenum, New York 1993)

4. O. A. Finklea: in A. J. Bard, I. Rubenstein (Eds.): *Electroanalytical Chemistry*, Vol. 19 (Marcel Dekker, New York 1996) p. 110
5. R. Woods: J. Phys. Chem. **75**, 354 (1971)
6. J. R. Gardner, R. Woods: Aust. J. Chem. **30**, 981 (1977)
7. P. E. Richardson, C. S. O'Dell: J. Electrochem. Soc. **132**, 1350 (1985)
8. A. Kowal, A. Pomianowski: J. Electroanal. Chem. **46**, 411 (1973)
9. C. S. O'Dell, R. K. Dooley, G. W. Walker, P. E. Richardson: in P. E. Richardson, S. Srinivasan, R. Woods (Eds.) *Proc. Int. Symp. Electrochemistry in Mineral and Metal Processing* (Electrochem. Soc., Pennington 1984) PV 84–10, p. 81
10. P. E. Richardson, J. V. Stout III, C. L. Proctor, G. W. Walker: Int. J. Miner. Process. **12**, 73 (1984)
11. C. Basilio, M. D. Pritzker, R.-H. Yoon: Preprint 85–86, 114th AIME Annual Meeting (AIME, New York 1985)
12. O'Dell, G. W. Walker, P. E. Richardson: J. Appl. Electrochem. **16**, 544 (1986)
13. R. Woods: in E. A. Mullar, G. Gonzalez, C. Barahona (Eds.): *Copper 87*, Vol. 2 (University of Chile, Santiago 1988) p. 121
14. R. Woods, C. A. Young, R.-H. Yoon: Int. J. Miner. Process. **30**, 17 (1990)
15. A. N. Buckley, R. Woods: Colloids Surf. A **104**, 295 (1995)
16. R. Woods, Z. Chen, R.-H. Yoon: Int. J. Miner. Process. **50**, 47 (1997)
17. Z. Szeglowski, J. Czarnecki, A. Kowal, A Pomianowski: Trans. IMM **86**, C115 (1977)
18. P. Talonen, J. Rastas, J. O. Leppinen: Surf. Interface Anal. **17**, 669 (1991)
19. R. Woods, C. I. Basilio, D. S. Kim, R.-H. Yoon: J. Electroanal. Chem. **328**, 179 (1992)
20. S. Chander, D. W. Fuerstenau: J. Electroanal. Chem. **56**, 217 (1974)
21. S. Chander, D. W. Fuerstenau: Int. J. Miner. Process. **2**, 333 (1975)
22. A. N. Buckley, R. Woods: J. Electroanal. Chem. **357**, 387 (1993)
23. G. A. Hope, K. Watling, R. Woods, unpublished data
24. R. Brdicka: Z. Elektrochem. **48**, 278 (1942)
25. R. Brdicka: Coll. Czech. Chem. Comm. **12**, 522 (1947)
26. D. M. Kolb: in H. Gerischer, C. W. Tobias (Eds.): *Advances in Electrochemistry and Electrochemical Engineering*, Vol. 11 (Wiley, New York 1978) p. 125
27. H. Schmalzried: Prog. Solid State Chem. **13**, 119 (1980)
28. R. Woods, D. S. Kim, C. I. Basilio, R.-H. Yoon: Colloids Surf. A **94**, 67 (1995)
29. R. Woods, C. I. Basilio, D. S. Kim, R.-H. Yoon: Colloids Surf. A **83**, 1 (1994)
30. R. Woods: Aust. J. Chem. **25**, 2329 (1972)
31. A. N. Buckley, R. Woods: Colloids Surf. A **89**, 71 (1994)
32. P. Nowak: Colloids Surf. A **76**, 65 (1993)
33. R. Woods: in J. O'M. Bockris, B. E. Conway, R. E. White (Eds.): *Modern Aspects of Electrochemistry*, Vol. 29 (Plenum, New York 1996) p. 401
34. K. L. Sutherland, I. W. Wark: *Principles of Flotation* (Aus. Inst. Mining Metallurgy, Melbourne 1955)
35. R. Woods, D. S. Kim, R.-H. Yoon: Int. J. Miner. Process. **39**, 101 (1993)
36. J. R. Gardner, R. Woods: Aust. J. Chem. **27**, 2139 (1974)
37. P. E. Richardson, G. W. Walker: In *Proc. XVth Int. Mineral Processing Congress*, Cannes, France, Vol. II (GEDIM, St. Etienne 1985) p. 198
38. J. B. Zachwieja, G. W. Walker, P. E. Richardson: Miner. Metall. Process. **4**, 146 (1987)

39. P. J. Guy, W. J. Trahar: Int. J. Miner. Process. **12**, 15 (1984)
40. N. W. Johnson, A. Jowett, G. W. Heyes: Trans. IMM **91**, C32 (1982)
41. S. R. Grano, J. Ralston, R. Smart: Int. J. Miner. Process. **30**, 69 (1990)
42. G. W. Heyes, W. J. Trahar: Int. J. Miner. Process. **6**, 229 (1979)
43. W. Barzyk, K. Malysa, A. Pomianowski: Int. J. Miner. Process. **8**, 17 (1981)
44. A. N. Buckley, R. Woods: Colloids Surf. A **53**, 33 (1991)
45. R. Woods, C. I. Basilio, D. S. Kim, R.-H. Yoon: Int. J. Miner. Process. **42**, 215 (1994)
46. J. O. Leppinen, C. I. Basilio, R. H. Yoon: in P. E. Richardson, R. Woods (Eds.): *Proc. Int. Symp. Electrochemistry in Mineral and Metal Processing II* (Electrochem. Soc., Pennington 1988) PV 88–21, p. 49
47. G. W. Heyes, W. J. Trahar: Int. J. Miner. Process. **4**, 317 (1977)
48. C. I. Basilio, D. S. Kim, R.-H. Yoon, D. R. Nagaraj: Miner. Eng. **5**, 397 (1992)
49. A. Ihs, K. Uvdal, B. Liedberg: Langmuir **9**, 733 (1993)
50. A. N. Buckley, T. J. Parks, A. M. Vassallo, R. Woods: Int. J. Miner. Process. **51**, 303 (1997)
51. R. Woods, G. A. Hope, G. M. Brown: Colloids Surf. A **137**, 329 (1998)
52. R. Woods, G. A. Hope, G. M. Brown: Colloids Surf. A **137**, 339 (1998)
53. R. Woods, G. A. Hope: Colloids Surf. A **146**, 63 (1999)
54. M. P. Seah, W. A. Dench: Surf. Interface Anal. **1**, 2 (1979)
55. P. M. A. Sherwood: Colloids Surf. A **134**, 221 (1998)
56. A. N. Buckley, R. Woods: Int. J. Miner. Process. **28**, 301 (1990)
57. A. V. Shchukarev, I. M. Kravets, A. N. Buckley, R. Woods: Int. J. Miner. Process. **44**, 99 (1994)
58. J. A. Mielczarski, J. B. Zachwieja, R.-H. Yoon: Preprint 90-174, SME Ann. Meeting, Salt Lake City, Utah, 1990 (Soc. Manufacturing Engineers, Dearborn 1990)
59. G. Contini, A. M. Marabini, K. Laajalehto, E. Suoninen: J. Colloid Interface Sci. **171**, 234 (1995)
60. K. Laajalehto: Private communication (1996)
61. J. A. Mielczarski, E. Mielczarski, J. Zachwieja, J. M. Cases: Langmuir **11**, 2787 (1995)
62. J. A. Mielczarski: Langmuir **13**, 878 (1997)
63. A. N. Buckley, R. Woods: J. Appl. Electrochem. **26**, 899 (1996)
64. A. N. Buckley, R. Woods: in R. Woods, F. M. Doyle, P. E. Richardson (Eds.): *Proc. 4th Int. Symp. Electrochemistry in Mineral and Metal Processing* (Electrochem. Soc., Pennington 1996) p. 1
65. I. J. Kartio, K. Laajalehto, E. J. Suoninen, A. N. Buckley, R. Woods: Colloids Surf. A **133**, 303 (1998)
66. A. N. Buckley, R. Woods: Appl. Surf. Sci. **17**, 401 (1984)
67. I. Kartio, K. Laajalehto, E. Suoninen: in R. Woods, F. M. Doyle, P. E. Richardson (Eds.): *Proc. 4th Int. Symp. Electrochemistry in Mineral and Metal Processing* (Electrochem. Soc., Pennington 1996) p. 13
68. E. Suoninen, K. Laajalehto, I. Kartio, S. Heimala, S. Jounela: in P. E. Richardson, R. Woods (Eds.): *Proc. 3rd Int. Symp. Electrochemistry in Mineral and Metal Processing* (Electrochem. Soc., Pennington 1992) p. 259
69. G. Wittstock, I. Kartio, D. Hirsch, S. Kunze, R. Szargan: Langmuir **12**, 5709 (1996)
70. A. N. Buckley, G. W. Walker: in K. S. E. Forssberg (Ed.): *Proc. XVI Int. Mineral Processing Congress* (Elsevier, Amsterdam 1988) p. 589

71. J. B. Zachwieja, J. J. McCarron, G. W. Walker, A. N. Buckley: J. Colloid Interface Sci. **132**, 462 (1989)
72. J. Mielczarski: J. Colloid Interface Sci. **120**, 201 (1987)
73. K. Laajalehto, P. Nowak, E. Suoninen: Int. J. Miner. Process. **37**, 123 (1993)
74. G. Fairthorne, J. S. Brinen, D. Fornasiero, D. R. Nagaraj, J. Ralston: Int. J. Miner. Process. **54**, 147 (1998)
75. W. J. Trahar: in M. H. Jones, J. T. Woodcock (Eds.): *Principles of Flotation*, The Wark Symposium (Aus. Institute Mining Metallurgy, Melbourne 1984) p. 117
76. G. D. Senior, W. J. Trahar: Int. J. Miner. Process. **33**, 321 (1991)
77. K. Reihs, R. Aguiar Colom, S. Gleditzsch, M. Deimel, B. Hagenhoff, A. Benninghoven: Appl. Surf. Sci. **84**, 107 (1995)
78. J. S. Brinen, F. Reich: Surf. Interface Anal. **18**, 448 (1992)
79. J. S. Brinen, D. R. Nagaraj: Surf. Interface Anal. **21**, 874 (1994)
80. D. Pletcher, R. C. Walsh: *Industrial Electrochemistry* (Blackie Academic, Glasgow 1993)
81. V. A. Ettel, B. V. Tilak: in J. O'M. Bockris, E. Yeager, B. E. Conway, R. E. White (Eds.): *Comprehensive Treatise of Electrochemistry* (Plenum, New York 1981) p. 327
82. C. J. Kraus, D. C. Kerby: in I. H. Warren (Ed.): *Application of Polarization Measurements in the Control of Metal Deposition* (Elsevier, Amsterdam 1984) p. 84
83. T. N. Andersen, R. C. Kerby, T. J. O'Keefe: J. Metals **37**(1), 36 (1985)
84. B. E. Langner, P. Stantke: in M. Koch, J. C. Taylor (Eds.): *Productivity and Technology in the Metallurgical Industries* (Minerals, Metals, Materials Soc., Warrendale 1989) p. 717
85. R. Winand, M. Degrex, V. Bastin: in W. C. Cooper, D. J. Kemp, G. E. Lagos, K. G. Tan (Eds.): *Copper 91* (Pergamon, New York 1991) p. 341
86. C. A. Davis, G. A. Hope: in D. Hall, Y. Kondo (Eds.): *Quality Management in Industrial Electrochemistry/1993* (Electrochem. Soc., Pennington 1993), PV 93-19, p. 172
87. T. J. O'Keefe, L. R. Hurst: J. Appl. Electrochem. **8**, 109 (1978)
88. D. R. Turner, G. R. Johnson: J. Electrochem. Soc. **109**, 798 (1962)
89. S. E. Afifi, A. A. Elsayed, A. E. Elsherief: J. Metals **39**(2), 36 (1987)
90. J.O'M. Bockris, M. Enyo: Trans. Faraday Soc. **58**, 1187 (1962)
91. G. A. Hope, G. M. Brown: in R. Woods, F. M. Doyle, P. E. Richardson (Eds.): *Proc. 4th Int. Symp. Electrochemistry in Mineral and Metal Processing* (Electrochem. Soc., Pennington 1996), PV 96-6, p. 429
92. I. J. Perry, J. C. Jenkins, Y. Okamoto: In *Proc. 110th Ann. Meeting AIME*, Chicago, Illinois, 1981 (AIME, New York 1981)
93. D. J. Kemp, O. Matwijenko, J. D. Scott: in W. C. Cooper, D. J. Kemp, G. E. Lagos, K. G. Tan (Eds.): *Copper 91* (Pergamon, New York 1991) p. 529
94. S. H. Macomber, T. E. Furtak: Chem. Phys. Lett. **90**, 59 (1982)
95. B. H. Loo: Chem. Phys. Lett. **89**, 346 (1982)
96. M. Fleischmann, I. R. Hill, G. Sundholm: J. Electroanal. Chem. **157**, 359 (1983)
97. A. El Hajbi, P. Chartier, G. Goetz-Grandmont, M. J. Leroy: J. Electroanal. Chem. **227**, 159 (1987)
98. Z. Q. Tian, Y. Z. Lian, M. Fleischmann: Electrochim. Acta **35**, 879 (1990)

99. H. Kim, J.-J. Kim: J. Raman Spectrosc. **24**, 77 (1993)
100. G. M. Brown, G. A. Hope, D. P. Schweinsberg, P. M. Fredericks: J. Electroanal. Chem. **380**, 161 (1995)
101. G. M. Brown, G. A. Hope: J. Electroanal. Chem. **413**, 153 (1996)
102. G. M. Brown, G. A. Hope: J. Electroanal. Chem. **382**, 179 (1995)
103. M. B. Ferrari, G.F, Gaspari: Cryst. Struct. Commun. **5**, 935 (1976)
104. R. C. Bott, G. A. Bowmaker, G. M. Brown, C. A. Davis, G. A. Hope, B. E. Jones: Inorg. Chem. **37**, 651 (1998)
105. G. M. Brown, G. A. Hope: J. Electroanal. Chem. **397**, 293 (1995)

SXPS and XANES Studies of Interface Reactions of Organic Molecules on Sulfide Semiconductors

Dirk Mayer, Karl Heinz Hallmeier, Dominic Zerulla, and Rüdiger Szargan

Universität Leipzig, Wilhelm-Ostwald-Institut für Physikalische und Theoretische Chemie, Linnéstr. 2, 04103 Leipzig, Germany
szargan,khall@rz.uni-leipzig.de

Abstract. The interaction of the organic heterocycle 2-mercaptobenzothiazole (MBT) with CdS($10\bar{1}0$) surfaces was investigated using SXPS and NK XANES spectra excited with synchrotron radiation. The molecules were found to be chemisorbed after deprotonation at the nitrogen atom. Angular dependent NK XANES measurements of the adsorbate complex indicated an upright and tilted orientation of the MBT molecular planes of the adsorbed MBT molecules in the [$\bar{1}2\bar{1}0$] direction of the crystal. In consequence of an interface reaction step an overlayer of the disulfide bis-(2-benzothiazolyl)disulfide (BBTD) was detected. This second reaction step was found to be induced by photoelectrical generation of electron–hole pairs in the semiconductor. A proposed reaction model includes the transport of the charge carriers to the semiconductor surface and subsequent charge transfer across the interface. The model was verified by adsorption experiments on differently doped semiconductor crystals which resulted in distinct differences in adsorbate composition.

1 Introduction

The controlled chemical and physical manipulation of organic adsorbates on sulfide surfaces in solutions may further develop the separation of natural complex sulfide mixtures by froth flotation and may create new applications in catalysis and sensor and solar techniques.

Much of the previous adsorption work deals with the bonding of thiolates and related compounds forming more or less self-assembled monolayers [1,2,3,4,5,6,7,8,9,10,11]. The adsorption of heterocyclic thiolate compounds introducing π-systems and specific functional groups into the molecular assembly of the adsorbate modifying also the redox behaviour of the system was investigated only recently [12,13,14]. With increasing progress in direct characterization of the solid/electrolyte interface by scanning microscopies (STM, AFM, SECM), by FTIR and electroanalytical in situ techniques, also interest in ex situ investigation of the chemical states of particular atoms as well as their in-depth distribution above and below the interface by means of electron [15,16,17] and X-ray absorption spectroscopies [18,19] is rising.

K. Wandelt, S. Thurgate (Eds.): Solid–Liquid Interfaces, Topics Appl. Phys. **85**, 97–112 (2003)

This chapter presents new results on the adsorption of 2-mercaptobenzothiazole (MBT=Hmbt) on CdS(10$\bar{1}$0). The investigation focuses on chemical binding, molecular orientation and oxidation of the adsorbate applying standard models of photoelectron attenuation and chemical shift of the core electron spectra (XPS) as well as the angle dependence of the X-ray absorption near edge structures (XANES). Highly resolving and surface sensitive techniques using monochromatized synchrotron radiation and sophisticated fitting procedures for decomposing the complex S 2*p* photoelectron spectra were used to separate the different spectral components of the adsorbate and the substrate. The influence of temperature and light intensity on the interface reaction rates is pointed out in order to elucidate the possibility for controlling and engineering surface properties by means of thermal and photoelectric effects.

2 Experimental

Photoelectron spectra were excited with tunable synchrotron radiation from BESSY (Berlin), the monochromator PM-5 and the HIRES photoelectron spectrometer. The XP-spectra were fitted with the peak-fitting program "Unifit for Windows" [20] using Gaussian–Lorentzian peak shape deconvolution. XANES spectra were derived from partial or total electron yield recorded with the HE-TGM 2 monochromator applying different retarding potentials in order to enhance the surface sensitivity of the measurements.

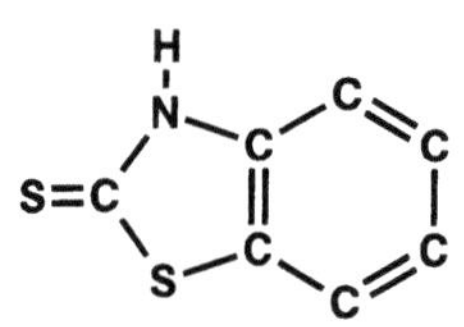

MBT=Hmbt

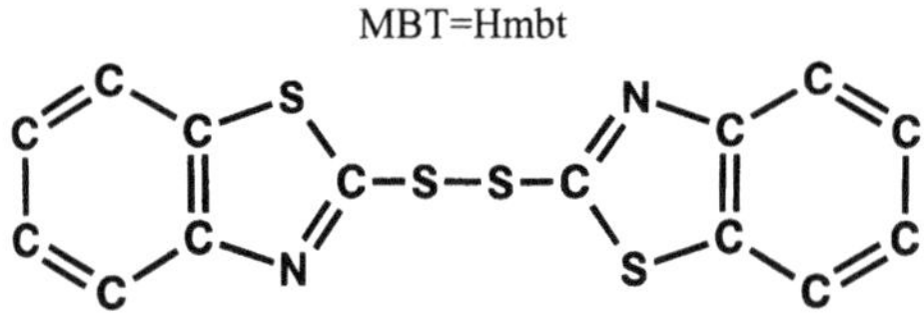

BBTD

Adsorption experiments were carried out using commercial lowly (10^8 Ωcm) and highly (1–10 Ωcm) doped CdS(10$\bar{1}$) wafers (Crystal GmbH, Berlin Germany). 2-mercaptobenzothiazole (MBT) and the oxidation product disulfide bis-(2-benzothiazolyl)disulfide (BBTD) were purchased from Aldrich Chemicals (Steinheim, Germany) in 98 % purity. The interface reactions were performed at clean and preoxidized crystal surfaces in aqueous 10^{-5} M and 10^{-4} M MBT solutions under nitrogen flux for different durations of reaction and temperatures. Because of the low solubility of the MBT in water addition of 5 % v/v methanol was necessary. After adsorption the sample was rinsed by a methanol/water (5 % v/v) solution and dried in a nitrogen flow. The CdS surfaces were oxidized by ozone produced in air by illuminating with UV light of a low-pressure mercury lamp. Elemental sulfur was deposited by etching the crystals in a 5 ‰ v/v Br_2/methanol solution.

3 Results and Discussion

The molecule of the chelating agent 2-mercaptobenzothiazole contains three reactive centers and exhibits a special affinity to metal atoms of different chalcogenide crystals [14,19].

3.1 Chemical Composition of the Adsorbate

The S 2*p* photoelectron spectra of the adsorbate system are composed by four components (Fig. 1) resulting from the substrate sulfur S^1 and two sulfur atoms S^2 and S^3 of the MBT.

The component S^2 with a binding energy of 162.9 eV can be attributed to the exocyclic sulfur atom of the chemisorbed mbt^- anion. The S 2*p* energy

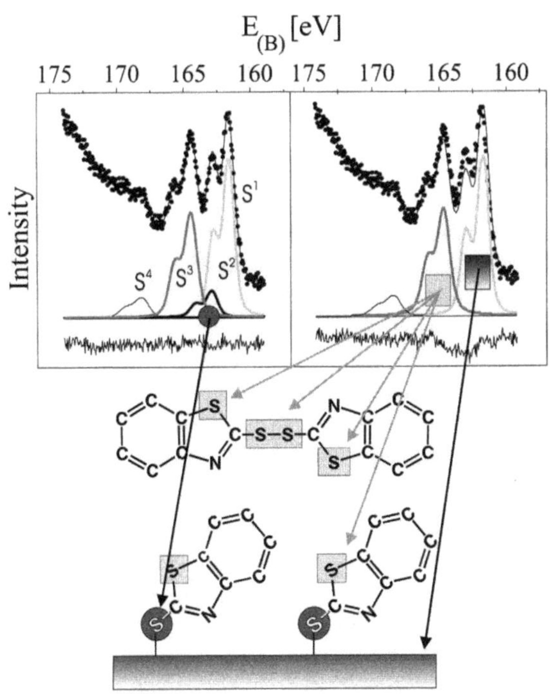

Fig. 1. S 2*p* photoelectron spectra of CdS(10$\bar{1}$0) after 10 min reaction time in 10^{-5}M MBT solution

of this atom is 0.8 eV shifted to higher energies with respect to the exocyclic sulfur of the pure MBT reflecting electronic charge donation from the sulfur to the cadmium atom at the semiconductor surface. The endocyclic sulfur atom of the MBT molecule should contribute to the S 2*p* signal S^3 in accord to the 1:1 atomic ratio. This component S^3, however, is much more intensive pointing at additional sulfur species containing (–S–) or (–S–S–) groups.

Therefore we conclude that bis-(2-benzothiazolyl)disulfide was formed by oxidation/dimerization. The four sulfur atoms from two (–S–) and one (–S–S–) groups (see above) giving only slightly shifted S 2*p* signals should contribute to the same spectral component S^3. Comparing the S 2*p* binding energies with results of the MBT adsorption on PbS (galena) and FeS_2 (pyrite) shows nearly the same energies for the S^3 components and distinct differences for S^2 [14,19]. Otherwise the distances between the binding energies of S^1 and S^2 differ just weakly by less than 0.3 eV for all three adsorbate systems. One may conclude the atomic potential at the exocyclic sulfur to be determined by the bonding interaction with the substrate:

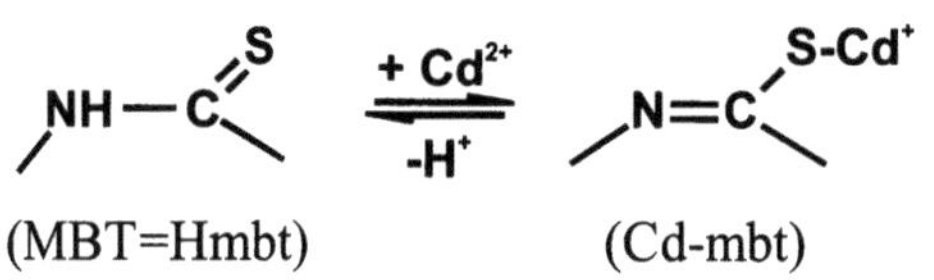

In contrast, the comparable energies for the S^3 components of the adsorbate system and the pure MBT indicate the absence of any interaction between the endocyclic sulfur and the substrates. The doublet S^4 at 168 eV is due to a very small fraction of SO_4^{2-} remained after ex situ preparation of the CdS wafer with UV/ozone and rinsing the soluble sulfate by water.

The chemisorption of the MBT molecule at the surface via the thiolate group is accompanied by a deprotonation of the nitrogen atom as indicated above. This deprotonation should strongly influence the N 1*s* and NK XANES features. Thus the N 1*s* spectra (Fig. 2) exhibit a low-energy signal at about 399 eV created by the sp^2 hybridized nitrogen atom (–N=) arising from the formation of a π bond between the nitrogen and the adjacent carbon atom as a consequence of the deprotonation. This π bond is indicated also by the NK XANES features which exhibit characteristic sp^2 preabsorption structures (see below).

The intensity of the (–N=) component rises with increasing MBT concentration due to increasing amount of reaction products. The additional component (–N̲H–) at 400.6 eV may originate from ubiquitous impurities in the atmosphere as a result of the ex situ preparation.

The NK XANES of the adsorbate resulting from the interaction between a 10^{-4} M MBT solution and a CdS wafer is similar to the spectrum of the $Cd(mbt)_2$ complex (Fig. 3) and clearly shifted with respect to the reference MBT because of the deprotonation.

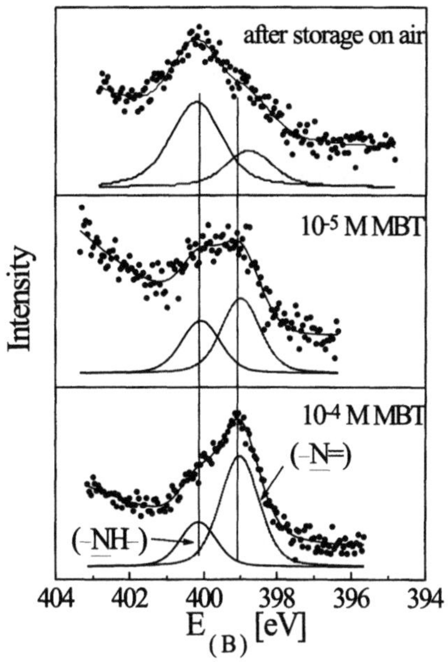

Fig. 2. N 1 s spectra of CdS(10$\bar{1}$0) after 10 min reaction time in 10^{-4}M and 10^{-5}M MBT solution

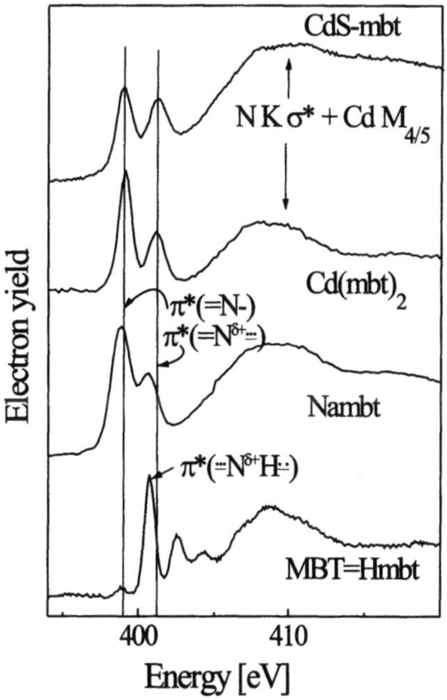

Fig. 3. NK XANES of Hmbt, Nambt and Cd(mbt)$_2$ in comparison with the spectrum of CdS(10$\bar{1}$0) after 10 min reaction time in 10^{-4} MBT solution

Two π resonances appear arising from $1\,s^{-1}2p\pi^*$ excitations at (–N=C–S) groups of the thiolate mbt^-. These characteristic resonances occur in molecules with (–N=C–) groups in the bridging position with conjugated π electrons and can be understood as transitions of N $1s$ electrons into two different antibonding $2p\pi^*$ states [18,19]. The splitting of the π resonances arises from conjugation effects as confirmed by Hartree–Fock calculations of unoccupied states in benzalaniline using the $Z{+}1$ approximation for the excited atom [21]. A high energy σ resonance above 408 eV can be attributed to a $1\,s^{-1}\sigma^*(2p)$ excited state located at the nitrogen atom.

3.2 Orientation of the Adsorbed Molecules

The orientation of the molecules on the surface is determined by bonding interactions between the molecule and the surface, by intermolecular interactions and by the dimensions of the molecules. By rotating the adsorbed mbt^- anion around the Cd–S and S–C bonds and varying simultanously the Cd–S–C angle nearly every molecular orientation can be obtained (Fig. 4, left).

One MBT molecule oriented in the $[\bar{1}2\bar{1}0]$ direction completely occupies two Cd sites. In the [0001] direction one molecule completely occupies one Cd atom and partially a second one. By rotating the molecule out of this orientation the second Cd atom becomes accessible for the adsorption of an additional molecule (see Fig. 4, right).

Such rotation can be caused by intermolecular interactions between adjacent molecules if the substrate dominated distance a between the molecular planes is larger than their van der Waals radii. The resulting effective molecular distance c is given by $\cos\alpha = c/a_x$.

The orientation of molecules with respect to a particular crystal direction can be determined by angle-dependent K XANES measurements varying the angles between the edge or surface normal of the crystal with respect to the polarization direction $\overline{\boldsymbol{E}}$ of the exciting synchrotron radiation. For this

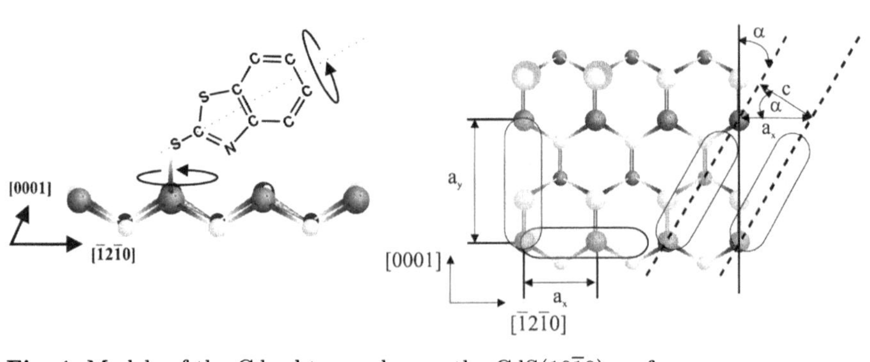

Fig. 4. Models of the Cd-mbt complex on the $CdS(10\bar{1}0)$ surface

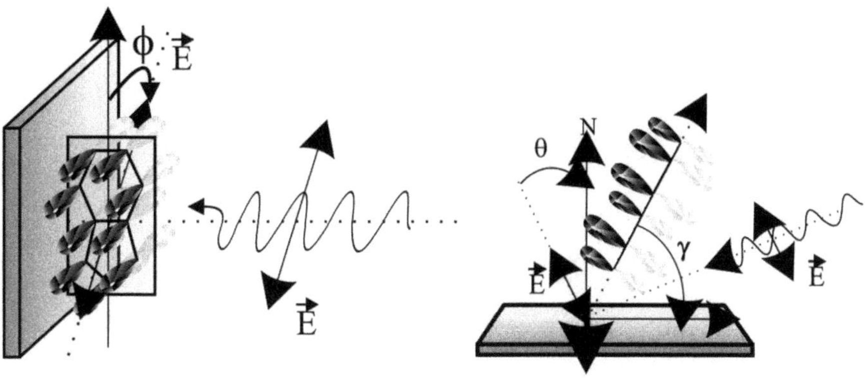

Fig. 5. Azimuthal (*left*) and polar (*right*) orientation of the molecular plane of MBT on the CdS(10$\bar{1}$0) surface with respect of the polarization vector $\overline{\boldsymbol{E}}$ of the exciting radiation

aim the azimuthal angle ϕ between a crystal edge and the direction of $\overline{\boldsymbol{E}}$ is changed by rotating the sample around the surface normal (Fig. 5, left).

The resulting resonance intensities at different azimuthal angles depend on the position of the respective $2p\pi$ molecular orbitals. The maximal transition moment and absorption cross-section is expected to have equal direction to the electric field vector of the exciting radiation $\overline{\boldsymbol{E}}$ and the $p\pi^*$ orbital. Figure 6 shows the azimuthal angle dependence of the NK XANES of the adsorbate sytem CdS-mbt with the angles 90°, and 0° indicating the $\overline{\boldsymbol{E}}$-direction along the [0001] and [$\bar{1}2\bar{1}0$] directions respectively. The normalized intensities of the first $2p\pi^*$ resonances with respect to the σ^* resonance are shown in the polar diagram giving a maximum at 90° and a minimum at 0°.

Since antibonding $2p\pi^*$ states are oriented perpendicularly with respect to the σ frame of the molecular plane, the molecular plane is directed in the [$\bar{1}2\bar{1}0$] direction. Thus every molecule occupies two Cd atoms at the surface. This adsorbate configuration, only slightly deviating from the [$\bar{1}2\bar{1}0$] direction, was predicted by DFT calculations giving a minimum of total energy including bonding interactions between the nitrogen atoms of the mbt$^-$ and Cd surface sites (see Fig. 4, left) [22].

The NK XANES exhibits nearly the same resonance intensities after rotating the sample by about 180° (Fig. 6, middle and lower part) indicating the conservation of the mirror plane of the molecule during the core excitation.

After determination of the azimuthal orientation of the adsorbed molecule also the determination of the tilt angle γ of the molecular plane with respect to the surface is possible. For this aim the polar angle θ between the surface normal N of the crystal and the direction of $\overline{\boldsymbol{E}}$ is changed by tilting the sample relative to the beam direction (Fig. 5, right). Comparing the experimental intensity ratios at the π resonance at normal ($\theta = 90°$) and tilted incidence with theoretical curves gives the tilt angle of the adsorbed molecule [18].

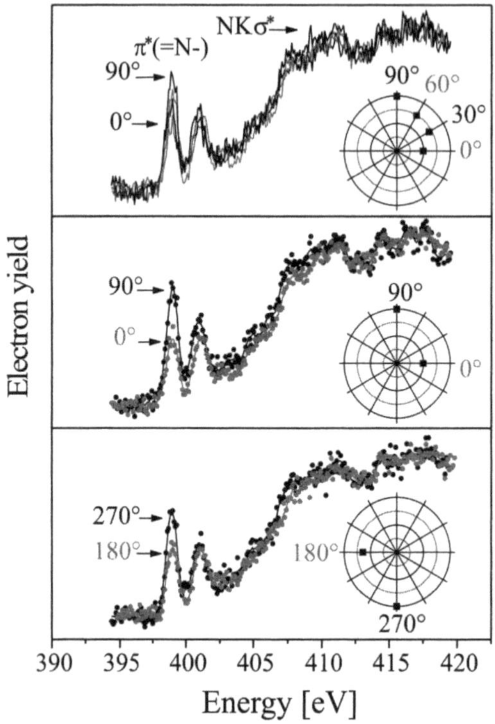

Fig. 6. Variation of the NK XANES of Cd-mbt on CdS(10$\bar{1}$0) with varying azimuthal angle ϕ

Figures 7 and 8 show the experimental polar angle dependence of CK and NK XANES, respectively, of the adsorbate system CdS-mbt. The π resonance in the preabsorption region is stronger at normal X-ray incidence with a field vector $\boldsymbol{E}$ giving a large projection in the direction of the π^* orbital of an almost upright standing molecule.

The decreasing intensity in the π^* region with decreasing angle of incidence θ gives an indication for a tilted orientation of the molecular planes of the adsorbed molecules relative to the crystal surface. Analyzing the intensity ratios comparing to that predicted by theory [18] a tilt angle of $\gamma = 63° \pm 5°$ between the molecular plane and the surface was obtained. Such an upright adsorbate constitution in the [$\bar{1}2\bar{1}0$] direction proves the assumption of a bidentate coordination of MBT with the exocyclic sulfur and the nitrogen of MBT to two adjacent Cd atoms in the CdS surface, as predicted by DFT calculations [22].

3.3 Thermal Activation of the Adsorption

Treating the CdS surface with a 10^{-4} M MBT solution a strong influence of the temperature on the absorption rate was found (Fig. 8). For a more

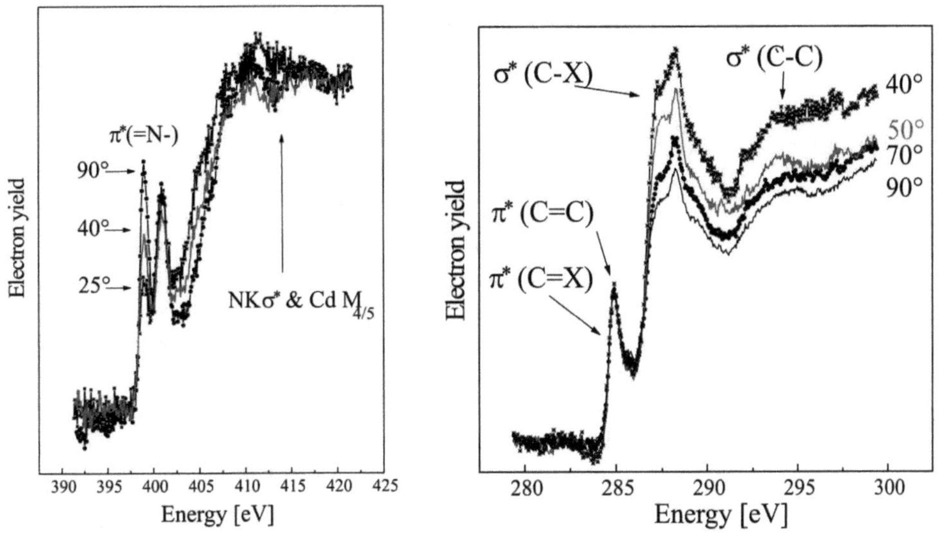

Fig. 7. Variation of NK XANES *(left)* and CK XANES *(right)* of Cd-mbt on CdS($10\bar{1}0$) with varying polar angle θ

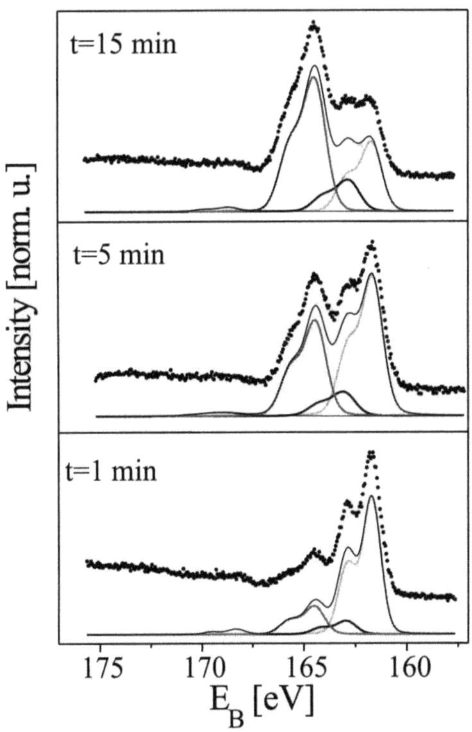

Fig. 8. S $2p$ spectra of CdS($10\bar{1}0$) after different reaction times in 10^{-5} M MBT solution at 45 °C

detailed kinetic interpretation of the absorption (k_a) and oxidation (k_b) processes

$$\begin{aligned} \mathrm{Hmbt} + [\mathrm{Cd}^+]_{\mathrm{CdS}} &\xrightarrow{k_a} [\mathrm{Cd}-\mathrm{mbt}]_{\mathrm{CdS}} + \mathrm{H}^+ \\ &\xrightarrow{k_b} (1/2)\mathrm{BBTD} + [\mathrm{Cd}^+]_{\mathrm{CdS}} + (1/2)\mathrm{H}_2 \end{aligned} \qquad (1)$$

the S $2p$ intensity ratios of S^1, S^2 and S^3 components have been used to calculate the fractions of the (–S–) and (–S–S–) chemical states of the BBTD and the (–S–) and ($-S^-$) states of the adsorbate Cd-mbt [23] presented in Fig. 9 in the form of fractions of a monolayer coverage. Estimating roughly a zero reaction order for the starting period with approximately constant numbers of free surface sites a linear approximation of the growing coverage with time seems meaningful giving rate constants of $k_a = 0.051/\,\mathrm{min}(25\,°\mathrm{C})$ and $0.224/\,\mathrm{min}(45\,°\mathrm{C})$. From the ratio of these rate constants an activation barrier $E_A = 58.3\,\mathrm{kJ/\,mol}$ was obtained. The thermal activation of MBT adsorption also controls the generation of the oxidation product BBTD (see Fig. 10, upper part) if photon induced electronic excitations in the semiconducting CdS substrate provides enough free energy for surpassing the energy barrier.

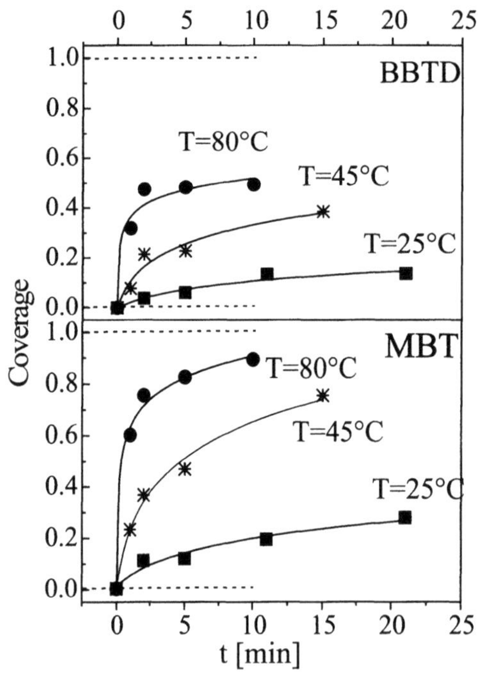

Fig. 9. Variation of the concentration of different sulfur bonding states of the MBT and BBTD molecules on CdS(10$\bar{1}$0) depending on the reaction time in 10^{-4} M MBT solution at different temperatures

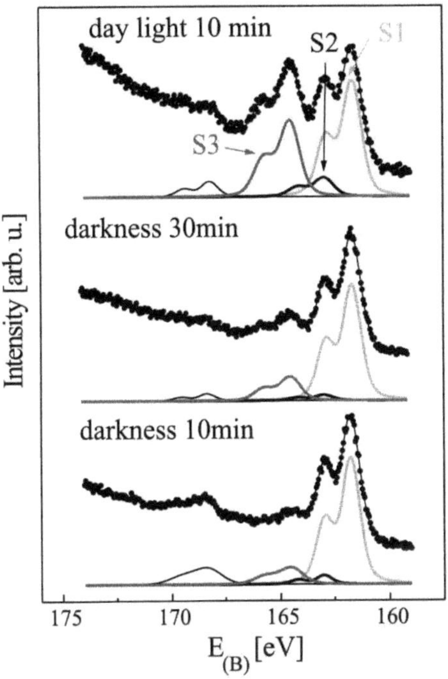

Fig. 10. S 2p spectra of CdS(10$\bar{1}$0) after reaction in 10^{-5} M MBT solution indicating the influence of daylight on the reaction rate

3.4 Photooxidation of the Adsorbate

The semiconductor CdS with a direct band gap of 2.4 eV absorbs light in the visible spectral range. This process may influence the chemical reactions at the semiconductor/electrolyte interface. This effect is clearly indicated comparing the S 2p spectra of Fig. 10 reflecting the adsorbate composition on CdS samples treated at daylight and darkness with a solution of MBT at room temperature. In contrast to the treatment in darkness the investigation of illuminated samples gave a high contribution of (–S–S–) adsorbate S^3 signals. Obviously the free energy of the dimerization/oxidation by water reduction known from the standard redox potentials giving a barrier of 0.615 eV (see Fig. 11) is provided by photon induced electronic excitation of the CdS substrate.

The reaction model in Fig. 12 based on a charge transfer model developed by *Gerischer* [24] includes possible steps of the photodimerization reaction in the bulk substrate, at the surface, in the ad-layer and in solution.

In the left part of the diagram the formation and recombination of electron-hole pairs after absorption of light are presented. Trapping in band gap states and transfer of arriving charge carriers across the surface is indicated in the middle part of the figure. The adsorbed MBT with a redox potential

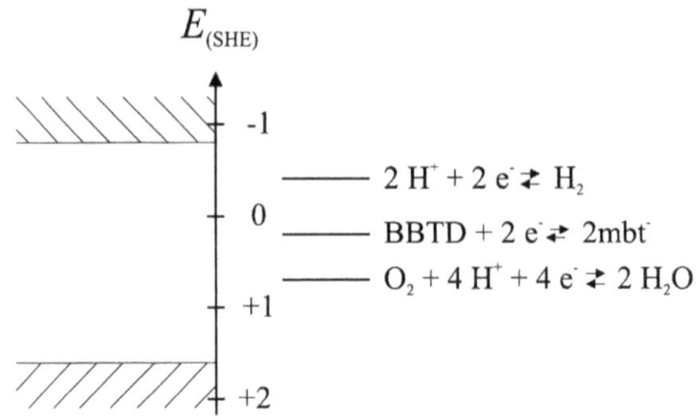

Fig. 11. Redox potentials for MBT, oxygen and water scaled schematically with the band edges of CdS

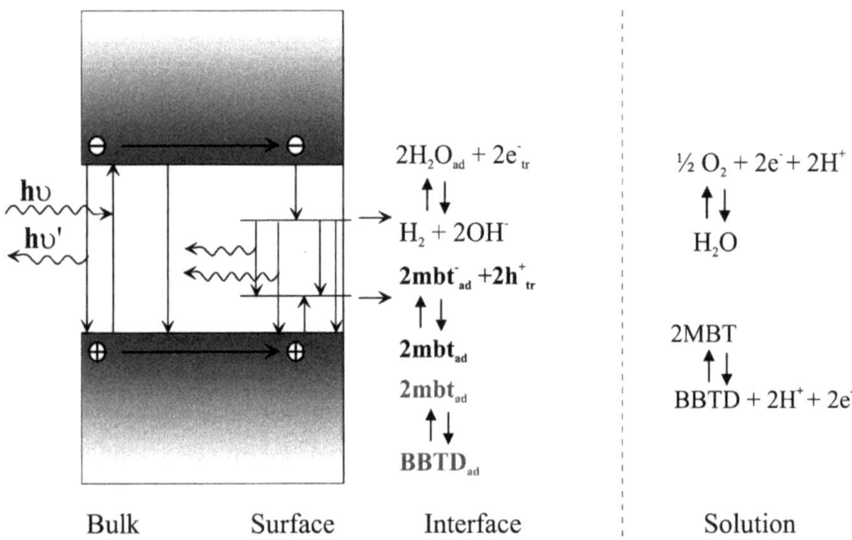

Fig. 12. Reaction model for the oxidation of MBT in the adsorbate system CdS/MBT: photon induced electronic excitation and recombination processes in the CdS substrate and charge transfer across the surface (*left*); water reduction, MBT oxidation and dimerization in the ad-layer (*middle*) and in bulk solution (*right*)

of 0.205 V (SHE) for a 10^{-4}M MBT solution at pH 7 provides hole acceptor states well above the valence band edge at 1.6 eV. The hole transfer initiates the dimerization giving BBTD. The reduction of water might be considered as corresponding redox process for maintaining the charge carrier balance. The redox potential of water at pH 7 amounts -0.41 V well below the conduction band edge at -0.8 V [25] enabling electron transfer across the surface. Formation of hydrogen has also been confirmed during charge transfer reactions with a set of organic compounds on colloidal CdS and ZnS [26].

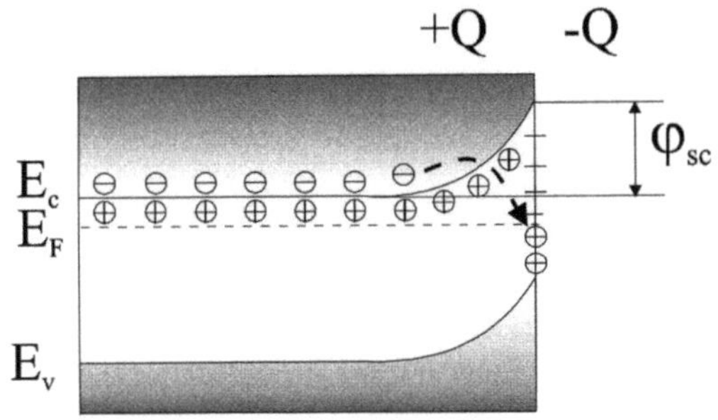

Fig. 13. Model of the space-charge region with the potential barrier φ_{sc} at the semiconductor surface

In order to minimize the influence of adsorbed O_2 on the oxidation the solutions were saturated with nitrogen. Nevertheless oxygen reduction may contribute weakly to the BBTD production because a small increase with time of the S $2p$ S^3 signal containing the (–S–S–) group after treatment of the sample in darkness has been detected (see Fig. 10).

The number of charge carriers and the charge transfer rate across the surface determining the oxidation rate is controlled by the band edge position close to the surface. In contrast to the simplified band pattern presented above, the majority carrier migration from the bulk to unoccupied states at the surface generates a dipole layer accompanied by an upbending of the bands and a depletion of electrons in the near-surface region of the n-type CdS (Fig. 13).

The resulting surface barrier φ_{sc} may restrict or even prevent the transport of charge carriers to the surface. Photon ($h\nu > E_g$) generated electron hole pairs separated by the electric field of the space charge region give minority carriers (holes) at the surface ready for the oxidation of the adsorbed mbt^- donor. In contrast to this branch of the charge transfer the electrons have to pass the surface barrier before arriving at the adsorbed acceptor molecules preferentially by tunneling. The probability of this process increases with decreasing φ_{sc} and x_0. The decrease of φ_{sc} by photon induced electron excitation explains the strong increase of the Cd-mbt oxidation rate during illuminating with daylight.

The interpretation of the daylight effect is confirmed by comparing the adsorbate oxidation rate on highly and lowly doped n-CdS. The S $2p$ spectra show a distinctly higher intensity of the component S^3 from BBTD for the highly doped CdS (Fig. 14). Obviously a higher charge transfer rate occurs in the semiconductor with the larger number N of ionized donors. This phenomenon can be understood taking into account the dependence of the depletion extension region being $x_0^2 \sim 1/N$ explaining the higher tunneling probability for the higher dopant concentration.

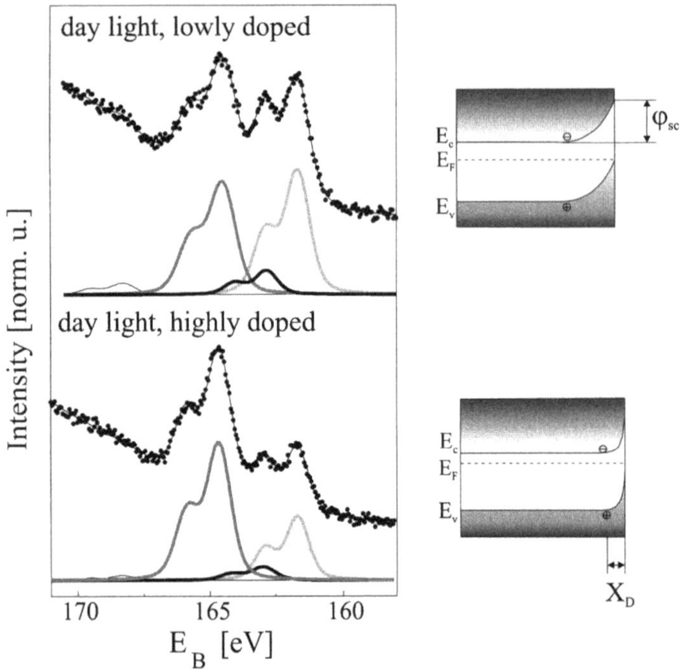

Fig. 14. S 2*p* spectra of MBT adsorbate demonstrating the influence of doping on the photon induced oxidation of MBT adsorbate on CdS from 10 min treatment in 10^{-4}M MBT solution at 25 °C

4 Conclusions

The reaction of MBT with the n-typ CdS(10$\bar{1}$0) surface gives a layer containing a Cd-mbt adsorbate complex and bis(2-benzothiazolyl)disulfide (BBTD). The MBT molecules are chemisorbed via the thiolate group accompanied by a deprotonation of the nitrogen atom. In consequence of a second interface reaction step the disulfide bis-(2-benzothiazolyl)disulfide (BBTD) is formed.

Angular dependent NK XANES measurements of the adsorbate complex indicate an orientation of the molecular planes in the [$\bar{1}2\bar{1}$0] direction of the crystal and a tilted orientation of upright standing molecules confirming the bidentate coordination via the exocyclic sulfur and the nitrogen of MBT to two adjacent Cd atoms in the CdS surface. As a result of temperature dependent adsorption experiments an activation barrier of 58.3 kJ/ mol was obtained.

Photon induced electron–hole pair generation influences the oxidation of the Cd–mbt adsorbate at the solid/electrolyte interface causing a strong increase of the electron transfer rate across the surface when illuminating the sample by daylight. The thermal activated chemisorption of MBT in the chemisorption/oxidation/dimerization sequence becomes rate controlling during daylight exposition increasing the BBTD coverage rate.

The proposed model is supported by the investigation of lowly and highly doped CdS samples indicating an increased amount of the disulfide species in the latter case. Obviously reactions at the interface predominate in comparison to redox processes in the solution.

Acknowledgements

This work was supported by BMBF contract No. 05 625 OLA5. The authors are grateful to the BESSY staff, especially Dr. W. Braun, Dr. Ch. Jung, C. Hellwig and M. Mast.

References

1. A. Ulman: *An Introduction to Ultrathin Organic Films: from Langmuir-Blodgett to Self-Assembly* (Academic Press, New York 1991)
2. L. H. Dubois, B. R. Zegarski, R. G. Nuzzo: J. Am. Chem. Soc. **112**, 570 (1990)
3. D. Zerulla, I. Uhlig, R. Szargan, T. Chassé: Surf. Sci. **402–404**, 604 (1998)
4. R. Szargan, S. Karthe, E. Suoninen: Appl. Surf. Sci. **55**, 227 (1992)
5. J. Mielczarsky, E. Suoninen: Surf. Interface Anal. **6**, 34 (1984)
6. A. N. Buckley, R. Woods: Int. J. Miner. Process. **28**, 301 (1990)
7. I. Persson: J. Coord. Chem. **32**, 261 (1994)
8. M. D. Porter, T. B. Bright, D. L. Allara, C. E. D. Chidsey: J. Am. Chem. Soc. **109**, 3559 (1987)
9. J. Hautman, M. L. Klein: J. Chem. Phys. **91**, 4994 (1989)
10. M. Grunze: Phys. Scr. B **149**, 711 (1993)
11. P. Harder, M. Grunze, R. Dahint, G. M. Whitesides, P. E. Laibinis: J. Phys. Chem. B **102**, 426 (1998)
12. E. Umbach, K. Glockler, M. Sokolowski: Surf. Sci. **404**, 20 (1998)
13. C. M. Whelan, M. R. Smith, C. J. Barnes, N. M. D. Brown, C. A. Anderson: Appl. Surf. Sci. **134**, 144 (1998)
14. A. Schaufuß, P. Roßbach, I. Uhlig, R. Szargan: Fresenius J. Anal. Chem. **358**, 262 (1997)
15. K. Siegbahn, C. Nordling, A. Fahlmann: *ESCA: Atomic, Molecular and Solid State Structure by Means of Electron Spectroscopy* (Almqvist Wiksell, Uppsala 1967)
16. C. R. Brundle, A. D. Baker (Eds): *Electron Spectroscopy: Theory, Techniques, and Applications*, Vol. 1 (Academic Press, London 1977)
17. D. Briggs, M. P. Seah (Eds): *Practical Surface Analysis*, Vol. 1, *Auger and X-ray Photoelectron Spectroscopy* (Salle Sauerländer, Frankfurt 1990)
18. J. Stöhr: *NEXAFS Spectroscopy*, Springer Ser. Surf. Sci. **25** (Springer, Berlin, Heidelberg 1992)
19. K. H. Hallmeier, D. Mayer, R. Szargan: J. Electron Spectrosc. **96**, 245 (1998)
20. R. Hesse, T. Chassé, R. Szargan: Fresenius J. Anal. Chem. **365**, 48 (1999)
21. C. Hennig, K.-H. Hallmeier, I. Uhlig, S. Irle, W. H. E. Schwarz, C. Jung, C. Hellwig, A. Bach, M. Möbius, L. Beyer, R. Szargan: BESSY-Jahresbericht, 172 (1993)

22. D. Mayer, K. H. Hallmeier, B. Flemmig, J. Reinhold, R. Szargan: BESSY-Jahresbericht 341 (1998)
23. D. Mayer: Adsorbatanalyse an chemisch modifizierten Cadmiumsulfidoberflächen. Thesis, Universität Leipzig (1999)
24. H. Gerischer: Z. Phys. Chem. N. F. **27**, 48 (1961)
25. J. R. White, A. J. Bard: J. Phys. Chem. **89**, 1947 (1985)
26. J. Kisch: J. Prakt. Chem. **336**, 635 (1994)

Recovery of Gold from Its Ores: An STM Investigation of Adsorption of the Aurocyanide onto Highly Orientated Pyrolytic Graphite

Eddy Poinen and Steve M. Thurgate

Department of Physics, Murdoch University
90th South Street, Murdoch, Western Australia
thurgate@central.murdoch.edu.com

Abstract. The development of new ways of recovering gold from low-grade ores has been of great importance to the gold mining industry throughout the world. In Australia, as a consequence of the occurrence of such ores, the developments have been even more significant. The Carbon-In-Pulp (CIP) process relies on the selective adsorption of the gold dicyanide ion from a slurry of crushed ore. A detailed explanation of how this occurs has yet to be offered, despite the widespread application of this technique for more than 30 years.

In this work, we report on the history of the development of the Carbon-In-Pulp (C-I-P) process. We then describe a series of experiments where we have used in situ Scanning Tunneling Microscopy (STM) to investigate the adsorption of gold dicyanide on Highly Orientated Pyrolytic Graphite (HOPG). In these studies, we determined the effects on adsorption of the addition of calcium ions to the solution.

From these experiments, we propose that the adsorption onto graphite occurs as a consequence of intercalation of calcium ions between the carbon sheets. The consequent modification of the electron distribution in the carbon sheets makes possible the adsorption of gold dicyanide ion.

1 Overview

Gold is mined in all five continents and throughout history has had a crucial impact on the economic well-being of nations. In the past 30 years there has been a technological revolution that has lowered operating costs of mining gold [1]. The effect of this new technology has been to make deposits that were previously uneconomic, viable. This has had a particularly large impact on the gold industry in Australia. While the new technologies have been optimized to high levels of efficiency, a microscopic understanding of all the of parts this process is yet to be achieved.

The new technologies rely on dissolution and adsorption processes from complex slurries of crushed ores. The surface science problems implicit here involve understanding the interaction of the liquor with the metallic gold during the dissolution process and the selective adsorption onto the surface of the adsorbing carbon. In this work we have used the technique of in situ STM to investigate the second of these processes, the adsorption of gold dicyanide onto graphite.

K. Wandelt, S. Thurgate (Eds.): Solid–Liquid Interfaces, Topics Appl. Phys. **85**, 113–138 (2003)

Some understanding of the significance of these innovations can be gained from the change in gold output in Australia in this period. Australia is the third largest producer in the world, producing 314 tonnes in 1997–98. The output from Australian mines has increased steadily over the past two decades from about 17 metric tons in the early 1980s to about 253 metric tons in 1995 [2].

The current technologies are very efficient and large scale and can be used to extract gold from low-grade "oxide" ores. In broad detail, concerning the processes used, the gold-bearing ore is treated with a caustic cyanide solution to form the aurocyanide ion complex $Au(CN)_2^-$. Activated carbon is pumped into the pulp formed to absorb the soluble aurocyanide ion. The gold loaded activated carbon is then filtered from the pulp. The aurocyanide complex is desorbed from the carbon by placing it in a caustic solution and raising the temperature to about 90 °C.

The interaction of the gold complex with the carbon substrate in the presence of several other ions and organic molecules is strong, even though the concentration of other species can be several orders of magnitude higher than that of the gold complex. The recovery of the aurocyanide complex using activated carbon can be as high as 97.5 % [3]. The underlying reasons for this strong affinity between activated carbon and gold have not been well understood, despite the fact that gold has been adsorbed onto charcoal in separation processes since 1894.

In recent years the adsorption of gold onto carbon has been investigated using scanning probe microscopy. These techniques allow for atomically resolved data to be gained in situ. In this chapter we report on the use of STM to examine the interaction between the gold complex and a graphite substrate in solution, and on the role of impurity atoms in the slurry.

2 Applications of Gold

As a material, gold has a number of properties that make it unique among the elements. These have made it useful in numerous applications ranging from traditional areas such as jewellery, dentistry and alloys to the electronics and aerospace industries. Gold's rich shining appearance combines with its unique physical and chemical properties to make it a most attractive metal for jewellers. In its native form it has a soft yellow lustre, the only element to have this colour. It has the highest malleability of any element, a relatively low melting point (1064 °C), a specific gravity of 19.3 at 20 °C [4], and can be alloyed to other metals to produce alloys of different properties and hues. Its malleability has allowed jewellers to beat it to a very thin film, which has been, and is still, used to coat many artefacts and even buildings.

Gold has been known to have curative properties. In the thirteenth century, in the form of aurum potabile, gold was administered as a cure for several diseases [5]. Ailments such as rheumatoid arthritis were treated with gold (I) thiol compounds. For over 50 years, sodium thiomalate has been used in the treatment of rheumatoid arthritis [6,7]. This gold therapy is generally found successful in cases of advancing rheumatoid arthritis which does not respond to other treatments. The treatment of such ailments with gold compounds is termed chrysotheraphy. It has recently been suggested that the aurocyanide ion may be used as a treatment for AIDS. The gold complex can penetrate cells easily and has a low toxicity level [8].

The catalytic effect of gold compounds has been investigated by several researchers. Gold is able to catalyse certain reactions, such as the conversion of CO into CO_2. The high-temperature treatment of gold catalyst supported on TiO_2 is an order of magnitude more active towards the CO conversion into CO_2, than gold supported on SiO_2 [9]. Palladium-gold has been found to be a catalyst for the production of vinyl acetate from ethylene, acetic acid and oxygen. Gold powders have been shown to be active catalysts in the hydrogenation process of ethylene [10].

Gold has several applications in the aerospace industry, where its characteristics of inertness and good reflectivity are valuable. Gold was used as a film coating for the Apollo 11 spacecraft, which landed the first man on the Moon. Gold is the final coating of light-weight mirrors to be used in space [11,12]. Its high reflectivity of infrared makes it useful as a coating of certain space satellites, as a heat shield.

The high thermal capacity, electrical conductivity and corrosion resistance of gold have led to its use in the emerging electronics industry. Although silver has the best conductivity ($1.59\,\mu\Omega\,cm$) of all metals, diffusion of Ag into the silicon substrate and the possibility of corrosion are problems. Gold ($2.46\,\mu\Omega\,cm$) is therefore more suited to the microelectronics industry [13]. Its main use is in the form of fine gold wires (typically 33 µm diameter) to electrically connect bond pads on semiconductor devices either to their headers or to tracks in hybrid assemblies.

Gold is used in fuel cells as well as in photography. Because of its softness it can also be used in bearings to reduce friction. Components that operate at high temperatures need lubricants that can also perform well under these extreme conditions. Films of gold/chromium are used as lubricants to coat ceramic devices operating at elevated temperatures [14]. Gold has been used as a sensor for monitoring noxious vapours (especially mercury vapours) in the workplace. A new technique proposed by *Thundat* et al. uses small silicon nitride microlevers coated with gold to monitor mercury levels with picogram mass resolution [15].

3 Gold in Nature

In nature, gold is found either as a free metal or in the form of tellurides. It is widely distributed and on average has a concentration of 3–4 ppb in the earth's crust. It is also found in seawater at a concentration of 2×10^{-11} M, but gold concentrations are heavily dependent on location. At a concentration of 0.04 ppm gold is even found in the Sun [5].

Gold can occur in several different forms: very fine flakes of Au, chemically combined with tellurium, and as microscopic flakes in iron pyrites (iron sulfide). Gold is also often found as very small inclusion or ionic substitution in association with quartz or arsenopyrite, stibnite, chalcopyrite and pyrite. Sometimes gold has a small percentage of silver associated to it. The metallic form of gold is generally found in placer or alluvial deposits in river systems and coastal areas. The high specific gravity of gold ($19.3\,\mathrm{g\,cm^{-3}}$) as compared to the surrounding rocks (2.5–$3.5\,\mathrm{g\,cm^{-3}}$) means that gold becomes concentrated in river systems in areas of low flow rates. This form of gold is easy to extract using simple technology: gold pans, sluice boxes, and dredges. The gold rushes throughout the world were focused mainly on this type of deposit [16].

3.1 Gold in Australia

Convicts cutting a road near Bathurst, New South Wales, found the first gold deposits in Australia in 1814. Gold nuggets were found by Edward Hargraves near Bathurst in 1851. This sparked a gold rush which brought thousands of migrants to the penal colony. In the state of Victoria there were many more gold rushes, notably to Clunes, Bendigo and Ballarat. The largest nugget ever found ("Welcome Stranger", 71 kg) was found in 1869 at the base of a tree by Deason and Oates. Gold was also found in the states of Queensland, South Australia, Tasmania, Western Australia and the Northern Territory.

4 Processing of Gold

The easiest form of gold recovery is the gravity concentration method that traps the heavier gold particles as they are conveyed down a concentration column by water. A somewhat better method is the use of mercury to amalgamate with gold and form a spongy Au/Hg amalgam. Gold is then recovered by distilling the amalgam and smelting the gold. This process has been in use since Roman times [12].

Gold, with an electronic configuration of $[\mathrm{Xe}]4f^{14}5d^{10}6s^1$, is a group IB element. It is considered to be a noble metal because of its lack of reactivity towards common reagents. Gold has different oxidation states: Au(I), Au(II), Au(III), and Au(V). Au(I) and Au(III) form stable compounds of gold. The oxidation of gold in solution does not occur readily, although dissolution of the metal can be achieved in the presence of complexation agents and an oxidizing agent. A leaching agent can be used to complex the gold out of its ores. Cyanide, thiourea and chloride ions are good complexation agents in dissolving gold and forming a stable gold complex. The dicyanoaurate ion formed, $Au(CN)_2^-$, is a very stable Au(I) complex with an overall formation constant,

$\beta_2 = 10^{38}$. The reaction of the cyanide ions with the gold is accompanied by the reduction of oxygen in solution [5].

For more than a century cyanide has been used in gold recovery, though there are still some problems associated with this technology. Low leaching rates (24–72 h), the high toxicity of cyanide, environmental restrictions and low effectiveness for treatment of refractory ores have been cited as hurdles with cyanide treatment of gold ores [17].

In general, the hydrometallurgy of an ore consists of three main steps: leaching, solid purification/concentration, and recovery. The same processes are applied to the treatment of gold ores, the actual process used being largely dependent on the type of ore. There are usually pretreatment methods such as pressure oxidation and bacterial oxidation to release the gold and allow it to be accessible to the leaching agent. The general scheme for hydrometallurgical processes in gold is shown in Fig. 1 [18].

Through the leaching step, the gold is complexed with a complexation agent, usually cyanide ions. The complexed gold is then purified and concentrated by using either resins or activated carbon. The next process is the refining procedure. The gold solution is treated in an electrowinning cell or precipitated/cemented with zinc metal. At this point, the gold may not be pure and may contain a fraction of other metals such as silver and platinum. The impure gold is generally termed a gold doré. Purification of the gold doré is done by either chlorine-refining or electro-refining [19].

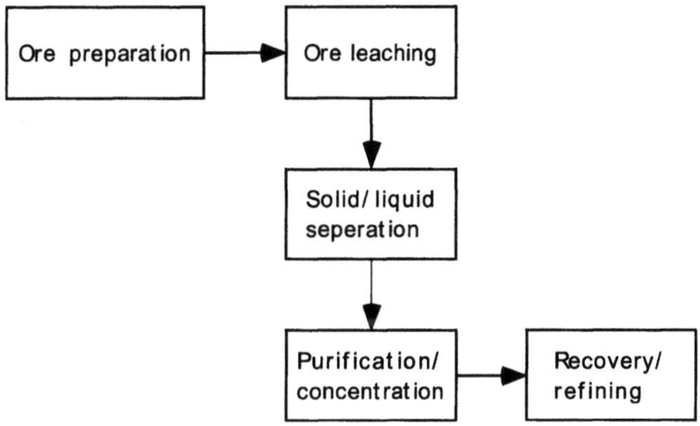

Fig. 1. Generalized hydrometallurgical processes in the treatment of an ore to produce either a metal or a compound

4.1 The Cyanidation Step

The best method of extraction of the gold from its auriferous ore is the cyanidation method in which the ore is finely crushed and then cyanide is added with some caustic solution. There is a need to keep the solution caustic, (pH above 10) to prevent the formation of the highly volatile and poisonous gas, HCN. The pK_a of the weak acid HCN is about 9.31. Hence, only half of the total cyanide exists as CN^- ions with the other half as HCN at pH 9.31. The ratio of CN^- rises to about 90 % if the pH is raised to pH 10.2 [20]. The mixture of the crushed ore and the cyanide solution forms a pulp.

An investigation into the dissolution of gold with cyanide in an alkaline medium was made by *Kudryk* and *Kellogg* in 1954 [21]. They proposed that the mechanism by which the gold dissolved, and the parallel cathodic reduction of the dissolved oxygen, was electrochemical in nature. The overall dissolution reaction is represented by (1):

$$4Au + 8CN^- + O_2 \rightarrow 4Au(CN)_2^- + 4OH^- . \quad (1)$$

This is also known as Elsner's equation. The process of dissolution of gold in an aerated alkaline solution is more adequately described by the following equations [22]:

$$2Au + 4CN^- + O_2 + 2H_2O \leftrightarrow 2Au(CN)_2^- + H_2O_2 + 2OH^- \quad (2)$$

$$2Au + 4CN^- + H_2O_2 \leftrightarrow 2Au(CN)_2^- + 2OH^- . \quad (3)$$

This oxidation reaction is thought to occur in three steps as shown in Fig. 2. The first step is thought to be the formation of the adsorbed species $AuCN^-$:

$$Au + CN^- \leftrightarrow AuCN_{ads}{}^- \quad (4)$$

$$AuCN_{ads}{}^- \rightarrow AuCN_{ads} + e^- \quad (5)$$

$$AuCN_{ads} + CN^- \leftrightarrow Au(CN)_2^- . \quad (6)$$

The second reaction is thought to be the rate-determining step [23,24]. There is still no general agreement on the exact mechanism at play in the dissolution and passivation of the gold in aerated cyanide solution and more research into this system is needed [25].

The dissolution of the gold is accompanied by the parallel cathodic reduction reaction of the dissolved gold. The cathodic reaction proposed by *Kudryk* and *Kellogg* is illustrated by

$$O_2 + 2H_2O + 4e^- \rightarrow 4OH^- . \quad (7)$$

The overall reaction can be understood by the mixed potentials and the dissolution of gold can also be regarded as a corrosion cell (Fig. 2).

4.2 Concentration

The solution obtained at the end of the cyanidation step is a dilute (1–20 g/ton) solution. To extract the gold more efficiently a concentrated high-grade solution is needed. This is done with the use of either synthetic

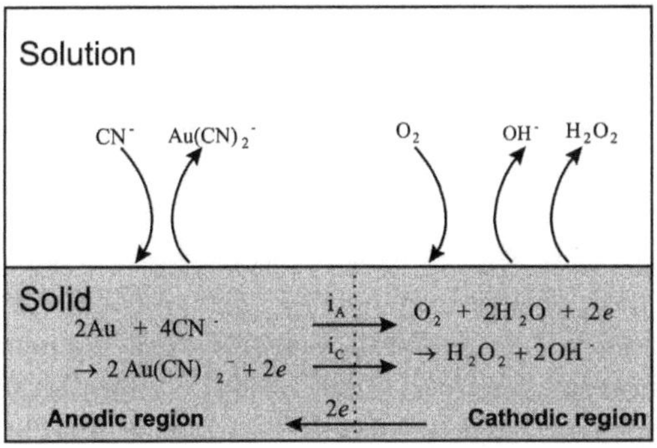

Fig. 2. The gold dissolution process in cyanide solution can be considered as a corrosion cell with anodic and cathodic regions [20]

extractants or activated carbon. The dilute solution is placed in contact with the extractant. Once loaded, the extractant is removed from the circuit and the gold species are desorbed into a smaller volume of solution. Ion exchange resins can be used as an extractant but activated carbon is the most widely used product for gold recovery. Once the gold species have been stripped the extractant can be regenerated and sent back to the circuit. This makes the process economical.

4.3 The CIP/CIL Process

The cyanidization process is carried out on ore that contains microscopic flakes of gold. If the gold is associated with sulfides, the ore must be roasted at 650 °C to convert the iron sulfide into iron oxide containing gold specks.

The ore is finely crushed and is dissolved in an alkaline cyanide solution. The solution strength is about 0.02 % in cyanide at a pH of 10. The resulting product is a pulp which is mechanically agitated to promote the dissolution of the gold (O_2 is added as air). This is called the leaching process. The pulp is then fed into similar size tanks known as adsorption tanks. Granules of activated carbon are added to these tanks for adsorbing the gold cyanide species. This is termed the Carbon-In-Pulp (CIP) process.

In some cases adsorption tanks are not used in the circuit and the activated carbon granules are injected straight into the leaching tanks. This is called the Carbon-In-Leach (CIL) process. The carbon is moved countercurrent to the pulp which is moved from tank to tank by gravity. The loaded carbon granules are removed from the tanks and washed.

The gold-loaded carbon is then placed in a hot caustic solution ($\approx$ 95 °C) with some sodium cyanide (NaCN). There are a few variants to the procedure just described, but essentially it is the basic type used. In the hot caustic solution the gold complex is desorbed from the activated carbon and the solution which is derived from this washing stage is called the pregnant solution. It is then shifted to the electrolysis cell whereby a process known as electrowinning is performed.

5 Activated Carbon

Activated carbon is a generic term that applies to a broad class of carbon-based compounds that typically have a large surface area ($2000\,m^2/g$), due to its highly developed porous structure. Activated carbon cannot be defined by a single physical structure or chemical analysis but is essentially graphitic in nature. As a result of its absorbent nature, activated carbon is mainly used in industry in separation, purification and extraction. In fact, it has been used since Napoleonic times [26], when activated carbon was used for the purification of sugar made from sugar beet. The source of raw material for activated carbon compounds can be as varied as wood, lignite, coal and apricot pips.

Nowadays, activated carbon is used in many fields other than the gold mining industry: in water purification systems, for the purification of raw materials or intermediates in the pharmaceutical industry, for the purification of gases and air in chemical processes, in air conditioning and for gas respirators.

5.1 Activated Carbon in the Gold Industry

Activated carbon is extensively used in industry for the purpose of recovery, separation and purification of compounds/elements in gas and liquid phases. The main use for activated carbon worldwide is directly related to the purification and treatment of water. Many varieties of activated carbon are presently available and each type has suitable properties in each process involved and the environment in which it is to be used. This section is focused on the type of activated carbon used in the gold recovery processes. Activated carbon has revolutionized the gold industry throughout the world and has made it possible to extract gold from the dirt rejected during previous extraction processes. In the following section a short history of the use of activated carbon in gold processing is presented, from the time when chlorine was used to dissolve the gold ore, to the time when cyanide became the method of choice for the recovery of gold.

Prior to the introduction of cyanide as a leaching agent in gold ore dissolution in 1890, chlorine was the lixiviant used. The gold dissolved to produce the soluble $AuCl_4^-$ species. The gold was then recovered by adding carbon or charcoal. In contact with the carbon, the gold chloride was reduced to metallic gold. The loaded charcoal was then burnt to yield the gold [27]. In 1894, Johnson patented the use of wood-based charcoal for the recovery of gold from cyanide leached solution. It had been previously known that gold could be recovered by wood charcoal from chloride leached solution. Charcoal was produced by heating wood to red heat, quenching with water, followed by grinding. The recovery of gold using charcoal was not efficient as the gold loading capacity was low due to low porosity. The gold was recovered by burning the wood to ashes and smelting. The ashes contained only 10–20 % gold. This made the use of charcoal unprofitable in the early 20th century [28].

With the successful introduction of cyanide leaching in gold processing in the late 19th century, chloride technology was gradually replaced. The cyanidation process rapidly gained popularity and soon became the premier method for the extraction of gold. After the dissolution step the gold could be recovered by the Merill-Crowe process. The gold was cemented/precipitated with the addition of zinc dust to the clarified leached solution. The process is very efficient and simple, and thus for a long period of time there were few changes or major innovations in gold plant practices.

Since 1900, there have been several attempts to investigate a carbon-based technology in gold processing. In the early 1950s, *Zadra* et al. [29] from the US Bureau of Mines, developed much of the technology on which the present-day CIP process is based. The procedure made use of activated carbon instead of charcoal to adsorb the gold cyanide complex after the leaching phase. The introduction of activated carbon increased the gold-loading capacity of the adsorption phase. In addition, after the removal of the gold from the carbon, the latter can be regenerated by reactivation and returned to the circuit. The first carbon-based plant operation was set up in the Witwatersrand gold mines in South Africa and the results were impressive. This success prompted several operators of new mines to investigate the use of carbon-based technology.

If the CIP process was adopted at a plant, the adsorbent could be reused and therefore the process proved economical. Given the above factors, the CIP process was rapidly accepted as a process for successfully treating low-grade ores. The CIP process was versatile in that it could be applied to any type of feed, and achieved superior recoveries at lower capital costs than the conventional procedure. The easy dissemination of the carbon technology made it adaptable to different environments/mines. The success of the carbon technology has spurred the development of other carbon adsorbent based technology such as Carbon-In-Leach (CIL) and Carbon-In-Column (CIC).

5.2 Physical Properties of Activated Carbon

Activated carbon is thought to be essentially similar to graphite but with a higher degree of disorder. X-ray diffraction studies have supported this view [30,31]. The current view is that activated carbon consists of two basic structures: small regions of elementary crystallites composed of roughly parallel layers of graphitic carbon, and a disordered cross-linked lattice of carbon hexagons [28].

In activated carbon, the dimensions of the elementary crystallites are dependent on the temperature treatment, but they typically vary between

9 and 12 Å in height and 20–23 Å in width. In a recent investigation by *Adams* [32], activated carbons from different sources were investigated with X-ray diffraction techniques. It was found that,on average, a microcrystal in activated carbon consists of two disordered planes about 12 aromatic rings in diameter, the distance between the graphitic planes being 3.7 Å, larger than in graphite. This is similar to the results achieved by *Marsden* [20] who found the crystallite structure of activated carbon to be approximately three layers in thickness, and about nine carbon hexagons in length.

When the outer surface area is compared to the internal surface area, it is found that the latter is extremely large (1000–2000 m^2/g). This is mainly due to the second step of the activation process. A material that has a wide range of pore sizes as well as shapes is generated. In Fig. 3 it can be seen that the coconut-activated carbon has different basic structures in terms of pore sizes and shapes.

It is not possible to have a definite pore size and shape dimension to characterize activated carbon. *Dubinin* [33] classified the pore structure of activated carbon in the following manner: macropore (pore size greater than 2000 Å and micropore with a porse size of less than 16 Å. Any pore sizes within this range of values are regarded as mesopores. This classification was based on changes in gas or vapour adsorption mechanism with pore size. It is also expected that different raw materials will result in differences in the contribution to the above size characteristics.

Adams [32] investigated the same carbon compound activated in different atmospheres. Activation in the presence of steam was found to promote micropore volume formation whereas activation in CO_2 had the reverse effect. In addition, pore sizes have a significant effect on the kinetics of the adsorption mechanism.

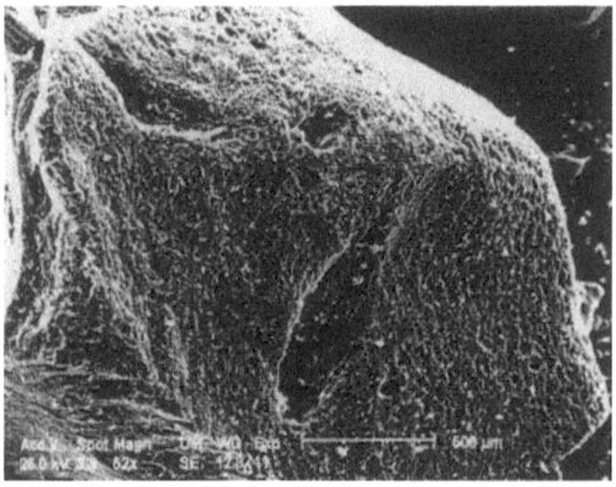

Fig. 3. SEM picture of a coconut-based activated carbon with distinct pore structure and sizes

5.3 Chemical Properties of Activated Carbon

The chemical nature of the surface of activated carbon also plays an important role in its adsorption properties. These properties are not well understood, however the chemical nature of the surface can be generally classified into three main groups:

(1) Edge and dislocation at the edge of the crystallites, which results in the formation of residual valencies.
(2) The presence of chemically bonded elements such as oxygen and hydrogen bonded to the carbon. These may have been present in the original raw material or may have been formed after the activation phase.
(3) Presence of inorganic matter such as ash components or impregnating agents (these may either enhance the adsorption or affect it negatively).

Activated carbon can be classified into two categories: H-carbons and L-carbons. H-carbons are formed at temperatures above 700 °C, usually at about 1000 °C. They have the ability to absorb H^+ ions when in contact with water. L-carbons on the other hand are activated at 300–400 °C and when in contact with water, OH^- ions are absorbed preferentially. The L-carbon is a stronger solid acid than the H-carbon [20].

Carbons used in the recovery of gold are usually activated at temperatures of 600–700 °C and thus show both acidic and basic properties. The adsorption of acid or base is usually accompanied by a change in the surface charge of the carbon. Oxygen chemisorbs quite readily to the carbon surface forming carbon–oxygen (C–O) functional groups. The terms chemisorb oxygen, oxygen-containing functional groups and surface oxides are considered to be synonymous in this section. These functional groups affect surface reactions, wettability, electrical and catalytic properties of the surface. It has been estimated that 90 % of the oxygen present on the carbon surface is present as these functional groups.

With the aid of different analytical techniques, the identities of several groups have been elucidated. These include acidic functional groups such as lactones, carbonyls, phenols, carboxylic acid, quinones and fluorescein lactones groups [34]. The exact identities of the basic functional groups have not been clearly defined as yet.

6 Investigations of Gold Adsorption on Activated Carbon

Due to the success of activated carbon in the adsorption of the gold complex several studies have been performed to investigate the loading capacity of activated carbons and the major parameters influencing such adsorption. The kinetics studies are briefly presented in this section.

6.1 Kinetics Loading Capacity Studies

The loading rate is dependent on the number of sites available to the aurocyanide complex. *Fuersteneau* et al. [35] found that the rate of adsorption of gold cyanide on activated carbon is proportional to temperature, concentration of carbon, and initial gold concentration, but inversely proportional to the carbon size squared. They proposed that the rate is controlled by film diffusion and the activated energy was determined to be about 2.0 kcal/ mol. Equally, the pore structure of the activated carbon has an influence on the rate. Pore-volume distribution of different sized pores showed that the activated carbons most suitable for loading gold values are those which have mainly micropores. From batch experiments, researchers at Murdoch University (Labrooy model) [27] have proposed the following model:

$$\Delta\,[\mathrm{Au}]_{\mathrm{c}}^{t} = k[\mathrm{Au}]_{\mathrm{s}}^{n} t^{n}\,, \tag{8}$$

where $\Delta\,[\mathrm{Au}]_{\mathrm{c}}^{t}$ and $[\mathrm{Au}]_{\mathrm{s}}^{n}$ are gold concentrations on carbon and in solution at time t (hours) and n is an empirical constant ≈ 1.

Le Roux [36] investigated several kinetic models and confirmed that the Labrooy model based on two parameters was adequate to describe experimental data over a period of 8 h.

Tsuchida and *Muir* (at Murdoch University) [37,38] have performed potentiometric studies in the adsorption of the aurocyanide complex onto activated carbon. Additionally, they have investigated the effect of oxygen in this adsorption process. They suggest that the adsorption process involves ion exchange between the aurocyanide ion and OH^- ions present on the surface of the activated carbon. The adsorbed aurocyanide species is then degraded to AuCN on the surface of the activated carbon.

6.2 Spectroscopic Investigations

Over the years that activated carbon has been used in the gold industry, there have been numerous spectroscopic techniques applied to gold loading on activated carbon. These complementing technical tools have supplied several clues about the interaction of the aurocyanide with the activated carbon and have extended our knowledge about the adsorption process. The next section reviews the specific spectroscopy technique used and the information derived from it.

6.3 Mössbauer Spectroscopy

Gold-loaded activated carbon was analysed by Mössbauer spectroscopy. A study by *Cashion* et al. [39] found that the gold species that adsorbs onto the carbon surface is the gold complex $Au(CN)_2^-$. An investigation by *Kongolo* et al. [40] confirmed the earlier results of Cashion and found that at acidic pH (< 4) a species similar to AuCN was found on the carbon among other gold cyanide complexes. The only point of contention between the two studies was that Cashion proposed that the point of attachment of the gold complex to the carbon was with the terminal nitrogen atoms, through chemisorbed oxygen. The significant point from Mössbauer studies was that the gold species mainly adsorbed onto activated carbon was identified to be the aurocyanide complex $Au(CN)_2^-$.

6.4 Infrared Spectroscopy

Infrared Spectroscopy has been considered as a technique for studying the formation of the activated carbon–gold species by monitoring the shift in the triple cyanide bond at (2080 cm^{-1}). *Sibrell* and *Miller* [41] have evaluated the technique and concluded that IR was not sensitive enough. A Fourier-transform infrared spectrophotometric investigation on loaded activated carbon was made by *Adams* [32]. He found that the aurocyanide species adsorbed onto activated carbon without any chemical change. This study could not however distinguish between ion pairs (aurocyanide + metal cation) and the aurocyanide ion loaded onto the carbon.

6.5 X-ray Photoelectron Spectroscopy

In X-ray Photoelectron Spectroscopy (XPS), the chemical environment of the element under study is reflected in an XPS spectrum. Based on the XPS results, *Klauber* [42] proposed the following model (Fig. 4) for the adsorption of the aurocyanide onto activated carbon.

There has been some controversy associated with some of the XPS results on activated carbon [42]. Differences in calibration techniques and referencing in the various laboratories conducting XPS studies on loaded activated carbon have been cited for the variation in results [43].

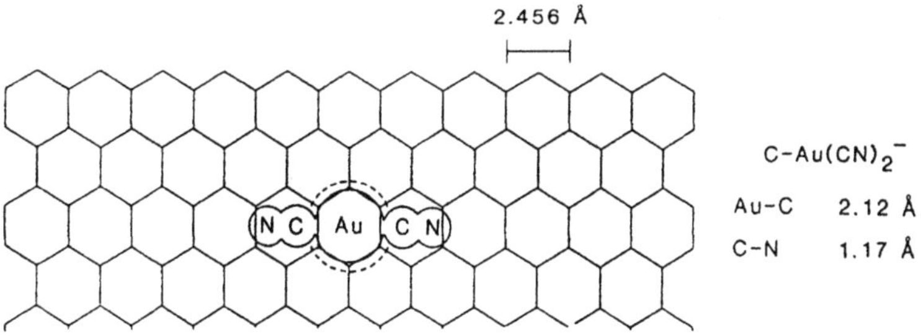

Fig. 4. Klauber's model for the adsorption of the aurocyanide ion complex onto activated carbon

6.5.1 Current Status

A number of researchers have proposed that under plant conditions of high ionic strength, the adsorption mechanism was different. They proposed that there must be a counter-cation to anchor the aurocyanide to the carbon surface [43,28]. The ion pair formation is depicted in the following equation:

$$\mathrm{M}^{n+} + n[\mathrm{Au(CN)}_2^-] \rightarrow \mathrm{M}^{n+}[Au(CN)_2^-]_n \,. \tag{9}$$

This view seemed logical as the positive charge could neutralize the negative charge on the aurocyanide ion complex and help the aurocyanide to attach itself to a negatively charged surface. The position and orientation of the counter-ion on the carbon surface was not discussed.

Recent combined XPS and IR investigation by *Ibrado* and *Fuersteneau* [43] have supported Klauber's results. They suggested that there is a partial donation of delocalized π-electrons from the graphitic plane of the activated carbon to the Au atom. The currently widely accepted view is that the aurocyanide complex ion is attached to the graphitic plane of the activated carbon without any chemical change.

7 Influence of Graphitic Structure

In the early part of this century, Green discovered that graphite did not have any strong affinity for gold, while charcoal and activated carbon were effective in adsorbing the gold cyanide [28]. On the other hand, recent studies have indicated that the graphitization process occurring during the activation process is the main factor contributing to the adsorption capacity of the activated carbon.

In gold loading studies, *Adams* [32] found that the loading on different activated carbons is proportional to the graphitic content of the activated carbon. Adams proposed that the quantity of gold adsorbed onto the surface depends on the degree of "graphitization", which is proportional to the amount of sp^2 bonded carbon. In studies of the gold loading activity of different carbonaceous compounds such as coals, carbon black, anthracites, lignites, graphites, and activated carbon, Ibrado and Fuersteneau have demonstrated that the crucial factor for gold loading is the graphitic nature of the substrate.

Miller and *Sibrell* [44] compared different carbon compounds in terms of their gold absorption capacity. They observed that the carbon compound with the higher degree of graphitization showed better gold adsorption capacity.

The only research to date on highly orientated pyrolytic graphite was undertaken by *Sibrell* and *Miller*[45]. In this investigation, labelled $\mathrm{Au(C^{14}N)_2^-}$ made from dissolution of Au in a $\mathrm{KC^{14}N}$ solution, was adsorbed onto the surface of graphite (HOPG). The study compared the adsorption of the aurocyanide complex either at the edge plane or at the basal plane. Sibrell and Miller proposed that the adsorption occurs at the edges.

8 Scanning Tunnelling Microscopy of Graphite

Given the conclusion that the gold complex adsorbs on graphitic planes and the ease in which graphitic carbon can be imaged with atomic resolution by STM, an in situ investigation of the system was carried out. In an initial attempt, the interaction of the aurocyanide complex and HOPG (grade ZYA) was examined with the ambient STM under solution. This work gave some interesting insights into this system [46].

The STM is a superb tool for surface studies and in many instances can achieve atomic resolution quite easily. The STM does not exactly follow the topology of the surface but rather the topology of the electronic distribution at the surface. This topology usually matches the topology of the surface, but not always. The HOPG surface is usually termed a 2D surface with no charge density waves.

The first images of graphite were made on HOPG in a UHV system by *Binnig* [47]. The first atomic resolution images of graphite made in ambient conditions were made at Stanford by *Park* and *Quate* [48]. From then on the ability of the STM and the SFM to image graphite on an atomic scale has been demonstrated many times. A typical atomically resolved image of the graphite surface obtained under an electrolyte from our laboratory is shown in Fig. 5.

First principles calculations by *Tomanek* et al. [49] have shown that the STM should observe a strong asymmetry in the tunnelling current between adjacent carbon atoms on the surface of hexagonal and rhombohedral types of graphite. The asymmetry was predicted to be independent at small voltage bias and it was found that the STM images preferentially carbon atoms that have no neighbouring carbon atom in the second layer. These sites are referred

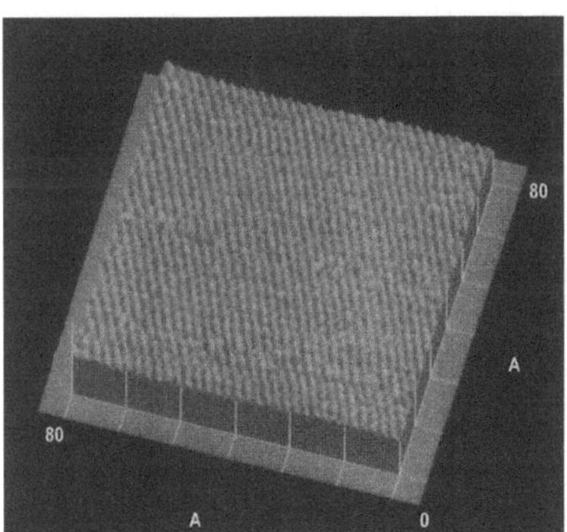

Fig. 5. STM image of HOPG under KOH/$KAu(CN)_2$, (pH 10 and 1000 ppm Au) electrolyte. It shows the characteristic hexagonal pattern of the graphite. Conditions for tunnelling were set at $i_t = 0.75$ nA and $V_t = -0.13$V

to as β sites. The α carbon atoms are not imaged under normal conditions with the STM [50].

9 Experimental

All the STM images were acquired using a Park Scientific Instruments Universal system [51] with the ambient STM head attachment. All STM measurements were performed in a closed cell. A small container filled with millipore MQ (10^{18} MW cm) water was placed in the cell to saturate the atmosphere in order to reduce the evaporation of the liquid in contact with the HOPG. In situ STM tips were made and covered with melted apiezon wax. The maximum faradaic current was determined to be in the range of 0.01 nA. For most of the in situ investigation the tips were freshly made prior to the experiment.

The Scanning Electron Microscopy (SEM) scans were performed on a Phillips XL 20 instrument. The microscope can take images in backscattered (BSE) and secondary (SE) modes. Some SEM images were made with an ISI-100 microscope used in the environmental mode (pressure $\sim$ 0.1 Torr). The SEM images taken with this microscope were in BSE mode.

The HOPG samples were supplied by Advanced Ceramics [52]. They were grade A and B material and had large areas of flat planes on the exposed surface. The aurocyanide ion was purchased as the potassium aurocyanide salt, $K[Au(CN)_2]$ [53]. A 1000 ppm Au solution was prepared using Millipore MQ water. The pH of the solution was adjusted to 10–11 with the addition of KOH in order to prevent the formation of hydrogen cyanide and to simulate plant conditions. The graphite substrate was prepared by peeling the surface layer using an adhesive tape. This procedure was performed at least five times to ensure that no contamination occurred from any previous study. Several drops of the solution to be studied were then placed on the hydrophobic graphite surface. The graphite sample was then placed in a clean, sealed glass enclosure. The time of contact between the solution and the graphite sample was between 24 and 45 h.

A range of data was examined in this study. In an initial study, samples were prepared by simply exposing the surface to the $[Au(CN)_2]$ solution for varying lengths of time, up to 30 h. Care was exercised to ensure that only the carbon surface was exposed to the solution. In a second set of experiments, $CaCl_2$ (1000 ppm Ca^{2+}) was added to the solution, and the effect monitored. Again, the exposure times were typically 30 and 45 h.

10 Results and Discussion of In Situ STM Investigations

10.1 $KAu(CN)_2$/KOH/HOPG

Large areas ($1.5 \times 1.5\,m^2$) were scanned and then the microscope was focused on a particular area. The carbon surface was found to be flat even after several hours in the solution. On further magnification, the surface showed the characteristic hexagonal structure of graphite, without any adsorption at the atomic/molecular level. This was found to be true even in defect regions or near step edges. Evidence of exfoliation of graphite layers was observed with extended contact time. However, there was very little evidence of any adsorption of the gold complex under these conditions. This would seem to contradict the proposal that gold adsorbs onto the basal plane of graphite as proposed by Klauber [42].

10.2 $KAu(CN)_2$ $KOH/CaCl_2$/HOPG

It is known in the industry that addition of Ca ions enhances the adsorption of gold on carbon. In light of this, calcium ions were added in the form of $CaCl_2$ at a concentration of 1000 ppm (Ca) in an equal volume to that of the aurocyanide complex solution 2000 ppm (Au).

Defect regions appeared on the surface of the graphite and these were studied extensively at higher magnification. Occasionally these regions showed small areas of gold adsorption. An example is presented in Fig. 6. Most of the surface remained flat with the presence of step and some defect regions. Further investigation of the defect shown (high region) is presented in Fig. 6b. The microscope was then focused further in the defect region and the imaging mode was changed to the constant-height mode. Typically the surface revealed the characteristic flat carbon structure (Fig. 6c) but occasionally there was evidence of gold adsorption. Figure 6d presents evidence of adsorption on the top sheet.

Similar data can be seen in Fig. 7. This scan was of a largely flat plane with a central step along the middle (Fig. 7a). The area marked by the arrow can be seen in Fig. 7b.

Steps were investigated as possible sites for adsorption of the gold complex. Some were found with little or no adsorption apparent (Fig. 8a), while in other areas it is clear that the adsorption has taken place on the top plane of a step as can be seen in Fig. 8b,c.

Another set of data is shown in Fig. 9. It shows a range of structures that are typical of how the gold was observed in these defective regions.

A combined SEM/X-ray fluorescence analysis (EDAX) was performed on the HOPG surface after adsorption with the $KAu(CN)_2/KOH/CaCl_2$ solution for 32 h. The analysis was preformed in an environmental cell at a pressure of 0.1 Torr. The SEM image is presented in Fig. 10a and the EDAX spectrum of the area showed by the arrow is depicted in Fig. 10b. The white area of the SEM picture was evidence of gold adsorption at the edges of graphitic sheets. The presence of gold and calcium was found in the spectrum recorded for the white areas (arrows).

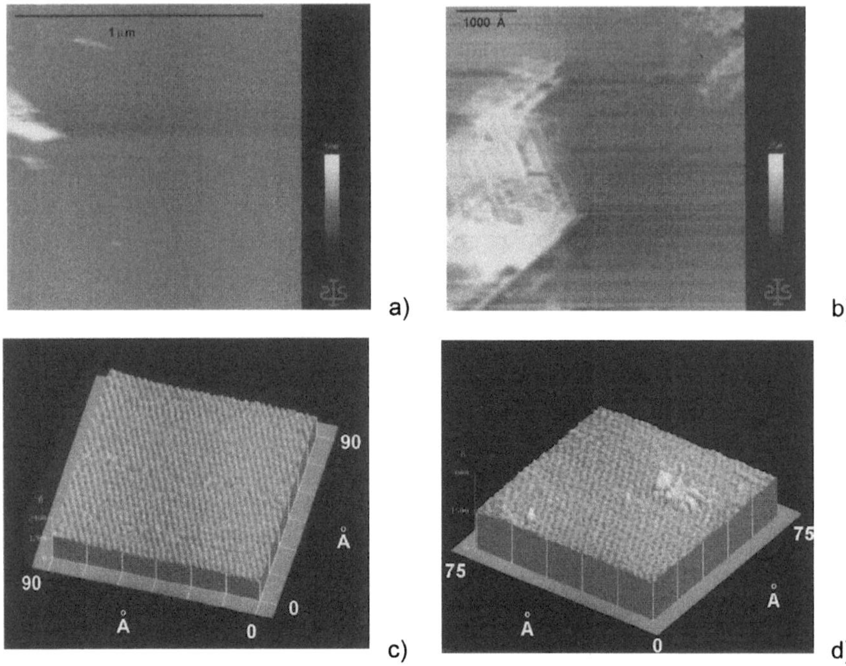

Fig. 6. Set of in situ STM scans of the HOPG surface under $KAu(CN)_2/KOH/CaCl_2$ electrolyte. (**a**) Extended flat areas were found with some defect areas. (**b**) Magnification of the defect (high) region. (**c**) The surface in the defect region showed typical carbon atomic structure. (**d**) Image showing a small area of gold adsorption on a graphite sheet

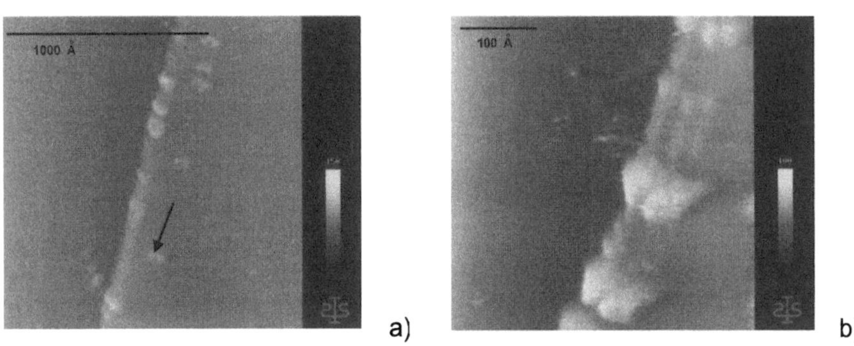

Fig. 7. Set of in situ STM scans of the HOPG surface under $KAu(CN)_2/KOH/CaCl_2$ electrolyte for 44 h. (**a**) Image of a HOPG sample showing a flat surface with a vertical step. (**b**) Magnified image of the raised region marked by the arrow in (**a**)

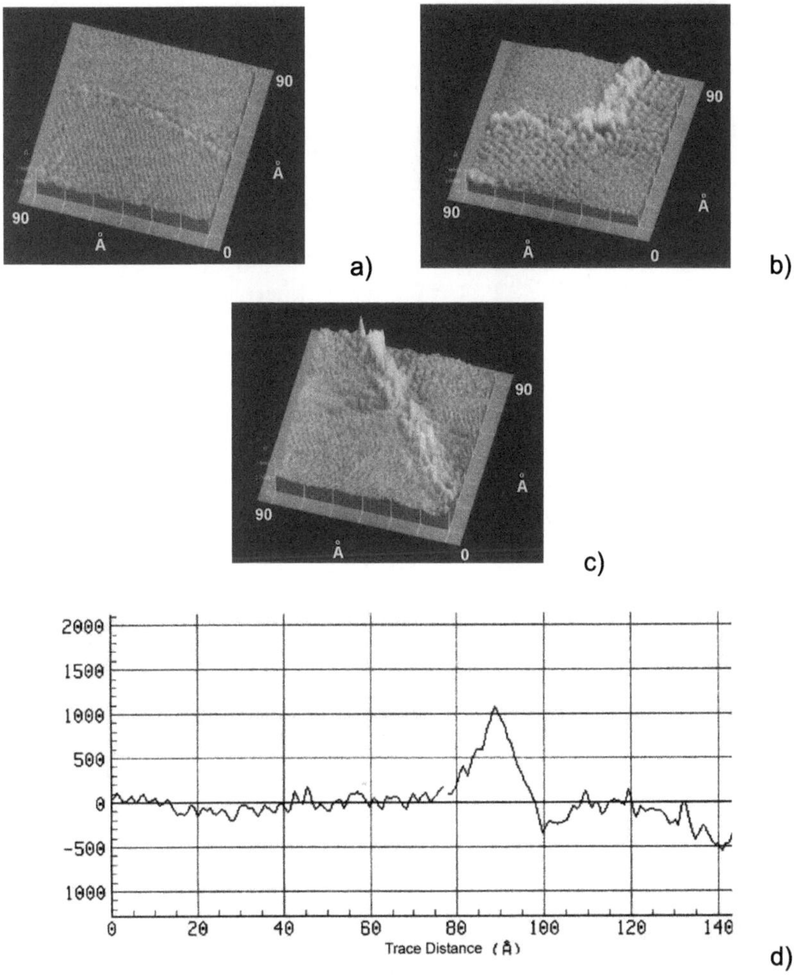

Fig. 8. In situ STM scans of a HOPG surface under $KAu(CN)_2/KOH/CaCl_2$ electrolyte. (**a**) Typical image of a step at atomic scale with little or no adsorption at the edge. (**b**) Image of a break in the top graphitic sheet with the step decorated by adsorption. The carbon structure on both top and underlying sheets is evident. (**c**) A step showing gold adsorption in the edge region. (**d**) Profile in the direction of the line drawn in panel (**c**)

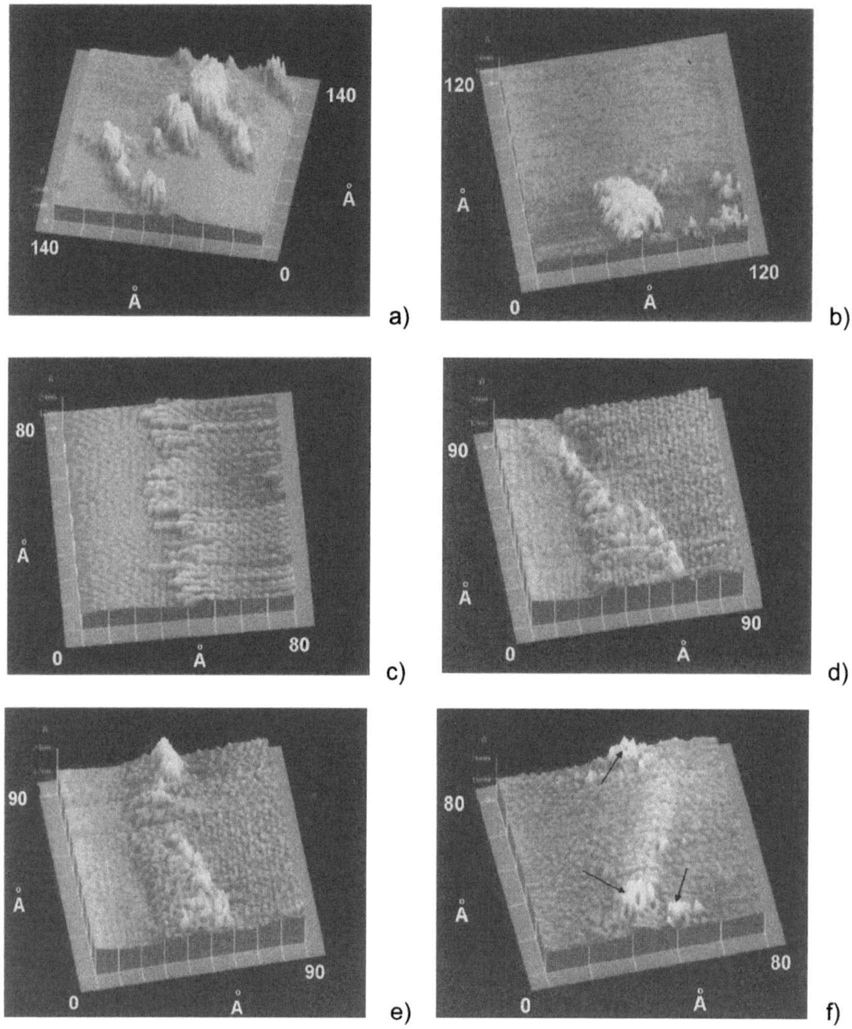

Fig. 9. Set of in situ STM scans of a HOPG surface under $KAu(CN)_2/KOH//CaCl_2$ electrolyte. (**a**) Gold adsorbed on a relatively flat area of the carbon substrate. (**b**) A mound type of gold. (**c**) Atomically resolved image of one carbon sheet overlying another. There is no gold adsorption obvious in this image. (**d**) This image shows a gold adsorbate region running across two sheets in a diagonal direction. (**e**) Another image in the same region. (**f**) Atomic scale image of the carbon surface with gold adsorbed

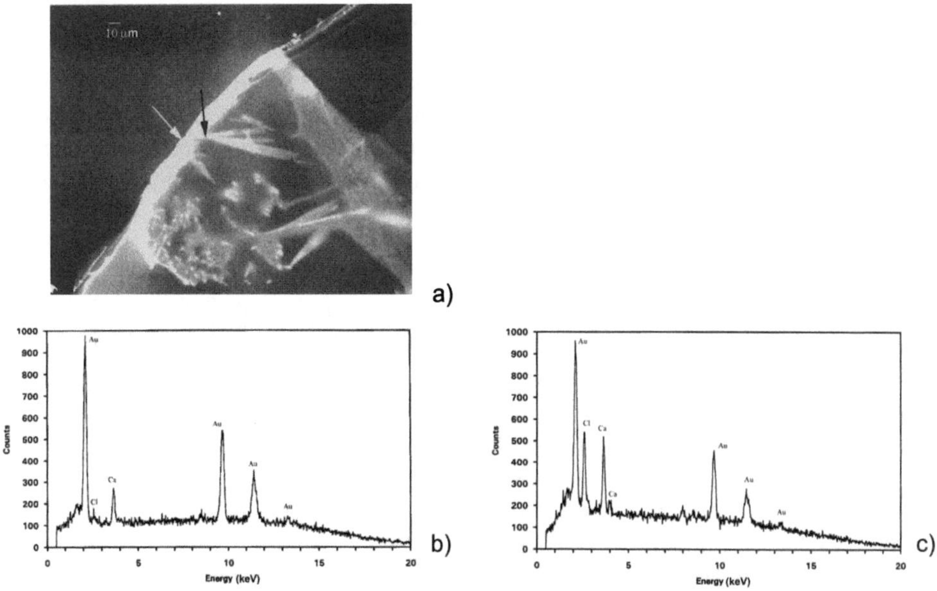

Fig. 10. SEM image of the HOPG surface after contact with the $KAu(CN)_2/KOH/CaCl_2$ electrolyte for more than 32 h. (**b**) and (**c**) EDAX scans of the areas showing the close association of Ca and Au, but no K, on an HOPG surface (*white arrow* is scan (**b**), *black arrow* denotes scan (**c**))

11 Discussions

It is known that graphite particles have a low adsorption characteristic for the aurocyanide complex [54], however, a number of studies have promoted the view that $Au(CN)_2^-$ adsorption in carbon compounds is strongly correlated to the amount of sp^2 carbon present [26,44]. *Groszek* et al. [55] compared the differential molar heats of adsorption on graphitized carbon (98 % basal plane sites) and coconut-activated carbon. They found that the energy changes were similar, being −21 kJ/ mol for graphitized carbon and −26 kJ/ mol for activated carbon. They concluded that the $Au(CN)_2^-$ complex has a strong affinity for the graphitic basal planes. *Miller* and *Sibrell* [44] have also concluded that the graphitic structure is indeed important for gold adsorption-concluding that at monolayer coverage the adsorption density of the $Au(CN)_2^-$ is expected to be about 3.3×10^{-10} mol/ cm^2.

To date, the only other gold adsorption work reported on highly pyrolytic orientated graphite was made by *Sibrell* and *Miller* [45]. Radioactive $Au(C^{14}N)_2$ was placed in contact with two types of graphitic substrate: a basal and an edge plane. They observed adsorption of the radioactive aurocyanide complex mainly on the edge plane.

11.1 In Situ STM Findings in the Presence of Ca^{2+}

It is widely known in gold processing plants that the addition of calcium to the pulp increases the level of aurocyanide ions loaded on to activated carbon [28,56,57]. When Ca^{2+} ions were added to the contact solution in the STM experiments, the probability of finding the gold adsorbate with the STM was increased.

These areas showing up as adsorbate in the STM scans were of significantly higher conductivity than that of the surrounding carbon. Electromigration STM studies of gold on carbon by *Besold* et al. [58] have shown that Au had a higher tunnelling characteristic than the surrounding carbon structure. The areas observed in the study were different from surface graphitic oxide as presented in the work of *Gewirth.*

In our work, the features associated with gold adsorption found on the surface were only found in defect areas and close to macroscopic steps. Higher-resolution STM scans in these defect areas revealed that most of the surface is graphitic carbon.

It is generally known that the HOPG surface provides extended areas of atomically flat surface. The graphite surface is hydrophobic and relatively free of contamination [60,61]. In addition, HOPG is a chemically inert surface and the surface has to be (activated) polarized anodically for the C–C bonds to break or be expanded [62].

The long contact times between the HOPG and the electrolyte before any significant adsorption can be observed strongly suggest that there is a time-dependent factor inherent to the adsorption of the gold complex on HOPG. This could be explained by the fact that the HOPG surface is hydrophobic and the extended contact time allowed the insertion of the liquid between the graphitic sheets. It is possible that the long exposure time could enhance the reaction of the carbon surface with the contact liquid (or with dissolved O_2) producing defect sites which subsequently act as sites for Ca intercalation.

The form of the gold adsorbate on the carbon surface is not so clear from the in situ STM study. Some images showed close to molecular resolution but these were not clear enough to investigate the unit cell structure of the gold adsorbate. The high-resolution STM images provide evidence that the clusters/mounds have an apparent structure but unlike the contrast that can be noticed between the carbon atomic background and the adsorbate, however the STM cannot distinguish between the calcium and the gold or the CN species. Although the SEM (BSE)/EDAX spectra have shown the association between Ca and the gold, their arrangement within the carbon sheets is actually not known.

Adams and *Fleming* [63] suggested that aurocyanide is extracted onto activated carbons as an ion pair such as under normal plant conditions. We have proposed that with elapsed time, solvated calcium ions are brought into the slowly expanding graphitic sheets prior to the adsorption of the aurocyanide onto the HOPG surface. The gold complex species on the surface could be of the form M_n^+–C–$[Au(CN)_2^-]_n$, where M represents the calcium metal ion.

11.2 Probable Mechanism of Interaction

The attraction of the aurocyanide complex to the graphite substrate is thought not to be a strong bond as the complex ion attachment in activated carbon is a physisorption process with no actual covalent/ionic bond formed. It is agreed that the attachment of the aurocyanide ion to activated carbon is not made via a purely electrostatic attraction because a neutral $Hg(CN)_2$ complex can successfully compete with the aurocyanide ions for sites on activated carbon [20].

Given the results obtained from this in situ study it is proposed that solvated calcium ions are intercalated within the graphite sheets. These Ca^{2+} ions (ionic radius 0.99 Å) would have a size (6–9 Å with one solvation sheet [64]) that could be slowly brought within the graphitic sheets only after the sheets had started to expand from the normal interlayer spacing of 3.35 Å. Intercalation within graphite is not a new phenomenon [50,65]. There are cases of electrochemical intercalation whereby molecules are known to intercalate from H_2SO_4 solution in contact with graphite [59]. Although the interlayer distance in graphite is 3.35 Å, in activated carbon it is about 3.7 Å but the latter has a more open and amorphous network structure. The long time needed for adsorption could be explained in terms of the expansion of the graphitic sheets with time by the insertion of the solvent into the graphite layers at defect points. The solvated calcium ions would be transported slowly into the graphitic structure because they would have to partially shed their solvation shell to get within the expanded sheets. There would be an interplay between the energy change from losing part of the solvation sheet and the energy change from being in between two graphitic sheets with a partially solvated sheet. This could well explain the fact that all of the in situ STM images of gold adsorption have been obtained near or at defects or steps. Within the layers the intercalated calcium would modify the electronic structure of the top sheet in such a way as to encourage the aurocyanide to attach itself to the graphite sheet.

The fact that the gold adsorbate/clusters are found in, and very close to, defect areas is an indication that the defects are important precursors to the adsorption process on HOPG. The solvated calcium ions could be trapped close to edges, due to C–O functional groups (formed at these edges) thus remaining in close proximity to the edge. In Figs. 8b,c edges are decorated by a gold adsorbate at the step on the edge of the upper layer.

12 Concluding Remarks

STM has been shown to be a useful tool for understanding the $Au(CN)_2^-$/ HOPG system. Imaging of the adsorbate at the molecular level has been achieved, and the studies have shown that the adsorption occurred close to defects and steps and in the presence of the Ca^{2+} ion.

It is proposed that the Ca^{2+} intercalates inhomogeneously within the graphite sheets at and close to defective regions. The aurocyanide complex then adsorbed after the intercalation of the partially solvated calcium ions. This mechanism is likely to be important in the adsorption of $Au(CN)_2^-$ on activated carbon in industrial plants.

References

1. C. A. Fleming: CIM Bull. **91**, 55 (1998)
2. C. Allen, T. Brearley, A. Clarke, J. Harman, P. Berry: *Australia in the World Gold Market* (ABARE Research Report. Canberra 1999) 8
3. J. Avraamides: *Perth Gold*, Randol Perth Int. Gold Conf. 1988 (Randol International, Golden, CO) 261
4. R. C. Weast (Ed.): *CRC Handbook of Chemistry and Physics* (Chemical Rubber, Cleveland 1975)
5. R. J. Puddephatt: *The Chemistry of Gold* (Elsevier, Amsterdam 1978)
6. W. F. Kean: J. Pharm. Sci. **77**, 1033 (1988)
7. B. M. Sutton: Gold Bull. **19**, 15 (1986)
8. S. L. Best, P. J. Sadler: Gold Bull. **29**, 87 (1996)
9. S. D. Lin, M. Bollinger, M. A. Vannice: Catal. Lett. **17**, 245 (1993)
10. J. Schwank: Gold Bull. **16**, 103 (1983)
11. Pat. Appl. EP642040, Gold Bull. **28**, 95 (1995)
12. D. W. de Havelland: *Gold and Ghosts* (Perth Hesperian Press, Perth 1985)
13. T. T. Kodas, M. J. Hampden-Smith (Eds.): *The Chemistry of Metal CVD* (VCH, New York 1994)
14. P. A. Benoy, C. Dellacorte: Surf. Coat. Technol. **62**, 454 (1993)
15. T. Thundat, E. W. Wachter, S. L. Sharp, R. J. Warmack: Appl. Phys. Lett. **66**, 1695 (1995)
16. C. J. Paterson: in A. Arbriter, K. N. Han (Eds.) *Gold: Advances in Precious Metals Recovery* (Gordon and Breach, New York 1990) p. 49
17. M. T. Van Meersbergen, L. Lorenzen, J. S. J. Van Deventer: Min. Eng. **6**, 1067 (1993)
18. S. G. Hodge, M. G. Ryan, J. T. Woodcock, J. K. Hamilton (Eds): Australasian Mining and Metallurgy: The Sir Maurice Mawby Memorial Volume, Monography 19, Parkville, **2** 1098 (1993)
19. R. B. Bhappu: Miner. Proc. Extractive Metall. Rev. **6**, 67 (1990)
20. J. Marsden, I. House: *The Chemistry of Gold Extraction* (Ellis Horwood, New York 1992) p. 266
21. V. Kudryk, H. H. Kellog: J. Met. **6**, 541 (1954)
22. C. P. Thurgood, D. K. Kirk, F. R. Foulkes, W. F. Graydon: J. Electrochem. Soc. **128**, 1680 (1981)
23. D. W. Kirk, F. R. Foulkes: J. Electrochem. Soc. **127**, 1993 (1980)
24. D. W. Kirk, F. R. Foulkes, W. F. Graydon: J. Electrochem. Soc. **127**, 1962 (1980)

25. M. J. Nicol, C. A. Fleming, R. L. Paul: in G. G. Stanley (Ed.): *The Extractive Metallurgy of Gold in South Africa*, S. A. Inst. Min. Metall. Johanesburg, **1** 831 (1987)
26. A. S. Ibrado, D. W. Fuerstenau: Hydromet. 30 (1992) 243.
27. La Brooy: *Mineral Chemistry M357: Gold Extraction* (Murdoch University 1996) p. 3
28. G. J. McDougall, R. D. Hancock: Gold Bull. **14**, 138 (1981)
29. J. B. Zadra, A. L. Engel, H. J. Heinen: RI 4843, U.S. Bureau of Mines (1952)
30. J. S. Mattson, H. B. Mark: *Activated Carbon* (Marcel Dekker, New York 1971)
31. J. C. Bokros: in P. C. Walker (Ed.): *Chemistry and Physics of Carbon* (Marcel Dekker, New York 1969) Vol. 5
32. M. D. Adams: Tech. Dig. XVIII Int. Min. Proc. Cong., Sydney (1993) p. 1175
33. M. M. Dubinin: *Chemistry and Physics of Carbon* (Marcel Dekker, New York 1966)
34. G. J. McDougall, R. D. Hancock: Min. Sci. Eng. **12**, 85 (1980)
    N. Tsuchida, D. M. Muir: Metall. Trans. B **17**, 529 (1986)
35. M. C. Fuersteneau, C. O. Nebo, J. R. Keslo, R. M. Zaragoza: Min. Metall. Proc. **4**, 177 (1987)
36. J. D. Le Roux, A. W. Bryson, B. D. Young: J. S. Afr. Inst. Min. Metall. **91**, 95 (1991)
37. N. Tsuchida, D. M. Muir: Metall. Trans. B **17**, 523 (1986)
38. N. Tsuchida, D. M. Muir: Metall. Trans. B **17**, 529 (1986)
39. J. D. Cashion, A. C. McGrath, H. Volz, J. S. Hall: Trans. Inst. Min. Metall. Sect. C **97**, 129 (1988)
40. K. Kongolo, A. Bahr, J. Friedl, F. E. Wagner: Metallurgical Transaction B **21**, 239 (1990)
41. P. L. Sibrell, J. D. Miller: *World Gold 1991* (Cairns, Queensland 1991) p. 25
42. C. Klauber: Langmuir **7**, 2153 (1991)
43. A. S. Ibrado, D. W. Fuerstenau: Min. Eng. **8**, 441 (1995)
44. J. D. Miller, P. L. Sibrell: in D. R. Gaskells (Ed.): *The Nature of Gold Adsorption from Cyanide Solutions by Carbon* (TMS, New Orleans 1991) p. 647
45. P. L. Sibrell, J. D. Miller: Min. Metall. Proc. **9**, 189 (1992)
46. G. Kirton: Hons. Thesis, Murdoch University (1994)
47. G. Binnig H. Fuchs, Ch. Gerber, H. Rohrer, E. Stoll, E. Tosatti: Europhys. Lett. **1**, 31 (1986)
48. S. L. Park, C. F. Quate: Appl. Phys. Lett. **48**, 112 (1986)
49. D. Tomanek, S. G. Louie, H. J. Mamin, D. W. Abraham, R. E. Thomson, E. Ganz, J. Clarke: Phys. Rev. B **35**, 7790 (1987)
50. O. J. Vohler, F. V. Strum, E. Wege: Carbon Materials, in G. L. Trigg (Ed.): *Encyclopedia Appl. Phys.* (VCH Publishers, New York 1992) Vol. 3, p. 21
51. Park Scientific Instruments, 1171 Borregas Ave., Sunnyvale, CA 94089, USA
52. Advanced Ceramics, Union Carbide
53. Johnson Matthey, Alfa, Postfach 65 40, D-76045 Karlsruhe, Germany
54. M. D. Adams: React. Polym. **21**, 159 (1993)
55. A. J. Groszek, S. Partyka, D. Cot: Carbon **29**, 821 (1991)
56. R. J. Davidson: J. S. Afr. Inst. Min. Metall. **75**, 67 (1974)
57. K. Kongolo, C. Kinabo, A. Bahr: Hydromet. **44**, 191 (1997)
58. J. Besold, R Kunze, N. Matz: J. Vac. Sci. Technol. B **12** (1994)
59. A. A. Gewirth, A. J. Bard: J. Phys. Chem. **92**, 5563 (1988)

60. E. Ganz, K. Sattler, J. Clarke: J. Vac. Sci. Technol. A **6**, 419 (1988)
61. R. S. Robinson, K. Sternitzke, M. T. McDermott, R. L. McCreery: J. Electrochem. Soc. **138**, 2412 (1991)
62. B. Zhang, E. Wang: J. Electroanal. Chem. **388**, 207 (1995)
63. M. D. Adams, C. A. Fleming: Metall. Trans. B **20**, 315 (1989)
64. T. Radnick, private communication
65. M. S. Dresselhaus, G. Dresselhaus: Intercalation Compounds, in G. L. Trigg (Ed.): *Encyclopedia Appl. Phys.* (VCH Publishers, New York 1994) Vol. 8, p. 133

Part II

Anion Adsorption

Atomic Structure of Cu(111) Surfaces in Dilute Sulfuric Acid Solution

Peter Broeckmann, Michael Wilms, Matthias Arenz,
Alexander Spänig, and Klaus Wandelt

Institut für Physikalische und Theoretische Chemie, Universität Bonn,
Wegelerstraße 12, 53115 Bonn, Germany
k.wandelt@uni-bonn.de

Abstract. The whole range of processes occurring at a single crystal Cu(111) electrode in a sulfuric acid electrolyte (5 mMH_2SO_4) between hydrogen evolution and copper dissolution has been studied using a combination of Cyclic Voltammetry (CV), in situ Scanning Tunneling Microscopy (STM), and in situ Fourier Transform Infrared Spectroscopy (FTIR) as well as ex situ characterization by means of Auger Electron Spectroscopy (AES), X-ray Photoelectron Spectroscopy (XPS) and Low Energy Electron Diffraction (LEED). In particular, in situ STM measurements with atomic resolution, carried out under various conditions, namely in the potentiostatic, potentiodynamic and quasi-spectroscopic mode, provide very detailed and direct information about the following interfacial properties and processes: the adsorption and desorption kinetics of the sulfate anions; the structure of the sulfate adlayer and the reconstruction of the copper surface underneath including drastic morphological changes of the electrode surface; the absolute adsorption geometry of individual SO_4^{2-} anions; the anodic copper corrosion and redeposition as well as the formation of cationic adsorbate (probably hydronium) layers in the regime of hydrogen evolution. Altogether the results demonstrate that already a relatively simple system such as Cu(111) in sulfuric acid solution may exhibit a surprisingly high degree of complexity.

1 Introduction

Electrochemical processes at or with electrode surfaces, like electrolysis, electrocatalysis, electrocorrosion, electrodeposition, etc. include the adsorption of ions as an elementary reaction step (Fig. 1). Unlike comparable reactions between a solid surface and a gas phase, however, electrochemical processes always proceed in the presence of the respective counterions. It is therefore important to investigate a certain reaction, e.g. metal deposition, in the presence of different anions in order to establish the influence of these counterions on the reaction mechanism in general and the properties of the electrode surface in particular. "Classical" electrochemical measurements of quantities like charge, current, capacity, etc., yield only information averaging over the whole (or unknown parts of the) electrode surface [1]. Only the more recent development of new spectroscopic and microscopic techniques enables these

K. Wandelt, S. Thurgate (Eds.): Solid–Liquid Interfaces, Topics Appl. Phys. **85**, 141–197 (2003)

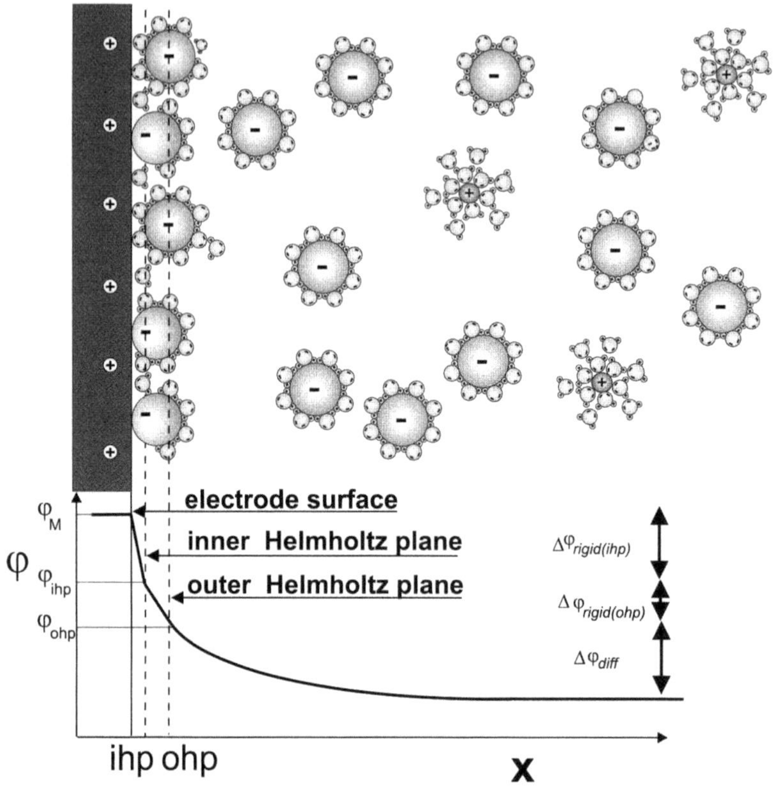

Fig. 1. Model of the "electrochemical double layer". The centers of gravity of the negative charge of partially or full hydrated anions adsorbed on a positively polarized metal (ϕ_M) electrode define the "inner" or "outer Helmholtz plane", respectively

solid–liquid interfaces to be studied in situ with similar resolution and precision as it has become standard in surface science under ultra high vacuum (UHV) conditions. In particular different versions of infrared spectroscopy like Fourier Transform Infrared Spectroscopy (FTIR) [2] and Sum Frequency Generation [3], and Scanning Probe Techniques like Scanning Tunneling Microscopy (STM) and Atomic Force Microscopy (AFM) [4,5,6] provide detailed insight on the atomic and molecular level [7,8,9]. Quite often, however, these in situ studies need to be complemented by ex situ measurements preferably under UHV conditions taking advantage of the great variety of surface science techniques. Very good overviews of this modern surface science approach to interfacial electrochemistry combining in situ and ex situ techniques are given in [10] and [11].

As a very basic and prototypical example for anion adsorption we describe in this chapter investigations of those processes which occur in the electrochemical double layer (Fig. 1) at a single crystal Cu(111) electrode in a sulfuric acid electrolyte. The experiments are carried out using in situ FTIR spectroscopy and various modes of Scanning Tunneling Microscopy (STM) as well as ex situ surface science techniques like Auger Electron Spectroscopy (AES), X-ray Photoelectron Spectroscopy (XPS), Low Energy Electron Diffraction (LEED), etc. The choice of copper as electrode material was motivated among others by the fact that copper plays an increasing role in modern microelectronics chip production [12,13].

2 Experimental

All STM measurements presented in this chapter were performed with a newly designed and home-built Electrochemical Scanning Tunneling Microscope (EC-STM), which combines a specially modified Besocke type STM [14,15] with an electrochemical flow cell [16]. In contrast to commercial instruments the new design (see Fig. 2) offers, in brief, the following favorable features

- The small, rigid and compact construction reduces vibrational disturbances, and the radially symmetric arrangement of the same piezoceramics for the scanner and the piezo-legs for coarse approach guarantees a first-order thermal drift compensation [14].
- The EC-STM operates within a closed aluminum chamber, which can be filled with an inert gas and which serves as further protection against external noise, thermal gradients, electromagnetic radiation, etc. All functions of the STM are fully software controlled including safety measures against tip-sample crashs.
- The electrochemical cell (Figs. 2–4) accommodates, unlike most other instruments, a realistic volume of the electrolyte of about 2.5 ml. This enables a direct comparison of EC-STM and electrochemical measurements, e.g. cyclic voltammetry, in the same cell. The copper sample (working electrode) is pressed against a hole in the bottom of the electrochemical flow cell in such a way that only the (111) surface is in contact with the electrolyte (Fig. 3). A Pt wire and a reversible hydrogen electrode (RHE) serve as counter-electrode and reference electrode, respectively.
- By means of a newly designed bipotentiostat, carefully adapted to the control electronics of the STM, it is even possible to take potentiodynamic images, i.e. the potential of the sample versus the reference electrode (Fig. 3) can be varied during imaging without changing the tip-potential versus the reference electrode, thereby protecting the tip against faradaic processes. The STM electronics has also provisions for doing local tunneling spectroscopy.
- The complete set-up of the EC-STM including the electrochemical periphery is displayed in Fig. 4. The electrochemical solutions can be intensively

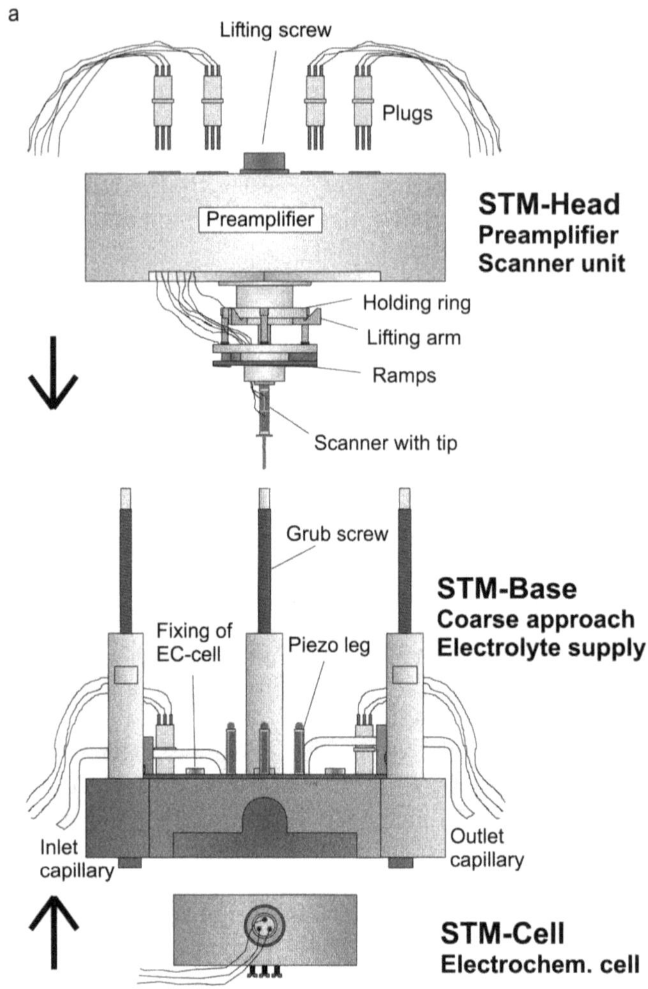

Fig. 2. Sketch of our home-built electrochemical scanning tunneling microscope (STM); (**a**) disassembled, (**b**) assembled [16]

outgassed by a flow of purified inert gas (argon) before they are pumped into the electrochemical cell. The solution in the electrochemical cell can even be continuously exchanged without lifting the tip and opening the aluminum chamber, i.e. imaging under potential control is possible while exchanging the electrolyte.

- The STM tips used were electrochemically etched from a 0.25-mm-diameter tungsten wire in 2 M KOH solution and subsequently isolated by passing them through a drop of simple nail polish.

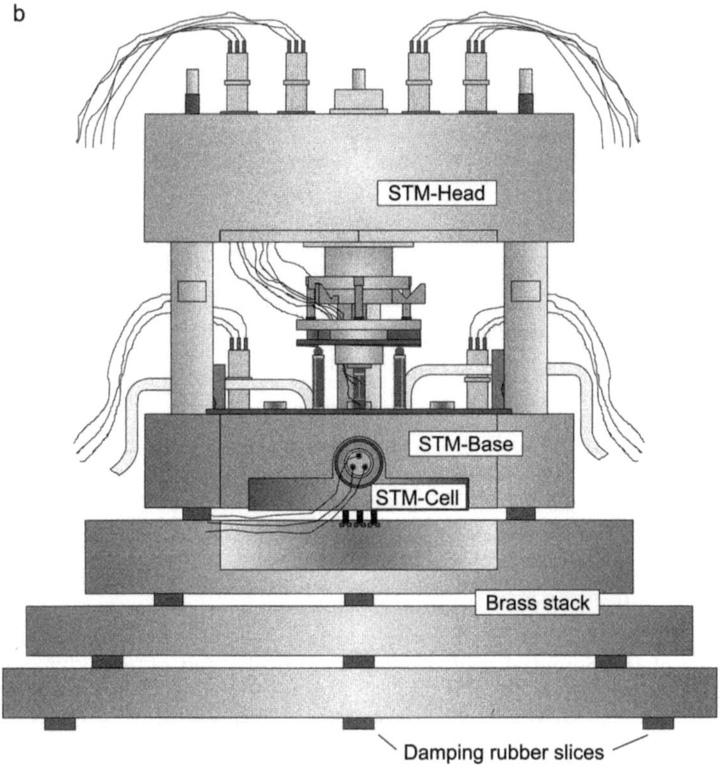

Fig. 2. continued

A very detailed description of the whole instrument, including the design of the mechanical set-up, the combined electronics of STM-control and bipotentiostat, and the electrochemical periphery can be found in [16].

For in situ FTIR measurements a Nicolet magna 560 spectrometer was available with a nitrogen cooled MCT detector. All IR measurements were performed in a glove box enclosing the spectroelectrochemical cell. The whole glove box was again permanently purged with purified argon to avoid oxygen contamination. The spectroelectrochemical cell is designed for an external reflection mode in a thin liquid layer configuration. Therefore the cell is coupled at its bottom with a CaF_2 prism beveled at 60° from the surface normal as shown in Fig. 5 [17]. The spectra were recorded with a resolution of $8\,cm^{-1}$ using p-polarized light.

The Cu(111) sample was cut from a single-crystal rod, oriented by Laue backdiffraction, and first mechanically polished using a diamond paste with different grain sizes down to 0.25 µm. The resulting surface orientation was within 0.5° of the (111) plane. Before each STM measurement, the single crystal was electropolished in a separate cell for about 40 s in 50 % orthophosphoric acid at an anodic potential of 2 V. After that the sample was first

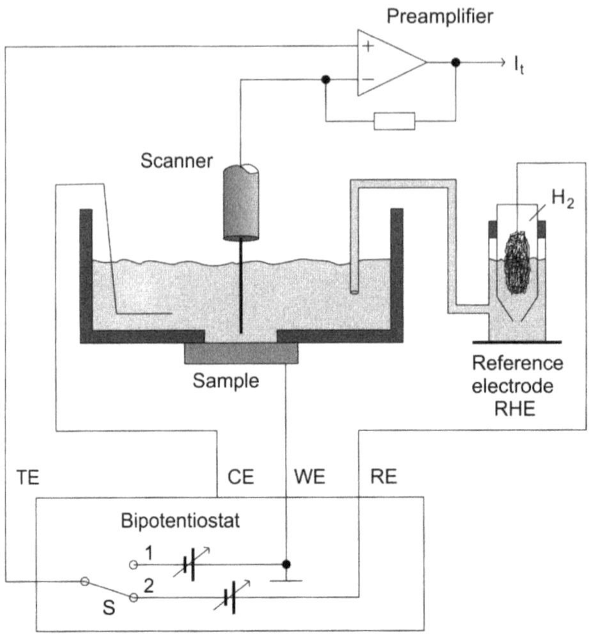

Fig. 3. Principle of the potential control for potentiostatic and potentiodynamic imgaging (TE = tip electrode; CE = counterelectrode; WE = working electrode; RE = reference electrode; RHE = reversible hydrogen electrode; S = switch; I_t = tunneling current)

rinsed with pure Milli-Q water and then, covered with a droplet of water left to protect the surface, mounted into the electrochemical STM cell. For the FTIR measurements the Cu(111) surface was prepared under UHV conditions by Ar ion sputtering (1 keV, 30 min) and annealing (900 K, 10 min) until no more contaminations could be detected in Auger spectra and a sharp (1×1) LEED pattern was observed. The thus prepared sample was then transferred without contact to air (Fig. 5) into the electrochemical cell for the in situ spectroscopic measurements. Identical voltammograms in sulfuric acid were observed independent of whether the sample was prepared by electrochemical etching in orthophosphoric acid or in UHV. All measurements were carried out in 5 mHH_2SO_4 solution prepared from high purity water (Milli-Q purification system, $> 18\,\mathrm{M}\,\Omega\,\mathrm{cm}$), and reagent grade or better chemicals were used. All electrochemical potentials of the working electrode (sample) are referenced against the reversible hydrogen electrode (RHE).

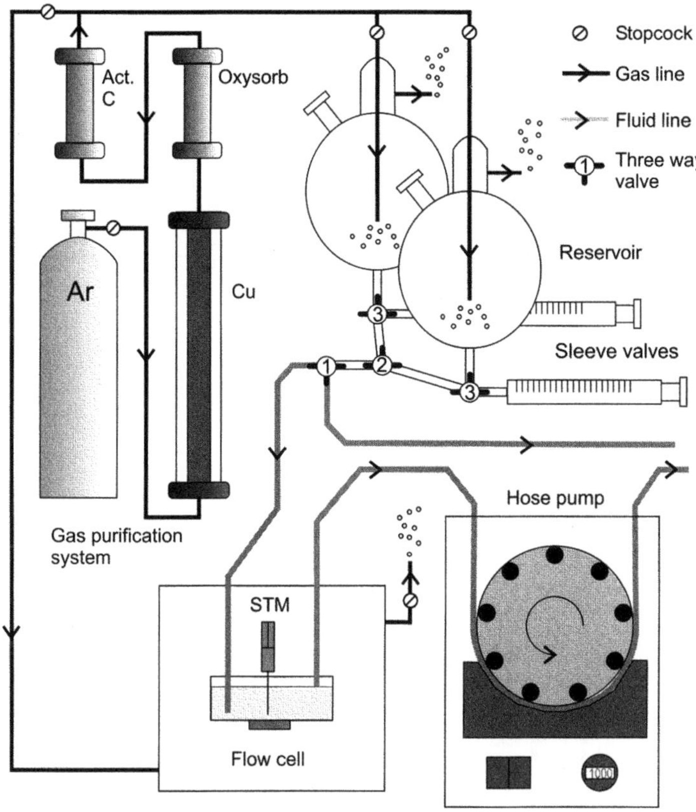

Fig. 4. Electrolyte supply and gas purification system of the electrochemical STM set-up [16]

3 Sulfate Adsorption

3.1 Electrochemical and Spectroscopic Characterization

Figure 6 shows a typical cyclic voltammogram (CV) of Cu(111) in 5 mM H_2SO_4 recorded with a scan rate of 10 mV/s within the electrochemical cell of the EC-STM (Fig. 3). The cathodic limit of this curve is defined by hydrogen evolution at the free copper surface, while the anodic limit is determined by the copper dissolution. Further characteristics in the anodic and cathodic run of this CV are features which are correlated with the adsorption and desorption of the sulfate anions, respectively. As described in the following, this latter assigment is supported by in situ potentiodynamic STM measurements as well as ex situ chemical analysis of the surface composition using X-ray photoelectron spectroscopy (XPS). Further support stems from in situ FTIR studies reported in Sect. 3.2. The large separation (hysteresis) between the potential ranges of anion adsorption (positive potential sweep) and des-

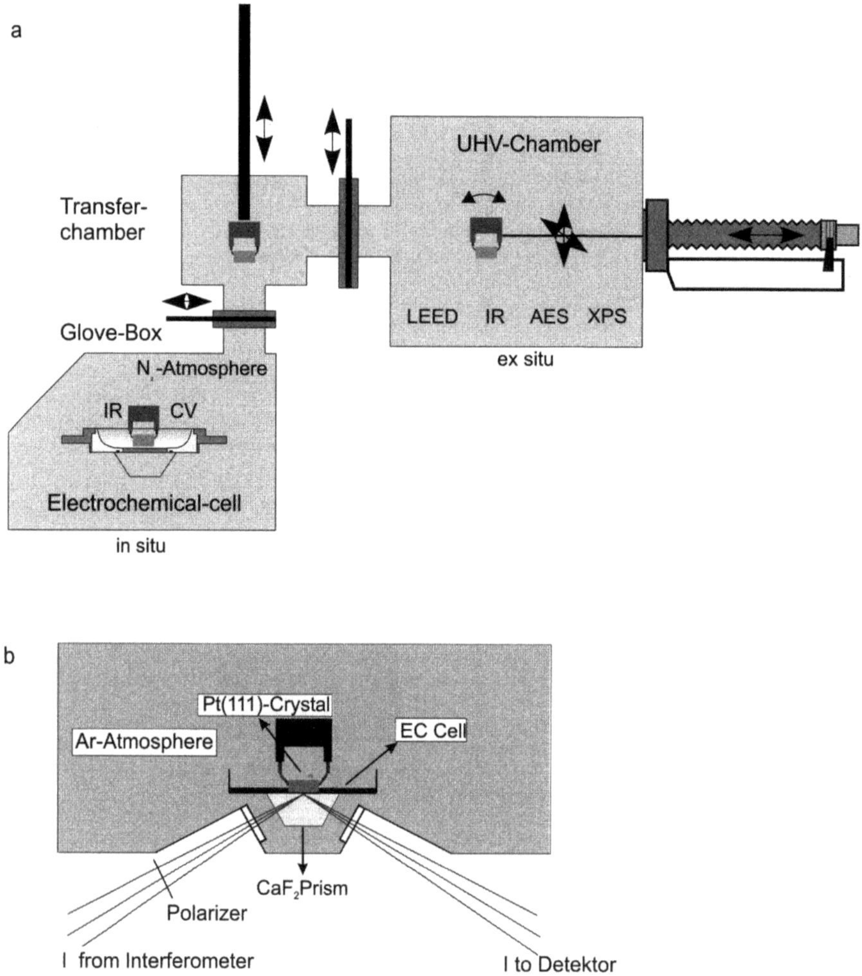

Fig. 5. (**a**) Sketch of the transfer chamber for in situ and ex situ FTIR measurements. The UHV chamber contains also provisions for surface analysis with Low Energy Electron Diffraction (LEED), Auger Electron Spectroscopy (AES), X-ray Photoelectron Spectroscopy (XPS), etc. A close-up of the thin layer configuration of the spectroelectrochemical cell is shown in (**b**)

orption (negative potential sweep), respectively, points to a strong kinetic hindrance of at least one of the two processes.

The occurrence of sulfate adsorption and desorption can best be visualized by a potentiodynamic STM image. Figure 7 was recorded simultaneously with a cyclic voltammogram and represents line scans (from left to right) at different electrode potentials. The whole span of the image in the slow scan direction (from top to bottom) corresponds to a full cycle of the CV.

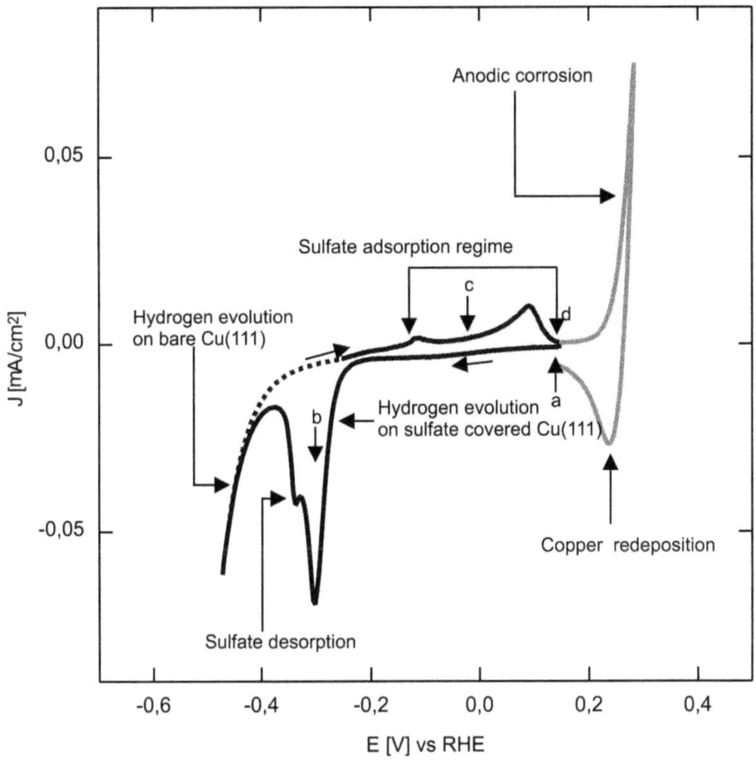

Fig. 6. Typical steady-state cyclic voltammogram of a Cu(111) electrode in 5 mM H_2SO_4; scan rate $dE/dt = 10\,mV/s$; the marks a–d refer to Fig. 7. For an explanation of the *dashed line* see text

The timing and the recording speed were chosen such that the scan of the STM image starts at the same time as the cycle of the voltammogram (at point a in Fig. 6) and ends exactly at the end of the cycle in point d. In the range between points a and b a well-ordered hexagonal structure is seen to cover four surface terraces (I–IV). At the potential corresponding to point b, i.e. at the onset of sulfate desorption, this structure disappears and does not reappear until readsorption occurs near point c. Note that the image in Fig. 7 is a superposition of spatial and temporal changes of the surface structure. The structural changes occurring at points b and c in the STM image are thus only a qualitative visualization of phase changes at the surface, which however, are clearly correlated with the desorption and adsorption peaks in the CV, respectively.

The chemical identity of the adlayer producing the hexagonal structure in Fig. 7 was verified by XPS. In a transfer chamber similar to the one shown in Fig. 5a the sample was emersed from the electrolyte under potential control and transferred into an ultra-high vacuum (UHV) chamber without being

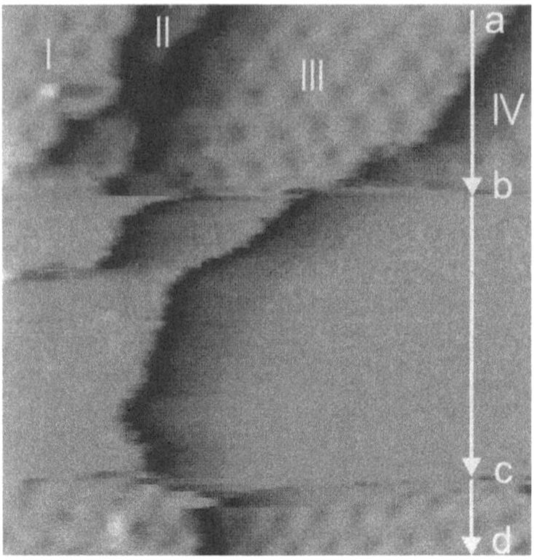

Fig. 7. Potentiodynamic STM image recorded during a full cycle of the voltammogram shown in Fig. 6a,b,c,d correspond to potentials marked in Fig. 6. Four surface terraces (I–IV) are imaged. Image size 39 nm × 39 nm; bias voltage $U_b = -10\,\text{mV}$; electrode potential $E = -215\,\text{mV}$ vs. RHE

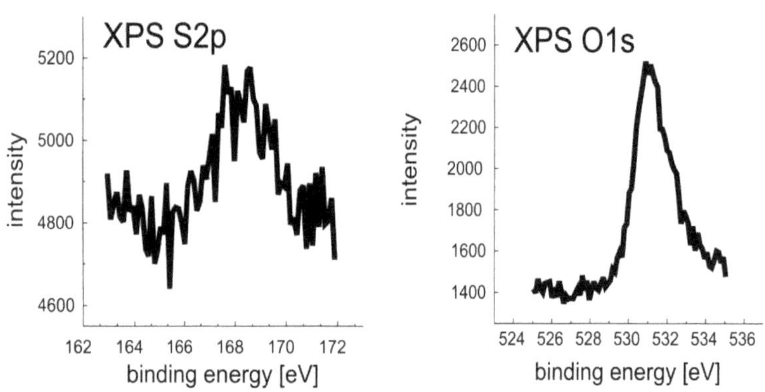

Fig. 8. XPS spectra of the emersed sulfate-covered Cu(111) surface. The relative peak intensities and binding energies are consistent with sulfate

exposed to air. The emersion at 150 mV preserved the adlayer outside the electrolyte. In the XPS spectra sulfur and oxygen were found (Fig. 8) and the binding energies of the S(2p) and O(1s) lines are indicative of adsorbed sulfate. Copper oxide formation could be excluded. The XPS data, however, do not allow one to distinguish between sulfate and bisulfate. The controversial discussion about whether the adsorbate is sulfate or bisulfate [18,19,20,21] is still not fully settled, but probably obsolete in the light of the results presented in Sect. 3.1 of this chapter. For the sake of simplicity, in the following we therefore refer to sulfate only.

Coming back to the detailed structure of the CV in Fig. 6 it is obvious that even after subtraction of a Butler–Volmer fit (dashed line) for the hydrogen evolution current on the bare Cu surface the cathodic desorption peak is significantly larger than the anodic adsorption feature. This is due to an additional hydrogen evolution current from the *sulfate-covered* Cu surface caused by the so-called Frumkin effect [22]: The presence of the adsorbed sulfate anions increases the H_3O^+ concentration in the vicinity of the electrode, which in turn reduces the overpotential for the hydrogen reduction reaction. As a result in the cathodic run one actually first observes an exponentially increasing hydrogen reduction current from the sulfate-covered electrode (first flank of the cathodic feature). The subsequent onset of sulfate desorption leads to a decrease of this hydrogen evolution and, hence, to the occurrence of the first (largest) peak in the cathodic current feature. The following increase of SO_4^{2-} desorption then produces the second peak until finally the hydrogen evolution current from the *sulfate-free* Cu surface begins to dominate.

The assignment of only the second peak in the desorption feature as being actually due to sulfate desorption can be verified by a simple experiment. If the potential is only cycled in the range between 150 mV and −300 mV one observes the exponential increase of the cathodic current between −250 mV and −300 mV but no adsorption features in the reverse anodic run because the surface remained sulfate-covered. Inasmuch as the potential range is extended to more and more negative potentials in this experiment the true sulfate desorption peak becomes visible and, as a consequence, the re-adsorption peak in the anodic run grows. This correlation proves the cathodic double peak as being a superposition of an initial hydrogen reduction current (on the sulfate-covered electrode) and the actual sulfate desorption process. Similar behavior was also found with Cu(111) in dilute hydrochloric acid as described in the following chapter by Stuhlmann et al. in this book.

3.2 Atomic Structure of the Bare Cu(111) Electrode

After desorption of the sulfate adlayer the free copper surface could be imaged with atomic resolution. Figure 9 shows the typical hexagonal lattice of the Cu(111) surface. In order to gain high statistical reliability on the values of interatomic distances and angles of surface structures it is useful to calculate pair-correlation functions. Since such pair-correlations represent an average over a relatively large number of unit cells, they allow surface structures to be derived with much greater precision than would be possible by manually locating individual surface particles and their neighborhood. For every lattice point a computer algorithm calculates the distances between this lattice point and all other lattice points within a certain distance. The frequency (pair density) of each distance is then plotted versus this distance (pair–distance correlation). Likewise the pair–angle correlation represents the frequency with which angles between connecting lines from all lattice points

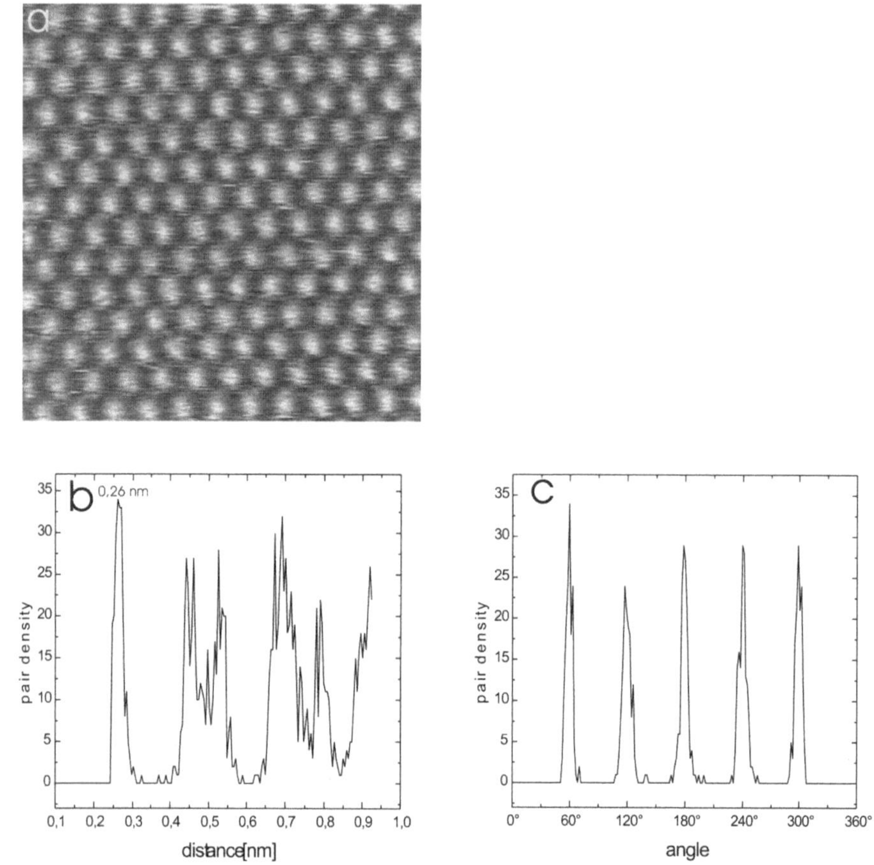

Fig. 9. In situ STM image (**a**) of the sulfate-free Cu(111) electrode measured potentiostatically in solution. Image size 3 nm × 3 nm; bias voltage $U_b = -10\,\text{mV}$; tunneling current $I_t = 30\,\text{nA}$; electrode potential $E = -215\,\text{mV}$ vs. RHE; positive potential sweep. (**b**) and (**c**) represent pair–distance and pair–angle distributions calculated from panel (**a**). The nearest-neighbor distance of $a = 0.26\,\text{nm}$ agrees with the interatomic Cu–Cu separation in the Cu(111) surface (see text)

to all nearest-neighbor lattice points occur. The pair–distance and the pair–angle function calculated from Fig. 9a of the bare Cu(111) electrode are shown in Figs. 9b,c respectively. As expected only multiples of 60° show up and the interatomic distance of 0.26 ± 0.01 nm agrees very well with the literature value of 0.256 nm [23].

Such pair-correlation functions immediately make clear if a hexagonal structure is distorted. A hexagonal lattice stretched along one main symmetry axis, for example, yields the specific pair–angle correlation shown in Fig. 10a. The corresponding pair–distance correlation consists of two peaks where the peak for the smaller distance has twice the height of that for the larger distance. The situation is inverted for a hexagonal lattice compressed along one symmetry axis (Fig. 10b). In this case the heights of the distance peaks as well as the heights of the angle peaks (except 180°) are exchanged compared to the stretched hexagon (Fig. 10a). These considerations will become useful for the discussion of the sulfate adsorption structure presented in the following.

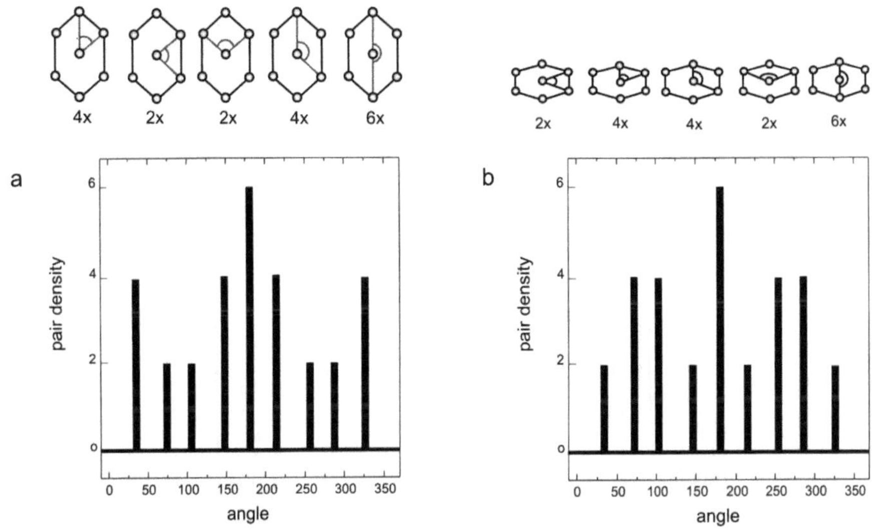

Fig. 10. Schematic pair–angle distributions of an elongated (**a**) and a compressed (**b**) hexagonal pattern. Note the distinct difference in pair densities

3.3 Atomic Structure of the Adsorbate Layer

After changing the sample potential to anodic values and passing the adsorption peak in the CV (Fig. 6) the formation of the sulfate adlayer takes place (Fig. 7). A potentiostatic image of the SO_4^{2-}-induced structure is displayed in Fig. 11a. Line scans (Fig. 11b) along the three main axes firstly show periodicities in the nanometer range, namely 2.7 ± 0.2 nm and 3.45 ± 0.2 nm, respectively [24]. Consequently, the observed structure is not an atomic structure but rather a long-range height modulation (z-corrugation) of 0.04 nm with Moiré character. From now on this structure will therefore be termed in short a "Moiré structure". Secondly, Fig. 11b reveals an anisotropy of the Moiré structure; the periodicity is not the same in all three directions. At first sight this anisotropy appears rather surprising since the substrate lattice has an ideal hexagonal structure. But the pair-correlation functions unambiguously confirm a hexagonal pattern which is stretched along one symmetry axis as suggested by the models in Fig. 10. The first peak (2.7 ± 0.2 nm) in

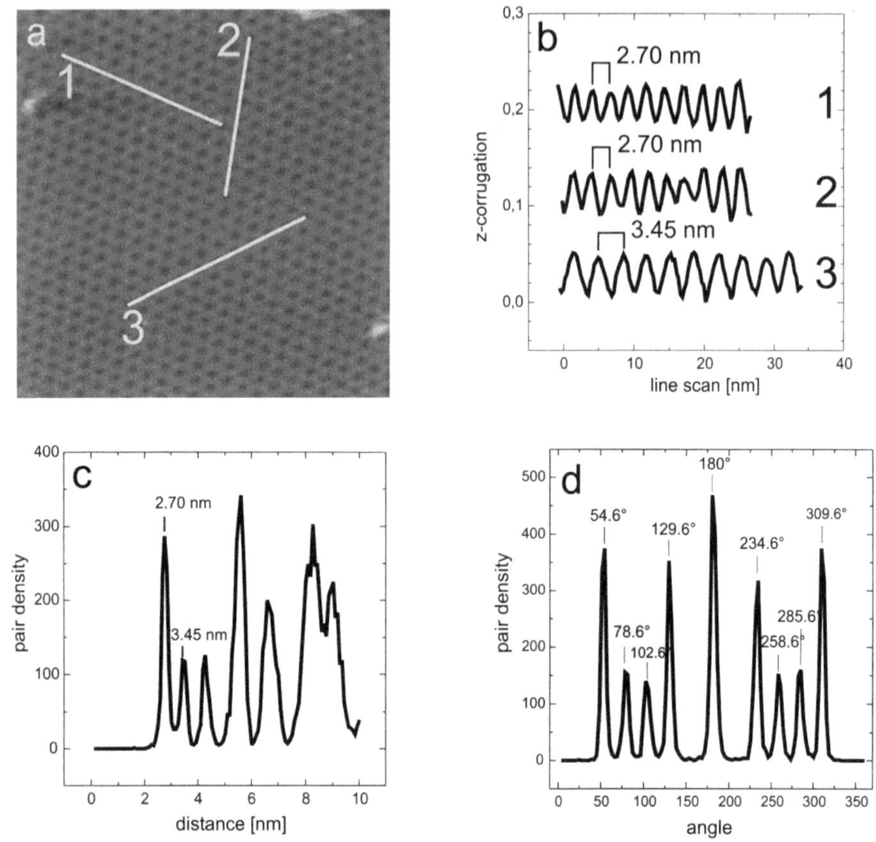

Fig. 11. (**a**) Moiré superstructure of the sulfate adlayer; image size 64 nm × 64 nm; bias voltage $U_b = 195\,\mathrm{mV}$; tunneling current $I_t = 1\,\mathrm{nA}$; electrode potential $E = -440\,\mathrm{mV}$. (**b**) Line scans along the white lines in panel (**a**). (**c**) Pair–distance and (**d**) pair–angle correlation calculated from (**a**) between units of the Moiré structure. Compare (**d**) with the schematic correlation function shown in Fig. 10

the pair–distance correlation function (Fig. 11c) appears with double intensity compared to the second peak (3.45 ± 0.2 nm). The third peak (without value) belongs to the shortest distance of the second coordination sphere, etc. The pair–angle correlation function (Fig. 11d) shows qualitatively the same angle distribution as shown in Fig. 10a and also confirms the deviation from an ideal hexagonal pattern. As an immediate consequence of this anisotropy, different domains of the Moiré structure rotated by 120° with respect to each other can be found on the surface (see Fig. 12).

A close-up of the structure of one such domain is displayed in Fig. 13a [24,25]. On this atomic scale the individual adsorbed particles become visible. The brighter dots correspond to the sulfate molecules; they are arranged in rows with an intermolecular spacing of 0.47 nm and a maximum long range

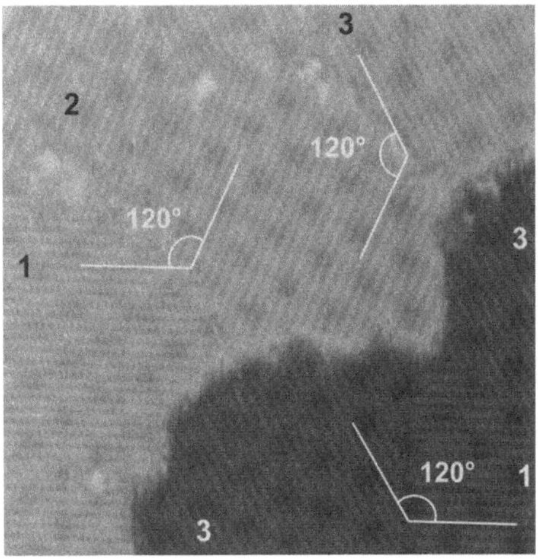

Fig. 12. Rotational domains (1, 2, 3) of the sulfate-induced Moiré structure; image size 26 nm×26 nm; bias voltage $U_b = 100\,\text{mV}$; tunneling current $I_t = 1\,\text{nA}$; electrode potential $E = 210\,\text{mV}$

height modulation of 0.04 nm (see Fig. 11a). The distance between sulfate particles in adjacent rows is 0.71 nm (see Fig. 13a). The corresponding pair–angle function in Fig. 13c indicates again the non-hexagonal structure of the sulfate adlayer. Weaker but clear dots are observed between the sulfate rows. As discussed below these are assigned to coadsorbed water molecules forming characteristic zig-zag chains. Considering the high affinity of sulfate anions to water molecules in aqueous solutions it is reasonable to assume that these anions do not completely lose their solvation sphere upon adsorption on the electrode surface. It is also plausible that the coadsorbed water molecules by virtue of hydrogen-bridge bonds play an important role for the stabilization of the adsorbed sulfate adlayer and the unusual anisotropic spacing between the negatively charged anions. The existence of these hydrogen-bridge bonds probably also dissolves the question as to whether the adsorbed particles are sulfate or bisulfate. Adsorbed halogenides such as chloride and bromide do not show this phenomenon of water coadsorption [26,27], which can be understood in view of their lower heats of solvation.

Based on the intermolecular distances and angles read off Fig. 13 it is possible to construct a structure model for the sulfate adlayer [24]. This model, together with one unit cell, is depicted in Fig. 14. This structure, in fact, is very similar to those found for sulfate layers on other fcc(111) transition metal electrodes (Au, Pt, Pd, Ir, Rh) in dilute sulfuric acid electrolytes [18,20,21,28,29,30]. All these adlayers are based on a $\begin{pmatrix} 2 & 1 \\ \bar{1} & 2 \end{pmatrix}$ unit cell. In the literature this adlattice is – not quite correctly – also known as

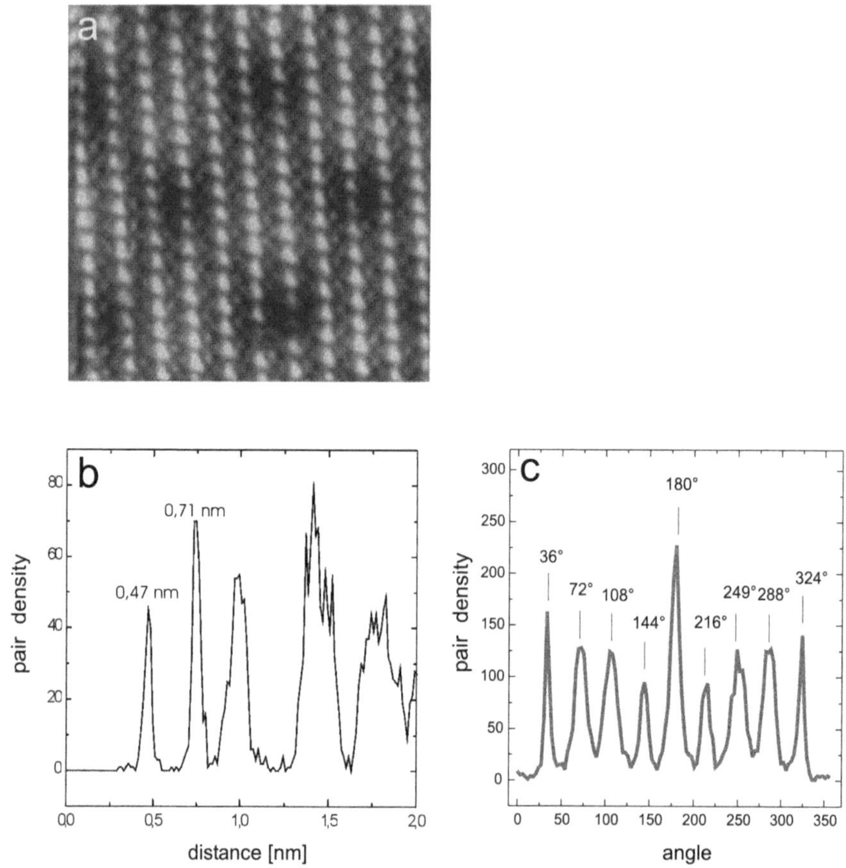

Fig. 13. (**a**) STM image of the coadsorption layer of sulfate anions (large bright dots) and water molecules (zig-zag chains of weak dots between the sulfate rows); note also the long-range modulation in the image due to the Moiré superstructure. (**b**) Pair–distance and (**c**) pair–angle distribution calculated from panel (**a**)

a $(\sqrt{3} \times \sqrt{7})$R30°-structure. For the sake of convenience we will therefore call this structure a $\sqrt{3} \times \sqrt{7}$-structure from here on.

Also the weaker spots between the sulfate rows are observed on these other sulfate-covered electrode surfaces. *Wan* et al. [20] proposed a model for sulfate on Rh(111) in which the coadsorbed water species form a bilayer chain. In their high-resolution STM images each second water molecule appears brighter. For copper a very similar model is proposed [31] with two kinds of water molecules within the unit cell (see Fig. 14). For the case of a sulfate-covered Au(111) electrode the chemical identity of the coadsorbed species was actually proved to be water by *Ataka* et al. [32] using Surface Enhanced Infrared Absorption Spectroscopy.

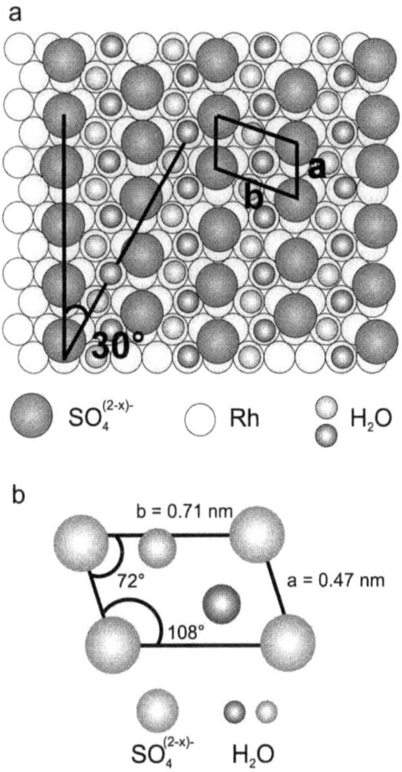

Fig. 14. Model of the SO_4^{2-}/H_2O coadsorbate structure on Rh(111) according to *Wan* et al. [20]. Close-packed sulfate rows are separated by zig-zag chains of water molecules. Note the similarity to the measured structure in Fig. 13a

Comparing the sulfate adlayer on Cu with those on the other fcc(111) surfaces, however, one fundamental difference becomes evident: While the sulfate structures on the other surfaces are simply commensurate only on copper the additional long-range Moiré-type modulation is observed [24,25,33].

Two strategies are possible to explain the observed Moiré effect: In the first scenario the interaction between the sulfate molecules and the copper substrate is weak and the interactions within the adsorbed layer lead to the formation of an incommensurate structure with varying adsorption sites for individual sulfate molecules. This explanation draws on similar cases of Moiré structures known from UHV adsorption systems.

On the other hand the sulfate $\sqrt{3} \times \sqrt{7}$ structures on the other fcc(111) electrode surfaces are fully commensurate, even though in the sequence of Au, Pt, Pd, Ir and Rh the interatomic distance shrinks considerably (see Table 1). For the adsorbed sulfate layers the lattice vector of the respective $\sqrt{3} \times \sqrt{7}$-unit cell shrinks accordingly, reaching $a = 0.46$ nm and $b = 0.74$ nm on Rh(111) [20]. Based on the even shorter Cu–Cu distance of 0.256 nm on

Table 1. Measured and calculated structure parameters for sulfate adlayers on the various noble and transition metals. The measured values originate from the following references: Au(111) [18,21]; Pt(111) [19,28]; Pd(111) [29]; Ir(111) [30]; Rh(111) [20]; Cu(111) [25]

Substrate	Substrate lattice constant	Sulfate unit cell	Calculated adlayer lattice constant Calculated angle	Measured adlayer lattice constant Measured angle
Au(111)	$a = 2.89$ Å	$(\sqrt{3} \times \sqrt{7})$-unit cell	$a = 5.00$ Å $b = 7.65$ Å $\gamma = 71°$	$a = 5.00$ Å $b = 7.60$ Å $\gamma = 71°$
Pt(111)	$a = 2.77$ Å	$(\sqrt{3} \times \sqrt{7})$-unit cell	$a = 4.80$ Å $b = 7.33$ Å $\gamma = 71°$	$a = 4.50 \pm 0.2$ Å $b = 7.10 \pm 0.2$ Å $\gamma = 71 \pm 1°$
Pd(111)	$a = 2.75$ Å	$(\sqrt{3} \times \sqrt{7})$-unit cell	$a = 4.76$ Å $b = 7.28$ Å $\gamma = 71°$	$a = 4.8$ Å $b = 7.3$ Å $\gamma = 71°$
Ir(111)	$a = 2.72$ Å	$(\sqrt{3} \times \sqrt{7})$-unit cell	$a = 4.71$ Å $b = 7.20$ Å $\gamma = 71°$	$a = 4.7$ Å $b = 7.1$ Å $\gamma = 71°$
Rh(111)	$a = 2.68$ Å	$(\sqrt{3} \times \sqrt{7})$-unit cell	$a = 4.64$ Å $b = 7.09$ Å $\gamma = 71°$	$a = 4.60 \pm 0.2$ Å $b = 7.40 \ldots 7.50$ Å $\gamma = 72°$
CU(111)	$a = 2.56$ Å	?	$a = 4.43$ Å $b = 6.77$ Å $\gamma = 71°$	$a = 4.7 \pm 0.1$ Å $b = 7.1 \pm 0.1$ Å $\gamma = 72 \pm 1°$ Moiré $a' = 2.80 \pm 0.2$ nm $b' = 3.60 \pm 0.2$ nm

the Cu(111) surface compared to 0.268 nm for Rh(111) the intermolecular distance within the sulfate rows on Cu(111) should shrink even further to $a = 0.443$ nm, in order to make the sulfate overlayer again fully commensurate with the Cu(111) substrate (see Table 1). This, however, is not the case. The *measured* value $a = 0.47$ nm (see Figs. 13 and 14) is significantly larger. Yet, as will be shown below, the sulfate overlayer is *not* incommensurate with the copper substrate. This apparent discrepancy as well as the appearance of the Moiré superstructure on Cu(111) (and not on the other fcc(111) surfaces) is explained by the following second scenario.

The second scenario is based on the assumption that a very strong interaction between the sulfate anions and the copper substrate causes a reconstruction, namely an expansion, of the topmost substrate layer such that the sulfate adlayer fits commensurably on this *expanded* topmost copper layer with Cu–Cu distances similar to Rh. As suggested by the sequence of interatomic distances Au > Pt > Pd > Ir > Rh (Table 1) it seems that for Rh the maximum of adlayer density has been reached. A further reduction of the substrate lattice constant in going from Rh to Cu is no longer accompanied by a corresponding adlayer compression, but in turn leads to an expansion of the topmost Cu layer. The Moiré superstructure, so far unique for sulfate on Cu(111), is then caused in this model by a misfit between the first expanded and the second non-expanded copper layer. Even though, at first glance, this second scenario appears rather unusual it will be shown in the following that this model is correct.

Originally [16,26,34] the proposal of a sulfate-induced surface expansion emerged from the observation of a significant mass transport out of the copper surface during the Moiré formation. This mass transport manifested itself by the nucleation and growth of many small copper islands on the sulfate-covered Cu terraces as well as by a dilatation of terraces at steps. Vice versa, after the desorption of sulfate and the disappearance of the Moiré structure one observes not only the disappearance of the islands but also the emergence of a large number of defects, namely vacancy islands and small pits [24,26]. All these effects are, however, only indirect but consistent indications for the supposed surface expansion; their influence on the gross morphology of the sulfate-covered Cu(111) electrode will be dealt with in greater detail in Sect. 3.6. Here we will first concentrate on a *direct* proof for the sulfate-induced expansion of the topmost copper layer, which is based on a comparison of EC-STM images taken with different tunneling parameters. This technique was used earlier to determine the absolute adsorption site of, e.g. chloride anions on Cu(111) [26]. STM images showing contributions from both the chloride and the copper lattice were subjected to a Fourier transformation procedure and bandpass filtered. After separate backtransformation of the individual contributions of copper and chloride a superposition of both structures (originating from one and the same STM image) clearly showed the threefold hollows of the Cu(111) surface being the chloride adsorption sites.

The same approach for the sulfate Cu(111)-system is demonstrated in Fig. 15. The STM images shown are part of a whole series recorded successively at the same sample potential ($E = 100\,\mathrm{mV}$) but using different tunneling parameters. Between two corresponding images the tunneling current is changed from 10 nA to 2 nA. In Fig. 15a the characteristics of the long-range Moiré modulation including the distorted hexagonal arrangement of the adsorbed molecules are visible as expected for this electrode potential. This molecular resolution, however, disappears in Fig. 15b, while the dimensions of the Moiré unit cell remain unchanged. The same experiment was repeated in Fig. 15c,d on an atomic scale. Again the disappearance of the adsorbed sulfate structure can be observed by changing the tunneling current from 10 nA to 2 nA. Instead a hexagonal structure with a much smaller lattice vector becomes visible. It is remarkable that this structure also shows exactly the same distorted long-range Moiré periodicity as in Fig. 15c. In order to emphasize more clearly the exact structural relationship between both structures a unit cell of the sulfate adlayer from Fig. 15c is superimposed with the hexagonal structure in Fig. 15d. This procedure reveals indeed a ($\sqrt{3} \times \sqrt{7}$)-like coincidence symmetry for the sulfate unit cell. Hence, it is very suggestive to interpret Figs. 15b,d as the STM images from the reconstructed copper surface (underneath sulfate) proposed in the second scenario. However, it must be pointed out that the underlying reconstructed (expanded) copper substrate as well as the sulfate structure do not exactly match a perfect hexagonal symmetry, but slightly deviate therefrom. Because the thermal drift is negligible in the sequence of STM images shown in Figs. 15c,d these small distortions become determinable. For the reconstructed (sulfate-covered) copper layer in Fig. 15d lattice constants of $a' = 0.26 \pm 0.01$ nm and $b' = 0.273 \pm 0.01$ nm are determined. Both lattice vectors enclose an angle of $56° \pm 1°$, i.e. 4° off the anlge of 60°. Hence, one can suppose that the slight distortion of the long-range Moiré superstructure originates from this distortion of the reconstructed first copper layer.

Further evidence for a commensurate sulfate structure on a reconstructed first copper layer can be obtained by using tunneling currents between the extreme limits of Figs. 15a,b. These STM images contain contributions from both the adsorbate and the substrate. From such a mixed STM image, a so-called power spectrum, derived from a two-dimensional Fourier transformation, of the original STM image is shown in Fig. 15e. Bright spots in the power spectrum represent lattice points in "reciprocal space" similar to a LEED picture in Low Energy Electron Diffraction in UHV. The complexity of the spectrum demonstrates the presence and the superposition of three different lattices in the real-space image which can now be analyzed separately. Starting with the lower frequencies the spots of the long-range Moiré periodicity are clearly visible near the center of the power spectrum (Fig. 15e). At higher frequencies spots are seen forming a distorted hexagonal arrangement which is assigned to the sulfate-derived contribution. Finally, in the high-frequency domain a hexagonal symmetry can be observed which is assigned to the reconstructed copper layer. A detailed analysis of this power spectrum reveals that these brighter spots assigned to the reconstructed copper exactly coincide with linear combinations of the spots assigned to the sulfate adsorbate. Using a LEED simulation program the reciprocal lattices of the sulfate adlayer as well as of the reconstructed copper layer are calculated in Fig. 15f under the assumption of a pseudo ($\sqrt{3} \times \sqrt{7}$) unit-cell coincident with the distorted copper layer as shown in Fig. 15c,d. All features of the original power spectrum in Fig. 15e are reproduced with the exception of the spots

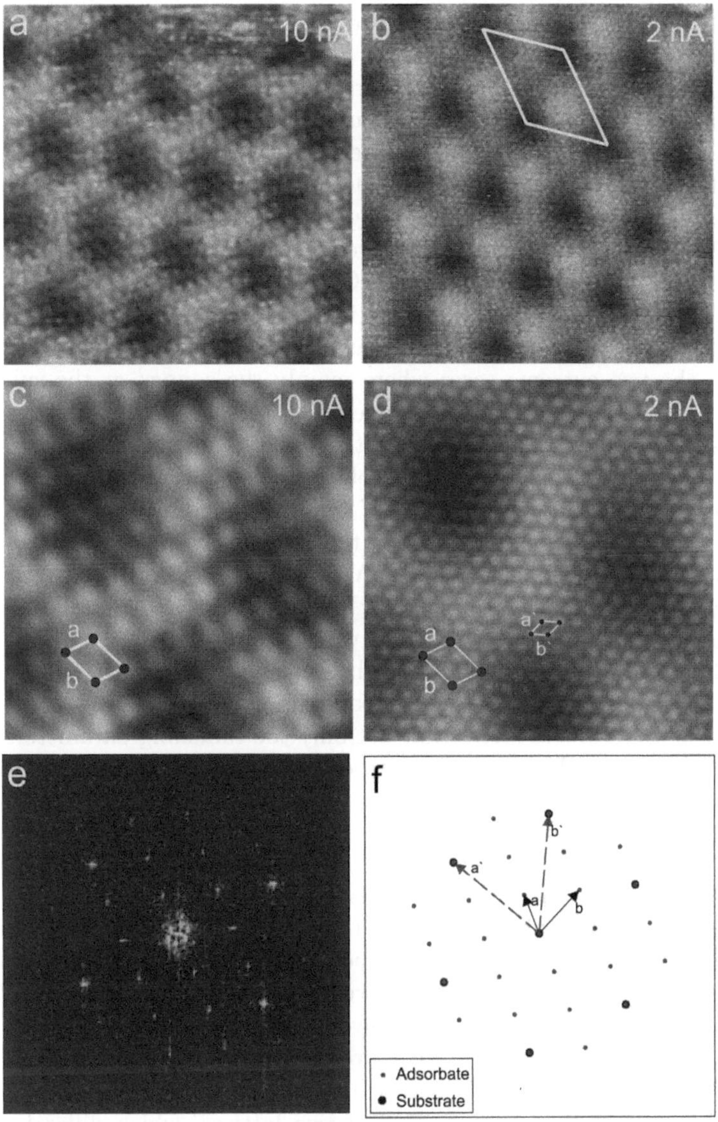

Fig. 15. Dependence of the sulfate-induced structure on Cu(111) as a function of tunneling conditions. The tunneling parameters of the STM images are: (**a**) image size 13.7 nm × 13.7 nm; bias voltage $U_b = -2\,mV$; tunneling current $I_t = 10\,nA$. (**b**) 13.7 nm × 13.7 nm; $U_b = -2\,mV$; $I_t = 2\,nA$. (**c**) 5.9 nm × 5.9 nm; $U_b = -2\,mV$; $I_t = 10\,nA$. (**d**) 5.9 nm × 5.9 nm; $U_b = -2\,mV$; $I_t = 2\,nA$. In all cases the electrode potential is $E = 100\,mV$ vs. RHE, negative potential sweep. Panel (**e**) represents a power spectrum of an STM image showing all three periodic contributions, namely the Moiré structure (near the center), the sulfate structure and the copper structure (outer hexagon). Panel (**f**) shows a simulation of the power spectrum without the Moiré structure (see text)

assigned to the Moiré periodicity which, however, were deliberateley left out of consideration for this procedure.

It must be pointed out that power spectra calculated from STM images of the sulfate adlayer taken with "normal" tunneling parameters as in Fig. 13 do not show the additional bright spots in the high-frequency range assigned to the reconstructed first copper layer. In fact, the given STM data reveal a very complex dependence of the STM imaging properties on a number of parameters: Apart from the well-known dependence on the *bias voltage* as already shown in [13] for a chloride-covered Cu(111) electrode, the imaging properties also significantly change by a variation of the *tunneling current* and therefore, of course, by a variation of the tip–sample distance. Figures 15a–d represent only the two extreme cases where mainly the copper *or* the sulfate contributions dominate. Choosing tunneling parameters between these extreme limits a continuous transition from the sulfate-dominated structure to the copper-dominated structure can be observed (see Sect. 3.2). An interesting detail is the actual constellation of the tunneling currents used to generate Figs. 15a–d. While the structure of the reconstructed copper shows up at low values of the tunneling current (2 nA, large tip–sample distance) the corresponding adsorbate image is obtained at higher values (10 nA, small tip–sample distance). This unexpected behavior, which seems to suggest a longer range of the Cu electron wave functions into the electrolyte than of the sulfate electron wave functions, awaits theoretical explanation.

However, it is necessary to stress that not only the tunneling parameters but also the potential at the working electrode plays an important role for these kinds of experiments. A variation of the working potential by just some 30 mV made such an approach impossible so that only the "usual" Moiré structure as in Fig. 13 could be imaged.

Summarizing so far, at positive electrode potentials sulfate anions are coadsorbed with water molecules on the Cu(111) surface forming a $(\sqrt{3}\times\sqrt{7})$-structure which is in registry with a reconstructed, i.e. expanded and slightly distorted, first Cu layer. The mismatch between this reconstructed first Cu layer and the unreconstructed second Cu layer is the reason for the occurrence of the Moiré superstructure. A structure model of this sulfate-induced reconstruction of the surface layer is sketched in the side view in Fig. 16a.

A Moiré arises from the superposition of two lattices, which differ by a lattice mismatch Δa and are rotated by an angle α, to form a third superlattice, namely the Moiré. This Moiré superlattice depends critically on Δa and α and can, thus, provide more precise structure parameters than a mere inspection of the STM images [24]. From the relative positions of particles in equivalent sites imaged at the same height, conditions can be formulated for the coincidence between the expanded first and the unreconstructed second Cu layer. The imaged anions are taken as markers for $\sqrt{3}$ and $\sqrt{7}$ positions on the expanded Cu layer. From images like the one shown in Fig. 13a it is obvious that between two depressions of the Moiré structure one has to travel

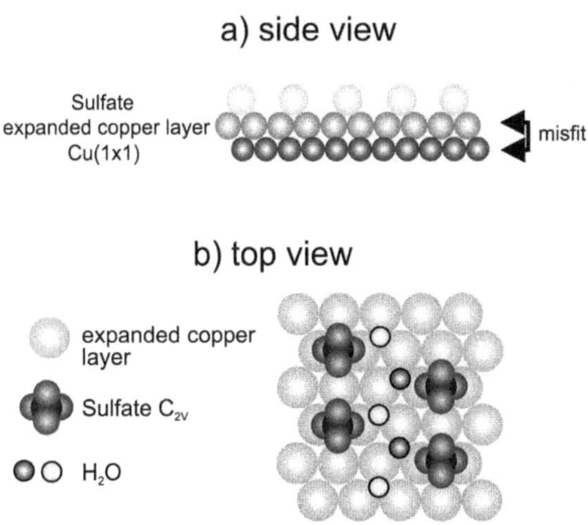

Fig. 16. Models of the sulfate-induced surface structure. (**a**) Side view showing the expansion of the first copper layer; (**b**) top view illustrating the twofold symmetry of the SO_4^{2-} adsorption geometry as deduced from Fig. 18

ten steps along a sulfate row in the $\sqrt{3}$ direction and one step in the perpendicular direction to reach the next close packed sulfate row. This means that the distance between two coincidence positions with the substrate in terms of the expanded Cu lattice constant a_{reconstr} is (see Fig. 14):

$$d = \sqrt{(10\sqrt{3})^2 + (5/2)^2} a_{\text{unreconstr}} \, . \tag{1}$$

The fact that the Moiré depressions are not perfectly aligned along the $\sqrt{3}$ direction of the sulfate layer leads to the conclusion that not only are the lattice constants of substrate and adsorbate different, but both lattices are also rotated with respect to each other. A priori this rotation angle as well as the stretching factor of the lattice constant are not known. A necessary condition for a repetition of equivalent lattice sites at distances d is that d can also be expressed by a combination of lattice vectors of the unreconstructed second Cu layer:

$$d = \sqrt{(n\sqrt{3})^2 + (m)^2} \tag{2}$$

where n and m are both either integers or half-integers.

Equating d from (1) and (2) yields possible values of n and m as well as the lattice expansion $a_{\text{reconstr}}/a_{\text{unreconstr}}$ and the rotation angle α between both lattices:

$$\alpha = \arctan \frac{m}{n\sqrt{3}} - \arctan \frac{5/2}{10\sqrt{3}} \, . \tag{3}$$

If a_{reconstr} is treated as a freely adjustable parameter, an infinite number of n, m pairs can be found; but with the reasonable restriction of that the ratio

$a_{\mathrm{reconstr}}/a_{\mathrm{unreconstr}}$ should only be a few percent, the number of solutions is limited. An additional restriction is imposed by the fact that the Moiré pattern is only very slightly rotated with respect to the $\sqrt{3}$ direction of the adsorbate as described above. Indeed, this condition reduces the number of possible solutions of (1) and (2) to two, namely:

$$\text{(i)}\quad m=\frac{1}{2},\quad n=10\frac{1}{2},\quad a_{\mathrm{reconstr}}/a_{\mathrm{unreconstr}}=1.040,\ \text{and}\ \alpha=-6.64^\circ$$

$$\text{(ii)}\quad m=3\frac{1}{2},\quad n=10\frac{1}{2},\quad a_{\mathrm{reconstr}}/a_{\mathrm{unreconstr}}=1.058,\ \text{and}\ \alpha=2.68^\circ$$

Taking $a_{\mathrm{unreconstr}} = 0.256$ nm for the second Cu layer we arrive at $\sqrt{3}a_{\mathrm{reconstr}} = 0.461$ nm and $\sqrt{7}a_{\mathrm{reconstr}} = 0.704$ nm in case (i) and $\sqrt{3}a_{\mathrm{reconstr}} = 0.469$ nm and $\sqrt{7}a_{\mathrm{reconstr}} = 0.717$ nm in case (ii). In both cases the calculated values are within the error limits of those obtained from direct inspection of the STM images, but the values for case (i) are closer to those measured directly between the sulfate particles. An expansion of the lattice constant by 4 % implies an increase of the area occupied by a fixed number of atoms by 8 % or, correspondingly, by a displacement of 8 % of atoms out of a fixed area. This is, indeed, verified in Sect. 3.6, further supporting the expansion model. Also the angle of 6.64° resulting from solution (i) above agrees reasonably with the large-scale images shown in Figs. 12 and 15.

3.4 Adsorption Site Configuration

So far we have only determined the symmetries and lattice parameters of the two-dimensional adsorbate and substrate layers. The obvious method of choice to determine the exact adsorption site and the molecular orientation of the sulfate anions would be in situ IR spectroscopy and in fact a lot of work has already been done in this respect [21,32,35]. In these studies, however, controversy exists about the symmetry of the adsorbed sulfate anions, which is reduced from T_d of the free anion to either C_{3v} or C_{2v} upon adsorption. For C_{2v} symmetry four surface-active infrared frequencies would be expected, two symmetric stretching modes $\nu_\mathrm{s}(SO_2)$ and two symmetric bending modes $\delta_\mathrm{s}(SO_2)$. In turn, for C_{3v} symmetry three bands are to be expected, $\nu_\mathrm{s}(SO)$, $\nu_\mathrm{s}(SO_3)$ and $\delta_\mathrm{s}(SO_3)$. Earlier in situ studies of sulfate adsorption on other fcc(111) surfaces, however, were hampered by the fact that not all of these frequencies could be observed making the assignment of the existing bands very difficult. For instance, *Edens* et al. found one strong potential dependent band in the region at 1150–1220 cm^{-1} and one weak band at 958 cm^{-1} on Au(111), and proposed sulfate being bound to the surface via a single oxygen atom in C_{3v} symmetry (monodentate coordination) [21,36]. *Ataka* et al. obtained the same results as Edens et al. on Au(111) but explained them by a three-fold coordination via three oxygen atoms of the sulfate in C_{3v} symmetry (tridendate) [32]. Finally, *Iwasita* et al. [35] investigated the sulfate adsorption on Pt(111) and found one band at 1220–1280 cm^{-1} with a weak shoulder near $1180\,cm^{-1}$. These results were interpreted by a C_{3v} adsorption symmetry with sulfate being coordinated to the surface via three oxygen atoms. The latter authors also studied the sulfate adsorption on Pt(100) and Pt(110), and found in both cases two bands, namely at $1203–1236\,cm^{-1}$ and $1105–1112\,cm^{-1}$, respectively. They assigned the higher frequency to the symmetric SO_2-stretching vibration of the uncoordinated oxygen atoms and the lower frequency to the SO_2-stretching mode of the coordinated oxygen atoms leading to a C_{2v} bridging (bidendate) coordination [37].

Our measurements on Cu(111) also show only one vibrational mode in the frequency range of $1205–1225\,cm^{-1}$ (see Fig. 17). This value is slightly higher than the high-frequency band observed on Au(111) and lower than the one on Pt(111) but close to the corresponding frequencies observed on Pt(100) [37]. For both (111) substrates the previous authors [21,32] suggested a C_{3v} adsorption symmetry for the sulfate anion on the basis of spectral data available for sulfate complexes, especially for cobalt sulfato complexes which were reported by *Nakamoto* et al. [38]. In these complexes the frequencies for the symmetric S-O stretching vibration shift to higher values with increasing coordination of the sulfate molecule. Even though coordination via three oxygen atoms is actually not present in these sulfato complexes the observed higher frequencies for *adsorbed* sulfate led the previous authors to conlude on a tridendate adsorption state, as an extrapolation of the observed shift in the cobalt complexes. But not only is the adsorbate coordination responsible for a frequency shift in sulfato complexes, also the nature of the metal atom in the center of these complexes has a significant influence. For example bidendate sulfato complexes of Cu, Ni and Co also show a shift of the S-O

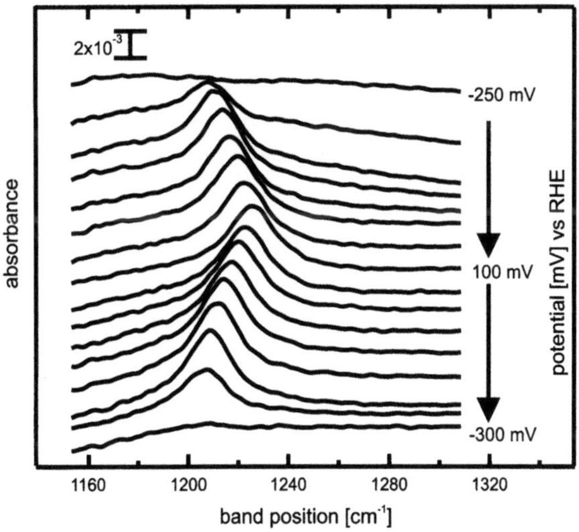

Fig. 17. Series of infrared spectra recorded at different electrode potentials between $-300\,mV$ and $100\,mV$ vs. RHE; background potential $-380\,mV$ vs. RHE

stretching vibration to higher frequencies [38] as the chemical identity of the core atom changes, although they exhibit the same bridging coordination via two oxygen atoms. This calls for caution when trying to determine the sulfate adsorption geometry on the basis of IR spectroscopic data alone.

More clear-cut information about the exact adsorption site and the geometry of the adsorption complex could actually again be obtained from high-resolution in situ STM measurements. Figure 18a shows a potentio-

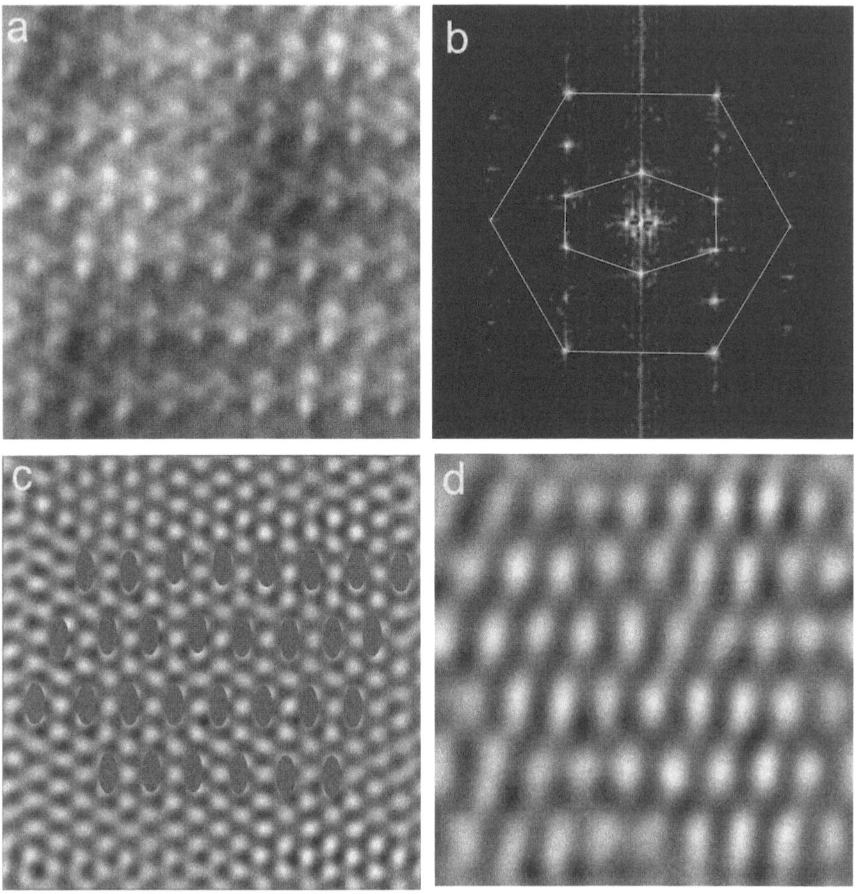

Fig. 18. (**a**) Moiré structure of the sulfate adlayer at anodic potentials; image size 4.4 nm × 4.4 nm; bias voltage $U_b = -130\,mV$; tunneling current $I_t = 40\,nA$; electrode potential $E = -120\,mV$ vs. RHE, negative potential sweep. (**b**) Power spectrum calculated from (**a**). (**c**) Separately backtransformed contribution of the copper substrate [outer hexagon in (**b**)]. (**d**) Separately backtransformed contribution of the sulfate adsorbate [inner hexagon in (**b**)]. A superposition of (**d**) and (**c**) (black dots) discloses the twofold symmetric adsorption site of the sulfate anions (see Fig. 16b)

static STM image of the sulfate-covered Cu(111) surface with high atomic resolution monitored at a potential of −120 mV. The tunneling parameters were chosen such that features of the adsorbate as well as of the substrate are visible in the same image. By use of a Fourier transformation procedure both contributions in Fig. 18a can easily be separated as demonstrated in Fig. 18b. The contributions originating from the copper substrate and the sulfate adlayer, respectively, are connected by lines in the two-dimensional Fourier spectrum. Both of these contributions are then separately backtransformed into real space using a band-pass filter. The resultant lattices are shown in Fig. 18c for the substrate and in Fig. 18d for the adsorbate structure. Since both lattices are retrieved from the very same STM image (Fig. 18a) their superposition unambiguously reveals a twofold bridging adsorption site for the sulfate molecules with a resulting C_{2v}-symmetry (Fig. 18c), in clear contrast to all earlier conclusions on the sulfate adsorption site symmetry based on IR spectroscopic measurements. The sulfate anions are suggested to be bonded via two oxygen atoms to two Cu atoms as sketched in the top view of Fig. 16b. Recent STM experiments of sulfate adsorption on Au(111) have also confirmed a twofold bridge position as the preferred sulfate adsorption site [39].

3.5 Adsorption/Desorption Kinetics

The clear separation of the potential ranges of sulfate adsorption and sulfate desorption in Fig. 6, respectively, indicates a strong hindrance of one or both of these processes. This fact allows one to control the rate of Moiré formation by choosing a certain constant adsorption potential [24,26]. In Fig. 19 a series of STM images shows the slow Moiré formation process [24]. Coming from a negative potential where no sulfate was present on the surface the potential was stepped to the constant potential of −70 mV and the same surface area was imaged with a delay time of 2 min between successive images. The first image (Fig. 19a) shows two flat terraces of Cu(111) with a monatomic step; no Moiré structure can be seen yet. In the second image (b) at the upper right and lower left corner small areas of the Moiré structure emerge which grow from the step edge onto the *upper* terrace, as indicated by the profile in Fig. 19f taken along the white line in Fig. 19b. They appear darker (lower) than the surrounding terrace due to their different electronic properties (see below). Note also that the same Moiré structure is already formed in these very small initial patches as it finally covers the whole surface (Fig. 19e). The negative charge of the sulfate anions and the resulting repulsive interaction should prevent this early aggregation. This behavior contradicts the well-known phenomenon of electrocompression which leads to a continuous compression of a complete and commensurate anion monolayer with increasing positive electrode potential like e.g. in the case of adsorbed iodide on Au(111) [40,41]. The unusual behavior of sulfate on Cu(111), instead, points to the particular stability of the Moiré structure and the screening effect of

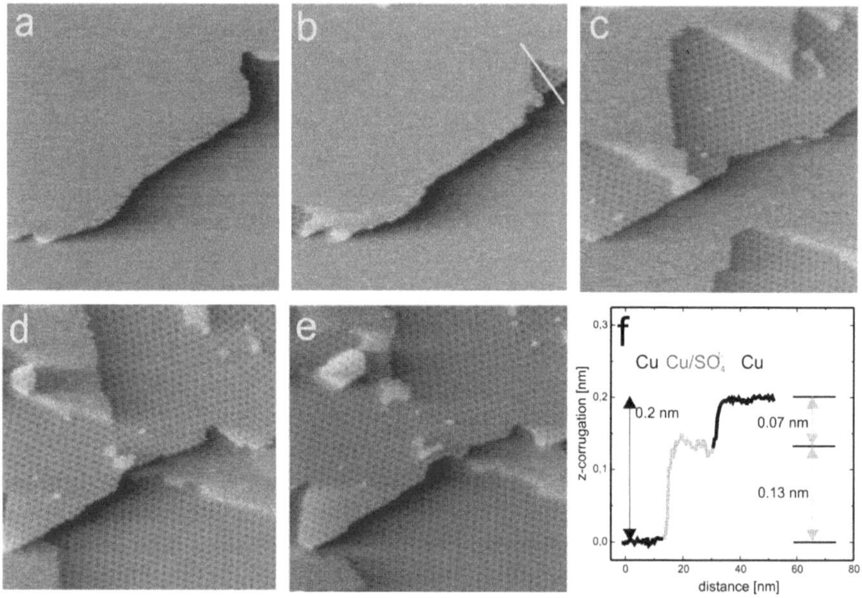

Fig. 19. (**a**)–(**e**) Series of STM images showing the sulfate-induced Moiré formation process. Tunneling parameters for all images: Size 101 nm × 101 nm; bias voltage $U_b = 169\,\text{mV}$; tunneling current $I_t = 1\,\text{nA}$; electrode potential $E = -70\,\text{mV}$ vs. RHE. Measuring time for the whole series was 10 minutes. Note the formation of a (Moiré covered) island in the upper left corner of (**d**) and (**e**). Panel (**f**) shows a height profile along the white line in (**b**) disclosing the onset of Moiré formation at the upper edge of a step (see text)

the coadsorbed water molecules between the sulfate anions. The following images (c)–(e) show the steady growth of this structure from the step edge onto the upper terrace. Starting in image c the Moiré structure is also growing in from the next step to the right (the latter not seen in the image). Furthermore at the end of the STM series in Figs. 19d–e the nucleation and growth of an additional Moiré covered island can be observed in the upper left corner. This island formation during the adsorption process is a consequence of the displacement of copper atoms out of the first copper layer which accompanies the reconstruction (expansion) of this layer. A more detailed description of these morphological changes during the reconstruction process is given in Sect. 3.6.

The STM images indicate the presence of the adsorbed sulfate only through the emergence of the Moiré and the $\sqrt{3} \times \sqrt{7}$-structure. They do not give clear indication whether the Moiré formation is preceeded by a disordered, probably mobile adsorption state of the sulfate anions. Possibly the rather (and increasingly) fuzzy appearance of the large Moiré-free terraces in images b and c in Fig. 19 may be a hint to the existence of such a mobile adsorption state.

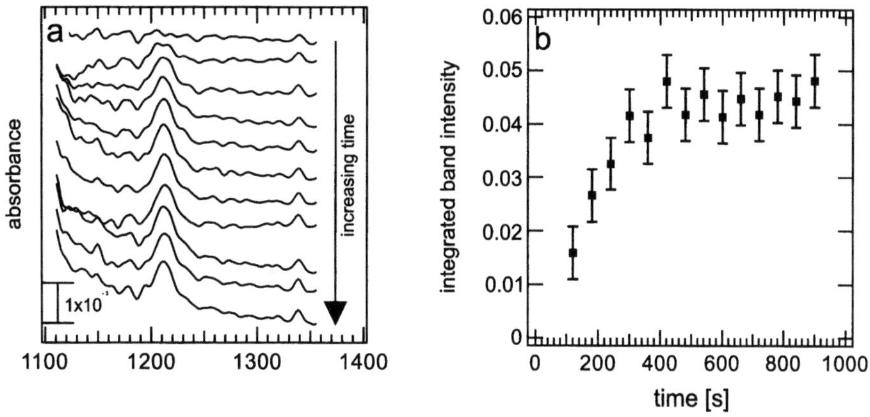

Fig. 20. (**a**) Series of infrared spectra recorded after potential steps from −350 mV to −200 mV vs. RHE as a function of time, background spectrum at −350 mV vs. RHE. (**b**) Integrated absorption band intensities from (**a**) plotted against adsorption time

In order to search for such a disordered stage of sulfate adsorption, series of IR spectra were recorded at potentials placing special attention on the potential region *before* the Moiré formation process sets in. The infrared spectra shown in Fig. 20a were recorded after a potential step from −350 mV, where definitely no sulfate is present on the surface, to a potential of −200 mV. At this fixed potential spectra were then recorded continuously for about 15 minutes with a total recording time of 60 s per spectrum in order to achieve a good signal-to-noise ratio [42].

In the chosen potential region clearly no formation of the Moiré structure is imaged by STM yet. However, in the infrared spectra the formation of a small absorption band at about 1210 cm^{-1} can already be observed. This absorption band belongs to the symmetric S-O stretching vibration of adsorbed sulfate [43,44]. In Fig. 20b the integrated intensity of this absorption band is shown as a function of adsorption time. The intensity, i.e. the sulfate coverage, grows continuously up to about 6 minutes adsorption time, and then saturates within the accuracy of the measurements.

In order to cover the complete adsorption/desorption regime, series of IR spectra were registered by changing the electrode potential in steps of 50 mV from −300 mV to 100 mV and back to −300 mV. The background spectrum (of the sulfate-free surface) was taken at −380 mV, and the recording time was 4 minutes per spectrum. The evolution of the sulfate absorption band and the well-known Stark shift are clearly seen in Fig. 17. In Fig. 21 the relative integrated intensity (a) and the band position (b) are plotted as a function of the applied potential. The Stark shift has the same slope, 55 cm^{-1}/ V, as previously reported [43], and Fig. 21a manifests a clear hysteresis in the change of the integrated intensity in the anodic direction (adsorption) ver-

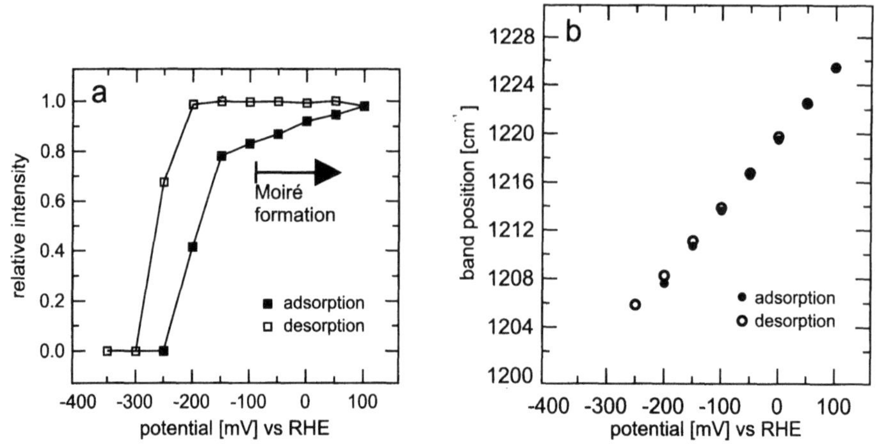

Fig. 21. (**a**) Integrated band intensities of the absorption bands in Fig. 17 plotted against the electrode potential. The onset potential of Moiré formation is indicated by the arrow. (**b**) Potential dependence of the S–O stretching vibration frequency (STARK shift)

sus cathodic direction (desorption). In the anodic run the sulfate adsorption starts at a potential between −250 mV and −200 mV and the coverage continues to increase somewhat over the whole range towards positive potentials. Conversely, in the cathodic run the desorption process sets in at potentials clearly lower than −200 mV, and sulfate is still present on the surface between −200 mV and −300 mV.

These results prove that the sulfate adsorption starts at potentials considerably more negative than does the Moiré formation. In this initial stage the adsorbed sulfate molecules seem to be very mobile on the surface and can therefore not be imaged directly by STM (except by the unspecific fuzzyness in images like 19b and c mentioned above).

Turning our attention now to the cathodic run of the IR series shown in Fig. 21 it is apparent that the desorption process is not kinetically hindered; in contrast to the adsorption process the complete desorption occurs in a narrow potential window. This finding is in full agreement with the decay of the Moiré structure and the sulfate desorption behavior as observed by STM [26]. The series of images in Fig. 22 is representative for this desorption behavior. In Fig. 22a the surface is almost completely covered by the Moiré structure. The brighter areas on the terraces correspond to defects and disordered regions in the sulfate adlayers (see also Fig. 19). As can be seen in the next images (Fig. 22b–i) the decay of the Moiré structure starts at these defects as well as at the lower step edges and proceeds to the upper step edges of the respective terrace until the Moiré has totally vanished. The persistance of small Moiré patches until the very end of the desorption process again points to the unusual stability of this structure. Hence, by comparing the STM and FTIR

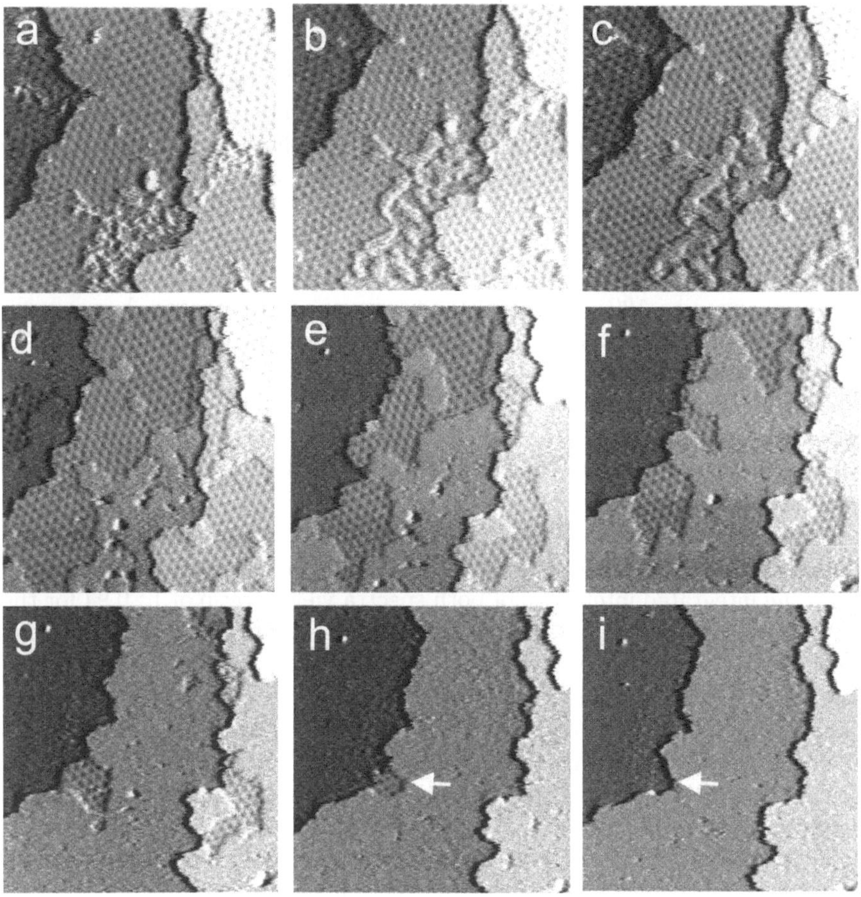

Fig. 22a–i. Series of STM images (differentiated) showing the decay of the Moiré adlayer; total recording time 15 minutes. Note the persistance of small Moiré patches (at upper step edges) until very small remnant sulfate coverages (panel h). Tunneling parameters for all images: Size 84 nm × 84 nm; bias voltage $U_b = 40$ mV; tunneling current $I_t = 1$ nA; electrode potential $E = 210$ mV vs. RHE

data one can conclude that, unlike the adsorption process, in the cathodic run the sulfate molecules are directly desorbed from the surface after lifting the reconstruction. The Moiré formation is a kinetically hindered slow process due to the massive mass transport and reordering in the copper surface, whereas the desorption occurs in a relative small potential window. (The slow healing process of the copper surface *following* the sulfate desorption (see Fig. 26) has no more influence on the sulfate desorption itself.)

Due to the strong kinetic hinderance of the Moiré formation the adsorption rate and, hence, the value of the electrode potential at which the adsorption takes place should have a significant influence on the final order of the sulfate overlayer. This can be verified by STM measurements which were done after a sudden potential change (potential step) [24,42]. Figure 23 shows an image of the sulfate-covered surface after such a fast adsorption process. Only parts of the surface are covered with an ordered Moiré structure, whereas the rest of the surface is covered with a disordered sulfate adlayer. Since under these circumstances STM is again unsuited to quantify the adsorbate coverage infrared spectroscopy can help. If the adsorption procedure is conducted such that the potential is changed slowly from $-350\,\mathrm{mV}$ (where no sulfate is present at the surface) to $-70\,\mathrm{mV}$, held there for 10 minutes and then slowly changed to $100\,\mathrm{mV}$ in the anodic region, STM images show a well-ordered sulfate adlayer. The corresponding infrared spectrum is presented in Fig. 24 (upper spectrum). If, however, starting from the same potential of $-350\,\mathrm{mV}$ as before, a sudden potential *step* to $100\,\mathrm{mV}$ is applied a disordered structure results as shown in Fig. 23. The related IR spectrum taken after a pause of only 10 s is also displayed in Fig. 24 (lower spectrum). Both spectra obtained after these different adsorption procedures show the same band position which suggests the same local electric field at the sites of individual sulfate molecules and, hence, essentially the same adsorbate density. Furthermore the integrated intensities of both spectra yield a ratio of 1 : 0.94 for the highly ordered versus the less ordered sulfate adlayer. Since the adsorption time for the barely ordered adlayer was much shorter than the adsorption time for the ordered adlayer the intensity ratio of 1 : 0.94 is even the upper limit for the difference in coverage of both adlayers. It must therefore be concluded that only the order of the sulfate adlayer, i.e. the Moiré formation, depends on the adsorption rate but the coverage basically does not. This is very plausible in the light of the expansion of the first Cu layer and the concomitant displacement of Cu atoms, which leads to a severe restructuring of the surface morphology as described in the following section.

3.6 Surface Morphology

The sulfate-induced expansion of the first copper layer and the concomitant displacement of Cu atoms is not only the explanation for the slow growth kinetics of the Moiré structure but also leads to a variety of morphological changes at the surface [24,45]. These morphological changes, in turn, provide further support for the previously made assumption of a very strong sulfate–copper interaction as being the driving force for the surface restructuring. This was already concluded from the existence of very small Moiré patches at low sulfate coverages as seen in Figs. 19b and 22h.

Figure 25 gives a clear impression of these intensive changes; upon adsorption of sulfate the surface gets very rough and is dotted with many small islands. Conversely, upon rapid desorption the surface is left with a high density of vacancy islands and holes (Fig. 26a), which heal out as time goes by (Fig. 26b,c). The island formation is a consequence of atomic displacement from the expanding first copper layer. The holes and vacancy islands are

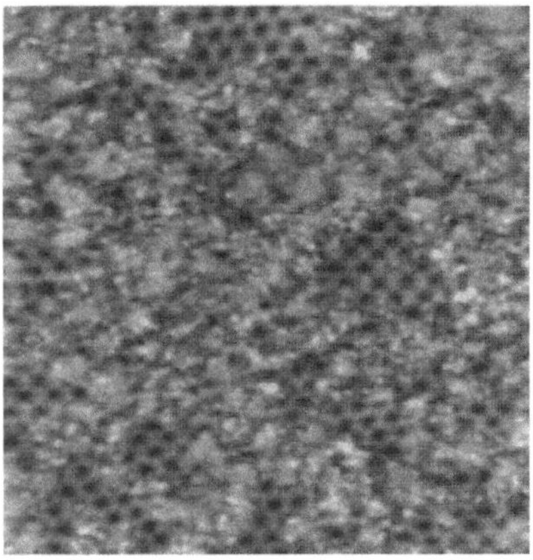

Fig. 23. Less-ordered sulfate adlayer after potential step induced adsorption. Image size 64 nm × 64 nm; bias voltage $U_b = -100\,\mathrm{mV}$; tunneling current $I_t = 1\,\mathrm{nA}$; electrode potential (after step) $E = 210\,\mathrm{mV}$ vs. RHE

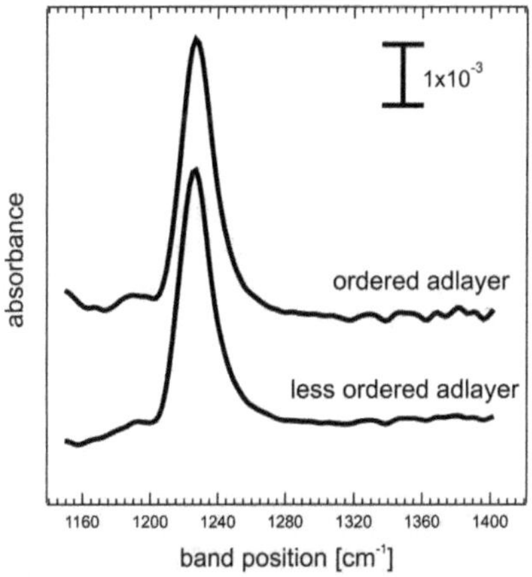

Fig. 24. Infrared spectrum of a well-ordered sulfate Moiré structure at 100 mV vs. RHE and of a less-ordered sulfate Moiré structure after a potential step from −350 mV to 100 mV vs. RHE, respectively. Background spectra were registered at −350 mV vs. RHE in both cases, and the difference in structural order is reflected by Figs. 11a and 23

a result of the sudden shrinkage of the previously expanded Cu layer to its normal structure.

A quantitative measure of the mass transport accompanying the sulfate-induced expansion of the first Cu layer can be obtained from an STM image covering a large surface area. Figure 27a shows an adsorbate-covered surface area of 373 nm × 373 nm. On this scale the Moiré structure is no longer visible. Figure 27b shows the same area after desorption of the sulfate. A compar-

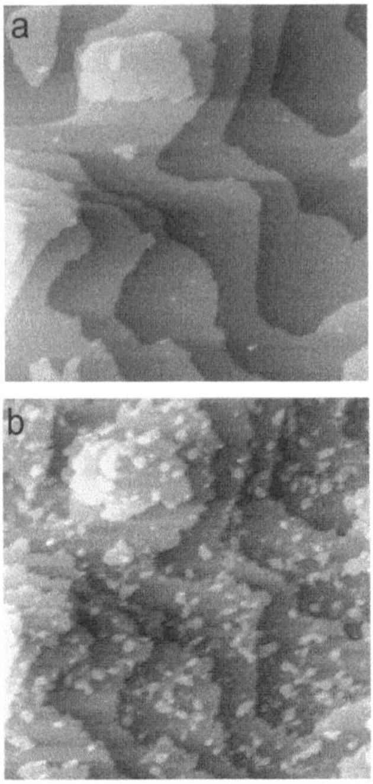

Fig. 25. Morphological changes induced by sulfate adsorption on Cu(111) in 5 mMH_2SO_4 solution. (**a**) Before sulfate adsorption; image size 340 nm × 340 nm; bias voltage $U_b = -20$ mV; tunneling current $I_t = 1$ nA; electrode potential $E = -266$ mV vs. RHE, positive potential sweep. (**b**) After sulfate adsorption; 340 nm × 340 nm; $U_b = -20$ mV; $I_t = 1$ nA; $E = 155$ mV vs. RHE, positive potential sweep

ison of both pictures reveals that the total Cu area is about 5 % larger in Fig. 27a, that is with sulfate being adsorbed [24]. This value of 5 % enlargement is somewhat smaller than the value of 8 % estimated from the structure parameters of the Moiré layer in Sect. 3.1, but the discrepancy is certainly within the accuracy limits.

It appears obvious that the sulfate-induced dilatation of the Cu surface layer and the related relaxation of compressive strain is easiest at step edges. This explains why the Moiré formation process starts at the *upper* edge of steps and proceeds onto the upper terrace, which can expand towards the lower terrace. This can clearly be seen in Figs. 19b,f; the Moiré structure is on the *upper* terrace. This channel of strain relief, however, becomes blocked as the Moiré patches increase. The further growth of the Moiré structure then requires the actual displacement of Cu atoms out of the first Cu layer and either their incorporation at existing step edges or their coalescence into nuclei followed by island growth. This is seen in the upper left corner of

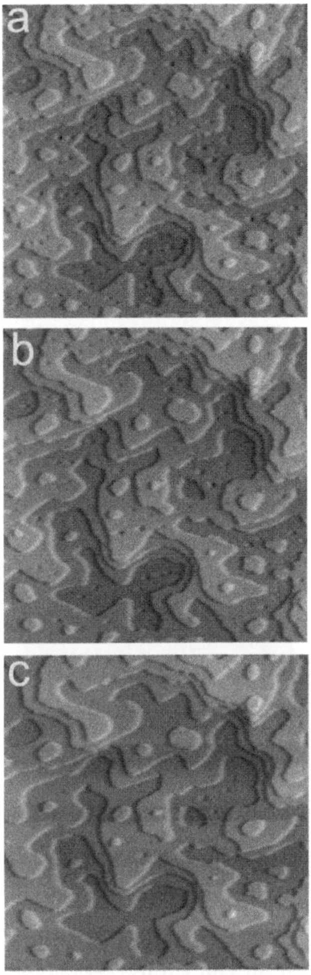

Fig. 26a–c. Morphology of the Cu(111) surface after desorption of the sulfate adlayer as a function of time; recording time for all three images 15 minutes. Tunneling parameters for all images: Size 256 nm × 256 nm; bias voltage $U_b = -41$ mV; tunneling current $I_t = 10$ nA; electrode potential $E = -210$ mV vs. RHE, positive potential sweep

Figs. 19d,e. Inasmuch as the Moiré pattern spreads out, the bright island grows.

A closer look at the evolution of this island at constant adsorption potential of −130 mV is shown in Fig. 28 and discloses further details of the growth process as well as of the imaging properties. Three different regions can be distinguished in these STM images. Firstly, the Moiré-covered island which is imaged brightest and which grows successively from Figs. 28a–d. Secondly, the dark and structured areas in Figs. 28a–d, which represent the

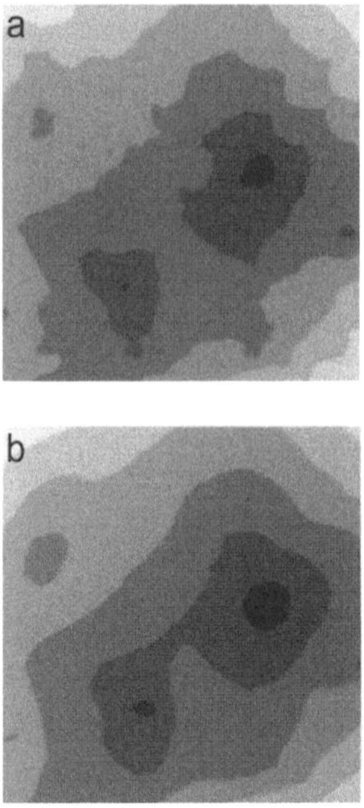

Fig. 27. Comparison of the Cu(111) surface topography before (**a**) and after (**b**) sulfate desorption. Tunneling parameters in both images: Size 337 nm × 337 nm; bias voltage U_b = 100 mV; tunneling current I_t = 1 nA. Electrode potentials (**a**) 160 mV and (**b**) −200 mV vs. RHE

reconstructed part of a copper terrace as indicated by the clearly visible Moiré superstructure. Thirdly, the greyish and diffuse areas which result from the unreconstructed part of the same terrace covered with disordered sulfate as proven in Sect. 3.5. This latter disordered layer of probably mobile sulfate is always imaged brighter than the Moiré structure on the same terrace, independent of the tunneling parameters. On the basis of pure geometrical arguments the ordered sulfate adsorbate on the expanded copper underlayer should be higher than the corresponding areas of disordered sulfate on non-reconstructed copper and should thus be imaged somewhat brighter. The fact that this is not the case points to an electronic effect, namely a reduced density of states at the Fermi level within the reconstructed surface layer which is actually imaged with STM in the constant current mode. However this reasoning needs to be confirmed by future theoretical calculations.

Coming back to the growth of the (bright) island in Figs. 28a–d it can be seen that the island grows in units of the Moiré structure at the expense of the disordered surrounding adlayer. Inasmuch as the disordered sulfate-covered copper patch turns in an ordered Moiré structure, thereby setting free Cu atoms, these displaced Cu atoms are incorporated at the edge of the island. Note that from image b to d thc island grows only towards the disordered

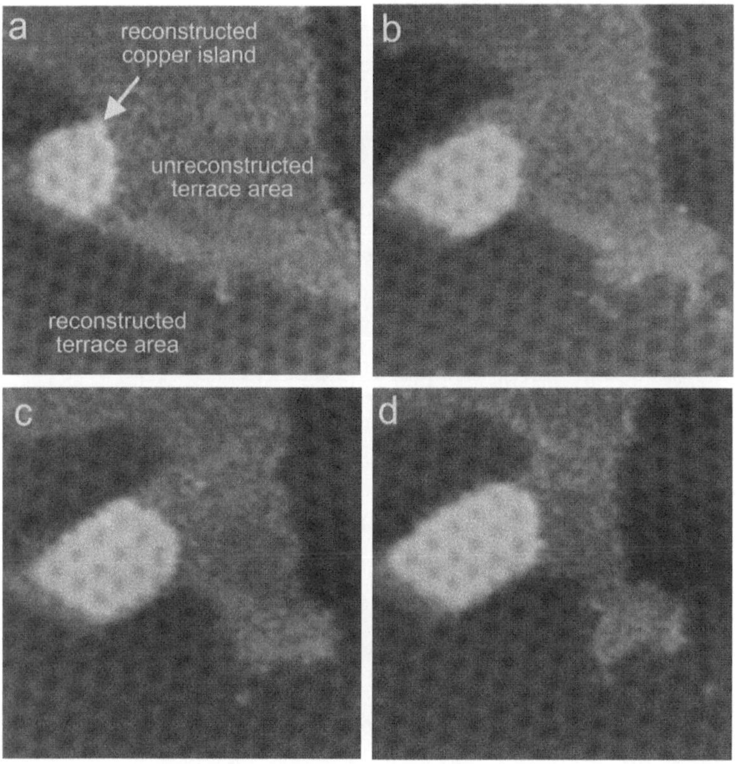

Fig. 28a–d. Direct correlation between the sulfate-induced formation of the Moiré structure and the nucleation and growth of a copper island. Note the growth of the island while the unreconstructed terrace area converts into the Moiré structure. Tunneling parameters in all images: Size 42.4 nm × 42.4 nm; bias voltage U_b = 169 mV; tunneling current I_t = 1 nA; electrode potential E = −129 mV vs. RHE, positive potential sweep

surface region. Furthermore nucleation and growth of islands has never been observed on an already Moiré-covered area. The reservoir of Cu atoms contributing to the growth of an island are the adjacent not yet reconstructed surface regions with disordered sulfate. Obviously, the small disordered halo on the left side of the island in Fig. 28a provides just enough Cu atoms upon reconstruction to add one (and only one) more Moiré unit at the left corner of the island (compare a and b). This leads us to another clear and very interesting finding: The islands grow only in units of the Moiré structure; only complete Moiré units are stable and are added to an existing island. This is emphasized by Fig. 29 which presents a whole collection of islands of different size and shape. The smallest island corresponds to just one single Moiré unit (Fig. 29d, lower left corner), and all islands are "quantized" in shape and size in units of the Moiré structure. This underlines the strong

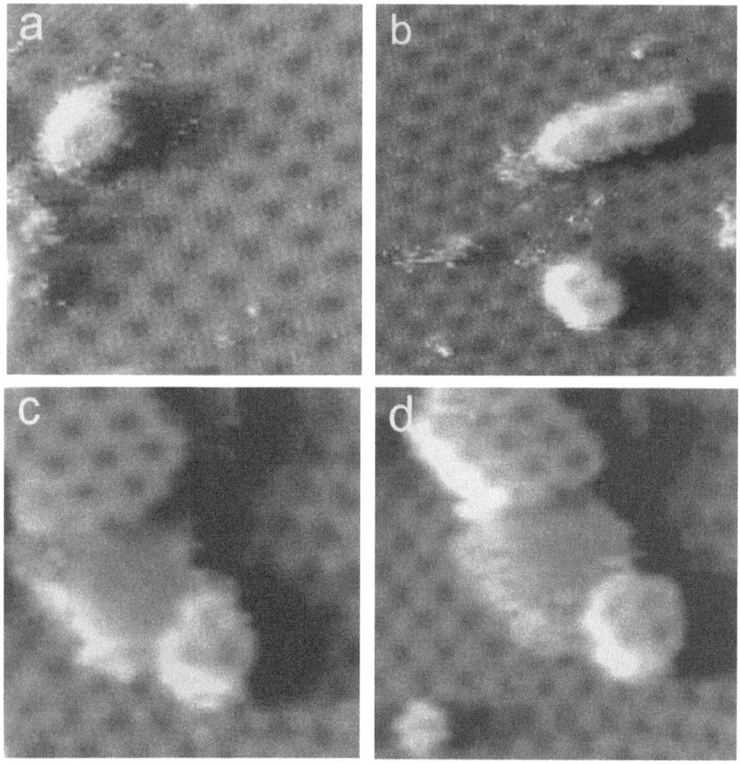

Fig. 29. Sulfate-stabilized copper islands on the Cu(111) electrode. Note the "quantization" of all islands in units of the Moiré structure. Imaging parameters: (**a**) Size 23 nm × 23 nm; bias voltage $U_b = -70\,mV$; tunneling current $I_t = 1\,nA$; electrode potential $E = 140\,mV$ vs. RHE. (**b**) 31 nm × 31 nm; $U_b = -70\,mV$; $I_t = 1\,nA$; $E = 140\,mV$ vs. RHE. (**c**) 19 nm × 19 nm; $U_b = -70\,mV$; $I_t = 1\,nA$; $E = -80\,mV$ vs. RHE. (**d**) 21 nm × 21 nm; $U_b = -70\,mV$; $I_t = 1\,nA$; $E = 140\,mV$ vs. RHE. All E-values refer to a positive potential sweep

sulfate–copper interaction and makes clear that one Moiré unit is the critical nucleus. Incomplete Moiré units have never been detected[1].

The stability of the Moiré unit as a whole also manifests itself by looking closely at step edges of the Moiré-covered surface. Steps on the sulfate-covered Cu surface are oriented parallel to the symmetry axes of the *Moiré structure* (Fig. 30) and not parallel to close-packed adsorbate rows (Fig. 31). Kinks correspond to the end of a chain of complete Moiré units, again demonstrating that the Moiré unit is the stable building bloc.

[1] A similar situation is known from the MBE growth of the Si(111) 7 × 7 structure under UHV conditions; its growth proceeds also only in full (7 × 7)-units [see e.g. B. Voigtländer, T. Weber: Phys. Rev. Lett. **77**, 3861 (1996)]

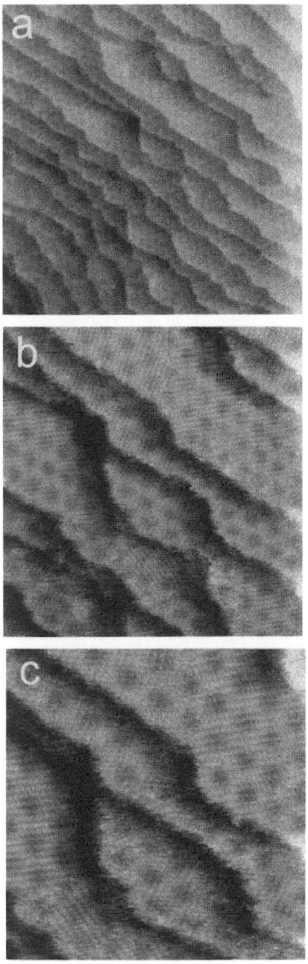

Fig. 30. Step structure and orientation on a sulfate-covered Cu(111) surface in 5 mH H_2SO_4 solution at a potential of $E = 140\,\text{mV}$ vs. RHE. Steps and kinks are formed by Moiré units (see text). Imaging parameters: (**a**) Size 102 nm×102 nm; bias voltage $U_b = 166\,\text{mV}$; tunneling current $I_t = 1\,\text{nA}$. (**b**) 43 nm×43 nm; $U_b = 166\,\text{mV}$; $I_t = 1\,\text{nA}$. (**c**) 29 nm × 29 nm; $U_b = 166\,\text{mV}$; $I_t = 1\,\text{nA}$

This type of "faceted" step differs from halogenide stabilized copper steps which are found to be preferentially oriented parallel to close-packed adsorbate rows [26,27,46,47,48,49,50]. Also the formation mechanisms leading to this sulfate-induced "step faceting" are significantly different. The halogenide stabilized steps can be seen as the result of local corrosion and redeposition processes taking place at positive electrode potentials near the onset of copper dissolution [26,32,46,50]. Under these conditions the anionic adlayer is already completely closed. Halogenides such as chloride and bromide are well known to enhance the surface mobility due to their strong complexing tendency lead-

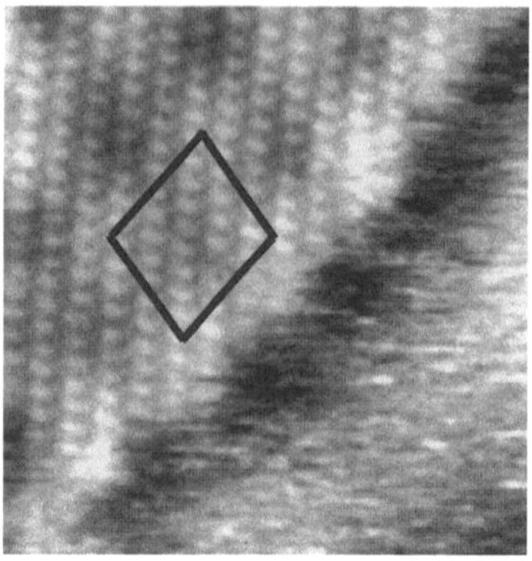

Fig. 31. Sulfate-stabilized copper step on the Cu(111) electrode in 5 mMH_2SO_4 solution. Note that only the upper terrace is covered by the Moiré structure but not the lower one (compare Fig. 19), and that the step runs parallel to the rows of Moiré units (not of the sulfate rows); one Moiré unit is marked. Imaging parameters: Size 13 nm × 13 nm; bias voltage $U_b = -70$ mV; tunneling current $I_t = 1$ nA; electrode potential $E = -80$ mV vs. RHE, positive potential sweep

ing to a rapid healing of surface defects. This phenomenon is also known from other substrates in the presence of halogenides and has been called "electrochemical annealing" [51]. The massive mass transport that is required for the anion-induced morphological surface restructuring takes place via surface diffusion or via diffusion through the solution phase depending on the applied potential [26,52].

In contrast to the situation in halogenid-containing electrolytes the "faceted" copper steps in the well-ordered sulfate adlayer on Cu(111) are a result of the growth of copper terraces and islands along preferential directions during the reconstruction process. A sulfate-enhanced restructuring of the surface morphology at positive electrode potentials as observed in the case of fully halogenide covered surfaces no longer takes place once the Moiré phase is completely closed. In contrast to the halogenides, adsorbed sulfate drastically decreases the surface mobility of copper. One reason for this effect is probably the low complexing tendency of sulfate anions. The role of adsorbed sulfate can be seen as an inhibitor and not as a promoter of dynamic processes. Probably adsorbed sulfate molecules block the kink sites at lower step edges which are the active reaction sites for local corrosion and redeposition reactions [46,50].

In general the length scale of surface reorganization is much larger in the case of the halogenide-covered surfaces than found for the sulfate-covered Cu(111) surface. In the case of halogenides straight steps of up to several hundred nanometers long and aligned parallel to the main symmetry axes of the adlayer have been observed [27,46,47,50]. Instead, on the sulfate-covered Cu(111) surface the step alignment is more a local phenomenon which is restricted to a few nanometers only. As a consequence the surface morphology of a sulfate-covered copper surface (see Fig. 25b and Fig. 30a) does not reveal characteristic symmetries on a larger scale from which one could conclude the adsorbate lattice symmetry on the atomic scale, in contrast to the situation with Cu(100) and Cu(111) surfaces in the presence of chloride and bromide [27,46,47,50].

Another interesting difference in the morphological behavior is the formation of multisteps in the presence of adsorbed halogenides which has not been observed for the sulfate-covered Cu(111) surface.

It must be stressed that the morphological changes occurring during the sulfate-induced reconstruction of the Cu(111) surface depend on several parameters. The most important one is the sulfate adsorption rate and, as a consequence, the Moiré formation rate. After a very slow sulfate adsorption experiment, the electrode surface is not significantly roughened in contrast to the experiment shown in Fig. 25. The low number of copper atoms which is displaced (per second) from the topmost copper layer during a slow reconstruction process can diffuse over the terraces and is eventually incorporated at the next step edge. Hence, mainly the copper steps are affected in this case which leads to a kind of step flow mechanism. Only if the diffusion of copper material to the next step edge is strongly hindered one does observe the nucleation of new copper islands on terraces. This happens if a fast-growing Moiré structure blocks the next lower step edges or if the growing Moiré encloses partially or completely an area which is not yet reconstructed. In these cases one can observe the nucleation and growth of copper islands even at low Moiré formation rates. Such an exceptional case was demonstrated in Fig. 28. It underlines the importance of local effects in these morphological changes mentioned above.

As pointed out before and demonstrated in Fig. 26 sulfate *desorption* again has a great impact on the morphology of the surface left behind. Since the reconstruction of the copper surface at positive potentials is stable only in the presence of the ordered sulfate/water overlayer a decay of this structure followed by the sulfate desorption automatically leads to the lifting of the surface reconstruction and a restoration of the normal Cu(111) density. This restoration is accompanied by the inverse processes which lead to the reconstruction: At first and on a short length scale the previously expanded copper layer just shrinks thereby creating many vacancy islands. Subsequentely transport of copper material back into the first Cu layer sets in, and as a consequence small copper islands disappear, larger islands shrink, terraces recede and step edges become smoother. This is a very slow process; even after 15 minutes (Fig. 26a–c) this healing process is not completed and some vacancy islands are still visible in Fig. 26a.

A sequence of adsorption and desorption cycles, and in particular the slow processes of surface reconstruction and island formation on the one hand, and of the healing after lifting the reconstruction and creation of vacancies

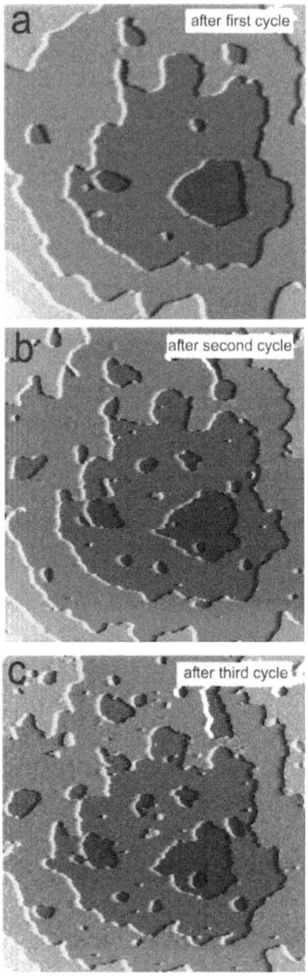

Fig. 32. STM images (differentiated) showing the surface roughening induced by repeated cycling of the electrode potential. Imaging parameters: (**a**)–(**c**) Size 353 nm × 353 nm; bias voltage $U_b = -130\,mV$; tunneling current $I_t = 10\,nA$; electrode potential $E = 300\,mV$ vs. RHE, positive potential sweep

on the other, eventually leads to a significant roughening of the surface. Figure 32 displays three images demonstrating this roughening effect. Between two STM image one complete potential cycle was carried out with a scan rate of 10 mV/ s, during which the tunneling tip was drawn back out of the tunneling region in order to exclude any influence from the tip on the dynamics of the surface processes. Already after three potential cycles the surface morphology is significantly rougher due to the appearance of new islands and holes and by changes of the step edges. After the three potential cycles the step edges are much less smooth than at the beginning of the experiment [45].

3.7 Anodic Corrosion

In order to have a full characterization of the electrochemical properties of an electrode not only are the potential ranges in which anion adsorption and desorption take place important but also the anodic and cathodic limits of the CV in the specific electrolyte. However, under these extreme electrochemical conditions the use of the in situ STM technique becomes more difficult because of the large faradaic currents due to chemical reactions starting at the electrode surfaces. Here we will first concentrate on the anodic corrosion regime. The next section describes the phenomena which occur in the potential range of hydrogen evolution.

For the Cu(100) electrode surface in acidic media *Vogt* et al. [46] investigated very carefully the influence of several parameters on the mechanism of copper corrosion such as the etch rate, the presence of impurities at the surface and the atomic structures. Especially the initial stage of a slow copper corrosion was a subject of this study. For a dilute sulfuric acid electrolyte as well as for a dilute hydrochloric solution a step-flow etching is observed for the incipient copper corrosion on the Cu(100) surface. The same authors also reported more recently [53] a step-flow mechanism at the early stages of copper corrosion for a sulfate-covered Cu(111) electrode.

Figure 33 shows the topography of a Cu(111) surface after applying large etching rates. Current densities of about $0.05\,\mathrm{mA/cm^2}$ were applied for several seconds. Under these more drastic etching conditions the step-flow mechanism of the early stages of corrosion turns rapidly into a pit etch mechanism. The STM images in Fig. 33 were taken in the cathodic potential sweep after the corrosion experiment. Clearly visible is the massive formation of etch pits. Pit depths of up to 2 nm can be measured. Another interesting feature is seen in Fig. 33b: Even on this smaller scale the sulfate Moiré structure expected for these potentials is no longer observed. The whole surface appears very rough and disordered *after* the dissolution and redeposition procedure. Single monatomically high copper steps can be seen in Fig. 33a, but all terraces are covered with undefined amorphous-like particles (Fig. 33b). No atomic resolution is obtained on this modified copper surface although the atomic structure of the sulfate Moiré adlayer was clearly seen *before* the corrosion experiment.

Similar behavior was observed for fast copper dissolution in hydrochloric solution [46]. In this case the disordered phase was explained by the formation of amorphous Cu_xCl_y which replaces the ordered c(2×2)-Cl adlattice. One can suppose that a similar formation of amorphous copper sulfate may be responsible for the features seen in Fig. 33b. However, this is a hypothesis which calls for ex situ measurements and spectroscopic verification of the exact chemical identity of the observed surface phase.

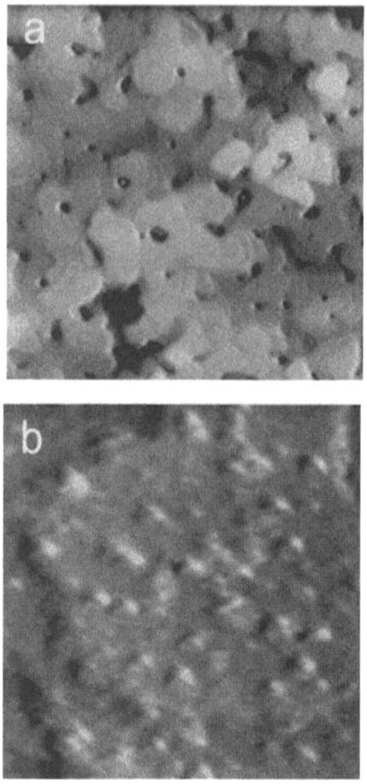

Fig. 33. (**a**) Massive surface roughening and formation of etch pits after anodic copper corrosion (current density 0.05 mA/ cm^2) and redeposition in 5 mMH_2SO_4 solution. The close-up (**b**) shows no more ordered Moiré structure. Imaging parameters: (**a**) Size 540 nm × 540 nm; bias voltage U_b = 40 mV; tunneling current I_t = 10 nA. (**b**) 44 nm × 44 nm; U_b = 40 mV; I_t = 10 nA. The electrode potentials is E = 170 mV vs. RHE, negative potential sweep, in both images

3.8 Cationic Structures

After desorption of the sulfate layer there is only a narrow potential range in which the bare Cu(111) structure can be seen (Fig. 9). This situation, however, is not stable; at sufficiently negative potential the adsorption of a further species sets in [54] leading to the formation of an ordered adsorbate layer (Fig. 34). This adsorption and adlayer formation process becomes faster with increasingly negative electrode potential. First, single particles are observed at $E = -220$ mV which in a further step aggregate and form small adsorbate clusters as seen in Fig. 34a. Figure 34b shows that already at submonolayer coverage this new adsorbate phase has formed a triangular island, and that there is a clear symmetry relation between the adsorbate and the substrate lattice, namely, that the close-packed adsorbate rows are aligned parallel to the main symmetry axes of the copper substrate (Fig. 34b). There is no measurable rotation between the close-packed rows of the adsorbate and the substrate, respectively. Changing the electrode potential to even more negative values the adlayer formation proceeds until the surface is almost fully covered with this new adsorbate layer. Only locally can one find small defects

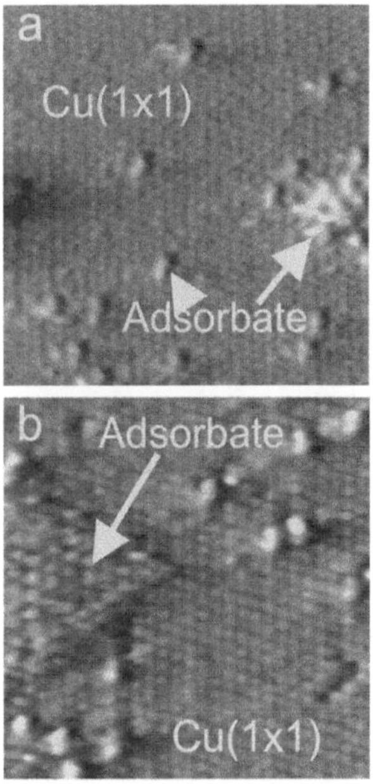

Fig. 34. Adsorption of a cationic species (probably hydronium; see text) in the regime of hydrogen evolution on the sulfate-free, unreconstructed (1×1) Cu(111) surface. Imaging parameters: (**a**) Size 9.1 nm $\times$ 9.1 nm; bias voltage $U_b = -180$ mV; $I_t = 20$ nA. (**b**) 7.6 nm$\times$7.6 nm; $U_b = -180$ mV; $I_t = 50$ nA. The electrode potential is $E = -220$ mV vs. RHE, negative potential sweep, in both cases

or vacancy islands within this adlayer as demonstrated in Fig. 35. Very often, these vacancy islands show a hexagonal or triangular shape with adsorbate step edges preferentially aligned parallel to the adsorbate main symmetry axes. On the atomic scale these step edges appear rather smooth and defect free. Interestingly, single adsorbate particles which exist within such an adsorbate vacancy island are imaged significantly brighter and bigger than the corresponding particles within the ordered adlayer (Fig. 35).

A more detailed analysis of the atomic structure of this new adsorbate reveals that the particles are arranged in an ideal hexagonal symmetry with a nearest adsorbate–adsorbate distance of $a = 0.34 \pm 0.015$ nm. Apart from the atomic corrugation this adlayer shows an additional long-range height modulation indicating that the adsorbate adlayer is not commensurate with the underlying copper substrate. This observation is supported by a line-scan along one of the close-packed adsorbate rows (Fig. 35b,c) showing a vertical modulation of the adsorbate particle positions of ± 0.008 nm with a periodicity corresponding to three interatomic distances within the close-packed adsorbate rows. Furthermore, the main symmetry axes of the new superstructure and the close-packed adsorbate rows are aligned along the same crystallographic direction. This reinforces that there is no rotation between

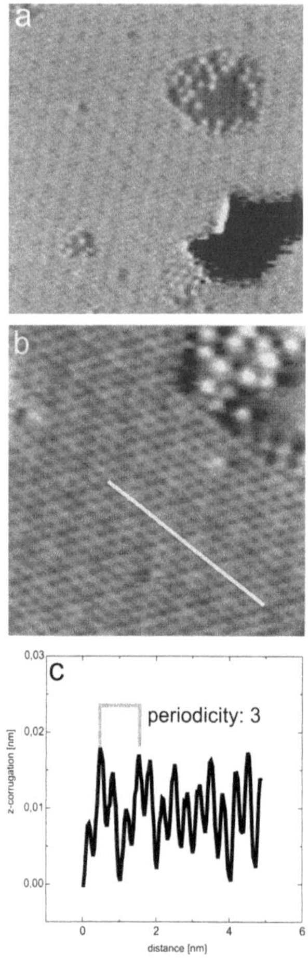

Fig. 35. Nearly completed cationic adsorbate layer in the regime of hydrogen evolution (negative potentials). Imaging parameters: (**a**) Size 15 nm × 15 nm, bias voltage $U_b = -180\,\text{mV}$; tunneling current $I_t = 10\,\text{nA}$. (**b**) 7 nm × 7 nm; $U_b = -180\,\text{mV}$; $I_t = 10\,\text{nA}$. The electrode potential is $E = -230\,\text{mV}$, negative potential sweep, in both images. Panel (**c**) shows the z-corrugation and the long range periodicity along the white line in (**b**)

the adsorbate and the substrate main symmetry axes as already read from Fig. 34b. Consequently, rotational domains of this adlayer have never been observed even on very large terraces. A surface reconstruction as explanation for the observed height modulation as in the case of the sulfate-induced Moiré pattern at positive electrode potentials (Fig. 16) clearly does not apply here. A noticeable change of the surface morphology, which is expected for a reconstruction process, also does not take place during the adlayer formation at negative potentials.

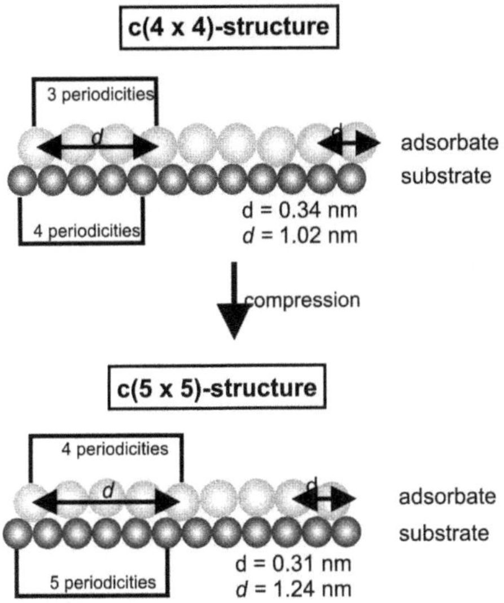

Fig. 36. Model for the electrocompression of the cationic adlayer observed on Cu(111) in 5 mMH_2SO_4 at negative potentials in the regime of hydrogen evolution

Considering all this information the adsorbate structure can be described by a simple c(4×4)-superstructure with an adsorbate unit cell which is three times larger ($a' = 1.04$ nm) than the nearest interatomic distance within the adlayer (see also Fig. 36). The observed height modulation can be explained in terms of the lattice mismatch between the adsorbate and the substrate, and the resultant occupation of inequivalent adsorption sites [54].

The observed height modulation of about 0.008 nm in Fig. 35b is rather small compared to the sulfate-induced Moiré structure (Fig. 11) with a height modulation of about 0.04 nm. In this latter case the imaging height of the Moiré structure did not depend on the chosen bias voltage or tunneling current. Quite in contrast to this, the imaging properties and, in particular, the imaging height of the new adsorbate structure at negative potentials change drastically by varying the tunneling conditions. Figure 37 gives an example of the complex correlation between tunneling current, bias voltage and the resulting STM imaging properties. All images in Fig. 37 are of the same image size. From Fig. 37a–d only the tunneling current and thereby the tip–sample distance has been changed. At the largest tip-sample distance in Fig. 37a only the long range (4×4) unit cell as a whole is imaged with an imaging height of about 0.04 nm. By changing the tunneling current stepwise from 1 nA to 30 nA the atomic adsorbate structure becomes more and more visible showing the characteristic periodicity of three along the adsorbate main symmetry axes. However, using tunneling currents of 30 nA (Fig. 37d) the surface appears less corrugated than at tunneling currents of 1 nA (Fig. 37a). Considering standard tunneling theories [55] the inverse behavior would be

Fig. 37. Imaging properties of the cationic adsorbate structure shown in Fig. 35. The imaging parameters are (**a**) bias voltage $U_b = -11\,\text{mV}$; tunneling current $I_t = 1\,\text{nA}$; (**b**) $U_b = -11\,\text{mV}$, $I_t = 2\,\text{nA}$; (**c**) $U_b = -11\,\text{mV}$; $I_t = 3\,\text{nA}$; (**d**) $U_b = -11\,\text{mV}$; $I_t = 30\,\text{nA}$. The size of all images is $4.4\,\text{nm} \times 4.4\,\text{nm}$ and the electrodepotential is $E = -230\,\text{mV}$, negative potential sweep, throughout

expected. Usually the imaging height increases with decreasing tip-sample distance and, hence, increasing tunneling currents. These tunneling effects may have to do with the presence of water in the tunneling gap, which is well known to strongly influence the tunneling process and possible tunneling channels [56,57,58]. However, a detailed understanding about the tunneling processes at solid/liquid interfaces is still lacking.

A change of the electrode potential to more negative values into the regime of massive hydrogen evolution (Fig. 38: $-420\,\text{mV}$, $j = -0.03\,\text{mA/cm}^2$) leads to an isotropic compression of the adlayer resulting in an 8.8 % decrease of the nearest interatomic distance from originally 0.34 nm (Fig. 35) to finally 0.31 nm (Fig. 38). As a consequence, the unit cell of the superstructure increases from a length of 1.02 nm to 1.24 nm. A line-scan along one of the main symmetry axes reveals (Fig. 38a,d) that now coincidence is reached every four interadsorbate distances (on five substrate interatomic separa-

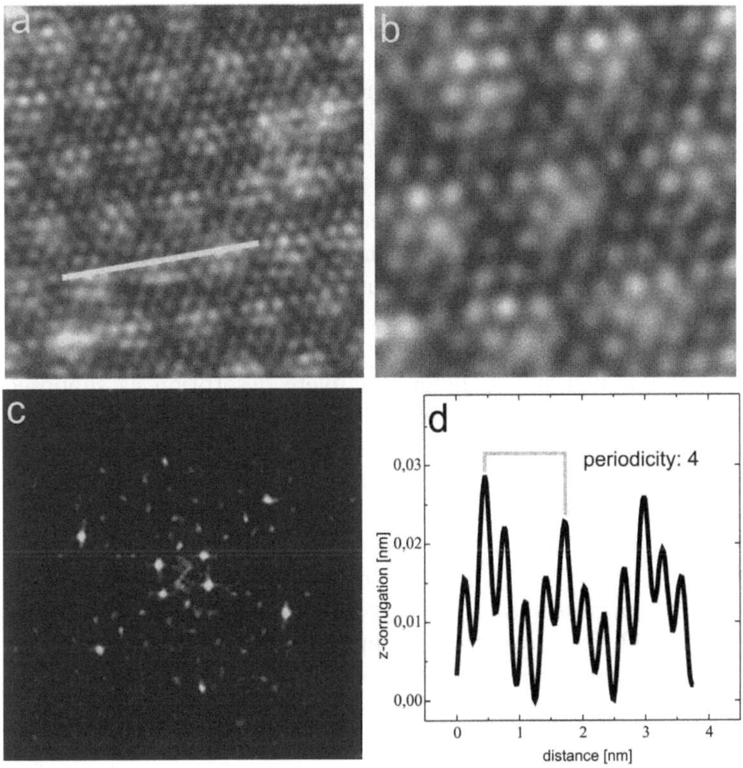

Fig. 38. STM images of a more compressed cationic adlayer (than the one in Fig. 35) in the regime of massive hydrogen evolution at even more negative potentials. The size of both STM images is (**a**) 5.2 nm × 5.2 nm, and (**b**) 2.6 nm × 2.6 nm. All other conditions are the same for both images: Bias voltage $U_b = 10\,\text{mV}$; tunneling current $I_t = 10\,\text{nA}$, and electrode potential $E = -420\,\text{mV}$, negative potential sweep. Panel (**c**) shows the power spectrum of this more compressed adlayer, while panel (**d**) displays the z-corrugation along the white line in image (**a**)

tions) instead of three (on four) for the less compressed phase (Fig. 35b). A schematic model of this electrocompression process is given in Fig. 36. This electrocompression phenomenon is well known for anionic structures on several electrode surfaces where a compression is observed for *increasing* electrode potentials and not for *decreasing* potentials as described here. Prominent examples for such anion compression on Cu(111) are highly ordered bromide structures [27,59] which show a characteristic uniaxial adlayer compression with increasing electrode potential.

Interestingly, no desorption of the cathodic adlayer can be observed by STM even at extremely negative potentals. This adsorbate remains on the surface even at extremely high current densities during massive hydrogen evolution reaction, always showing the same structures. Only the imaging properties change comparable to the effect observed for the less compressed structure (Fig. 37).

Under these conditions problems with the STM tip arise due to the massive formation of hydrogen bubbles at the surface. Whenever a hydrogen bubble has formed between the tip and the surface the tip potential control is lost leading to a tip crash and the end of the experiment.

Finally, it remains to identify the chemical nature of the adsorbed species which gives rise to the observed adlayer at negative potentials. As described in Sect. 3.5 in situ FTIR data have shown that at these negative potentials the sulfate is completely desorbed, and the decay of the sulfate-induced Moiré structure is directly correlated with the sulfate desorption. An intermediate state of adsorbed sulfate before complete desorption as supposed by *Brisard* et al. [60] can therefore be excluded. These infrared data [43] are also consistent with the STM results presented here, showing the bare copper surface (Fig. 34) *before* the formation of the new adlayer at more negative potentials. A further argument against persistance and rearrangement of sulfate molecules at these potentials is the fact that this kind of adsorbate structures (Figs. 36 and 37) has also been observed on Cu(111) in many other acidic electrolytes like toluensulfuric acid [61], hydrobromic acid [27] and recently also in perchloric acid [62]. From these comparative studies it becomes evident that the adsorbate species at negative potentials just before and during hydrogen evolution is independent of the anionic species in the acidic electrolytes used.

Contamination being present at the surface at negative potentials after desorption of strongly adsorbing anions could be excluded by means of XPS measurements after transferring the copper sample from the electrochemical environment into UHV. It is therefore concluded that this species adsorbed at negative potentials is due to a constituent of the solvent, i.e. water. Since STM is not a chemically sensitive method one can only speculate about the exact chemical identity of this species. First of all, one can think of the adsorption of OH species. However, by comparing the results presented in this chapter with STM measurements of Cu(111) in alkaline electrolytes by *Maurice* et al. [63] where the reversible adsorption of OH species at potentials negative of the Cu_2O formation regime were studied one can exclude the formation of a similar adlayer in acidic electrolytes at negative potentials. Although also forming a hexagonal superstructure on Cu(111) the OH adlayer in *alkaline* solution differs in symmetry and dimensions from the ordered adlayers observed here during the hydrogen reaction in *acidic* electrolyte. Furthermore, the adlayer formation mechanisms in both cases are completely different. While in the alkaline solution the nucleation and growth of the adlayer is accompanied by a mass transport out of the outermost copper layer due to an adsorbate induced reconstruction [63], such substrate restructuring is not observed here. Also in the alkaline solution different rotational domains were observed in contrast to the present situation in the acidic electrolyte at negative potentials. And finally, in alkaline solution the coverage of the OH species increases with *increasing* potentials, while here the adsorbate coverage increases with *decreasing* potentials. All this supports the notion that a *cationic* species may be involved in the adlayer formation in acidic electrolytes at negative potentials. However, it is unlikely that naked H_3O^+ cations adsorb at the electrode surface. It is well known that these "oxonium" cations in solution are strongly shielded by a hydration shell resulting in clusters from the $H_{2n+1}O_n^+$-type like $H_5O_2^+$, $H_7O_3^+$ and $H_9O_4^+$. Such "hydronium" clusters have been found by means of High Resolution Electron Energy Loss Spectroscopy (HREELS) to be present at Pt(110) surfaces after coadsorption of water and hydrogen under UHV conditions [64,65,66] (non-situ experiments) but not on Cu(111) [67,68]. In situ data from these species on metal surfaces at negative electrode potentials are rare. For Pt(111) in 0.5 M H_2SO_4 *Shingaya* et al. [44] suggest the coadsorption of hydrogen, water and hydronium cations at negative electrode potentials in order to explain in situ IR spectra obtained at a negative electrode polarization. Corresponding in situ experiments for Cu(111) are still lacking.

However, even without knowing the exact chemical identity of the observed adsorbate one can imagine that the presence of this cationic adsorbate has a strong impact on the electrochemical reactivity of the Cu(111) surface concerning reactions such as hydrogen evolution (HER) or the oxygen reduction (ORR) taking place at these negative potentials. *Brisard* et al. [60] explained the different reactivity of Cu(111) and Cu(100) as regards the ORR and the HER mainly on the basis of sulfate anion effects. In contrast to this explanation it is more likely that the observed ordered cation layer is responsible for the different reactivity of Cu(111) and Cu(100), because in the latter case a similar ordered superstructure is not found; only the bare Cu(100) lattice could be imaged in dilute sulfuric acid at negative potentials. The same is also true for Cu(110) [69].

3.9 Metastable Structures

In this last section some metastable structures shall be reported which have been observed occasionally and which seem to be intermediate states between the disordered mobile sulfate adsorbate and the stable Moiré structure. The specific conditions under which these metastable structures appear are not yet clear. But for the sake of completeness and as a stimulation for further investigations of this SO_4^{2-}/Cu(111) system they shall at least be shown here.

In very few images like Fig. 39a (upper part) isolated small areas of flat structures without Moiré periodicity could be seen. A higher-resolution image of such a flat area (Fig. 39b) reveals an ordered molecular structure (unlike the diffuse regions in Fig. 28). The angle distribution (Fig. 39d), which shows angles of multiples of 60° with nearly equal pair density, confirms perfect hexagonal symmetry. The intermolecular distance of 0.43 nm (Fig. 39c) is similar to the distance between the sulfate ions in the close-packed rows of

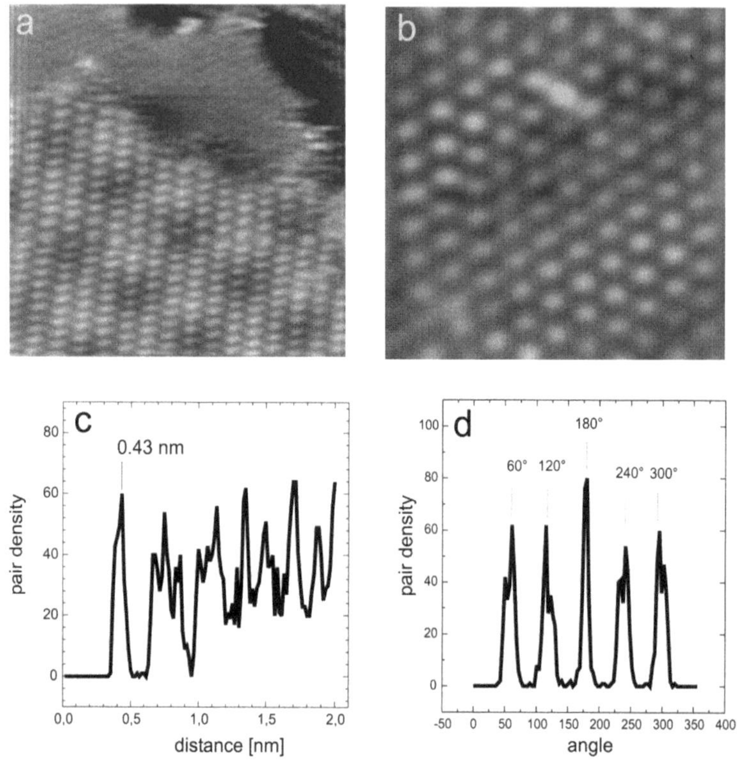

Fig. 39. (**a**) Moiré-free, but adsorbate covered regions adjacent to the Moiré structure; image size 8.2 nm × 8.2 nm; bias voltage U_b = 100 mV; tunneling current I_t = 1 nA. (**b**) Close-up of this adsorbate patch; 4.1 nm × 4.1 nm; U_b = 100 mV; I_t = 1 nA. In both images the electrode potential is E = 160 mV vs. RHE. (**c**) Pair–distance correlation and (**d**) pair–angle correlation of the adlayer structure shown in (**b**)

the Moiré structure (Fig. 13). It appears, however, unlikely that this ideal hexagonal structure is due to adsorbed sulfate particles because of the great repulsive interaction within a close-packed structure of such anions. It appears more likely that these are adsorbed water molecules [24].

In some other cases structures as displayed in Fig. 40a–d could be observed coexisting with the Moiré and the close-packed structure shown in Fig. 39b. The four images in Fig. 40a–d were recorded with time intervals of 2 minutes at a constant potential where the adsorption process was slow enough to be followed. All aspects of the previously described adlayer growth can be recapitulated, especially the nucleation and growth of islands and the persistance of Moiré-free areas at domain boundaries, etc. This series, however, reveals another new structure. After patches of the Moiré structure have already emerged two domains of a stripe pattern on different terraces shows up (Fig. 40b). The stripes in each domain are exactly parallel and the distance between them amounts to 4 nm. As the Moiré structure grows the stripe pattern recedes until the stripes have completely vanished. It is remarkable that the stripes never touch the Moiré. Magnification of the stripes (Fig. 41) reveals a complicated structure. The sharpness and the corrugation of 0.09 nm leads to the assumption that the stripes are related rather to a coadsorbate of sulfate/water than to a reconstruction of the surface. Other features in this image are the frizzy and flat structures between the sharp stripes. In the potential range cathodic of the desorption peak where the adlayer is desorbed and far away from the adsorption potential range yet another kind of unstable structure could be observed in isolated cases to occur on the flat terraces. Figure 42 displays a snapshot of these structures which are characterized by groups of parallel single or double lines. They enclose angles of 30°, 60° or 90°. These structures were so unstable that two successive images, one recorded immediately after the other, showed differ-

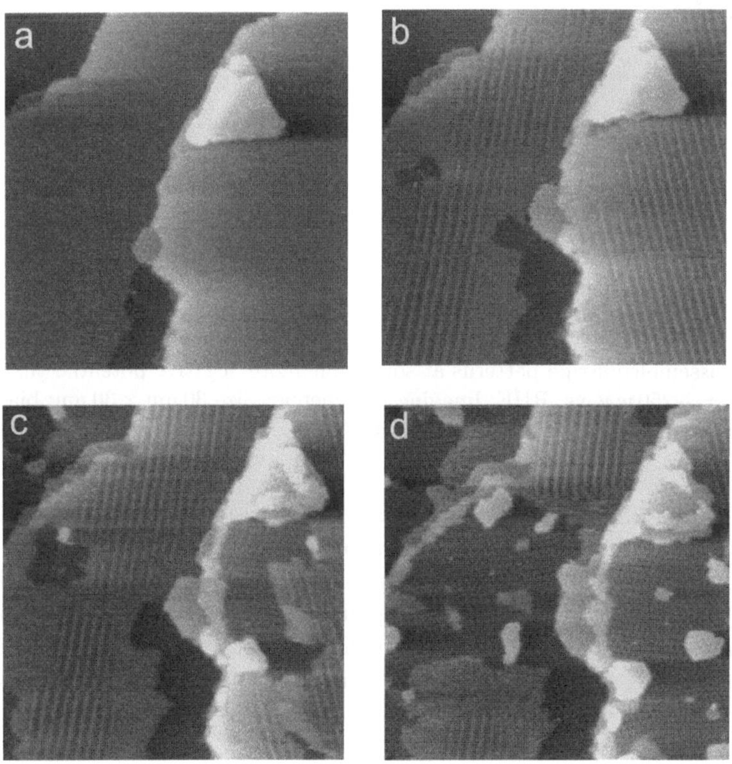

Fig. 40. Series of STM images showing metastable stripe patterns during the sulfate adlayer formation. Imaging parameters (**a**)–(**d**): Size 117 nm × 117 nm; bias voltage U_b = 327 mV; tunneling current I_t = 1 nA; electrode potential E = −110 mV vs. RHE. Recording time for the whole series 8 minutes

Fig. 41. High-resolution STM image of the stripes shown in Fig. 40b-d. Imaging parameters: size 14 nm × 14 nm; bias voltage U_b = 169 nm; tunneling current I_t = 1 nA; electrode potential $E = -130$ mV vs. RHE

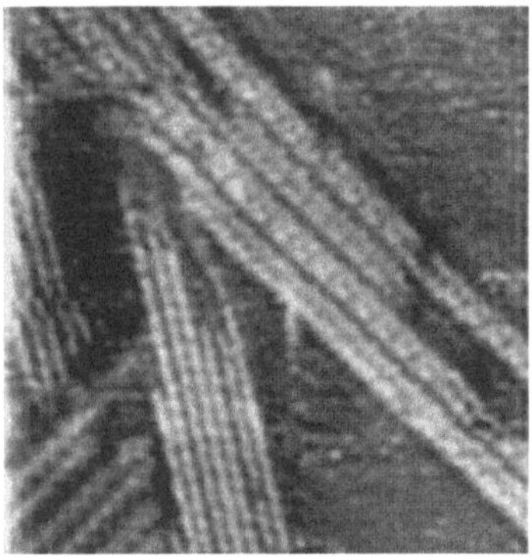

Fig. 42. Self-assembled stripe patterns at somewhat more negative potential than in Fig. 41: $E = -250$ mV vs. RHE. Imaging parameters: Size 30 nm × 30 nm; bias voltage U_b = 540 mV; tunneling current I_t = 1 nA

ent groups of lines. Since this potential range is very close to the onset of hydrogen evolution these stripes may be a precursor to the cationic structures described in Sect. 3.8. As in the case of the stripe pattern mentioned above, this phenomenon cannot yet be explained but was presented here in order to complete the survey of surface structures detected for Cu(111) in aqueous 5 mMH_2SO_4 solution in the whole potential range between hydrogen evolution and copper dissolution.

References

1. P. A. Christensen, A. Hamnett: *Techniques and Mechanisms in Electrochemistry* (Blackie Academic, Glasgow 1994)
2. T. Iwasita, F. C. Nart: In situ Infrared Spectroscopy at Electrochemical Interfaces, Progr. Surf. Sci. **55**, 271 (1997)
3. B. Pettinger, Ch. Bilger, J. Lipkowski: in this volume
4. H.-J. Güntherodt, R. Wiesendanger (Eds.): *Scanning Tunneling Microscopy I* (Springer Verlag, Berlin, Heidelberg 1991)
5. R. Wiesendanger, H.-J. Güntherodt (Eds.): *Scanning Tunneling Microscopy II* (Springer Verlag, Berlin, Heidelberg 1992)
6. K. Itaya: in N. Masuko, T. Osaka, Y. Fukunaka (Eds.): *New Trends and Approaches in Electrochemical Technology* (Kodansha, Tokyo, VCH, New York 1993)
7. D. M. Kolb: *Initial Stages of Metal Depositon: An Atomic View*, Schering Lect. Publ. 1 (Schering, Berlin 1991)
8. K. Itaya: In situ Scanning Tunneling Microscopy in Electrolyte Solutions, Progr. Surf. Sci. **58**, 121 (1998)
9. E. Buderski, G. Staikov, W. J. Lorenz: *Electrochemical Phase Formation and Growth* (VCH, Weinheim 1996)
10. S. Trasatti, K. Wandelt (Eds.): *Surface Science and Electrochemistry*, Surface Sci. 335 (Elsevier Science, Amsterdam 1995)
11. A. Wieckowski (Ed.): *Interfacial Electrochemistry* (Marcel Dekker, New York 1999)
12. http://www.research.ibm.com/journal/rd/425/intro.html
13. http://www.chips.ibm.com/gallery/
14. K. Besocke: Surf. Sci. **181**, 145 (1987)
15. T. Michely: PhD thesis, IGV Forschungszentrum Jülich, Bonn (1991)
16. M. Wilms, M. Kruft, G. Bermes, K. Wandelt: Rev. Sci. Instrum. **70**, 3641 (1999)
17. M. Lennartz, M. Arenz, C. Stuhlmann, K. Wandelt: Surface Sci. **461**, 98 (2000)
18. O. M. Magnussen, J. Hageböck, J. Hotlos, R. J. Behm: Faraday Discuss. Chem. Soc. **94**, 329 (1992)
19. A. M. Funtikov, U. Linke, U. Stimming, R. Vogel: Surface Sci. **324**, L343 (1995)
20. L.-J. Wan, S.-L. Jau, K. Itaya: J. Phys. Chem. **99**, 9507 (1995)
21. G. J. Edens, X. Gao, M. Weaver: J. Electroanal. Chem. **375**, 357 (1994)
22. A. Frumkin: Z. Phys. Chem. A **164**, 121 (1933)
23. Landolt-Börnstein: *Group 3* Vol. 6 (Springer, Berlin, Heidelberg 1971)
24. M. Wilms, P. Broekmann, C. Stuhlmann, K. Wandelt: Surface Sci. **416**, 121 (1998)
25. P. Broekmann, M. Wilms, K. Wandelt: Surf. Rev. Lett. **6**, 907 (1999)
26. P. Broekmann, M. Wilms, M. Kruft, C. Stuhlmann, K. Wandelt: J. Electroanal. Chem. **467**, 307 (1999)
27. A. Spänig, P. Broekmann, K. Wandelt: in preparation, P. Broekmann, PhD-thesis, Bonn, 2000 (see [31] below)
28. A. M. Funtikov, U. Stimming, R. Vogel: J. Electroanal. Chem. **428**, 147 (1997)
29. L.-J. Wan, T. Suzuki, K. Sashikata, J. Okada, J. Inukai, K. Itaya: J. Electronal. Chem. **484**, 189 (2000)
30. L.-J. Wan, M. Hara, J. Inukai, K. Itaya: J. Phys. Chem. **103**, 6978 (1999)

31. P. Broekmann: Ph.D. Thesis, Bonn, 2000
32. K.-I. Ataka, M. Osawa: Langmuir **14**, 951 (1998)
33. W. H. Li, R. Nichols: J. Electroanal. Chem. **456**, 153 (1998)
34. M. Wilms: PhD-thesis, University of Bonn (1999)
35. F. C. Nart, T. Iwasita, M. Weber: Electrochim. Acta **39**, 961 (1994)
36. T. Iwasita, F. C. Nart: Surf. Sci. Rep. **99**, 322 (1998)
37. T. Iwasita, F. C. Nart, A. Rodes, E. Pastor, M. Weber: Electrochim. Acta **40**, 53 (1995)
38. K. Nakamoto: *Infrared and Raman Spectra of Inorganic and Coordination Compounds*, 4th ed. (Wiley, New York 1986)
39. T. Wandlowski: private communication
40. T. Yamada, N. Batina, K. Itaya: Surface Sci. **335**, 204 (1995)
41. T. Yamada, N. Batina, K. Itaya: J. Phys. Chem. **99**, 8817 (1995)
42. M. Arenz, P. Broekmann, M. Lennartz, E. Vogler, K. Wandelt: Physica Status Solidi (a), **187**, 63 (2001)
43. M. Lennartz, P. Broekmann, M. Arenz, C. Stuhlmann, K. Wandelt: Surface Sci. **442**, 215 (1999)
44. Y. Shingaya, M. Ito: J. Electroanal. Chem. **467**, 299 (1999)
45. P. Broekmann, M. Wilms, A. Spänig, K. Wandelt: Progr. Surf. Sci. **67**, 59 (2001)
46. M. R. Vogt, A. Lachenwitzer, O. M. Magnussen, R. J. Behm: Surface Sci. **399**, 49 (1998)
47. P. Broekmann, M. Anastasescu, W. Lisowski, K. Wandelt: J. Electroanal. Chem. **500**, 241 (2001)
48. D. W. Suggs, A. J. Bard: J. Am. Chem. Soc. **116**, 10725 (1994)
49. D. W. Suggs, A. J. Bard: J. Phys. Chem. **99**, 8349 (1995)
50. M. R. Vogt, F. A. Möller, C. M. Schilz, O. M. Magnussen, R. J. Behm: Surface Sci. 367 (1997) L33
51. D. M. Kolb: Progr. Surf. Sci. **51**, 109 (1996)
52. S. Baier, M. Giesen, H. Ibach: Verhandl. DPG (VI) **35**, 723 (2000)
53. M. R. Vogt, W. Polewska, O. M. Magnussen, R. J. Behm: Verhandl. DPG (VI) **34**, 915 (1999)
54. P. Broekmann, M. Wilms, A. Spänig, K. Wandelt: in *Proceedings of the Symposium on Thin Films: Preparation, Characterization, Applications*, San Diego, California, April 1–5, 2001 (Kluwer Academic/Plenum Publishers, San Diego 2001)
55. C. J. Chen: *Introduction to Scanning Tunneling Microscopy* (Oxford University Press, Oxford 1993)
56. J. Halbritter, G. Rephuhn, S. Vinzelberg, G. Staikov, W. J. Lorenz: Electrochim. Acta **40**, 1385 (1995)
57. G. Nagy: Electrochim. Acta **40**, 1417 (1995)
58. G. Nagy: J. Electroanal. Chem. **409**, 19 (1996)
59. J. Inukai, Y. Osawa, K. Itaya: J. Phys. Chem. B **102**, 10034 (1998)
60. G. Brisard, N. Bertrand, P. N. Ross, N. Markovic: J. Electroanal. Chem. **480**, 219 (2000)
61. E. Vogler, P. Broekmann, M. Arenz, M. Lennartz, K. Wandelt: in preparation
62. B. Obliers, P. Broekmann, K. Wandelt: in preparation
63. V. Maurice. H.-H. Strehblow, P. Marcus: Surface Sci. **458**, 185 (2000)
64. T. F. Wagner, T. E. Moylan: Surface Sci. **182**, 125 (1987)

65. T. F. Wagner, T. E. Moylan: Surface Sci. **206**, 187 (1988)
66. N. Chen, P. Blowers, R. I. Masel: Surface Sci. **419**, 150 (1999)
67. J. Schott, D. Lackey, J. K. Sass: Surface Sci. **238**, L478 (1990)
68. D. Lackey, J. Schott, J. K. Sass, S. I. Woo, F. T. Wagner: Chem. Phys. Lett. **184**, 277 (1991)
69. M. Anastasescu, B. Obliers, P. Broekmann, K. Wandelt: in preparation; see also [31]

Chloride Adsorption on Cu(111) Electrodes: Electrochemical Behavior and UHV Transfer Experiments

Christopher Stuhlmann, Bernd Wohlmann, Zin Park, Michael Kruft, Peter Broekmann, and Klaus Wandelt

Institut für Physikalische und Theoretische Chemie der Universität Bonn
Wegelerstr. 12, 53115 Bonn, Germany
k.wandelt@uni-bonn.de

Abstract. The adsorption of chloride on Cu(111) in dilute HCl solutions was studied by in situ STM and UHV transfer experiments. X-ray Photoelectron Spectroscopy (XPS), Ultraviolet Photoelectron Spectroscopy (UPS) and Ion Scattering Spectroscopy (ISS) were applied ex situ. The cyclic voltammogram features two peaks which were unequivocally assigned to the adsorption and desorption of chloride. The transfer experiments are hampered by the presence of a large hydrogen evolution current. This faradaic process leads to the adsorption of an extra amount of chloride upon emersion from the electrolyte. This unwanted effect can be suppressed by the use of very dilute electrolytes or, even better, mixed electrolytes. The adsorbed chloride enhances the kinetics of the hydrogen evolution reaction. The influence of chloride on the reaction rate is demonstrated by a detailed analysis of the voltammogram, using a simple model for the simulation of voltammograms.

1 Introduction

In recent years, our understanding of the so-called specific adsorption of anions has been substantially fostered by the use of in situ structural techniques, in particular of in situ Scanning Tunneling Microscopy (STM). The role of anions in the reconstruction of Au(100) and Au(111) has been studied [1,2,3], and it was found that the transition between the reconstructed and the unreconstructed surface is triggered by changes of the anion coverage. A more complex case is the formation of underpotentially deposited (upd) Cu layers on Au(111) and Pt(111) where a large number of different Cu structures is observed, depending on the anion in the solution and the applied potential [4,5,6,7].

The interaction between Cu and the adsorbed anions can, however, be elaborated more clearly if bulk Cu is used as the substrate instead of a upd layer. Like *Inukai* et al. [8], we undertook a series of studies on the adsorption of different anions on Cu(111) electrodes. However, starting from the fairly simple Cl/Cu(111) system we advanced to sulfate and sulfonate adsorption, which provide increasing levels of complexity.

K. Wandelt, S. Thurgate (Eds.): Solid–Liquid Interfaces, Topics Appl. Phys. **85**, 199–221 (2003)

These studies gain by the combination of in situ STM with UHV transfer experiments. The transfer experiments are, however, rather difficult with Cu(111) electrodes due to the presence of a substantial hydrogen evolution current in the potential region of interest. The additional problems introduced by this circumstance and possible workarounds are discussed in this chapter.

Another interesting aspect is the influence of anions on faradaic reactions. We will show in this chapter that adsorbed chloride enhances the reaction kinetics of the hydrogen evolution reaction. Since the bulk of our work on the Cl/Cu(111) system has already been published elsewhere [9,10,11,12], we will focus our attention on these two particular facets of an intriguing electrochemical system.

2 Experimental

The experiments were carried out in an ultrahigh vacuum (UHV) transfer apparatus which allows the sample to be transferred between UHV and the electrochemical cell without exposure to the air. The apparatus comprises two stainless steel recipients. The main chamber is used for sample analysis by UHV-based spectroscopies. It features a hemispherical analyzer (Specs, EA10 plus), which can be operated at either polarization. Thus photoelectron spectra as well as ion spectra can be recorded with the same analyzer. A commercial twin anode (Al and Mg) laboratory X-ray source (VG, XR2E) and a home-built He discharge lamp are used for the acquisition of X-ray Photoelectron Spectra (XPS) and Ultraviolet Photoelectron Spectra (UPS), respectively. The ion gun (Varian, model 981-2043) is operated with He gas for Low-Energy Ion Scattering Spectroscopy (LEISS). The same ion gun is used for sputtering the sample surface with heavier noble gas ions (Ne or Ar). The surface structure is characterized by a three-grid LEED optics (VSI, Er-LEED 150) and a quadrupole mass spectrometer (VG, Q7) serves as residual gas analyzer. The sample is mounted on a special sample holder which is attached to a precision xyz manipulator with dual rotation for exact sample positioning. The main chamber is pumped by an ion getter pump and a titanium sublimation pump. The base pressure is 8×10^{-11} mbar.

The main chamber is separated from the second, smaller recipient by a gate valve. The second recipient is used as a load-lock chamber for the electrochemical experiments. In the load-lock chamber the sample holder is attached to a second manipulator which provides sample rotation only. With the gate valve closed the load-lock chamber can be brought to atmospheric pressure with purified argon gas without breaking the vacuum in the main chamber. The electrochemical cell is then introduced through a second gate valve in the bottom of the load-lock chamber and the electrochemical experiments are performed with the sample face-down in a hanging meniscus configuration.

The electrochemical cell is a flow cell enabling electrolyte exchange without loss of potential control. During the electrochemical experiments the load-lock chamber is held at a slight overpressure of Ar to prevent backflow of atmospheric air through the sink of the cell. A Pt foil is used as the counter electrode and the reference electrode is a Hg/Hg_2SO_4 electrode. All potentials are given with respect to this reference. After the electrochemical experiment the sample is removed from the electrolyte under potential control and transferred back to the UHV.

The sample was cut from a Cu single crystal and oriented by Laue back-diffraction to within 1° of the {111} direction. The surface was mechanically polished with diamond paste of successively smaller grain sizes down to 0.25 μm. Finally the sample was cleaned in UHV by Ar^+ ion sputtering (1 keV, 3 μA/ cm^2) until no contaminations could be detected by XPS. After annealing to 700 °C a clear and sharp (1×1) LEED pattern was found. The last preparation step was repeated prior to each electrochemical measurement.

For the preparation of the electrolyte solutions ultrapure water (Millipore, resistivity > 18 MΩ cm) was used. All chemicals were reagent grade or better.

Argon gas was used both for venting the load-lock chamber and for deaeration of the electrolyte. The gas was cleaned by a dual stage gas purification system, consisting of a glass tube with a high surface area copper catalyst (Merck, BTS) and an oxysorb cartridge (Messer-Griesheim). This rather extensive cleaning method was necessary because the Cu surface is extremely sensitive to residual oxygen.

3 Results

3.1 Cyclic Voltammetry

After preparation in UHV the sample was transferred to the electrochemical cell and immersed under potential control. The acquisition of cyclic voltammograms was started immediately after the immersion. Figure 1 shows two voltammograms which were obtained in this manner in HCl solutions of different concentrations. The immersion potential was set positive of the anodic peak in both cases, and the potential scan rate was 10 mV/ s. Both curves show a pair of peaks. The anodic peak appears at a more positive potential in 1 mM HCl than in 10 mM HCl whereas the potential of the cathodic peak is almost the same for both concentrations. In the potential region positive of the anodic peak only double layer charging is observed. This range is limited by the onset of Cu dissolution at about −0.5 V. The upper potential limit was set at or below this value in all experiments so as to prevent excessive corrosion of the Cu surface. The portion of the voltammogram to the left of the anodic peak is dragged down by the beginning hydrogen evolution reaction. The shape of the voltammogram can be understood as a superposition

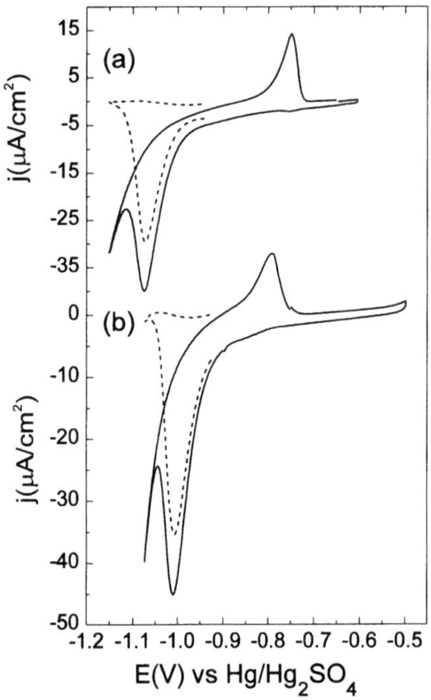

Fig. 1. Cyclic voltammogram of Cu(111) in 1 mM HCl (**a**) and in 10 mM HCl (**b**) *(solid curves)*. The potential scan rate is 10 mV/ s. The *dashed curves* are obtained by subtraction of the extrapolated hydrogen evolution current on the bare Cu(111) surface (see text for details)

of adsorption and desorption currents and of the faradaic current due to the hydrogen evolution reaction.

The anodic peak is assigned to the adsorption of Cl^- from the solution, according to the reaction equation:

$$Cu + Cl^- \rightarrow Cu\text{-}Cl_{ads} + e^- \, . \tag{1}$$

This interpretation is in agreement with the fact that the anodic peak shifts to higher potentials when the Cl^- concentration is reduced. The cathodic peak which is formed during the reverse scan is evoked by the desorption of the Cl^- which has been adsorbed during the anodic scan. A Cl^--free Cu surface is left behind in the potential region cathodic of this peak. Given this interpretation, the adsorption charge should be exactly counterbalanced by a corresponding desorption charge when the initial state of the surface is restored upon completion of the potential cycle. This means that the area under the anodic and the cathodic peak must be equal, quite contrary to what is observed in the voltammograms of Fig. 1: the cathodic peak is obviously larger than the anodic peak. This discrepancy is due to the superposition of the hydrogen evolution current. Since the desorption peak occurs at a more cathodic potential than the adsorption peak the contribution of the hydrogen evolution current to the former peak is larger.

A clearer distinction between faradaic contributions to the overall current and contributions due to adsorption and desorption is achieved by a variation of the potential scan rate v. This is demonstrated in Fig. 2 for a mixture of 1 mM HCl and 49 mM NaCl.

In the voltammogram recorded with $v = 10\,\mathrm{mV/s}$ a similar effect is seen as in the curves of Fig. 1, but due to the higher Cl^- concentration this effect is more pronounced here: not only is the size of the cathodic peak enhanced by an additional faradaic contribution, but the current undergoes a sign change in the anodic run at about −0.83 V. As the scan rate is reduced the overweight of the faradaic current over the adsorption current becomes even more pronounced leading to the gradual disappearance of the adsorption peak. At $v = 1\,\mathrm{mV/s}$ the cyclic voltammogram is entirely dominated by faradaic currents.

A more systematic study of the variation of v in solutions of different composition was hampered by the presence of a time effect, especially for the higher Cl^- concentrations. There was a systematic shift of the anodic and the cathodic peak as can be seen in Fig. 3. The cathodic peak shifts in the anodic direction and the anodic peak shifts in the cathodic direction, getting sharper and higher at the same time. On the anodic flank of the cathodic peak a shoulder forms which finally evolves into a small separate peak. The

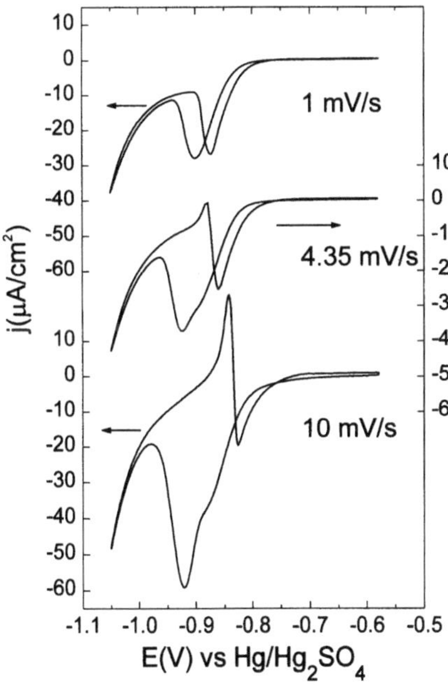

Fig. 2. Cyclic voltammogram of Cu(111) in 1 mM HCl + 50 mM NaCl, recorded at different scan rates v

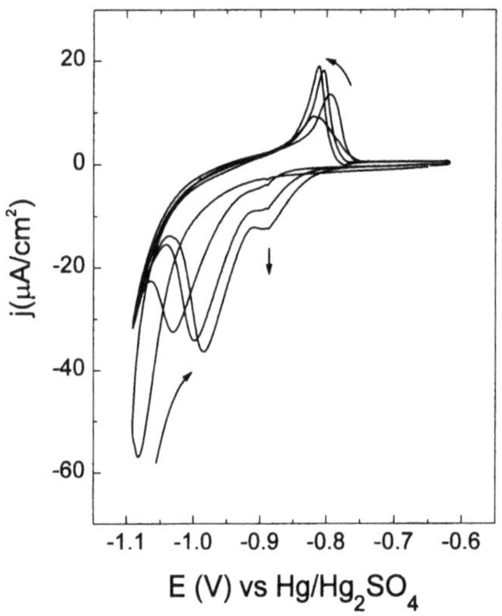

Fig. 3. Time evolution of the cyclic voltammogram of Cu(111) in 1 mM HCl/9 mM NaCl. A large number of voltammograms was recorded in sequence. Only every tenth voltammogram is shown for the sake of clarity and the running number is shown with each curve. The time evolution of the prominent features is indicated by the *arrows*. The scan rate was $v = 10\,\mathrm{mV/s}$ and thus the time elapsed between the first and the last curve was 50 min

time evolution is rapid in the beginning and it levels off later, but it still continues at a slow pace even after one hour in the electrolyte. This effect depends on the time the sample spends in contact with the electrolyte rather than on the potential applied. If the peak potentials are plotted as a function of time a continuous curve is obtained whether or not the potential is stopped at a fixed value for 5 min between two cycles at various intervals during the course of the experiment. The speed of the time evolution strongly depends on the Cl^- concentration. The changes are very slow in 1 mM HCl while in 100 mM HCl it is difficult to obtain meaningful results at all because the voltammogram is too unstable. There seems also to be an influence of the quality of sample preparation. The better the surface order as estimated from the sharpness of the LEED pattern, the slower is the change in the voltammograms. It is, however, not possible to give a more quantitative account of this notion.

3.2 UHV Transfer Experiments

After one or several potential cycles the sample was removed from the electrolyte solution and transferred to the UHV chamber. Figure 4b shows the Ultraviolet Photoelectron Spectrum (UPS) of the sample after emersion from 1 mM HCl at −0.94 V during the cathodic scan. According to the interpretation of the voltammogram presented in Fig. 1a this is a potential just

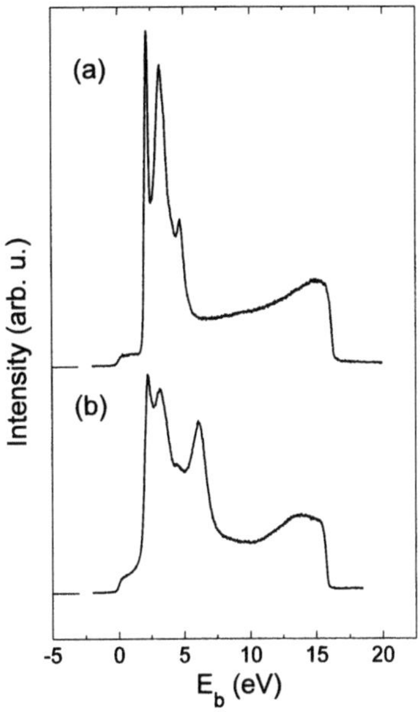

Fig. 4. (**a**) Ultraviolet photoelectron spectra (He I) of the clean Cu(111) electrode ($\vartheta_{\mathrm{in}} = \vartheta_{\mathrm{out}} = 45°$). (**b**) same as (**a**) after emersion from 1 mM HCl and transfer to UHV. The emersion potential was $E_{\mathrm{emers}} = -0.94\,\mathrm{V}$ during the cathodic sweep

before the onset of Cl^- desorption. A UP spectrum recorded for the clean surface under identical experimental conditions is shown in Fig. 4a for comparison. There are two very intense peaks at $E_{\mathrm{b}} = 2.2\,\mathrm{eV}$ and $E_{\mathrm{b}} = 3.2\,\mathrm{eV}$ and a weaker peak at $E_{\mathrm{b}} = 4.7\,\mathrm{eV}$ which are characteristic of the clean Cu(111) surface [13][1]. The two intense peaks are preserved after the emersion experiment though with reduced intensity. An additional peak appears at $E_{\mathrm{b}} = 6.1\,\mathrm{eV}$, which is clearly due to the adsorbate. The peak position is in agreement with our previous assignment to adsorbed Cl. The peak is caused by photoemission from the 3*p* orbital of Cl [14,15].

This assignment is confirmed by an experiment in which UP spectra and low-energy Ion Scattering Spectra (ISS) were recorded alternately. Since only those ions are detected by the hemispherical analyzer which are not neutralized during collision with the sample surface, the method is extremely surface sensitive. Because of the very high neutralization probability for ions penetrating into the bulk only the topmost layer contributes to the signal [16]. From the kinetic energy of the scattered ions the mass of the scatterers in the surface layer can be calculated. In the topmost spectrum of Fig. 5 (right panel) the expected positions for copper, chlorine and oxygen are marked.

Immediately after the transfer from the electrolyte the first UP spectrum was recorded (spectrum 1 ∈ Fig. 5). This spectrum is very similar to Fig. 4a. In particular, the additional emission peak at 6 eV is clearly seen. In the ion

[1] It should be noted that with the geometry of our UPS experiment the initial states of the electrons contributing to the spectra are close to the $\bar{M}$ point of the surface Brillouin zone

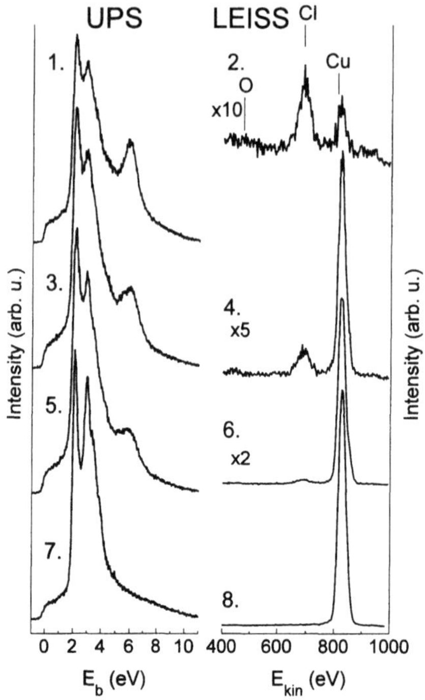

Fig. 5. UP spectra (He I) and low-energy ion scattering spectra of Cu(111) after emersion from 10 mM HCl at $E_{emers} = -0.7\,V$. For the ion scattering spectra He ions with a kinetic energy of 1 keV and a current of 0.3–1.8 μA were used. The spectra were normalized with respect to the sample current. The scattering angle was 120°. The spectra were recorded alternately in the order indicated by the numbers. The ion dose between the first and second row spectra was approximately 105 μA s, between the second and third row 94 μA s, between the third and last row 1740 μA s

scattering spectrum recorded immediately afterwards (spectrum 2 of Fig. 5) two peaks are found which indicate the presence of Cl and Cu in the topmost layer. Note that no oxygen is detected in the spectrum.

A side-effect of ion scattering spectroscopy is a mild surface sputtering. Adsorbates are therefore gradually removed during the measurement. In the second row spectra of Fig. 5 the Cl related features are significantly reduced compared to the first row. This trend continues also in the third row spectra. Finally, after prolonged ISS measurement, a clean Cu surface is restored (fourth row in Fig. 5). In the ion scattering spectrum only one peak is visible corresponding to the mass of Cu, and the UP spectrum is that of a clean Cu(111) surface, similar to that shown in Fig. 4a. The synchronous behavior of the 6 eV peak in the UP spectrum and of the Cl peak in the ion scattering spectrum confirms once more the assignment of the former to adsorbed Cl.

This picture is also consistent with the results of XPS measurements. The XP spectra show a strong Cu signal, a weaker Cl signal and virtually no oxygen. The different intensity ratio of the Cu and the Cl peaks in XPS and ISS is rationalized by the small photoemission cross-section of the Cl 2*p* orbital and by the different probing depth of the two methods. The XPS signal averages over at least five monolayers, and a rather weak signal is to be expected from a submonolayer coverage of an adsorbate. There is no hint from the XP spectra that Cl penetrates into the surface and a thick CuCl layer builds up after prolonged potential cycling. Such an effect should manifest itself in an increasing Cl/Cu ratio because of the relatively large probing depth of XPS.

The adsorption and desorption of Cl, which is clearly seen in the voltammograms, should also be reflected in the transfer experiments. The Cl coverage found by the different spectroscopies in UHV should depend on the potential at which the sample was withdrawn from the electrolyte. This is, however, not the case. Figure 6 displays the integrated peak intensities of the Cl 2*p* line found in an XPS experiment as a function of the emersion potential. The scatter of the data points is approximately equal to the experimental errors. A potential dependence of the Cl coverage on the surface can thus not be derived from this experiment by contrast to our interpretation of the voltammetric features. There are two possible explanations for this discrepancy: firstly, the peaks in the voltammogram may not be due to chloride adsorption and desorption. There is, however, clear evidence from in situ STM images (see below, Sect. 3.3) that the peaks are indeed due to chloride adsorption and desorption. The other possible explanation is that the amount of adsorbed Cl is not the same in the electrolyte and after the transfer. More specifically, since the maximum amount of Cl is found at all potentials *after* the transfer, an additional post-emersion adsorption of chloride would explain the observed behavior. The surplus of chloride adsorbed after the emersion proper obviously has to be supplied by the electrolyte film remaining on the surface. This assumption is easily checked by reducing the amount of chloride present in this film. Such a reduction is achieved by decreasing the Cl^- concentration in the electrolyte.

Figure 7 shows the result of an identical experiment with 10^{-4} M HCl instead of 10^{-3} M HCl. Indeed, the potential dependence of the Cl 2*p* intensity is now much closer to what is expected from the voltammogram. This approach has, however, the drawback that in such a dilute acid (pH = 4) copper oxide formation is thermodynamically favored, and the appearance of the O 1*s* signal at cathodic emersion potentials indicates that the Cl free surface is no longer pure Cu, but an oxide is formed once the protecting Cl overlayer disappears. This unwanted effect can be suppressed by using a mixed electrolyte with low pH *and* low Cl^-concentration. The results of such an approach are presented in Fig. 8. Note that now the S 1*s* signal of

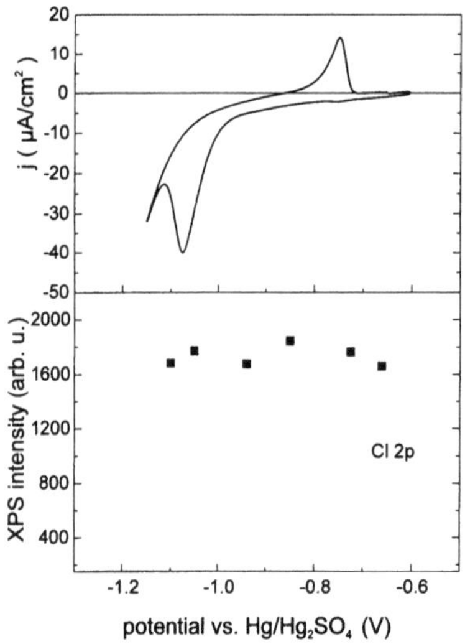

Fig. 6. Voltammogram of Cu(111) in 10^{-3} M HCl and integrated intensity of the XPS Cl $2p$ signal after transfer into UHV at the indicated potentials

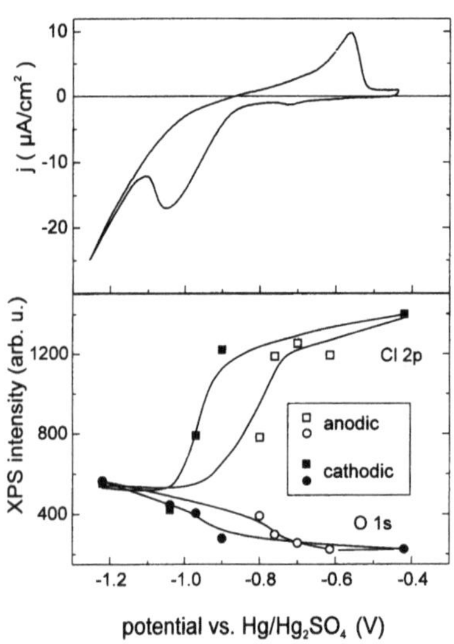

Fig. 7. Voltammogram of Cu(111) in 10^{-4} M HCl and integrated intensity of the XPS Cl $2p$ and the O $1s$ signals after transfer into UHV

adsorbed sulfate is observed at cathodic potentials. The implications of these experiments will be discussed in Sect. 4.2.

The occurrence of changes of the electrode state after the emersion can be checked directly by a simple experiment: the electrode is emersed from the electrolyte at a certain potential and re-immersed shortly afterwards at the same potential. The current transient after the re-immersion is recorded. If the electrode state after the emersion is exactly the same as it was in the electrolyte, then no current should be observed upon re-immersion, since in that case the outer circuit would not have even "noticed" the temporary

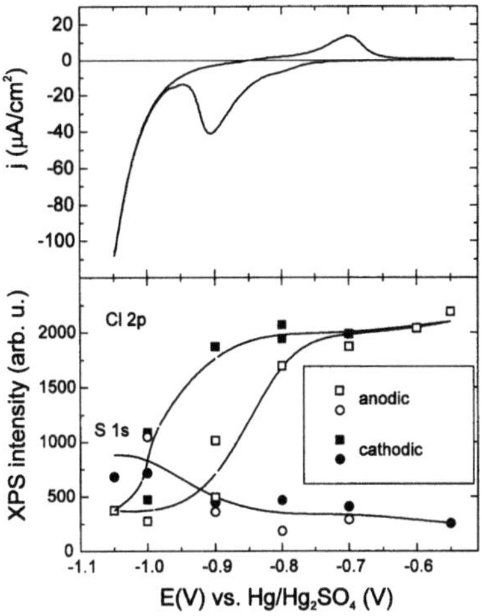

Fig. 8. Voltammogram of Cu(111) in $0.1\,\mathrm{N}\,H_2SO_4 + 10^{-4}$ mM HCl and integrated intensity of the XPS Cl $2p$ and the S $1s$ signals after transfer into UHV

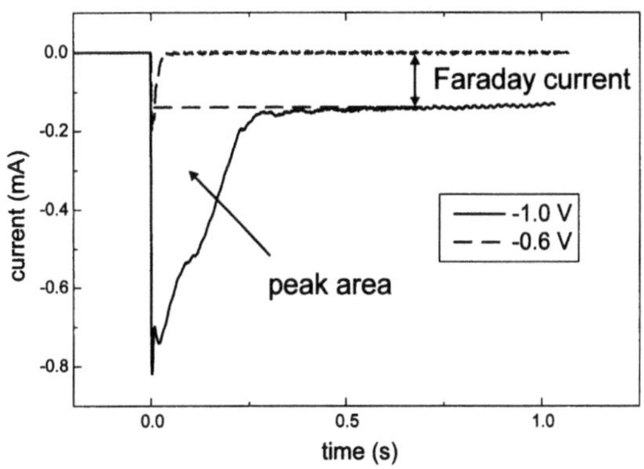

Fig. 9. Current transients recorded after emersion from the electrolyte at −0.6 V and −1.0 V and re-immersion at the same potential

absence of the electrode from the electrolyte. If post-emersion changes have occurred, however, these changes are reversed upon re-immersion, and corresponding currents should be observable in the transient. The results of such an experiment for 10^{-3} M HCl are shown in Fig. 9.

At an emersion potential of −0.6 V a nearly ideal situation is found with a zero current transient apart from a small capacitive charge at very short times. By contrast, after emersion and re-immersion at −1.0 V substantial currents are found over the entire time interval of the transient. The current can be split into two contributions. The Faraday current due to the hydrogen evolution, which is already quite palpable at this potential, does not change with time. It provides a constant offset on top of which a peaked curve is added. The latter curve is caused by the reversal of the post-emersion changes of the electrode state and is therefore the most interesting part in our context.

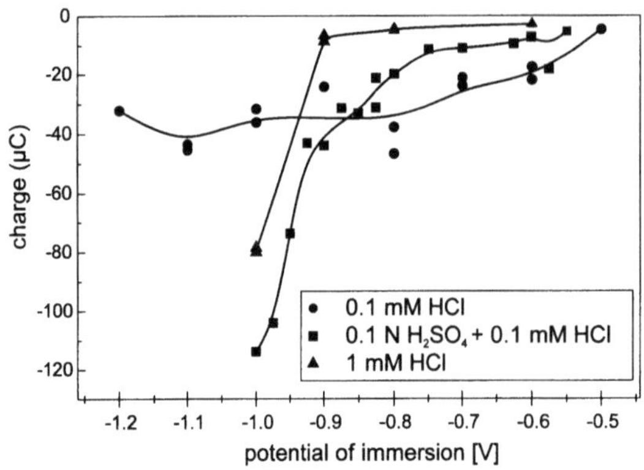

Fig. 10. Re-immersion charge determined from the peak area of curves as those shown in Fig. 9

The area under this part of the curve is plotted as a function of the emersion potential for all three electrolytes in Fig. 10.

It is obvious that the effect of post-emersion changes is the more pronounced the more cathodic the emersion potential is. The weakest effect is found for the most dilute electrolyte (10^{-4} M HCl).

3.3 Scanning Tunneling Microscopy

STM measurements of the Cl^- adsorption on Cu(111) have been discussed in great detail in our previous publications [9,12]. Therefore only a short summary of the results pertinent to the topic of this paper will be given here.

Figure 11 shows a series of potentiodynamic images for this system together with an assignment of the observed structures to the voltammetric features. In the potentiodynamic mode the potential of the sample is slowly scanned while the image is recorded. The slow scan direction of the STM images thus provides not only the usual geometric information, but it is at the same time a potential axis. The STM images of Fig. 11 clearly show a change of the substrate (1×1) structure to a superstructure with a larger unit mesh and back to the (1×1) structure, as the potential is scanned first in the anodic direction and then back in the cathodic direction. Corresponding regions in the voltammogram and in the STM images are marked by solid and dashed lines. It is clearly seen that the transition between the two structures occurs in the peaks of the voltammogram which can therefore be assigned to the adsorption and desorption of the Cl^- anions. A detailed inspection of the images reveals that the superstructure formed by the adsorbed Cl is a $(\sqrt{3} \times \sqrt{3})R30°$ structure. The adsorption site of the Cl was also determined by the following experiment. In the potential region where Cl is adsorbed images were taken with different values of the bias voltage U_b and constant tunneling current, in other words the tunneling resistance was varied.

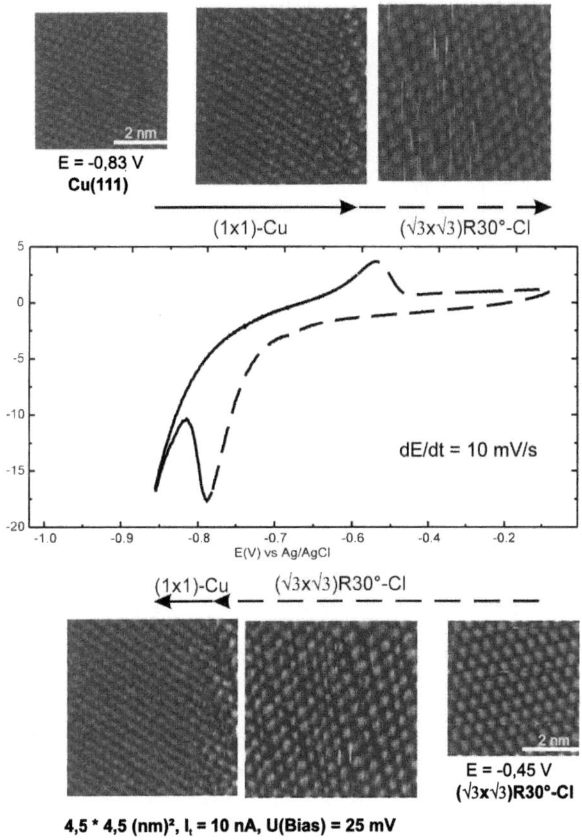

Fig. 11. Cyclic voltammogram and potentiodynamic STM image series of Cu(111) in 10 mM HCl. Image size: 4.5 nm × 4.5 nm, tunneling current $I_t = 10\,nA$

In Fig. 12 the result of this procedure is shown in the top row. At high tunneling resistances the adsorbate is imaged, similar to the images shown in Fig. 11. At low tunneling resistances (top row, right panel of Fig. 12) by contrast, the substrate structure is imaged. Note that the sample potential was not varied, so the adsorbate is present in both cases, but at low tunneling resistances the adsorbate becomes invisible, and the substrate is imaged *through* the adsorbate layer.

Finally, *both* the substrate and the adsorbate are detected in *one* image at intermediate values of the tunneling resistance. Such an image is shown in the center panel of the top row of Fig. 12. The contributions of substrate and adsorbate were disentangled from this rather complex image by a Fourier transform, followed by a separate backtransformation of the low-frequency part and the high-frequency part. The result of the calculation is shown in the bottom row of Fig. 12. If the leftmost panel and the rightmost panel of the bottom row (which were obtained from a single image) are placed on top

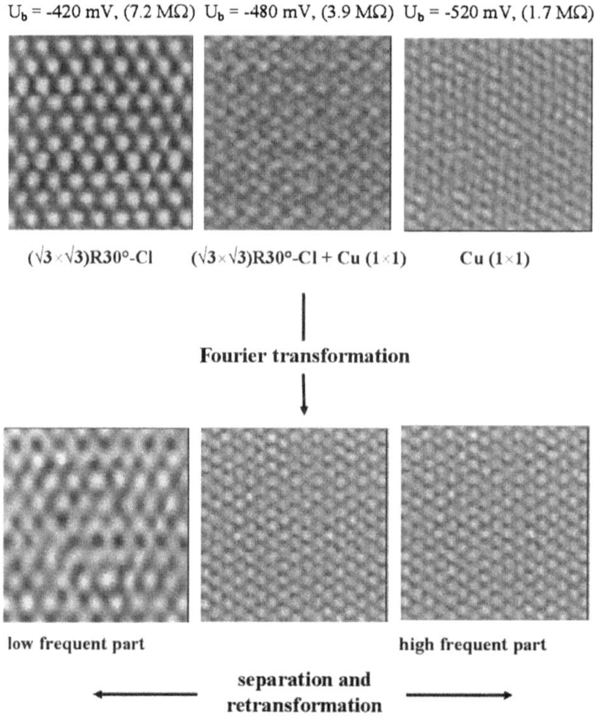

Fig. 12. Series of STM images *(top row)* of Cu(111) in 10 mN HCl recorded at the same potential $E = -550\,\text{mV}$, but with different tunneling resistances. (3.8 nm × 3.8 nm, $I_t = 18\,\text{nA}$). In the *bottom row* the substrate and adsorbate contributions to the *center image* of the *top row* are disentangled by a Fourier transformation (see text for details)

of each other, the adsorption site of the Cl is easily identified as a threefold hollow position.

4 Discussion

4.1 Influence of the Hydrogen Evolution Reaction on the Cyclic Voltammograms

The voltammograms of Fig. 1 display a pair of peaks which is typical of the adsorption and desorption of an ionic species from the electrolyte. This species has been identified as Cl^-. An analogous interpretation has been given by *Brisard* et al. [17] who found a similar voltammogram in 0.01 M $HClO_4$ containing small quantities of Cl^- (2×10^{-4} – 8×10^{-4} mol/l). Such an assignment is also confirmed by the UHV transfer experiments as well as by the STM images presented above. The significant difference of the peak areas of the anodic and the cathodic peak, however, requires further discussion.

The larger area of the cathodic peak has already been explained tentatively by the superposition of the hydrogen evolution current. In the following this effect will be discussed in more detail. The influence of the hydrogen evolution current can easily be eliminated from the voltammogram if its size is known for each potential. In this case it is simply subtracted from the total current of the voltammogram, leaving only the contributions of adsorption and desorption and of double layer charging. The problem would be greatly simplified if the hydrogen evolution current did not depend on the Cl coverage of the surface. It will turn out, however, that this simplifying assumption does not hold for Cu(111). The hydrogen evolution current for bare Cu(111) can be obtained from the leftmost portion of each voltammogram. A Cl free surface is found cathodic of the Cl^- desorption peak during the cathodic scan and also cathodic of the adsorption peak during the anodic scan where the latter spans a much larger potential range. The exact match of the current signal before and after scan reversal at the cathodic limit testifies that there is no diffusion limitation to the hydrogen evolution current. The current in this potential region is therefore essentially given by the stationary hydrogen evolution current (apart from a small contribution of double layer charging). It can be fitted by a Butler–Volmer relationship:

$$j_{\text{cat}}(E) = -j_0 \exp\left(-\frac{\alpha(E - E_0)F}{RT}\right) , \qquad (2)$$

where j_0 is the exchange current density, α is the transfer coefficient and F, R and T have their usual meaning. The equilibrium potential E_0 is calculated from the Nernst equation. For our reference electrode $E_0 = -0.809\,\text{V}$ and $E_0 = -0.750\,\text{V}$ is obtained in 1 mM HCl and 10 mM HCl, respectively. The potential region where the current data are fitted is sufficiently cathodic of E_0 so that only the cathodic partial current density j_{cat} needs to be considered in (2). If the current recorded during the anodic scan between the lower potential limit and about $-0.92\,\text{V}$ is fitted to (2) exchange current densities $j_0 = 0.029 \pm 0.014\,\mu\text{A cm}$ and $j_0 = 0.07 \pm 0.005\,\mu\text{A cm}$ and transfer coefficients $\alpha = 0.45 \pm 0.05$ and $\alpha = 0.5 \pm 0.02$ are obtained for 1 mM HCl and 10 mM HCl, respectively. These values are in reasonable agreement with literature data for polycrystalline Cu [18,19].

The hydrogen evolution current found by this procedure is now extrapolated over the entire potential range of the voltammogram and then subtracted from the measured data. The result of this calculation is shown as the dashed line in Fig. 1. For the sake of clarity the calculated curve is shown only where there is a substantial deviation from the measured curve (solid line). As one gets closer to the equilibrium potential the influence of the hydrogen evolution current becomes quite small and both curves coincide. Cathodic of about $-0.9\,\text{V}$ the dashed curve lies clearly above the solid curve. The area under the cathodic peak is noticeably reduced, but it is still obviously larger than that of the anodic peak. If we had completely removed all faradaic currents from the voltammogram the total charge under the dashed curve for the anodic and the cathodic scan should exactly cancel. Since this is not the case we have to drop our assumption that the hydrogen evolution current depends on the potential only, but not on the Cl coverage of the surface. A coverage-independent hydrogen evolution current also does not explain the typical features of the voltammograms shown in Fig. 2, particularly the reversal of current direction during the anodic scan. In order to correctly reproduce the observed shape of the voltammograms the model has to be refined by introducing a coverage-dependent exchange current density $j_0 = j_0(\theta_{\mathrm{Cl}})$, where θ_{Cl} is the Cl coverage of the surface.

The shape of the function $j_0(\theta_{\mathrm{Cl}})$ is by no means obvious. On the other hand, one would have to know $j_0(\theta_{\mathrm{Cl}})$ in advance to be able to disentangle the different contributions to the voltammogram. Therefore we do not attempt to extract $j_0(\theta_{\mathrm{Cl}})$ from our experimental data, but we take a different approach instead. Constructing a simple model, we show that it reproduces all the essential features of the measured voltammograms.

In our model we assume a rate k_a for Cl^- adsorption which depends on the potential as well as on the coverage:

$$k_a = k_0^a c_{\mathrm{b}}(1-\theta)\exp[\alpha_{\mathrm{ad}} f(E - E_{0\mathrm{ad}})]\exp(-\gamma\theta) , \tag{3}$$

where k_0^a is the rate constant, c_{b} is the bulk electrolyte concentration, θ is the relative surface coverage, i.e. the coverage ratioed by the maximum possible coverage $\theta_{\max}$. $E_{0\mathrm{ad}}$ is the equilibrium potential for adsorption, f is defined as $\frac{F}{RT}$ and α_{ad} is the transfer coefficient for adsorption. In our calculations α_{ad} is always set to $1/2$ for the sake of simplicity. While the expression $\alpha_{\mathrm{ad}} f(E - E_{0\mathrm{ad}})$ describes the potential dependence of the free energy of adsorption, its coverage dependence is accounted for by the argument $\gamma\theta$ of the second exponential. For the desorption rate we set:

$$k_d = k_0^d \theta \exp[-(1-\alpha_{\mathrm{ad}}) f(E - E_{0\mathrm{ad}})]\exp(\gamma\theta) \tag{4}$$

with the rate constant k_0^d and the other symbols having the same meaning as above. In equilibrium we have $k_a = k_d$ and it follows that:

$$\frac{\theta}{1-\theta} = \frac{k_0^a c_{\mathrm{b}}}{k_0^d}\exp[f(E - E_{0\mathrm{ad}})]\exp(-2\gamma\theta) . \tag{5}$$

This is the Frumkin isotherm where γ is related to the parameter g conventionally used in the Frumkin isotherm by the equation

$$\gamma = \frac{g\theta_{\max}}{RT} . \tag{6}$$

This model for adsorption and desorption in the a bsence of faradaic currents has been discussed in the literature [20].

We introduce additionally a coverage-dependent hydrogen evolution exchange current density $j_0(\theta)$ of the form:

$$j_0(\theta) = j_{0\mathrm{Cu}}(1-\theta) + j_{0\mathrm{Cl}}\theta . \tag{7}$$

In this equation $j_{0\mathrm{Cu}}$ represents j_0 for a Cl free Cu(111) surface whereas $j_{0\mathrm{Cl}}$ is the value of j_0 for a surface with maximum Cl coverage, i.e. $\theta = 1$. (Recall that θ is always given in units of $\theta_{\max}$ which is not necessarily a complete monolayer.) This linear function is the simplest possible form of $j_0(\theta)$. This is obviously the right choice if the Cl^- adsorption follows an island growth mechanism. There is no evidence from our experimental data that this is really the case. For Cl adsorption on Cu(111) in UHV, by contrast, a gradual compression of the adsorbate layer is observed as the Cl coverage is increased, at least for coverages above 1/3 of a monolayer [21]. Nevertheless we start our model calculations with (7). Even for this simple form of $j_0(\theta)$ the essential features of the voltammograms can be reproduced. Figure 13 displays a set of simulated voltammograms which are obtained from the model equations (3–7). The scan rate v for the voltammograms in Fig. 13a,b,c are set equal to those of the respective experimental voltammograms in Fig. 2. The remaining parameters of the model have the values $k_0^a c_\mathrm{b} = k_0^d = 0.3\,\mathrm{s}^{-1}, \gamma = -1.8, E_{0\mathrm{ad}} = -0.835\,\mathrm{V}$. The maximum coverage is $\theta_{\max} = \frac{1}{3}\Gamma_\mathrm{Cu}$ where Γ_Cu is the area density of Cu atoms in the (111) face $\Gamma_\mathrm{Cu} = 1.767 \times 10^{15}\,\mathrm{cm}^{-2}$. In each panel of Fig. 13 the contributions to the current density from adsorption and desorption j_θ and those from the hydrogen evolution reaction $j_\mathrm{h.e.}$ are shown separately as the dash-dotted and the dashed curve, respectively. The sum of both, $j_\mathrm{sum} = j_\theta + j_\mathrm{h.e.}$, is shown as the solid curve. This curve is to be compared with the experimental observations. j_θ is calculated according to the equation:

$$j_\theta = e\theta_{\max}\frac{\mathrm{d}\theta}{\mathrm{d}t} = e\theta_{\max}(k_a - k_d)\,, \tag{8}$$

where e is the elementary charge and k_a and k_d are given by equations (3) and (4). $j_\mathrm{h.e.}$ is calculated from the Butler–Volmer equation with j_0 given by (7) and $j_{0\mathrm{Cu}} = 0.8\,\mu\mathrm{A\,cm}, j_{0\mathrm{Cl}} = 15\,\mu\mathrm{A\,cm}$. In the model calculations the adsorption and desorption charge under the j_θ curve are exactly equal as it must be. There is a substantial separation between the peak positions in the 10 mV/s curve due to the relatively slow adsorption and desorption kinetics. This separation as well as the peak heights shrink as the scan rate is reduced in agreement with the results of *Angerstein-Kozlowska* et al. [20]. In our context, however, the behavior of $j_\mathrm{h.e.}$ is more interesting. The hydrogen evolution current displays a peak both in the cathodic and in the anodic scan. This peak is due to the interplay of the potential and the coverage dependence of $j_\mathrm{h.e.}$. At the start of the cathodic scan the surface is fully covered with Cl and the higher of the two j_0 values, $j_{0\mathrm{Cl}}$ applies. As the potential is lowered the hydrogen evolution current increases exponentially due to the exponential potential dependence in the Butler–Volmer equation. At sufficiently low potentials Cl^- desorption sets in, leading to a gradual decrease of $j_0(\theta)$. This effect overcompensates the increase of the exponential and $j_\mathrm{h.e.}$ becomes smaller again. Finally when the Cl^- is completely desorbed, the current increases again exponentially with $j_0(\theta) = j_{0\mathrm{Cu}}$. During the anodic scan the

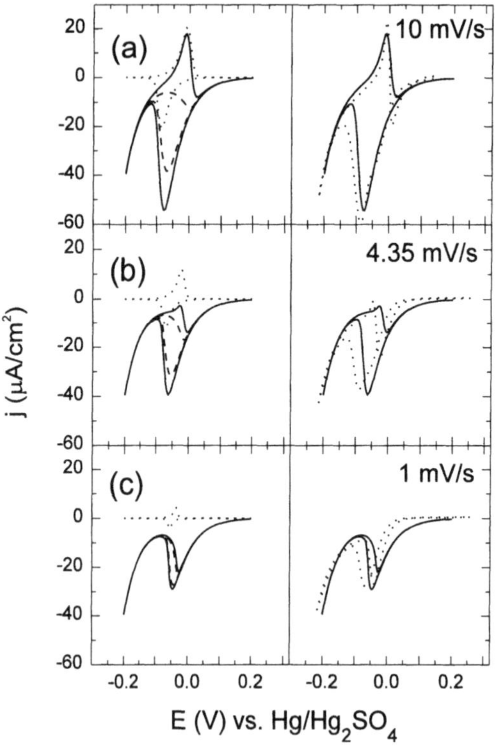

Fig. 13. (**a**)–(**c**) Calculated voltammograms, according to the model (3)–(8) (left panel of each row). The parameters are $k_0^a c_b = k_0^d = 0.3\,\mathrm{s}^{-1}, \gamma = -1.8, E_{0\mathrm{ad}} = -0.835\,\mathrm{V}, \theta_{\max} = \frac{1}{3}\Gamma_{\mathrm{Cu}}$. The potential scan rate v is indicated in each row. The individual contributions to the current density are marked by different line styles: j_θ (*dotted*), $j_{\mathrm{h.e.}}$ (*dashed*), solid line: $j_{\mathrm{sum}} = j_\theta + j_{\mathrm{h.e.}}$ (*solid*). In the right panel of each row j_{sum} (*solid*) is shown together with the experimental data of Fig. 2 (*dotted*) for comparison

same steps occur in reverse order. It is noteworthy that even for $v = 10\,\mathrm{mV/s}$ the contribution of $j_{\mathrm{h.e.}}$ to the cathodic peak is markedly larger than that of the Cl^- desorption j_θ. In the anodic scan the coaddition of the anodic peak in j_θ and the cathodic peak in $j_{\mathrm{h.e.}}$ forms the double peak structure which was also observed in the experiment. For $v = 4.35\,\mathrm{mV/s}$ the anodic peak of j_θ is smaller than for $v = 10\,\mathrm{mV/s}$. At the same time the peak of $j_{\mathrm{h.e.}}$ on the anodic scan becomes larger because the Cl^- adsorption starts at more cathodic potentials. The consequence of these two effects is that the cathodic part of the double peak structure in the anodic scan prevails over the anodic part in agreement with the experimental results. For the lowest scan rate, $v = 1\,\mathrm{mV/s}$, the same tendency results in the complete disappearance of the anodic part of the double peak and a cathodic peak appears in the anodic scan as well as in the cathodic scan.

In summary, the qualitative agreement between our simple model and the experiment is excellent. It should also be pointed out that the same parameter set (except for the value of v) has been used in all calculated voltammograms. All the typical features of the experimental voltammograms are explained by the model. Upon closer inspection it turns out, however, that there are some systematic deviations between the calculated and the measured data which are due to imperfections of the model. The cathodic peak in the simulations is somewhat narrower than in the experiment whereas the anodic double peak is markedly broader. This discrepancy is most pronounced for the higher scan rates, indicating that it has its origin in an oversimplification of the adsorption/desorption model. Firstly, the assumption $k_a = k_d$ is probably not justified. These two parameters influence the width of the adsorption and desorption peak, respectively. Secondly, the linear expression $\gamma\theta$ is not likely to completely describe the interactions within the adlayer.

The shoulder of the cathodic peak is also not reproduced by the model. It may be explained by a two-step desorption process. An alternative explanation is a nonlinear relationship $j_0(\theta)$. If j_0 decreases faster for coverages near θ_{max} a current dip occurs on the anodic flank of the desorption peak, giving rise to a shoulder or even a second peak.

Finally, the initial rise of the cathodic current is steeper in the experiment than in the simulations. This discrepancy could be corrected by the introduction of a coverage-dependent transfer coefficient $\alpha(\theta)$ for the hydrogen evolution current.

The possible improvements of the model sketched in the preceding paragraphs would require the addition of several new adjustable parameters. This would introduce new uncertainties, which are probably not outweighed by the new insight gained from such a refined model.

While our simple model explains the qualitative features of a single set of voltammograms, the origin of the observed time dependence of the voltammograms (Fig. 3) still remains unclear. In terms of our simple model the adsorption and desorption kinetics grow faster after prolonged cycling, and at the same time the enhancement of the hydrogen evolution reaction becomes more pronounced. There is no obvious reason for the improved kinetics of both processes from our experimental data. It should be mentioned, however, that long-range structural changes of the Cu(111) surface are observed in the STM images on a similar time scale. Particularly, the surface roughness is gradually reduced, small islands and vacancies on larger terraces disappear. At the same time the average terrace size grows and steps bunch together [12]. To our knowledge there is, however, no simple model which would allow one to predict the influence of such structural rearrangements on the observed kinetics of adsorption/desorption and faradaic processes.

4.2 Influence of the Emersion Step on the Transfer Experiment

The transfer experiments show that after emersion from a moderately concentrated HCl solution (10^{-3} M) the amount of adsorbed Cl does not necessarily represent the in situ situation (Fig. 6). This discrepancy was tentatively explained by post-emersion changes of the electrode state and a more quantitative approach was attempted in a re-immersion experiment (Fig. 10). Inspection of the re-immersion curves reveals immediately that the distortion of the in situ situation is most dramatic near the cathodic potential limit of the voltammogram. Indeed, the Cl^- coverage is expected to be zero in this potential region, but is still observed to be at the maximum level after the transfer. On the other hand, near the anodic potential limit the maximum Cl coverage is expected and observed, so the deviation from the in situ situation is small in accordance with the re-immersion transients (Fig. 10). The exponential increase of the re-immersion charge as the emersion potential is made more cathodic gives a strong hint that the observed effects are caused by the faradaic hydrogen evolution current. In this subsection we will elaborate these ideas in some detail.

The effects of the emersion step on a transfer experiment have already previously been discussed in the literature. Under electrochemical conditions the solution part of the electrode–electrolyte interface can be roughly subdivided into the bulk electrolyte and a region close to the electrode where the structure of the assembly of solvent, anions and cations is influenced by the electrode. This region, which is termed the double layer in conventional electrochemistry, is about 10 Å thick in fairly concentrated electrolytes. It is a constituent part of the electrode–electrolyte interface, and therefore it is of primary concern whether or not this region is transferred into UHV unaltered. A number of studies have addressed this question, and it was found that under favorable conditions the transfer of an intact double layer succeeds [22,23,24,25,26,27,28,29]. All these studies have in common that the emersion was performed at potentials in the so-called double layer region, i.e. at potentials where no faradaic currents are observed. The situation changes essentially in the presence of faradaic processes. Faradaic currents arise from electrochemical reaction with species from the bulk electrolyte. They can obviously not be sustained if the electrode is withdrawn from the electrolyte, since the supply of bulk species breaks down immediately. On the other hand, the interface is in a dynamic equilibrium which requires the reaction to proceed. If the reaction stops the state of the interface characterized by the double layer structure and the electrode potential has to change. In our system the faradaic reaction is hydrogen evolution, i.e. the reaction in the electrolyte is

$$2\,H_3O^+ + 2\,e^- \rightarrow H_2 + 2\,H_2O\,. \tag{9}$$

The reaction equation allows us to determine what changes have to appear in the double layer after the contact with the electrolyte has been lost. In the very first moment after the emersion the reaction proceeds as before since the dynamic equilibrium state of the double layer has not yet changed. H_3O^+ ions are supplied from the remaining electrolyte on the electrode surface, but electrons from the outer circuit are no longer available. Therefore, electrons are delivered by the chloride in the remaining electrolyte concomitant with the formation of an adsorbed Cl layer. The net reaction is now

$$2\,H_3O^+ + 2\,Cl^- \rightarrow H_2 + 2\,H_2O + 2\,Cl_{ads}\ , \tag{10}$$

explaining the post-emersion adsorption of additional chloride. It is also clear that this effect is stronger for higher hydrogen evolution reaction rates, i.e. for more cathodic potentials. An equivalent way of viewing this process is a relaxation of the electrode potential into the chloride adsorption region.

The above reaction is not directly observable because at the time when it occurs the outer circuit is open. An observation of the reverse process, which occurs upon re-immersion of the electrode, is, however, possible. If the emersion and re-immersion is performed at a potential where the adsorbed Cl is stable nothing happens after the re-immersion and the current transient is flat. If, however, the same experiment is done in a potential region where in situ no Cl is adsorbed then the additional Cl which was adsorbed after the emersion according to (10) is desorbed upon re-immersion, giving rise to a cathodic current in the current transient. This current is indeed observed in the transient for $-1.0\,V$ in Fig. 9.

In agreement with our explanation additional post-emersion adsorption of Cl^- is not observed for potentials where there are no faradaic currents in the voltammogram. Accordingly, there is virtually no charge under the re-immersion transients in this potential region.

As already mentioned, the additional Cl^- adsorption can be partly suppressed by diluting the electrolyte. However, it cannot be completely avoided by this method, as can be seen from the re-immersion curve for 10^{-4} M HCl in Fig. 10. The effect is somewhat weakened, but the curve appears to be rather ill-defined. Accordingly, there is still a significant amount of Cl on the surface after transfer at the cathodic potential limit (Fig. 7).

A much better result is achieved with the mixed electrolyte (Fig. 8). For this more concentrated electrolyte again strong post-emersion adsorption occurs, as gleaned from Fig. 10, but now the additionally adsorbed anions are supplied by the majority species in the electrolyte, which is sulfate in this case. This becomes apparent from the increasing S 1*s* signal in the cathodic potential limit (Fig. 8).

In principle the amounts of adsorbed species measured in the ex situ experiment can be corrected, using the re-immersion transient. From the transient the quantity of the surplus adsorbate can be calculated and this value should be subtracted from the measured one. This method does not work very well, however, since the values obtained from the transients are much too high. This is probably due to the fact that in the re-immersion experiment not only the front face of the sample is wetted, but there are also contributions from the very rough side faces. Unfortunately, in an immersion experiment the usual meniscus configuration cannot be used to limit the wetting to the well-defined front face of the sample. A "fudge factor" for the correction of this effect may be estimated from the faradaic contribution in the transients, which is also much larger than in the meniscus configuration. Such an estimate shows that the re-immersion charges have roughly the correct magnitude to be consistent with our interpretation of the voltammogram, but a more precise quantification is presently not possible.

5 Summary

In this chapter the adsorption of chloride on the Cu(111) surface was presented as a model system for a study combining in situ and ex situ methods in the same experiment. The structure of the adsorbate layer and its potential-dependent formation were observed by in situ STM whereas its chemical identity was found by ex situ spectroscopies. The influence of the adsorbate on the kinetics of the hydrogen evolution reaction was studied using the traditional electrochemical method of cyclic voltammetry in conjunction with a simple model calculation. Some peculiarities of UHV transfer experiments in the presence of faradaic currents were discussed as well as possible workarounds for the unavoidable problems of such an approach.

References

1. D. M. Kolb: In J. Lipkowski, P. N. Ross (Eds.): *Structure of Electrified Interfaces* (VCH, Weinheim, New York 1993) p. 65
2. X. Gao, A. Hamelin, M. J. Weaver: J. Chem. Phys. **95**, 6993 (1991)
3. X. Gao, A. Hamelin, M. J. Weaver: Phys. Rev. Lett. **67**, 618 (1991)
4. M. S. Zei, G. Qiao, G. Lehmpfuhl, D. M. Kolb: Ber. Bunsenges. Phys. Chem. **91**, 349 (1987)
5. N. Marković, P. N. Ross: Langmuir **9**, 580 (1993)
6. H. Matsumoto, J. Inukai, M. Ito: J. Electroanal. Chem. **379**, 223 (1994)
7. I. Oda, Y. Shingaya, H. Matsumoto, M. Ito: J. Electroanal. Chem. **409**, 95 (1996)
8. J. Inukai, Y. Osawa, K. Itaya: J. Phys. Chem. **B102**, 10034 (1998)
9. M. Kruft, B. Wohlmann, C. Stuhlmann, K. Wandelt: Surf. Sci. **377–379**, 601 (1997)
10. B. Wohlmann, Z. Park, M. Kruft, C. Stuhlmann, K. Wandelt: Colloids Surf. A **134**, 15 (1998)
11. M. Wilms, P. Broekmann, M. Kruft, C. Stuhlmann, K. Wandelt: Appl. Phys. A **66**, S 473 (1998)
12. P. Broekmann, M. Wilms, M. Kruft, C. Stuhlmann, K.Wandelt: J. Electroanal. Chem. **467**, 307 (1999)
13. D. Westphal, A. Goldmann: Surf. Sci. **131**, 113 (1983)
14. D. Westphal, A. Goldmann: Surf. Sci. **131**, 92 (1983)

15. K. K. Kleinherbes, A. Goldmann: Surf. Sci. **154**, 489 (1985)
16. H. Niehus, W. Heiland, E. Taglauer: Surf. Sci. Rep. **17**, 213 (1993)
17. G. Brisard, E. Zenati, H. Gasteiger, N. Marković, P. N. Ross: Langmuir **11**, 2221 (1995)
18. A. Hickling, F. W. Salt: Trans. Faraday Soc. **36**, 1226 (1940)
19. B. E. Conway, J. O'M. Bockris: J. Chem. Phys. **26**, 532 (1957), and references therein
20. H. Angerstein-Kozlowska, J. Klinger, B. E. Conway: J. Electroanal. Chem. **75**, 45 (1977)
21. W. K. Walter, D. E. Manolopoulos, R. G. Jones: Surf. Sci. **348**, 115 (1996)
22. W. N. Hansen, D. M. Kolb: J. Electroanal. Chem. **100**, 493 (1979)
23. W. N. Hansen: J. Electroanl. Chem. **150**, 133 (1983)
24. G. J. Hansen, W. N. Hansen: J. Electroanal. Chem. **150**, 193 (1983)
25. S. Trasatti: Pure Appl. Chem. **58**, 621 (1986)
26. R. Kötz, H. Neff, K. Müller: J. Electroanal. Chem. **215**, 331 (1986)
27. D. M. Kolb: Z. Phys. Chem. NF **154**, 179 (1987)
28. W. N. Hansen, G. J. Hansen: In M. P. Soriaga (Eds.): *Electrochemical Surface Science,* ACS Symp. Ser. **378**, 166 (1988)
29. Z. Samec, B. W. Johnson, K. Doblhofer: Surf. Sci. **264**, 440 (1992)

SHG Studies on Halide Adsorption at Au(111) Electrodes

Bruno Pettinger[1], Christoph Bilger[1], and Jacek Lipkowski[2]

[1] Fritz-Haber-Institut der Max-Planck-Gesellschaft, Faradayweg 4-6, 14195 Berlin (Dahlem), Germany
[2] Department of Chemistry and Biochemistry, University of Guelph, Guelph, Ontario, N1G 2W1, Canada
pettinger@fhi-berlin.mpg.de

Abstract. Second-order optical nonlinear spectroscopy such as Second Harmonic Generation (SHG) is inherently sensitive to interfacial regions. Using our recently developed method of Interference SHG Anisotropy (ISHGA), the anisotropy of the SHG field $E(2\omega, \varphi) \propto \sum_{n=0}^{m} k_n \cos[n\ (\varphi + \zeta_n)] = A + B\cos(\varphi + \alpha) + \ldots$ can be measured (i.e., the variation of the 2ω-field generated during the rotation of the sample around the surface normal). This field can be separated into its various sources, denoted as A, B, ..., permitting us to achieve information on the geometric and electronic structure of the interface and their changes upon potential and adsorption.

In this chapter we report on halide ion adsorption at Au(111) electrodes employing the ISHGA technique. The results show that even the isotropic contribution to SHG encodes a remarkable dependence on the various adlayer structures such as the formation of *disordered*, *ordered* and *compressed* adlayers in the case of specific adsorption. This contrasts with earlier assumptions that the isotropic part of the EM field (the A term) should be insensitive to structural changes of surface. Evidently, the isotropic property refers to its insensitivity with respect to the azimuthal rotation of the sample. But this does not rule out the possibility that the magnitude and complex phase of the A term depend on the interfacial structure and composition.

1 Introduction

Solely electrochemical investigations yielding information on potentials, currents and capacities have led to important insights into the electric double layer. However, such investigations alone are not sufficient to gain a detailed microscopic picture of the electrochemical processes occurring on atomic and molecular levels at the electrode/electrolyte boundary. Hence, these interfacial processes depend crucially on the local electronic and geometrical structures of the surface as well as on the adsorbate–substrate interactions. To understand and to describe electrochemical reactions, detailed information is required on the interfacial structures as well as on how they vary with potential and adsorption of molecules and ions.

For this purpose, a series of modern in situ methods was developed such as Scanning Tunneling Microscopy (STM), Atomic Force Microscopy (AFM)

K. Wandelt, S. Thurgate (Eds.): Solid–Liquid Interfaces, Topics Appl. Phys. **85**, 223–240 (2003)

and X-ray reflection and diffraction. They are able to reveal interfacial structures in situ with atomic resolution. Also in the optical field a number of in situ techniques has been developed. Widely used are, for example, infrared (IR) and Fourier Transform Infrared Spectroscopy (FTIR), but also Raman and Surface Enhanced Raman Spectroscopy (SERS), since they all belong to the class of vibrational spectroscopies which are powerful methods of identifying adsorbates and intermediates and to study their interaction with the substrate.

Second-order nonlinear optical spectroscopies were found to be particular suitable for in situ studies by two specific features: They exhibit (i) an inherent surface sensitivity, (ii) a radiation which is easily separable in space, frequency and time. If the interface is formed by bringing two centrosymmetric media into contact (such as fcc metals and electrolytes), the nonlinear radiation stems (nearly) solely from the interfacial region, because only there is the centrosymmetry lifted and, thus, SHG is permitted [1]. In addition, the radiation is emitted as a beam into a distinct direction; its frequency is the sum or difference frequency, $\omega = \omega_1 \pm \omega_2$, and its phase has a well-defined relationship with the exciting electromagnetic wave. In recent years, innumerable experimental and theoretical publications illustrate that the advent of nonlinear spectroscopies has opened new avenues for exciting research at interfaces.

In this chapter we report on halide ion adsorption at Au(111) electrodes using the recently developed method of Interference Second Harmonic Generation Anisotropy (ISHGA). The results achievable by this technique will show that even the isotropic contribution to SHG encodes a remarkable dependence on the various adlayer structures.[1]

2 SHG Anisotropy at Metal Electrodes

Electromagnetic waves induce in matter a polarization which has linear and, for sufficiently intense field strengths, also nonlinear components. The latter can lead to radiation at $\omega_1 + \omega_2$ and higher frequencies. If $\omega_1 = \omega_2 = \omega$, the second-order process is known as Second Harmonic Generation (SHG) emitting 2ω radiation [1].

2.1 SHG Anisotropy

At interfaces, the second-order polarization vector can be split into its components perpendicular and parallel to the surface [1]. In general, the former is

[1] This contrasts with earlier assumptions that the isotropic part (the A term) should be insensitive to structural changes of the surface. Evidently, the isotropic property refers to its insensitivity with respect to the azimuthal rotation of the sample; this does not rule out the possibility that the magnitude and complex phase of the A term depend on the interfacial structure.

sensitive to all effects which modify the electronic configuration perpendicular to the surface; among them are the electric field across the interface or the adsorption of molecules or ions. The latter probe the rotational symmetries of a surface of a crystalline sample which are related to the current surface structure(s); for electrodes they depend again on potential and adsorption, but in a different way. Keeping the zenith angle ϑ fixed, one can record the SHG intensity as a function of the polar angle φ,[2] i.e., one records the SHG anisotropy. Usually it is given as a plot of the SHG intensity vs. the azimuthal angle, φ. Evidently, at every angle φ, the SHG intensity results from a distinct superposition of the isotropic and anisotropic contributions to SHG. Thus, a characteristic anisotropy pattern can be related to a specific surface symmetry. In principle, the observed pattern can be analyzed, providing an easy way to monitor the (∞), one-, two- and threefold rotational symmetries of the sample and their possible changes in the course of experiments.[3]

One of the first reports on SHG anisotropy was given by *Heinz* and coworkers for a Si(111)-7 $\times$ 7 surface in UHV, clearly illustrating the rotational symmetry of the 7 $\times$ 7 reconstructed surface [3]. *Shannon* et al. were the first to observe SHG anisotropy at metal electrodes [4,5,6]. The authors correlated its pattern with interfacial symmetries and its variation by changes of the nonlinear susceptibility and surface structure.

These reports mark the rise of a new research field, SHG anisotropy at single-crystalline electrodes. One should be aware that at that time SHG could be counted among the rather few structural sensitive methods which could be applied in situ in electrochemistry. Among them were the technique of the emersion of electrodes and their subsequent transfer into a vacuum chamber to perform detailed surface structural studies [7], electroreflectance [8] and surface plasmon spectroscopy [9], infrared [10] and Raman spectroscopy [11] and, finally, SHG [12].

2.2 Earlier SHG Anisotropy Studies on Halide Adsorption

Friedrich et al. have shown that surface reconstruction at Au(111) and Au(100) electrodes alters the SHG anisotropy remarkably [13]. The authors monitored the transition from reconstructed to unreconstructed surfaces in situ by a change of the anisotropy patterns with potential. At negative potentials, the anisotropy displays a C_s symmetry expected for the ($\sqrt{3} \times 23$)

[2] In the literature, the rotational angle, φ, is often called the "azimuthal angle". Here, let us stick to the following conventions: The angle of rotation around the surface normal is denoted as the polar angle, φ, and the angle of incidence between the surface normal and the incident beam may be associated with the zenith angle ϑ.

[3] Note that (in the dipole approximation) SHG can only monitor contributions to the SH field by one- to three-fold rotational symmetries, related mirror planes and the isotropic component, sometimes denoted as the ∞-fold contribution.

superstructure, whereas at positive potentials a C_{3v} symmetry is observed expected for the non-reconstructed Au(111) surface. This Au(111)-(1×1) $\longleftrightarrow$ Au(111)-($\sqrt{3} \times 23$) transition is both potential and adsorption controlled and it is reversible [13,14,15,16,17].

In a number of papers *Pettinger* et al. reported on the influence of ionic and molecular adsorption on surface reconstruction and also on the isotropic part of the SH response at Au(111), Au(110) and Au(100) electrodes [15,16,17,18,19,20,21,22,23,24,25,26]. Here we will address only the case of ionic adsorption [15,16,17,18,19,20,21,22]. For details on SHG and molecular adsorption we refer to the literature [23,24,25,26].

Figure 1 presents two three-dimensional plots which illustrate how the Au(110) anisotropy pattern changes with potential in the absence and in the presence of bromide ions in solution. These ions are known to be desorbed at negative potentials and are adsorbed specifically at positive potentials. The left plot displays complicated patterns representing a superposition of various rotational symmetries. By moving the electrode potential from negative to positive values, these patterns gradually change. The right plot first shows a similar development of the anisotropy patterns at negative potentials. In the potential regime of bromide adsorption, however, a fast change of the patterns towards that of a twofold rotational symmetry can be clearly seen. This means that we monitor here a transition of a reconstructed (microfaceted) Au(110) into an unreconstructed Au(110) surface.

These anisotropy examples illustrate the enormous power of SHG for interfacial studies. Unfortunately, in this approach only SH intensities were measured. That means part of the information present in the SH field is lost. In addition, more information remains uncovered when using simple or insuf-

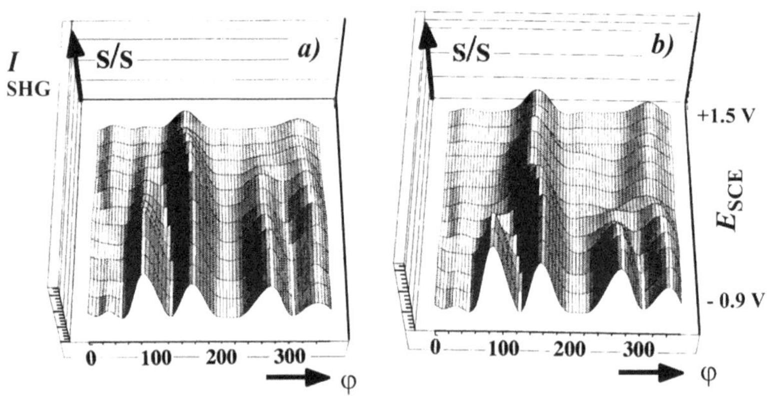

Fig. 1. Three-dimensional plots of the SHG anisotropy for a Au(110) electrode vs. the rotational angle φ and vs. the electrode potential E. (**a**) Pure 0.1 M $NaClO_4$; (**b**) 0.1 M $NaClO_4$ + 1 mM NaBr. The potential is stepped from -0.9V to $+1.2$V. (From [18])

ficient models in the analysis of the data (which, for instance, do not take into account the presence of steps and other irregularities as well as the existence of surface reconstruction domains). Wavelength and phase measurements can yield further, valuable and important information [27,28,29]. It is also important to separate the various isotropic and anisotropic contributions to SHG. In other words, one has to single out the individual sources of the SH radiation from an interfacial region without any ambiguity. As we will show in the next section, this goal requires the determination of the anisotropy of the SH field itself (rather than that of the SHG intensity). Once the anisotropy of the SH field is measured, it is a simple task to decompose it into its various symmetry contributions without using restricting assumptions (this is much in contrast to the analysis of the anisotropy of the SHG intensity, which can be difficult or even impossible, and if possible, the analysis can lead to ambiguous results).

2.3 Problems in Evaluation of Susceptibility Coefficients

It seems to be a simple and straightforward task to correlate the SHG anisotropy pattern of a sample with the presence and superposition of its rotational symmetries. If these symmetries change during the experiments, for instance, due to the formation or lifting of surface reconstruction, this leads to a different anisotropy pattern. In this way, structural changes can be monitored. Such effects have been reported already in the early study on the symmetry and disordering of a reconstructed Si(111)-(7 × 7) interface by *Tom* et al. [30]. The dependence on the azimuthal angle or on the rotation angle of a polarizer was explained by an angle-dependent susceptibility tensor [31,32]. In short, for a (111) surface and *p*-polarized incident beam and *p*-polarized emitted radiation, the SHG intensity vs azimuthal angle, φ, should be represented by the following relationship:

$$I_{pp} \propto |A + B\cos\varphi + C\cos 2\varphi + D\cos 3\varphi|^2 \; I_{\mathrm{L}}^2 \tag{1}$$

$$A = a_{333}^{(\infty)} + a_{311}^{(\infty)} + a_{113}^{(\infty)} \qquad B = a_{313}^{(1)} + a_{133}^{(1)} + a_{111}^{(1)}$$

$$C = a_{311}^{(2)} + a_{113}^{(2)} \qquad D = a_{111}^{(3)}$$

where I_{L} is the incident laser intensity and the coefficients A, B, C, D refer to the so-called isotropic and one-, two- and threefold amplitudes associated with the corresponding trigonometric terms $\cos n\varphi$, $n = \infty, 1, 2, 3$.[4] The subscript on the a_{ijk} terms indicate the ijk^{th} element of the susceptibility tensor $\chi^{(2)}$ of third rank. Evidently, even the A, B, C, D coefficients can be

[4] Note that the superscript (∞) is used in the literature to denote the isotropic nature of the associated terms. Since in the dipole approximation SHG can only resolve rotational symmetries up to a threefold one, it is more consistent to (re)define the isotropic case as $(\infty) \mapsto (0)$, which results in $\cos n\varphi = 1$. Thus, let us keep this definition in mind, but use the usual nomenclature.

composed of several terms $a_{ijk}^{(n)}$, where each term $a_{ijk}^{(n)} \cos n\varphi$ can be considered as a *basic source* to the total SH field. The superscript (n) indicates the n-fold rotational axis, where (∞) characterizes the isotropic nature of the term. Note that the coefficients A, B, C, D can be complex in general. Thus, up to eight terms can contribute to the SHG anisotropy pattern (if we only count the real and imaginary parts of A, B, C, D). Defining the SH field as

$$E(2\omega, \varphi) = A + B\cos\varphi + C\cos 2\varphi + D\cos 3\varphi\,, \tag{2}$$

we obtain

$$I_{pp} \propto \left(\mathrm{Re}\,(E(2\omega,\varphi))^2 + \mathrm{Im}\,(E(2\omega,\varphi))^2\right)\, I_{\mathrm{L}}^2\,. \tag{3}$$

If one expanded the above equation (3), one would have to sum up about 25 squared and mixed terms which constitute the overall anisotropy pattern. Not surprisingly, only for simple cases have the anisotropy patterns been analyzed with respect to all major symmetry contributions.

At silver(111) electrodes, for instance, there is no surface reconstruction. Thus, the coefficients B, C can be set to zero. Then the analysis according to (1) appears to be straightforward [2,33]. In the case of Au(111) electrodes, surface reconstruction can be present; fortunately, the B and C coefficients are usually small enough and, therefore, the coefficients A and $|D|$ can be determined together with the relative complex phase δ_{A-D} [23]. In all these cases, a priori information was used to simplify the expression (3), according to which the data were analyzed. Severe difficulties arise, however, when the anisotropy becomes rather complex and a priori information is not available. For example, the SHG anisotropy pattern of a Au(110) electrode has such a rich content that its analysis yielded unsatisfactory results, when using the usual concepts [19,34].

Furthermore, the data analysis itself bears general problems: (1) and (3) show that a SHG anisotropy curve can be described by a superposition of a few harmonic terms (in general, $n \leq 6$ or 8, but each term can be a complex quantity). In the first step, a Fourier analysis is straightforward as one gets all Fourier coefficients a_n and b_n. However, they are related to expressions containing squared terms and mixed products of the coefficients A, B, C, D; see (1) and (3). Thus, in order to evaluate these coefficients, one has to operate on a nonlinear system of equations, which can/must have multiple solutions. Indeed, either by solving this system of equations or by a fitting procedure, one can find multiple solutions, even for the case of a simple rotational symmetry, such as that of an unreconstructed, (111) oriented surface. A striking example is given in [35], showing that an anisotropy curve, which displays perfect $C_{3\mathrm{v}}$ symmetry, can be reproduced using four distinct sets of complex coefficients A, B, C, D. Thus, for a real SHG anisotropy experiment one may find various sets of coefficients, but the only one which describes the physical situation of the sample is unknown in advance.

3 Concept of Interference SHG Anisotropy

The above sketched problem of multiple solutions is due to the fact that we measure intensities of photons and not the field strengths of the emitted electromagnetic waves. In short, we measure $I^{\mathrm{SHG}} \propto \mid E\left(2\omega,\varphi\right)\mid^{2}$ and not the SH field $E\left(2\omega,\varphi\right)$.[5] In order to record the *anisotropy of the* SH *field*, we combine the methods of SHG interference and SHG anisotropy in a unique way which is sketched below. Using this approach one obtains experimental data which are necessarily linear in $E\left(2\omega,\varphi\right)$ [35]. Thus, a subsequent analysis leads straightforwardly and reliably to all coefficients and other, not yet discussed terms *without ambiguities.* In addition, this approach can be easily extended to the (up to now neglected) contributions of higher multipoles as well as to other optical spectroscopies.

To begin with, let us describe the variation of the SH field with the polar angle, φ, in a more generalized fashion:

$$E^{pp}\left(2\omega,\varphi\right)\propto A^{pp}+B^{pp}\cos\left(\varphi+\alpha\right)+C^{pp}\cos\left[2\left(\varphi+\beta\right]\right)$$
$$+D^{pp}\cos\left[3\left(\varphi+\gamma\right)\right]+\ldots, \quad (4)$$

where we introduce, offhand at this stage, angular shift terms α,β and γ. Their physical meaning will be discussed below. Note, analogous relationships hold for all polarization combinations.

Next, let us consider the combination of SHG interference and SHG anisotropy. For the interference one needs an additional SHG source, for instance, a quartz lamella. The first SHG wave is created by a quartz lamella and the second by the sample surface. Figure 2 illustrates the arrangement of the two SH sources: The quartz lamella can be placed additionally into the laser beam, in front of the sample and at a location x_1, where x_1 denotes its distance from the sample. The s or p polarized laser pulse passes the quartz lamella and creates the first SH wave; its field strength is denoted as E_q. Both the laser pulse and the first SH wave travel collinearly towards the sample and are reflected there. The laser pulse interacts with the sample surface and generates the second SH wave, denoted as $E\left(2\omega,\varphi\right)$. A filter, placed behind the sample, only blocks the fundamental laser beam. Both the first and second SH wave pass this filter and subsequently a polarizer. The latter projects out either the s or the p-polarized components of both 2ω radiations. A photomultiplier records their interference intensity during a polar angle scan of the sample (see Fig. 2).

Subsequently, the quartz lamella is moved to a different location x_2 (such that $\Delta x = x_2 - x_1 \cong 3.18$ cm). This means that the phase relationship changes by $\pi/2$ because of the different optical path lengths for the ω and 2ω beams in air $[n(\omega) \neq n(\omega)]$; this also causes a $\pi/2$ phase shift between the

[5] The second argument φ in $E\left(2\omega,\varphi\right)$ indicates that the generated SH field depends on the azimuthal angle φ of the crystalline sample.

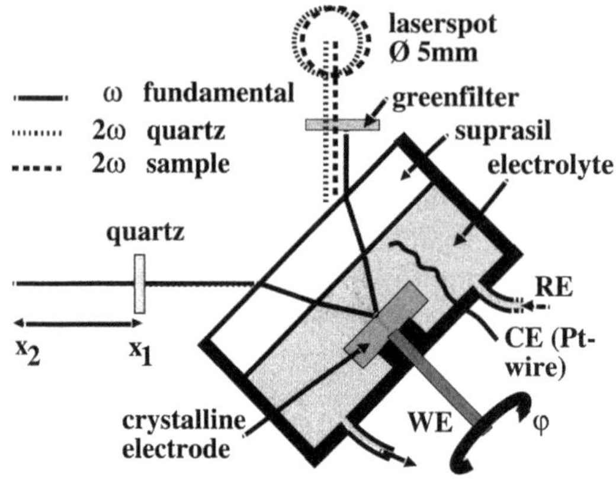

Fig. 2. Interference SHG anisotropy experiment. The first SHG wave is created by the quartz lamella (*dotted line*) located at x_1 or x_2; it propagates together with the fundamental beam toward the sample where the second SHG wave is created (*dashed line*). Upon reflection at the sample, both SHG waves travel on parallel routes and interfere at the observer point. (From [35])

two SHG waves; thus, $E_q \to iE_q$. For this configuration, the second interference SHG anisotropy curve is measured. Finally, the quartz lamella is moved off the beam, and a third, the standard SHG anisotropy curve, is recorded. Obviously, this approach requires one to record three anisotropy curves, one without the quartz lamella and two with the quartz lamella at locations x_1 and x_2. The intensities of the three anisotropy curves are described by the following most general expressions:

$$I_{pp}^{\mathrm{SHG}} \propto \left|\mathrm{Re}\left[E^{pp}(2\omega,\varphi)\right] + \mathrm{i}\ \mathrm{Im}\left[E^{pp}(2\omega,\varphi)\right]\right|^2 \tag{5}$$

$$I_{pp}^{\mathrm{ISHG}}(x_1) \propto \left|E_q + \mathrm{Re}\left[E^{pp}(2\omega,\varphi)\right] + \mathrm{i}\ \mathrm{Im}\left[E^{pp}(2\omega,\varphi)\right]\right|^2 \tag{6}$$

$$I_{pp}^{\mathrm{ISHG}}(x_2) \propto \left|\mathrm{Re}\left[E^{pp}(2\omega,\varphi)\right] + \mathrm{i}\ \left\{E_q + \mathrm{Im}\left[E^{pp}(2\omega,\varphi)\right]\right\}\right|^2 . \tag{7}$$

Depending on the location of the quartz lamella at x_1 or x_2, its SH field appears in (6) and (7) either via the terms E_q or iE_q. This concept can be used for all possible polarization combinations. Here, *pp* polarization is used.

The next step is to compute the difference curves, i.e., the difference of (6) and (5) and of (7) and (5); in addition, the SHG intensity of the quartz lamella alone, $E_q^2 \propto I_q$ (which is determined independently), is subtracted (for the sake of clarity we omit the polarization indices such as *pp*, *ps*, etc.). Finally, the difference data are divided by $2E_q$:

$$E^{\mathrm{SHG}}(x_j) = \frac{1}{2E_q}\left(I^{\mathrm{IHG}}(x_j) - I^{\mathrm{SHG}} - E_q^2\right); \quad j = 1,2 . \tag{8}$$

We arrive at

$$E^{\mathrm{SHG}}(x_1) \propto \mathrm{Re}\left[E(2\omega,\varphi)\right], \tag{9}$$

$$E^{\mathrm{SHG}}(x_2) \propto \mathrm{Im}\left[E(2\omega,\varphi)\right], \tag{10}$$

where $E^{\mathrm{SHG}}(x_j)$ are the experimental data which are proportional to the real and imaginary parts of the SH field $E(2\omega, \varphi)$, respectively. These $E^{\mathrm{SHG}}(x_j)$ data represent the experimentally measured anisotropy of the SH field. The composition of $E(2\omega, \varphi)$ is given by (4). Note that (9) and (10) hold for all cases, because in their derivation neither restrictions nor assumptions were made concerning the SH field $E(2\omega, \varphi)$. That means that also higher-order contributions to $E(2\omega, \varphi)$ are directly measurable by this approach, if they arise from n-fold rotational symmetries with $n > 3$ (see (4)); the only self-evident prerequisite is that their amplitudes are above the noise level [35,36,37]. As is well known, the complex phase of $E(2\omega, \varphi)$ (or of its constituents) is defined and measurable only against a reference phase, here relative to the phase of the SH wave from the quartz lamella at location x_1.

Since both curves, $\mathrm{Re}[E(2\omega, \varphi)]$ vs φ and $\mathrm{Im}[E(2\omega, \varphi)]$ vs φ, can be composed of an isotropic and a few harmonic terms, a Fourier analysis

$$
\begin{aligned}
a_{n,j} &= \frac{1}{\pi} \int_0^{2\pi} E^{\mathrm{SHG}}(x_j) \cos(n\varphi) \\
b_{n,j} &= \frac{1}{\pi} \int_0^{2\pi} E^{\mathrm{SHG}}(x_j) \sin(n\varphi)
\end{aligned}
\tag{11}
$$

yields the Fourier coefficients $a_{n,\,j}$ and $b_{n,\,j}$ with $j = 1$ or 2. They contain all symmetry information and therefore permit an easy evaluation of the full set of isotropic and anisotropic coefficients A , B, C, $D, \ldots$ and also the "geometric" angular offset terms α, β and γ. This is shown by the following set of equations:

$$
\begin{aligned}
A_r &= \tfrac{1}{2}\, a_{0,1} \\
k_{n,r} &= \{B_r, C_r, D_r, \ldots\} = \left.\sqrt{a_{n,1}^2 + b_{n,1}^2}\right|_{n=1,2,3,\ldots} \\
\zeta_{n,a} &= \{\alpha_a, \beta_a, \gamma_a, \ldots\} = \arctan\left(\tfrac{b_{n,1}}{a_{n,1}}\right)_{n=1,2,3,\ldots}
\end{aligned}
\tag{12}
$$

$$
\begin{aligned}
A_i &= \tfrac{1}{2}\, a_{0,2} \\
k_{n,i} &= \{B_i, C_i, D_i, \ldots\} = \left.\sqrt{a_{n,2}^2 + b_{n,2}^2}\right|_{n=1,2,3,\ldots} \\
\zeta_{n,b} &= \{\alpha_b, \beta_b, \gamma_b, \ldots\} = \arctan\left(\tfrac{b_{n,2}}{a_{n,2}}\right)_{n=1,2,3,\ldots}
\end{aligned}
\tag{13}
$$

where the polarization indices have been omitted, and again the dots mean a possible extension of this approach to higher rotational symmetries with $n > 3$. The angular offset terms α, β and γ have a clear *geometric* meaning, discussed in detail in [35,58].[6]

[6] Note that there are two sets of offset angles α_a, β_a, γ_a and α_b, β_b, γ_b as well as two sets of coefficients B_r, C_r, D_r and B_i, C_i, D_i. Whereas the latter two together constitute the set of the complex coefficients B, C, D, the concept of angular offset requires that each of the two sets contain real offset angles. Therefore, the indices are denoted as a and b.

The great advantage of this approach is that the thus determined susceptibility coefficients (together with the angular offset terms) reproduce the experimental data exceptionally well in all situations. In the following we are mainly concerned with the variation of the isotropic term A with potential and adsorption. The ISHGA approach is the only practicable method which allows one to separate isotropic and anisotropic contributions with sufficient accuracy.

4 Halide Ion Adsorption on Au(111)

The Au(111) electrode is structurally much simpler than the Au(110). This surface either shows the non-reconstructed state, denoted as Au(111)-(1×1), or the well-known $\sqrt{3} \times 23$ reconstruction. The latter occurs in all three equivalent $[1\bar{1}0]$ directions. This reconstruction is present at negatively charged surfaces and it is lifted at positively charged metal surfaces. There is a large body of literature on this reconstruction and how it is influenced by potential and adsorption. In the 1980s *Kolb* and coworkers used LEED to investigate this surface [38,39,40,41,42]; in the 1990s a variety of new methods became available for in situ studies of this reconstruction such as STM [43,44,45,46,47], X-ray [48,49,50,51,52,53] and SHG [13,16,17,23].

When using SHG, the reconstruction can be observed only if the reconstruction domains are distributed unequally. If this is not the case, the corresponding anisotropy pattern must exhibit a C_{3v} symmetry, although the surface is reconstructed. However, there are no ideal surfaces, i.e., there are always steps and other defects present. This results in slightly unequally distributed reconstruction domains, and therefore the SHG anisotropy shows weak additional onefold and twofold symmetry contributions (the B and C terms). However, this weakness spoiled (prior to ISHGA) a sufficiently precise evaluation of the various susceptibility coefficients. Nevertheless, specific adsorption of ions on Au(111) causes a significant decrease of these B and C terms which evidences the lifting of the surface reconstruction upon specific adsorption [13,16,17,23].

In addition, in these earlier publications the influence of adsorption on the isotropic term A was also investigated. Linear relationships of $A_r = \mathrm{Re}(A)$ with the metallic charge density, σ_M, have been reported in several cases but not for the imaginary part, $A_i = \mathrm{Im}(A)$. In these studies, however, the phase of the A term δ_A could not be determined independently, because only its phase relative to the phase of the threefold term D, $\delta_{A-D} = \delta_A - \delta_D$, could be evaluated. Since the absolute phase is meaningless, only the relative phases carry physical information and, therefore, δ_{A-D} is a good quantity as long as δ_D is a constant. However, if $\delta_D \neq$ const., we have the paradoxical result that $|A|$ and $|D|$ are correctly determined, but not the quantities $\mathrm{Re}(A)$ and $\mathrm{Im}(A)$, because A rotates in the complex plane with δ_D. Initially, the constancy of δ_D was postulated by assuming that the threefold amplitude D arises from the nonlinear response of the sample in the plane of the surface. Thus, D should not show a potential dependence. And in fact, in the absence of specific adsorption the $|D|$ term was found to be constant over a wide potential range, and this constancy was "transferred" to δ_D. Most interestingly, by employing the ISHGA method, a significant variation of δ_D was observed even in the absence of specific adsorption. Figure 3 shows the influence of the electrode potential on the D term, in the top and bottom panel arrangement for $|D|$, and δ_D. The four curves in each panel refer to distinct electrolytes. In the absence of specific adsorption, i.e., in pure perchlorate solution, the $|D|$ term is nearly potential independent. If halide ions are added to the electrolyte, the $|D|$ term decreases upon adsorption with an amount which increases in the sequence of Cl^-, Br^- and I^-. Figure 3 shows that the $|D|$ term drops by about 30%. These findings confirm results reported earlier [17]. The new results concerning the δ_D variation are presented in the bottom panel: The complex phase δ_D changes nearly linearly with potential in the absence of specific adsorption. In the case of specific

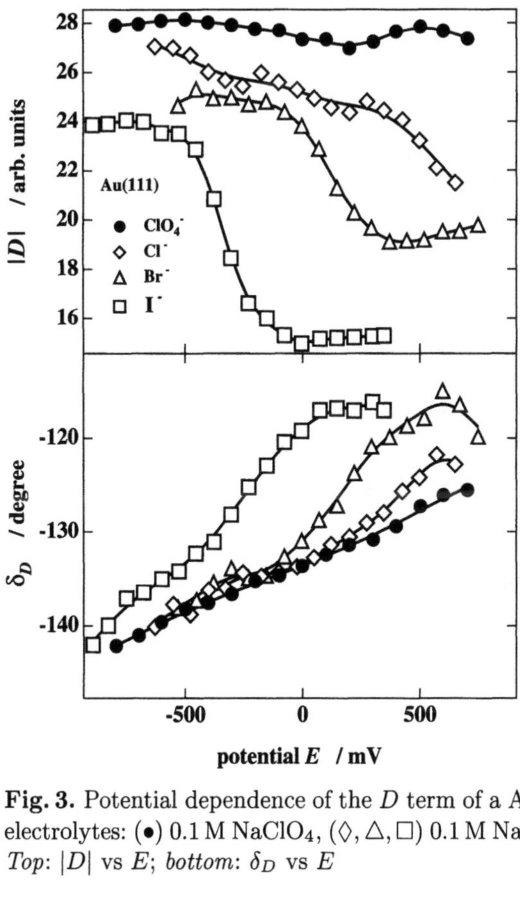

Fig. 3. Potential dependence of the D term of a Au(111) electrode in four different electrolytes: (•) 0.1 M $NaClO_4$, (◇, △, □) 0.1 M $NaClO_4$ + 1 mM NaX, X= Cl, Br, I. *Top*: $|D|$ vs E; *bottom*: δ_D vs E

adsorption of halide ions, the complex phase δ_D exhibits a kind of s-shaped variation in the potential region where the adsorption of the ions occurs.

Let us now consider the potential dependence of the A term as determined by the ISHGA method. It is displayed as $|A|$ and δ_A curves in the top and bottom panel of Fig. 4. For each halide ion the $|A|$ vs potential curve shows a broad peak. It has a maximum at potentials coinciding with the corresponding current density maximum seen in the CV curves upon adsorption. At more positive potentials the ions are still specifically adsorbed, but the $|A|$ curves decrease. This speaks either for a partial charge transfer from the ion to the metal (reducing the electric field in the inner double layer) or for a substantial change of the SHG resonance. The complex phases in Fig. 4b show a kind of s-shaped curve, with the rise coinciding roughly with the potential region of adsorption.

For Au(111) and the five electrolytes, the charge density vs potential relationships are well known. Using these relationships, the complex phase δ_A can be redrawn vs charge density, σ_M. Figure 5 replots the phase angles, given in Fig. 4b, as δ_A vs σ_M, for pure 0.1 M $NaClO_4$ and for 0.1 M $NaClO_4$ + 0.001 M NaX, X = Cl^-, Br^-, I^-, respectively. In this plot below 60 μC cm^{-2}, all five

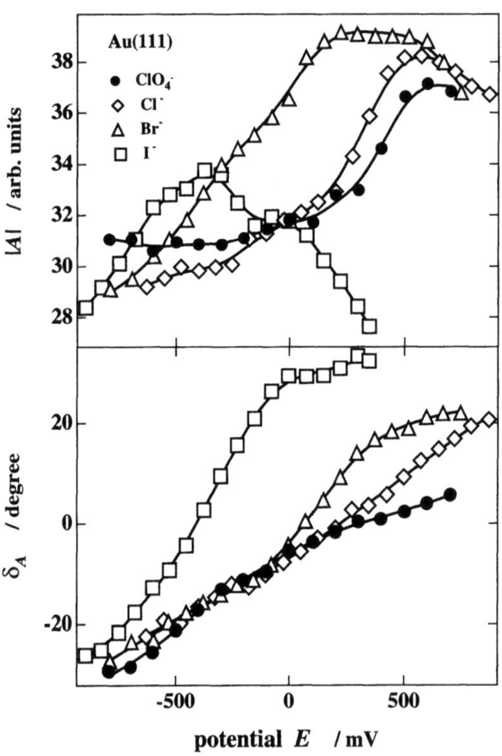

Fig. 4. Potential dependence of the A term of a Au(111) electrode for the same four different electrolytes as in Fig. 3. *Top panel*: $|A|$ vs E, *bottom panel*: δ_A vs E

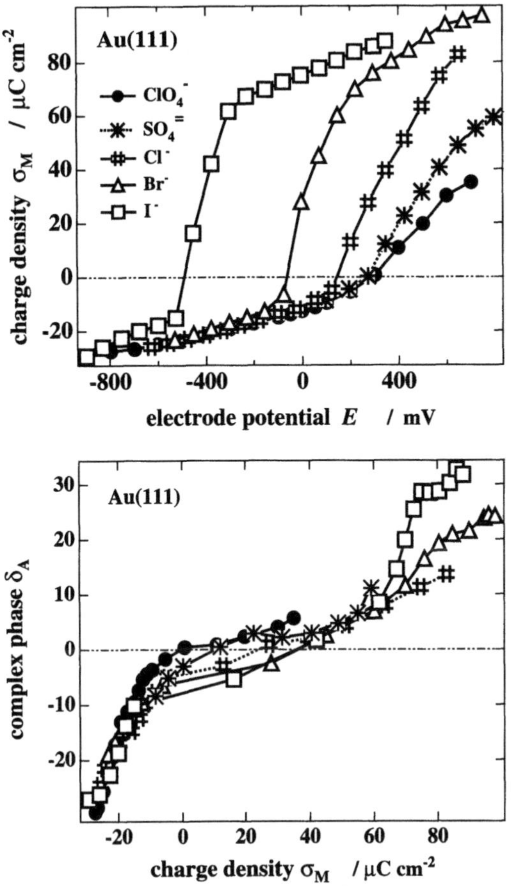

Fig. 5. *Top panel*: Charge density σ_{M} vs potential E for a Au(111) electrode in contact with the same four electrolytes as in Fig. 3. *Bottom panel*: Complex phase of the isotropic term, δ_A vs charge density, σ_{M}. δ_A data are the same as in Fig. 4

curves essentially follow the same curve. That means that in this regime they exhibit a common relationship and show nearly no dependence on the nature of the anion. This is much in contrast to the same plot but vs potential (given above in Fig. 4a), where the phase angles follow four different curves (thus showing a clear and distinct influence of the nature of the anion). These changes support the assumption that intraband transitions are involved in the SHG processes. Since the amount of phase angle shifts increases in the order of $ClO_4 \approx SO_4 < Cl^- < Br^- < I^-$, it parallels the increasing adsorption strength.

Obviously, the way the A-term is plotted highlights or conceals its dependence on the nature of the anions, on the electrode potential or on the charge density. This calls for a more general presentation of the A term which allows one to uncover (to some extent) the complex SHG response on varying adlayer structures encoded in this term. A plot of the A term in the complex plane, i.e., plot Im(A) vs Re(A) turned out to be a suitable presentation, although the dependences on potential or charge density are not directly visible.

The data given in Fig. 6 are the same as shown above as $|A|$ and δ_A vs potential. In order to illustrate common properties among the various curves, some of them are shifted in the complex plane. It is important to note that such shifts do not alter their forms nor rotate the curves in the complex plane. They solely shift the (0,0) position of the curves. The shifts were applied only to four of seven curves, to curves for iodide, chloride and the two curves for bromide (corresponding to different measurements on the same

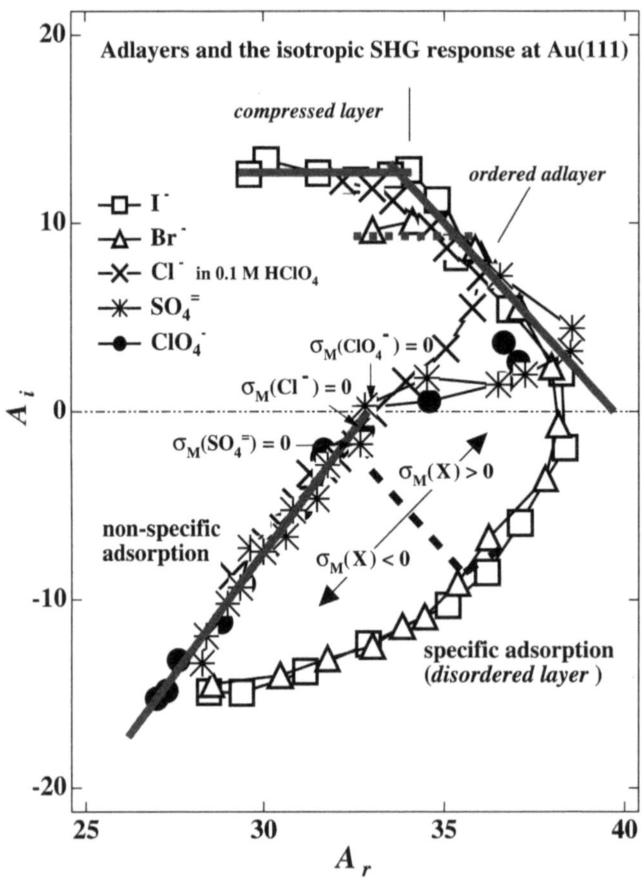

Fig. 6. Plot of the A term in the complex plane. Electrolytes: pure -•- : 0.1 M $NaClO_4$, 0.1 M $NaClO_4$ + 1 mM NaX, X = Cl (-#-), Br (-△- and -▲ -), I (-□-), SO_4 (-∗-) and 0.1 M $HClO_4$ +1mM NaCl (-×-). The *thick solid lines* are guides for the eye

system). That means, each of them is shifted by adding a specific constant complex number with two physically motivated objectives: (a) All curves should coincide at most negative potentials (that means, we require that in the potential region of nonspecific adsorption all curves should coincide), and (b) all curves for halide and sulfate ions should coincide in addition at the curve sections marked as *ordered adlayer*. In this complex plane plot the A_i term varies from about -19 to about $+12$, while the A_r term increases from $+26$ to $+38$ and decreases again to $+28$ (all rel. units).

Obviously, the A_i term exhibits the strongest dependence on potential or on charge density, while the A_r term varies in a minor way. Obviously, the phase variations shown above stem essentially from the A_i term (i.e., $\delta_A \propto A_i$). Most negative A_i values correspond to most negative potentials or charge densities and vice versa for the most positive A_i values.

Evidently, one can discern four different sections in the plot (Fig. 6) which are denoted a follows:

(i) The *nonspecific adsorption* region: This holds for ClO_4^- , $SO_4^=$ at all negative A_i values, for Cl^- only at rather negative A_i values, and for Cl^- in 0.1 M $HClO_4$ at all negative A_i values.

(ii) The *specific adsorption (disordered layer)* region: This includes all curve sections deviating from the straight lines shown in the graph. Most strikingly, the curves for bromide and iodide closely coincide (besides the final section at most positive A_i values).

(iii) The *ordered adlayer* region which represents the upwards oriented linear sections for positive A_i values: In this region the A_i values correspond to charge densities where ordered adlayers are formed ($> 60\,\mu C\,cm^{-2}$ for halide ions and $> 35\,\mu C\,cm^{-2}$ for sulfate ions) [54,55].

(iv) The *compressed layer* region: This corresponds to charge densities where compressed adlayers are established. Connected with that, there is no change in A_i but a decrease in A_r. A similar tendency is seen for bromide but not for chloride.

In the above mentioned linear curve sections, the plot of A in the complex plane shows a common behavior of A_i vs A_r for distinct ions. For bromide and iodide, the *total* A_i vs A_r curves coincide aside from the last section (once the above described shifts in the complex plane were applied). Linear sections can only occur when $A_i \propto A_r$, where $A_r = f(U, \sigma_M, \Gamma, \ldots)$ is a function of potential, charge density, etc. Then we have:

$$A_i = \alpha\ A_r + b = \alpha\ f(U, \sigma_M, \Gamma, \ldots) + b \tag{14}$$

where α is a proportionality factor and b is an appropriate offset, and the function $f(U, \sigma_M, \Gamma, \ldots)$ can be curved and may also depend on the nature of anions. The linearity only requires that $A_i \propto \alpha A_r$. Let us reconsider the complex phase δ_A: It is related to the arc tangent of the quotient A_i/A_r.

For all anions the complex phase vs charge density curves follow a common relationship below 60 μC cm^{-2}. Thus:

$$A_i \approx g(\sigma_{\mathrm{M}})\ A_r = g(\sigma_{\mathrm{M}})\ f_g(U, \sigma_{\mathrm{M}}, \Gamma, \ldots) \tag{15}$$

where the index g makes clear that we assign A_r to a more general function f_g compared to the function f valid for the above discussed linear sections of A_i vs A_r. Then, the complex phase, $\delta_A \approx \arctan[g(\sigma_{\mathrm{M}})]$, is mainly a function of σ_{M}. Contrary to that, the individual A_r and A_i terms dependent on potential, charge density, surface coverage, etc. in a complicated way.

Variations of the complex phase (of susceptibility terms) indicate in general that the electronic configuration of the surface is changed. Evidently, the electrode potential and/or the adsorption of ions and molecules can alter the interfacial electric field (acting on the electronic configuration of the metal surface) as well as a possible orbital mixing. The latter effect has to be ruled out, because for charge densities below 60 μC cm^{-2} there are only disordered layers which are rather mobile [56]. Hence, the phase variation is driven (mainly) by the electric field, since the latter is proportional to the charge density σ_{M}. Thus, $\delta_A = \delta_A(\sigma_{\mathrm{M}})$ obeys a common functional for all anions in the region of nonspecific adsorption or in the *disordered adlayer* region, but there is no common relationship for the δ_A curves for the *ordered adlayer* region [56].

5 Conclusion

Second-order optical nonlinear spectroscopy such as SHG is inherently sensitive to interfacial regions. Using Interference SHG Anisotropy (ISHGA) the anisotropy of the SHG field $E(2\omega, \varphi) \propto \sum_{n=0}^{m} k_n \cos[n\ (\varphi + \zeta_n)] = A + B\cos(\varphi + \alpha) + \ldots$ is observed. This field can be separated into its various sources, denoted as A, B, ..., permitting one to achieve information on the geometric and electronic structure of the interface and their changes upon potential and adsorption. We have shown this here in detail for a Au(111) electrode and its isotropic term A, which exhibits a remarkable dependence on potential and specific adsorption of halide and sulfate ions. When the A data, recorded as a function of the potential for various electrolytes, are plotted in the complex plane, one obtains curves which clearly show distinguishable regions for nonspecific adsorption and for the formation of *disordered*, *ordered* and *compressed* adlayers in the case of specific adsorption.

Acknowledgments

The authors are grateful to Professor G. Ertl for stimulating interest, various discussions and continuing support in these studies. This work was partly financed by the Deutsche Forschungsgemeinschaft.

References

1. Y. R. Shen: *The Principles of Nonlinear Optics* (Wiley, New York 1984)
2. G. L. Richmond: In: H. Gerischer, C. W. Tobias (Eds.) *Advances in Electrochemical Science and Engineering*, Vol. 2 (VCH, New York 1992) pp. 141–204
3. T. F. Heinz, M. M. T. Loy, W. A. Thomson: J. Vac. Sci. Technol. B **3**, 1467–1470 (1985)
4. V. L. Shannon, D. A. Koos, G. L. Richmond: J. Phys. Chem. **91**, 5548–5551 (1987)
5. V. L. Shannon, D. A. Koos, G. L. Richmond: J. Chem. Phys. **87**, 1440–1441 (1987)
6. V. L. Shannon, D. A. Koos, G. L. Richmond: Appl. Opt. **26**, 3579–3583 (1987)
7. D. M. Kolb, D. L. Rath, R. Wille, W. N. Hansen: Ber. Bunsenges. Phys. Chem. **87**, 1108–1113 (1983)
8. D. M. Kolb, R Koetz: Surf. Sci. **64**, 96–108 (1977)
9. A. Tadjeddine, D. M. Kolb, R Koetz: Surf. Sci. **101**, 277–288 (1980)
10. S. C. Chang, A. Hamelin, M. J. Weaver: J. Phys. Chem. **95**, 5560–5567 (1991)
11. R. K. Chang, T. E. Furtak (Eds.): *Surface Enhanced Raman Scattering* (Plenum Press, New York 1982)
12. G. L. Richmond, J. M. Robinson, V. L. Shannon: Prog. Surf. Sci. **28**, 1–70 (1988)
13. A. Friedrich, B. Pettinger, D. M. Kolb, G. Luepke, R. Steinhoff, G Marowsky: Chem. Phys. Lett. **163**, 123–128 (1989)
14. B. Pettinger, A. Friedrich, C Shannon: Electrochim. Acta **36**, 1829–1833 (1991)
15. B. Pettinger, A. Friedrich: In: M. P. Tosi, A. A. Kornyshev (Eds.): *Condensed Matter Physics Aspects of Electrochemistry (Working Party on Electrochemistry), Proc. Conf., 27. Aug. – 9. Sept. 1990, Trieste, Italy* (World Scientific, Singapore 1991) pp. 259–273
16. A. Friedrich, C. Shannon, B. Pettinger: Surf. Sci. **251–252**, 587–591 (1991)
17. B. Pettinger, J. Lipkowski, S. Mirwald, A Friedrich: J. Electroanal. Chem. **329**, 289–311 (1992)
18. B. Pettinger, S. Mirwald, J Lipkowski: Appl. Phys. A: Mater. Sci. Process. A **60**, 121–125 (1995)
19. B. Pettinger, J. Lipkowski, S. Mirwald: Electrochim. Acta **40**, 133–142 (1995)
20. Z. Shi, J. Lipkowski, S. Mirwald, B. Pettinger: J. Chem. Soc., Faraday Trans. **92**, 3737–3746 (1996)
21. S. Mirwald, B. Pettinger, J. Lipkowski: Surf. Sci. **335**, 264–272 (1995)
22. Z. Shi, J. Lipkowski, S. Mirwald, B. Pettinger: J. Electroanal. Chem. **396**, 115–124 (1995)
23. B. Pettinger, J. Lipkowski, S. Mirwald, A. Friedrich: Surf. Sci. **269–270**, 377–382 (1992)
24. B. Pettinger, S. Mirwald, J. Lipkowski. Ber. Bunsenges. Phys. Chem. **97**, 395–398 (1993)
25. J. Lipkowski, L. Stolberg, D. F. Yang, B. Pettinger, S. Mirwald, F. Henglein, D. M. Kolb: Molecular adsorption at metal electrodes, Electrochim. Acta **39**, 1045–1056 (1994)
26. D. Yang, D. Bizzotto, J. Lipkowski, B. Pettinger, S. Mirwald: J. Phys. Chem. **98**, 7083–7089 (1994)

27. R. Georgiadis, G. A. Neff, G. L. Richmond: J Chem. Phys. **92**, 4623–4625 (1990)
28. P. Guyot-Sionnest, A. Tadjeddine: J. Chem. Phys. **92**, 734–738 (1990)
29. E. K. L. Wong, G. L. Richmond: J. Chem. Phys. **99**, 5500–5507 (1993)
30. H. W. K. Tom, X. D. Zhu, Y. R. Shen, G. A. Somorjai: Surf. Sci. **167**, 167–176 (1986)
31. H. W. K. Tom: Ph.D. Thesis, University of California, Berkeley (1984)
32. J. E. Sipe, D. J. Moss, H. M. van Driel: Phys. Rev. B **35**, 1129–1141 (1987)
33. R. A. Bradley, R. Georgiadis, S. D. Kevan, G. L. Richmond: J. Chem. Phys. **99**, 5535 (1993)
34. C. D. Keefe, E. Revesz, M. Dionne, M Morin: Can. J. Chem. **75**, 449–455 (1997)
35. B. Pettinger, C. Bilger: Chem. Phys. Lett. **286**, 355–360 (1998)
36. C. Bilger, B. Pettinger: Chem. Phys. Lett. **294**, 425–433 (1998)
37. C. Bilger, B. Pettinger: J. Chem. Soc., Faraday Trans. **94**, 2795–2801 (1998)
38. D. M. Kolb, J. Schneider: Surf. Sci. **162**, 764–775 (1985)
39. D. M. Kolb, J. Schneider: Electrochim. Acta **31**, 929–936 (1986)
40. D. M. Kolb: Ber. Bunsenges. Phys. Chem. **92**, 1175–1187 (1988)
41. M. S. Zei, D. Scherson, G. Lehmpfuhl, D. M. Kolb: J. Electroanal. Chem. **229**, 99–105 (1987)
42. M. S. Zei, G. Lehmpfuhl, D. M. Kolb: Surf. Sci. **221**, 23–34 (1989)
43. N. J. Tao, S. M. Lindsay: J. Appl. Phys. **70**, 5141–5143 (1991)
44. M. H. Hoelzle, Th. Wandlowski, D. M. Kolb: Surf. Sci. **335**, 281–290 (1995)
45. M. H. Hoelzle, D. Krznaric, D. M. Kolb: J. Electroanal. Chem. **386**, 235–239 (1995)
46. M. Dietterle, T. Will, D. M. Kolb: Surf. Sci. **342**, 29–37 (1995)
47. D. M. Kolb: Prog. Surf. Sci. **51**, 109–173 (1996)
48. B. M. Ocko, A. Gibaud, J. Wang: J. Vac. Sci. Technol. **A10**, 3019–3031 (1992)
49. J. Wang, B. M. Ocko, A. J. Davenport, H. S. Isaacs: Science **255**, 1416–1418 (1992)
50. J. Wang, B. M. Ocko, A. J. Davenport, H. S. Isaacs: Phys. Rev. B **46**, 10321–10338 (1992)
51. B. M. Ocko, O. M. Magnussen, R. R. Adzic, J. X. Wang, Z. Shi, J. Lipkowski: J. Electroanal. Chem. **376**, 35–39 (1994)
52. T. Wandlowski, B. M. Ocko, O. M. Magnussen, S. Wu, J. Lipkowski: J. Electroanal. Chem. **409**, 155–164 (1996)
53. S. Wu, J. Lipkowski, O. M. Magnussen, B. M. Ocko, T. Wandlowski: J. Electroanal. Chem. **446**, 67–77 (1998)
54. B. M. Ocko, O. M. Magnussen, J. X. Wang, T. Wandlowski: Phys. Rev. B Condens. Matter **53**, R7654–R7657 (1996)
55. O. M. Magnussen, B. M. Ocko, J. X. Wang, R. R. Adzic: J. Phys. Chem. **100**, 5500–5508 (1996)
56. J. Lipkowski, Z. Shi, A. Chen, B. Pettinger, C. Bilger: Electrochim. Acta **43**, 2875–2888 (1998)
57. G. Beltramo, E. Santos, W. Schmickler: J. Electroanal. Chem. **447**, 71–80 (1998)
58. B. Pettinger, C. Bilger, G. Beltramo, E. Santos, W. Schmickler: Electrochim. Acta **44**, 897–901 (1998)

Part III

Electrochemical Surface Modification

Electrodeposited Magnetic Monolayers: In-Situ Studies of Magnetism and Structure

Werner Schindler

Institut für Hochfrequenztechnik und Quantenelektronik,
Universität Karlsruhe (TH), Kaiserstr. 12, 76128 Karlsruhe, Germany
werner.schindler@etec.uni-karlsruhe.de

Abstract. We show that ultrathin magnetic films of Co and Fe in the thickness range of a few atomic monolayers can be prepared by electrodeposition from an aqueous electrolyte. The high quality of these films is demonstrated by their magnetic and structural properties, which are equivalent to the properties of molecular beam grown Co and Fe films. In order to measure the intrinsic properties of films in this thickness range, in situ measurements are applied to the electrochemical cell, similarly to in situ measurements at ultrathin films in ultrahigh vacuum. Magneto-optical Kerr effect and surface X-ray diffraction are presented as examples to measure magnetism and structure, even at the same film. Thus, electrodeposition of magnetic monolayers turns out to be a useful alternative to the preparation of magnetic ultrathin films by ultrahigh vacuum techniques.

1 Motivation

Ultrathin magnetic films in the thickness range of a few atomic monolayers are mainly prepared and studied in ultrahigh vacuum using molecular beam growth, sputtering, or laser ablation techniques. Their properties are usually measured by a variety of ultrahigh vacuum compatible in situ analysis techniques, as, e.g., electron diffraction or electron spectroscopy.

In contrast, there is only little knowledge of the properties of electrodeposited ultrathin magnetic films, although electrodeposition can be performed under cleanliness conditions equivalent to ultrahigh vacuum conditions [1]. Early reports about unusual magnetic properties of electrodeposited Fe, Co, and Ni monolayers [2], which have been shown in the meantime to be caused by film impurities, may be a reason for the delay in considering electrochemistry as a clean film preparation method.

The investigation of only a few monolayer thick films requires the application of in situ measurements, in order to exclude oxidation or other environmental effects which would result in modifications of the intrinsic film properties. This is very important especially for ultrathin magnetic layers, which are known to be very sensitive to any impurities. Similarly to in situ analysis techniques attached to ultrahigh vacuum systems, a variety of in situ measurement techniques can be applied also to an electrochemical cell, even for simultaneous measurements on the same film, while maintaining

K. Wandelt, S. Thurgate (Eds.): Solid–Liquid Interfaces, Topics Appl. Phys. **85**, 243–258 (2003)

cleanliness conditions equivalent to ultrahigh vacuum conditions. Whereas in situ scanning tunneling microscopy (STM) has been widely applied in electrochemistry, particularly in growth studies of underpotential deposition of, e.g., Cu on Au(111) [3,4], other in situ techniques as, e.g., infrared spectroscopy [5,6], second harmonic generation [7], or X-ray diffraction [8,9], have not been as widely exploited. Unlike film growth, the physical and chemical properties of electrodeposited ultrathin films, such as electric, magnetic, or catalytic properties, are poorly known. However, they would be necessary to supplement simple growth studies, and even to increase the confidence in the electrodeposition technique.

We will present as examples of in situ measurements on ultrathin Co and Fe films (i) magneto-optical Kerr effect (MOKE), which has so far not been applied to an electrochemical cell in order to measure magnetic moments or coercivity of magnetic films with submonolayer resolution, and (ii) in situ surface X-ray diffraction (SXRD), which allows high-resolution studies of the film structure.

2 Experimental Details

Figure 1 shows schematically the electrochemical cell which has been developed for simultaneous MOKE and SXRD measurements [10]. The cell is rotatable by 360° about its horizontal axis (Fig. 1) to align different crystallographic directions of the substrate with respect to the external magnetic field direction or to the incident X-ray radiation. The cell fits into the gap of an electromagnet, which provides a magnetic field of up to 500 mT at the sample position. During the SXRD measurements, a magnetic field of up to 30 mT could be realized at the sample position. This allows one to record in situ complete magnetic hysteresis loops by applying an external magnetic field sweep. The field direction can be adjusted parallel or perpendicular to the crystal surface to measure the longitudinal and polar Kerr effect, or can be adjusted in arbitrary in-plane directions, to measure magnetic in-plane anisotropies.

Whereas the linearly polarized HeNe laser beam, as used in the MOKE measurements [11], penetrates water of several mm thickness, which results in a small, but constant, additional rotation of the polarization vector due to the Faraday effect of the electrolyte, SXRD requires a thin electrolyte layer configuration and a highly X-ray transparent Mylar window to reduce X-ray absorption to a minimum (Fig. 1). The thickness of the thin electrolyte layer between substrate surface and Mylar window is typically of the order of 10–20 μm [12]. The Mylar window can be inflated by applying an N_2 overpressure to the cell, resulting in an electrolyte layer thickness of approximately 2 mm between substrate and Mylar window, which hence avoids diffusion limitation of the ionic currents during film deposition and dissolution.

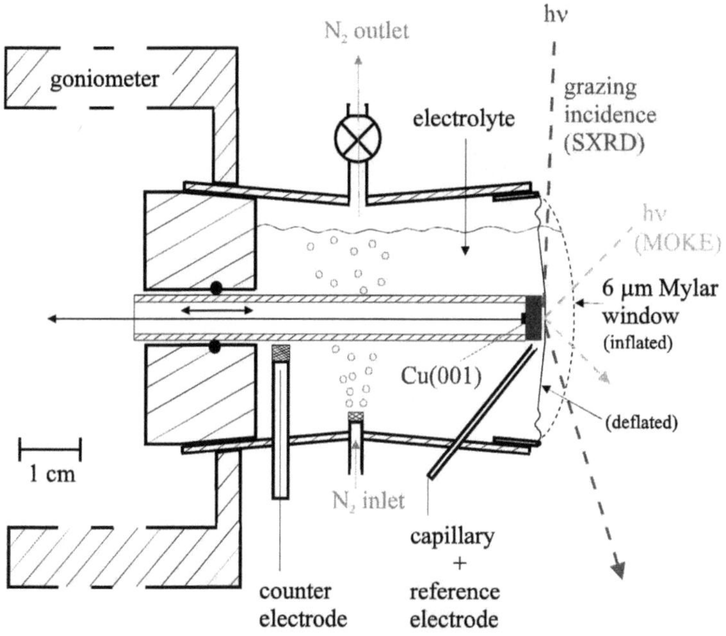

Fig. 1. Schematic drawing of the electrochemical cell as used for in situ magneto-optical Kerr effect measurements and in situ surface X-ray diffraction. The cell can be rotated by 360° about its horizontal axis and simultaneously deaerated with N_2. The Mylar window can be inflated to avoid diffusion limitation of the ionic currents during film deposition and dissolution

Oxygen is removed from the electrolyte by bubbling with 5 N (99.999 %) N_2 to avoid oxidation of the working electrode (WE, substrate) or of the deposits (Co or Fe), and to reduce the current in the double layer regime to mainly the capacitive current, caused by the voltage sweep which has to be applied to record the cyclic voltammogram (Fig. 2). Thus, ionic currents during film deposition and dissolution can be measured in the double layer range with a charge resolution of approximately $10 \mu C/cm^2$ (Fig. 2), which is equivalent to a thickness resolution of 0.02 ML. Co and Fe deposition from an aqueous electrolyte are accompanied by simultaneous H_2 evolution at the substrate, due to the negative Nernst potentials of Co and Fe. Therefore, only the anodic charge is taken as a measure for the film thickness, which represents exclusively the Co^{2+} or Fe^{2+} current resulting from the film dissolution. Nevertheless, the cathodic to anodic charge ratio is constant, and can be calibrated during several deposition/dissolution cycles. Film deposition is done in the presented experiments at an overpotential of approximately 120 mV.

MOKE and SXRD are performed at room temperature near the Nernst potential of Co/Co^{2+} and Fe/Fe^{2+}, respectively, as indicated by NP in Fig. 2.

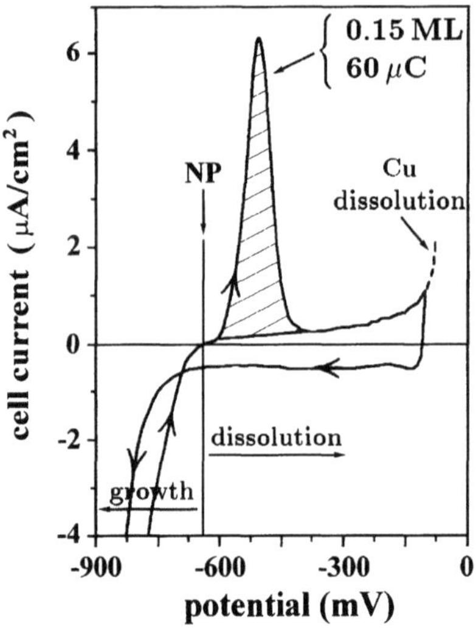

Fig. 2. Cyclic voltammogram of Co on Cu(001). In situ MOKE and SXRD measurements are done near the Nernst potential of Co/Co^{2+} (NP), where the films are stable with respect to their thickness

The WE potential is adjusted such that deposition and dissolution currents are balanced. Under these conditions, the films can be held stable with respect to their thickness for more than one hour [13].

Cu(001) single crystals are chosen as substrates, because the small in-plane lattice constant mismatch between fcc Cu and Co or Fe of less than 2% lets us expect epitaxial film growth. The Cu crystals are electropolished in 65% H_3PO_4 at + 1.8 V against a carbon electrode, carefully rinsed with ultrapure water (Milli-Q, Millipore Ltd.), and immediately transferred into the cell with the crystal surface protected against air by a drop of ultrapure water.

The SXRD measurements have been performed at the BW 2 beamline of Hasylab, Hamburg, using focused radiation of the 56 pole hybrid wiggler, and a photon energy of 8.5 keV, set 0.5 keV below the K edge of Cu. In order to achieve a high surface sensitivity, the X-ray angle of incidence has been chosen as 0.3° with respect to the Cu(001) crystal plane.

Further details of the cell design and experimental procedures have been published in references [1,10,14].

3 Co on Cu(001)

In contrast to other preparation techniques as, e.g., evaporation, sputtering or laser ablation, electrochemistry uniquely provides the opportunity to measure the film properties not only during deposition, but also during dissolution of the deposits [15]. This is demonstrated in Fig. 3 by a sequence of

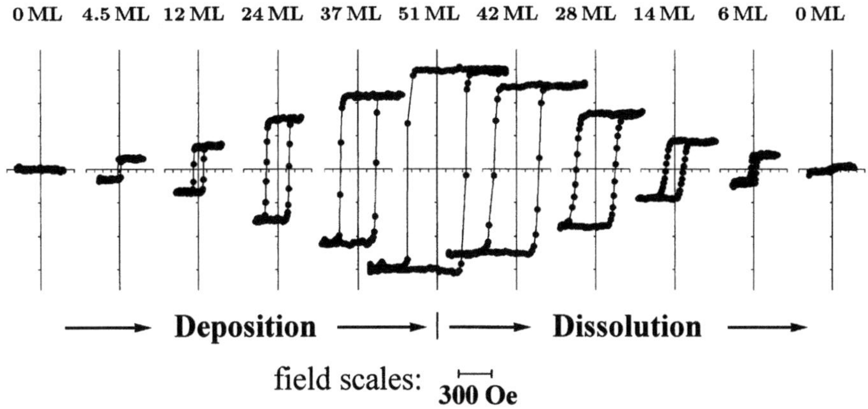

Fig. 3. Sequence of magnetic hysteresis loops during growth and dissolution of a single Co film

magnetic hysteresis loops, measured during growth and subsequent dissolution of a single Co film. Since the magnetic moment depends on coverage, film growth was interrupted during each MOKE measurement by switching the WE potential from the deposition regime at approximately 120 mV overpotential to the Nernst potential of Co/Co^{2+}, and for subsequent growth back into the deposition regime. The magnetic hysteresis at decreasing coverage was measured by dissolving a certain coverage between subsequent MOKE measurements, which were performed with the WE potential adjusted again around the Nernst potential of Co/Co^{2+}. Whereas the magnetic hysteresis loops during deposition are square-like, the wider loops during dissolution may be caused by a broader thickness distribution in the film across the measured surface area of approximately 1 mm^2, indicating a slightly inhomogeneous film dissolution.

The MOKE signal depends linearly on the coverage, which can be seen in more detail in Fig. 4a. The increase of the magnetic coercivity, which may be deduced from Fig. 3 to be nearly linear with coverage, in general depends strongly on the film microstructure. Even an increase of the deposition overpotential by 40–100 mV results in an increase of the coercivity by a factor 7 in films of the same thickness [13], which may be due to different film morphologies. This can be exploited to prepare easily magnetic materials with a tailored coercivity.

Since electrodeposition uniquely allows one to decrease the coverage without introducing defects or irreversible modifications into the films, studies of the film properties during decrease of the coverage become possible. This may be advantageous for example in studies of structural and magnetic phase transitions, as are observed, e.g., in ultrathin Fe films. Moreover, the reversibility may also be useful from a technical point of view in precisely tailoring magnetic moment or coercivity in single films or multilayer systems.

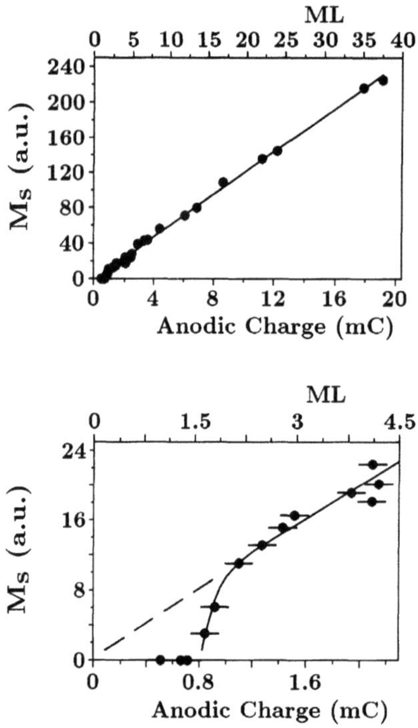

Fig. 4. Thickness dependence of the MOKE signal for Co on Cu(001)

As has been addressed in the introduction, the question of whether there are impurities in the electrodeposits is very important if films of high quality are required, or if the intrinsic material properties are to be investigated. Early magnetization measurements on electrodeposited Fe, Co, and Ni monolayers [2] showed at least two non-magnetic, i.e. "magnetically dead", layers not contributing to the magnetic moment of the films. In contrast to these findings, the proportionality of MOKE signal and coverage of our films between 2 and 40 ML (Fig. 4a) extrapolates to zero, hence indicating that each atomic layer contributes to the total magnetic moment, and that there are no "magnetically dead" layers [1]. The rapid decrease of the MOKE signal below 2 ML (Fig. 4b) neither indicaes any unusual phenomenon nor "magnetically dead" layers, but rather that the Curie temperature of films less than 2 ML thick decreases to below room temperature. This is in full agreement with the thickness dependence of the Curie temperature [16], and hence the dependence of the MOKE signal on the film thickness of evaporated Co films [17,18].

The electrodeposited films grow epitaxially in fcc structure on Cu(001), as can be inferred even from magnetic measurements. It is well known that the magnetic hysteresis of a single crystalline sample varies for magnetic fields applied along different crystallographic orientations due to different spin-orbit coupling [19]. Whereas the saturation magnetization remains nearly constant for all crystallographic directions, the remanence varies dramatically with orientation. Thus, by measuring the remanence of our films, easy and hard in-plane magnetization axes can be determined to be in [110] and [100] in-plane directions with respect to the Cu(001) substrate (Fig. 5). From this fourfold anisotropy in registry with the Cu(001) substrate, fcc growth of electrodeposited Co films can be inferred. The overall hard magnetization axis at all coverages is out-of-plane, i.e. along the [001] direction.

Structural properties can be investigated in more detail by in situ SXRD, which will be demonstrated in the following. Figure 6 shows the diffracted X-ray intensity around the fcc Cu [2, −2] in-plane diffraction peak of a 100 ML thick Co deposit on Cu(001). There are two peaks observed at [2, −2] and [2.022, −2.022], which correspond to fcc Cu and fcc Co, thus proving the epitaxial growth of fcc Co. Due to the surface sensitivity of SXRD at grazing incidence, the intensity of the Cu peak at [2, −2] is much smaller than the intensity of the fcc Co peak at [2.022, −2.022]. There is no in-plane distortion of the fcc Co unit cell, since the center of the Co peak is observed exactly in the [1, −1] direction. The Co in-plane lattice constant can be calculated from Fig. 6 to be 3.58 Å, i.e. approximately 1 % smaller than the Cu in-plane lattice constant (3.6147 Å). From the literature values of fcc Co one would have expected a lattice constant relaxation by 1.9 % to approximately 3.554 Å. Thus, in disagreement with current belief, even 100 ML thick Co films seem to have not yet reached their bulk lattice constants.

Since there is no in-plane lattice distortion of fcc Co, the evolution of the in-plane lattice constant with coverage can be observed by detector scans in the [1, −1] direction through the centers of the Cu and Co peaks, as is shown by the set of lineshapes for different Co coverages in Fig. 7. Since surface

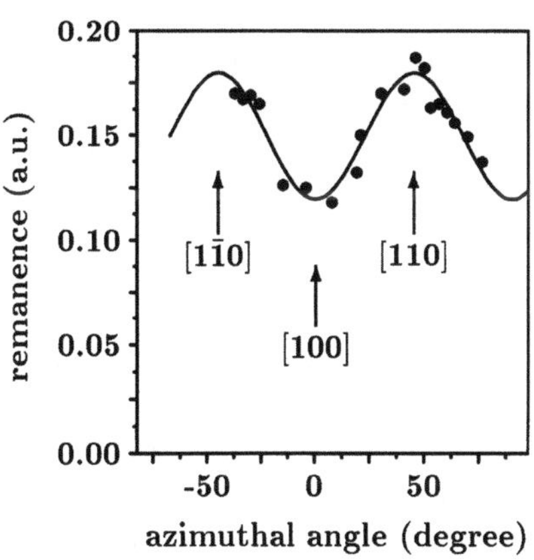

Fig. 5. Fourfold in-plane anisotropy of 13 ML Co on Cu(001). The easy magnetization axes are in the [110] in-plane directions, the hard magnetization axes in the [100] in-plane directions

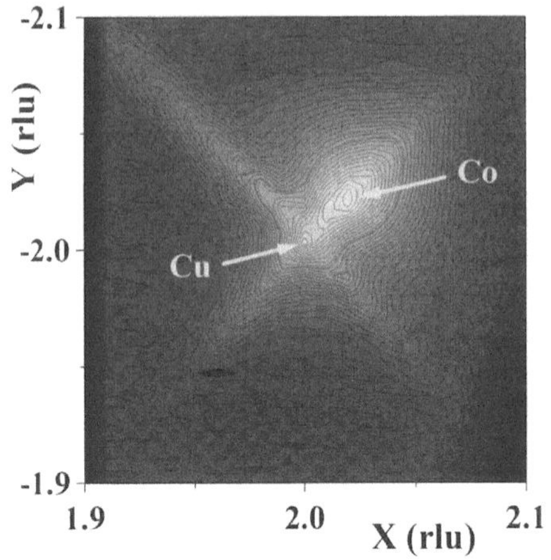

Fig. 6. XY scan around the [2, −2] in-plane Cu diffraction peak at a Co coverage of 100 ML. The two peaks correspond to fcc Cu and fcc Co, as indicated. The Co peak is shifted to [2.022, −2.022] due to the in-plane lattice constant relaxation in the Co film. (The increased intensity along the diagonals is caused by Cu islands of the substrate with a preferred step orientation in the [110] direction.)

X-ray diffraction provides a high resolution of better than 0.004 Å [10], even pseudomorphic and orthomorphic Co phases can be distinguished in the films as indicated in Fig. 7 by the dotted and dashed lines, respectively. This is an advantage compared to the widely used electron diffraction techniques (LEED, RHEED) in ultrahigh vacuum, which provide much less resolution than X-ray diffraction, and which are restricted to the first two ML next to the film surface. They cannot provide information about the structure of the buried substrate/film interface, or even deeper layers in the film.

From Fig. 7 the dependence of the pseudomorphic and orthomorphic peak positions as well as of the peak integrals on the Co coverage can be deduced. The first monolayers grow completely coherent with the Cu(001) surface. In-plane lattice constant relaxation starts around a thickness of 5 ML, which is observed as the onset of an orthomorphic peak in the X-ray diffraction lineshape (Fig. 7), or as onset of the orthomorphic peak integral (Fig. 8). This has been interpreted so far within the framework of classical theory by *van-der-Merwe* [20] as the onset of the formation of bulk dislocations in the films. Usually, this onset is called critical thickness for pseudomorphic film growth, and it is assumed that coherent (pseudomorphic) growth ceases abruptly at this particular thickness [21,22]. However, above 5 ML, the pseudomorphic peak integral in Fig. 8 neither levels off, nor starts to decrease with further increasing coverage, as was expected from classical theory. It even continues to grow up to approximately 15 ML, demonstrating that pseudomorphic film growth continues up to the 15th ML Simultaneously, the orthomorphic peak integral increases, until the transition to complete orthomorphic film

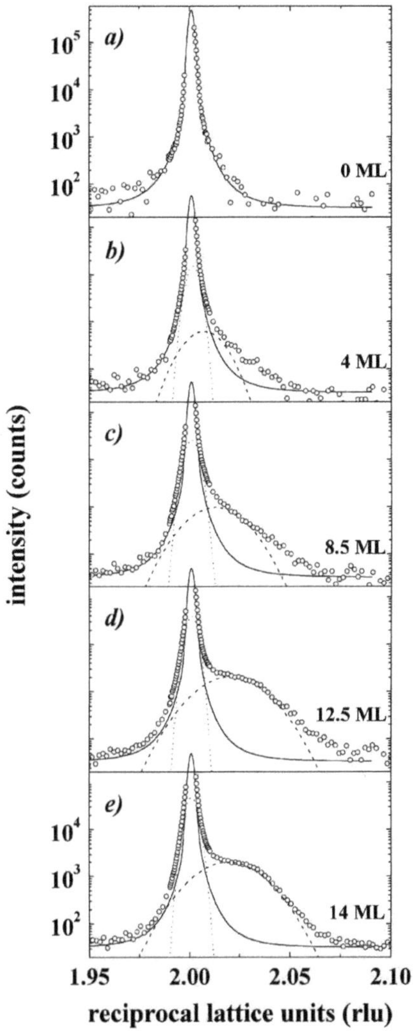

Fig. 7. (**a**)–(**e**) [2, −2] in-plane diffraction peaks of fcc Co on Cu(001) for different Co coverages: (○) data; *(solid line)* partial Gauss fit of Cu; *(dotted line)* partial Gauss fit of pseudomorphic peak; *(dashed line)* partial Gauss fit of orthomorphic peak; The x-axis scale is given in reciprocal lattice units (rlu), which are normalized to $q_{110}^{Cu} = 2\pi/a_{\text{in-plane}}$, i.e. $\mathrm{q(rlu)} = \mathrm{q}/\mathrm{q}_{110}^{Cu}$

growth occurs around 15 ML, indicated by the strong increase of the gradient of the orthomorphic peak integral (Fig. 8). Therefore, the critical thickness for pseudomorphic growth is approximately 15 ML, much larger than the onset of in-plane lattice constant relaxation around 5 ML. The formation of orthomorphic growth patches within the pseudomorphic Co matrix is an unusual relaxation mechanism of pseudomorphic strain, which has not

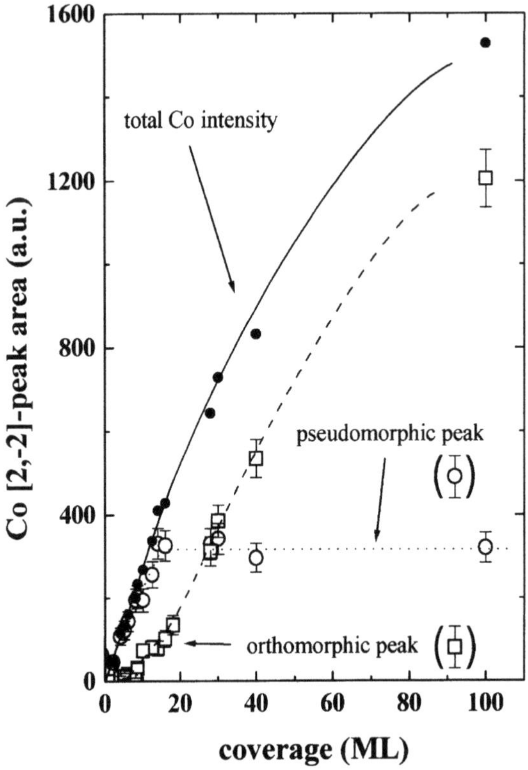

Fig. 8. Peak integral of the pseudomorphic (○, *dotted line*) and orthomorphic (□, *dashed line*) peak fits. Both peak integrals add up to the total integrated Co peak (•, *solid line*), which increases linearly at low coverages, and saturates for higher Co coverages due to the limited penetration depth of X-rays in grazing incidence condition

been observed in a heteroepitaxial growth system showing a lattice constant misfit of only 1%. In contrast to current belief, that Co/Cu(001) is a Frank-van-der-Merwe growth system at room temperature [21], the presented data would be fully consistent with the assumption that the onset of in-plane lattice constant relaxation at 5 ML is caused by the onset of the formation of Stranski–Krastanov islands within the pseudomorphic Co film [23].

4 Fe on Cu(001)

Unlike the growth of ultrathin Co films, the growth of evaporated Fe on Cu(001) displays a large variety of possible structures and features: The actual Fe phase depends on coverage [24,25], and even on the preparation method, as found by comparison of evaporated and laser ablated ultrathin Fe films [26,27]. Whereas the first monolayers start to grow in fcc structure, adopting the structure of the fcc Cu(001) substrate, a structural transition from fcc to bcc growth is observed around 10 ML, depending sensitively on temperature and surfactants at the substrate surface during growth. For example, a few Langmuir of CO adsorbates at the Cu(001) surface have been reported to enhance fcc growth up to several tens of ML [28]. This structural transition is accompanied by a spin-reorientation transition: Whereas the easy magnetization axis in the fcc growth regime of evaporated Fe films is out-of-plane, i.e. in the [001] direction, it switches to in-plane in the bcc growth regime.

Since this system is a very interesting example for how structural properties determine the magnetic properties of ultrathin films, both structure and corresponding magnetic properties have been measured simultaneously by in situ SXRD and MOKE, as shown in Fig. 9. Similarly to the SXRD measurements at Co films, the in-plane fcc Fe diffraction peaks are expected close to the corresponding fcc Cu diffraction peaks due to the small lattice constant mismatch. Evaporated bcc Fe grows on Cu(001) with the Fe(110) plane parallel to Cu(001), and the Fe [111] direction parallel to the Cu [110] direction [29]. Therefore, the in-plane diffraction peaks of bcc Fe(110) are expected to be separated from the Cu diffraction peaks. From previous LEED studies on evaporated films [29], the bcc Fe [011] in-plane diffraction peak would be expected at approximately [1.78, 0.26], in coordinates of the fcc Cu unit cell.

The first two columns of Fig. 9 summarize the results of SXRD at films of 2–14 ML thickness: Figure 9a–d show the X-ray lineshapes of the [2,0] in-plane diffraction peak, which is a superposition of the intensity of fcc Fe and that of fcc Cu. Figure 9e–g show the lineshapes of the ω-scan through the bcc Fe [011] peak at [1.78, 0.26]. Columns three and four show the magnetic hysteresis of the films, as measured in polar geometry which is sensitive to out-of-plane magnetization (Fig. 9h–k), and as measured in longitudinal geometry which is sensitive to in-plane magnetization (Fig. 9l–n), respectively.

Fcc Fe can be detected by the peak asymmetry of the fcc Cu peak, which has been fitted by a Gauss curve, and plotted as solid line in Fig. 9a–d. Bcc Fe peaks are observed at 7 and 14 ML film thickness (Fig. 9f,g), but not at coverages of 2 and 3.5 ML. The onset of bcc Fe peaks at 7 ML indicates, that around 7 ML a structural transition from fcc Fe to bcc Fe occurrs. Evaporated and pulsed laser ablated Fe/Cu(001) films show a continuous transition regime from fcc to bcc growth between 6 and 10 ML [30], in which range bcc precipitates are observed in coexistence with further fcc growth of the surrounding film matrix. The onset of bcc Fe peaks around 7 ML (Fig. 9f) in the electrodeposited films would be in good agreement with the onset of bcc precipitates at 6 ML in evaporated or pulsed laser deposited films [26,27].

From the SXRD data in Fig. 9, we cannot deduce if the bcc Fe intensity results from precipitates or from a laterally extended coverage. However, both peaks in the ω scans in Fig. 9f,g at [1.78, −0.26] and [1.78,0.26] originate from

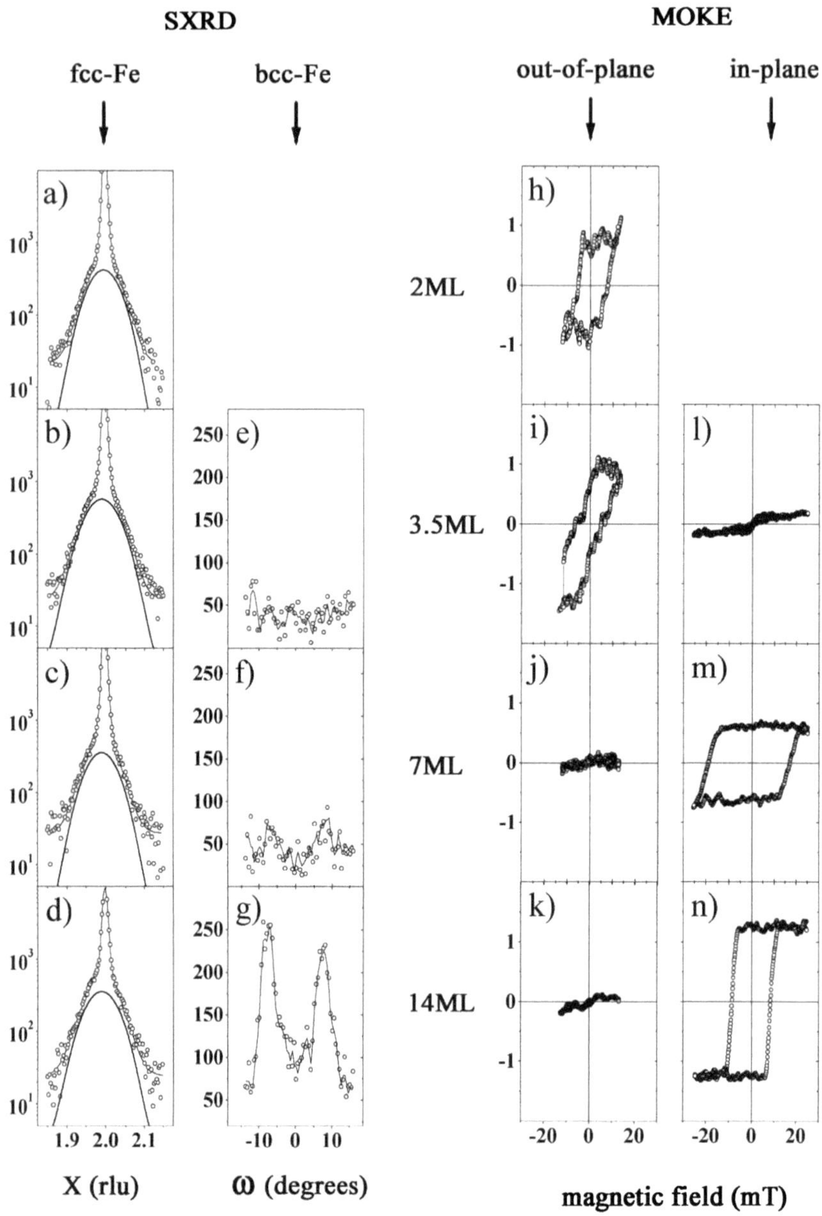

Fig. 9. fcc–bcc and spin reorientation transition of Fe on Cu(001), as monitored simultaneously by SXRD and MOKE. **(a)**–**(d)**: *solid curve*: fcc Fe intensity; **(e)**–**(g)**: bcc Fe peaks at [1.78, −0.26] and [1.78,0.26], in coordinates of the fcc Cu unit cell; **(h)**–**(k)**: polar hysteresis loops, corresponding to SXRD measurements **(a)**–**(d)**; **(l)**–**(n)**: longitudinal hysteresis loops, corresponding to SXRD measurements **(e)**–**(g)**

two bcc(110) domains, which are symmetrically to the Cu [100] direction. Due to the symmetry of the fcc substrate, there are two additional bcc(110) domains at $[-0.26, 1.78]$ and [0.26,1.78], i.e. rotated by 90°. All four domains are observed, and confirm the structural model of bcc Fe growth on Cu(001) as proposed by [29,30].

The easy axis of the film magnetization in films of 2 and 3.5 ML thickness is found out-of-plane (Fig. 9h,i), similarly to the magnetization of evaporated films in this thickness range [31]. At the onset of the growth of the bcc phase (Fig. 9f), the easy magnetization axis switches to in-plane (Fig. 9m). Whereas the coercivity of the out-of-plane magnetic hysteresis becomes larger than the available magnetic field sweep range in the measurements, the softening of the magnetic in-plane hysteresis above the spin reorientation transition can be observed by the decrease of the coercivity from 7 ML (Fig. 9m) to 14 ML (Fig. 9n). The spin reorientation transition coincides with the onset of the growth of bcc Fe around 7 ML. On the first view, this is quite different to the finding in evaporated films, that the easy magnetization axis remains out-of-plane until a significant amount of bcc precipitates is formed in the films, which causes the spin reorientation in evaporated films to occur not before a film thickness of 10 ML is reached [31,32].

However, so far we have not considered the amount of the bcc phase in films of 7 ML thickness. If we assume, that the fcc–bcc transition is completed at 14 ML (according to the fcc–bcc transition in evaporated films it is probably completed even at 10 ML), the bcc intensity at 14 ML (Fig. 9g) results from the whole film. Since this intensity is roughly four times larger than the bcc intensity at 7 ML (Fig. 9f), we can conclude that the bcc intensity at 7 ML corresponds to a coverage equivalent of approximately 3–4 ML. Thus, the fcc–bcc transition seems to be nearly completed even around 7 ML. Hence, the driving force to form the bcc phase seems to be stronger in electrodeposited films than in evaporated or laser ablated films, which show a fcc–bcc transition regime of 4 ML from 6 to 10 ML film thickness. The finding, that the spin reorientation transition occurs at a lower coverage than is observed in evaporated Fe films, may be explained by the H_2 evolution at the substrate during the Co film deposition, which may support growth of the bcc phase. Similarly, the fcc–bcc transition in evaporated Fe films has been found very recently to occur in the presence of adsorbed H_2 at the substrate surface also at a lower coverage than usually observed [33]. Thus, the growth and spin reorientation transition in electrodeposited Fe films turns out to be in very good agreement with the corresponding behaviour of evaporated Fe films.

The MOKE signal of Fe films up to 50 ML thickness is plotted in Fig. 10. The linear increase of the in-plane MOKE signal above the spin reorientation transition around 7 ML extrapolates to zero at zero coverage, indicating that even the first monolayers, which started to grow in fcc structure with the easy magnetization axis out-of-plane, reorientate and contribute to the in-plane magnetization. If the fcc layers would not reorientate, the magnetic moment

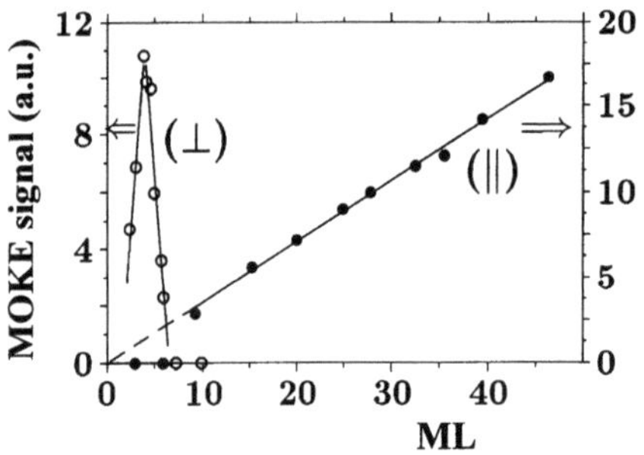

Fig. 10. Dependence of the MOKE signal M_S of Fe on the coverage. *Left scaling* (∘): easy magnetization axis out-of-plane; *right scaling* (•): easy magnetization axis in-plane, measured with the magnetic field applied in the [110] in-plane direction

of a 7 ML film could only be explained assuming a magnetic moment of the first bcc layer to be seven times larger than the magnetic moment of a single ML in the bulk material. The reorientation of the fcc layers is also in good agreement with the SXRD data of the fcc Fe peak at 14 ML coverage (Fig. 9d), which is smaller than the fcc Fe peaks at 2 and 3.5 ML (Fig. 9a,b).

5 Discussion

High-quality monolayers of Co and Fe can be electrodeposited even from an aqueous electrolyte, if cleanliness conditions in the electrochemical cell are equivalent to ultrahigh vacuum conditions. Cleanliness has been found to be the most crucial parameter to achieve structural and magnetic film properties, which are equivalent to the properties of evaporated or molecular beam grown films. The reproducible investigation of magnetic monolayers under these conditions becomes possible, if in situ measurement techniques are applied to an electrochemical cell. A variety of measurement techniques can be applied even simultaneously, as for example MOKE and SXRD. This opens up exciting prospects for the investigation of physical and chemical properties, and of their correlation to the structure of the films. So far, these possibilities have not been widely exploited, except for in situ STM. Moreover, the measurement of structural properties by SXRD provides high resolution, and allows for example even the investigation of the buried substrate/film interface of a heteroepitaxial system with only approximately 1% lattice constant mismatch, like Co/Cu(001). Systems showing such small lattice misfit have so far not been widely investigated [34,35]. Previous STM and LEED/RHEED measurements [22], which are limited to an information depth of approximately the first two monolayers next to the film surface, thus can be nicely supplemented. For example, it can be shown that the relaxation of heteroepitaxial, and even pseudomorphic films, may be more complicated than so far believed for systems showing a small lattice constant mismatch, like Co/Cu(001) [23].

The presented measurements on Fe/Cu(001) show the possibilities of studying even more complicated heteroepitaxial systems. The spin reorientation behaviour of electrodeposited Fe/Cu(001) is a nice example of the importance to study correlated film properties like structure and magnetism in simultaneous measurements, and under well-defined preparation conditions. The critical balance of surface energies and shape anisotropies, which determines the actual easy and hard axes of the film magnetization in this thickness range, is demonstrated by SXRD and MOKE measurements on the same Fe film. We find a spin reorientation behaviour of electrodeposited Fe films, which is fully consistent with the spin reorientation behaviour of evaporated [31] or laser ablated [27] Fe films. It may be an interesting question, how this spin reorientation depends on the deposition overpotential, since film morphology of electrochemically grown films is observed to change with the deposition overpotential [13]. According to the presented data, there is neither a stabilizing effect of the electrolytical double layer, nor of electrochemical adsorption, on the growth of fcc Fe detectable. Hydrogen evolution at the substrate surface during film deposition rather seems to cause the structural transition from fcc to bcc growth and the spin reorientation to occur at a lower coverage than so far observed in molecular beam grown films.

It turns out that electrodeposition of magnetic films neither needs to be restricted to the preparation of thick deposits, nor to technical applications. It may rather be useful also in the investigation of ultrathin film properties in the field of basic research if done properly under ultrapure conditions, which has become standard in ultrahigh vacuum technology for more than a decade.

Acknowledgements

The author would like to thank J. Kirschner for providing continuous support of this work, and Th. Koop for participating in the X-ray diffraction measurements. These have been done in collaboration with J. Zegenhagen, A. Kazimirov, and G. Scherb from MPI für Festkörperforschung, Stuttgart, R. Feidenhans'l and Th. Schultz from Risø National Laboratory, R. Johnson and O. Bunk from Universität Hamburg, Hamburg. Support with Cu crystals by H. Menge, and assistance by the staff of Hasylab and DESY at the synchrotron radiation laboratory is gratefully acknowledged.

References

1. W. Schindler, J. Kirschner: Phys. Rev. B **55**, R1989 (1997)
2. L. N. Liebermann, D. R. Fredkin, H. B. Shore: Phys. Rev. Lett. **22**, 539 (1969) L. N. Liebermann, J. Clinton, D. M. Edwards, J. Mathon: Phys. Rev. Lett. **25**, 232 (1970)
3. E. Budevski, G. Staikov, W. J. Lorenz: *Electrochemical Phase Formation and Growth* (VCH, Weinheim 1996)

4. M. H. Hölzle, V. Zwing, D. M. Kolb: Electrochim. Acta **40**, 1237 (1995)
5. T. Iwasita, N. C. Nart: Prog. Surf. Sci. **55**, 271 (1997)
6. S. M. Stole, D. D. Popenoe, M. D. Porter: in H. D. Abruna (Ed.): *Electrochemical Interfaces* (VCH, New York 1992), p. 341
7. G. L. Richmond: in H. D. Abruna (Ed.): *Electrochemical Interfaces* (VCH, New York 1992), p. 267
8. I. K. Robinson, D. J. Tweet: Rep. Prog. Phys. **55**, 599 (1992)
9. J. Zegenhagen, G. Materlik, J. P. Dirks, M. Schmäh: in *Synchrotron Techniques in Interfacial Electrochemistry* (Kluwer Academic, Dordrecht 1994), p. 349
10. Th. Koop, W. Schindler, A. Kazimirov, G. Scherb, J. Zegenhagen, Th. Schulz, R. Feidenhans'l, J. Kirschner: Rev. Sci. Instrum. **69**, 1840 (1998)
11. S. D. Bader: J. Magn. Magn. Mater. **100**, 4 (1991)
12. M. F. Toney: *Synchrotron Techniques in Interfacial Electrochemistry* (Kluwer Academic, Dordrecht 1994) p. 109
13. W. Schindler, Th. Koop, J. Kirschner: Z. Phys. Chem. **208**, 93 (1999)
14. W. Schindler, J. Kirschner: Rev. Sci. Instrum. **67**, 3578 (1996)
15. W. Schindler, Th. Koop, D. Hofmann, J. Kirschner: IEEE Trans. Mag. **34**, 963 (1998)
16. C. M. Schneider, P. Bressler, P. Schuster, J. Kirschner, J. J. de Miguel, R. Miranda: Phys. Rev. Lett. **64**, 1059 (1990)
17. P. Krams, F. Lauks, R. L. Stamps, B. Hillebrands, G. Güntherodt: Phys. Rev. Lett. **69**, 3674 (1992)
18. H. A. Wierenga, W. de Jong, M. W. J. Prins, Th. Rasing, R. Vollmer, A. Kirilyuk, H. Schwabe, J. Kirschner: Phys. Rev. Lett. **74**, 1462 (1995)
19. Ch. Kittel: *Introduction to Solid State Physics* (Wiley, New York 1996)
20. J. H. van der Merwe: J. Appl. Phys. **34**, 117 (1963)
21. W. A. Jesser, J. W. Matthews: Philos. Mag. **17**, 461 (1968)
22. J. Fassbender, U. May, B. Schirmer, R. M. Jungblut, B. Hillebrands, G. Güntherodt: Phys. Rev. Lett. **75**, 4476 (1995)
23. W. Schindler, Th. Koop, G. Scherb, A. Kazimirov, J. Zegenhagen, Th. Schultz, R. Feidenhans'l, J. Kirschner: Surf. Sci. **465**, L738 (2000)
24. S. Müller, P. Bayer, A. Kinne, C. Reischl, R. Metzler, K. Heinz: Surf. Sci. **331–333**, 723 (1995)
25. H. Landskron, G. Schmidt, K. Heinz, K. Müller, C. Stuhlmann, U. Beckers, M. Wuttig, H. Ibach: Surf. Sci. **256**, 115 (1991)
26. J. Shen, H. Jenniches, C. V. Mohan, J. Barthel, M. Klaua, P. Ohresser, J. Kirschner: Europhys. Lett. **43**, 349 (1998)
27. H. Jenniches, J. Shen, C. V. Mohan, S. S. Manoharan, J. Barthel, P. Ohresser, M. Klaua, J. Kirschner: Phys. Rev. B **59**, 1196 (1999)
28. A. Kirilyuk, J. Giergiel, J. Shen, J. Kirschner: Phys. Rev. B **52**, R11672 (1995)
29. P. Schmailzl, K. Schmidt, P. Bayer, R. Döll, K. Heinz: Surf. Sci. **312**, 73 (1994)
30. J. Giergiel, J. Kirschner, J. Landgraf, J. Shen, J. Woltersdorf: Surf. Sci. **310**, (1994)
31. J. Giergiel, J. Shen, J. Woltersdorf, A. Kirilyuk, J. Kirschner: Phys. Rev. B **52**, 8528 (1995)
32. J. Shen, C. Schmidthals, J. Woltersdorf, J. Kirschner: Surf. Sci. **407**, 90 (1998)
33. R. Vollmer: Phys. Rev. B **61**, 4146 (2000)
34. H. Ibach: Surf. Sci. Rep. **29**, 193 (1997)
35. R. Koch, J.Phys.: Condens. Matter **6**, 9519 (1994)

Structural Transitions in Organic Adlayers – A Molecular View

Thomas Dretschkow and Thomas Wandlowski

Institut für Grenzflächenforschung und Vakuumphysik, Forschungszentrum Jülich
Leo Brandt Str. 52425 Jülich, Germany
Th.Wandlowski@fz-juelich.de

Abstract. The structure and dynamics of phase formation of organic monolayers potentiostatically deposited on single-crystal electrode/aqueous electrolyte interfaces has been studied by combining classical electrochemical experiments with in situ Scanning Tunnelling Microscopy (STM). Depending on the potential, uracil, 5-methyluracil, 5,6-dimethyluracil and n, n'-bipyridines form highly ordered physisorbed and/or chemisorbed adlayers on Au(hkl)-electrodes, which were imaged with molecular resolution. The two-dimensional condensed physisorbed films of uracil derivatives display characteristic properties of a hydrogen-bonded network of planar oriented molecules. These pyrimidine bases deprotonate at sufficiently positive potentials and form highly ordered (commensurate, incommensurate) chemisorbed monolayers of perpendicularly oriented surface co-ordination complexes. The positions of the individual molecules and their long-range order are determined by the competing interactions between molecular and substrate co-ordination centres as well as lateral π-stacking. These motifs of molecular phase formation were extended and tested with the chelating bidentate ligand $2, 2'$-bipyridine and the non-chelating bridging ligand $4, 4'$-bipyridine adsorbed on Au(hkl)-electrodes. The dynamics of film formation and dissolution of the various ordered molecular phases and the substrate stability were investigated by in situ STM as a function of electrode potential, temperature and supporting electrolyte. Evidence was found for (hole) nucleation- and growth-controlled (dissolution) formation for uracil and n, n'-bipyridine adlayers at positive charge densities. Several metastable structures were observed during these transitions. Their appearance and characteristic properties depend strongly on the applied potential perturbation, and are accompanied with major changes of the electrode surface. The chapter concludes with a method to selectively nanostructure an Au(100)-electrode with $2, 2'$-bipyridine exploiting the uneven local charge distribution between reconstructed (hex) and unreconstructed (1×1) surface patches.

1 Introduction

Monolayer films on solid surfaces are of interest for fundamental as well as technological reasons [1,2]. Organic molecules are used as corrosion inhibitors [3], and are often employed as additives in plating baths [4]. Thin organic films have the potential to modify the chemical and structural properties of surfaces, which is of considerable importance for electronic and optical

K. Wandelt, S. Thurgate (Eds.): Solid–Liquid Interfaces, Topics Appl. Phys. **85**, 259–321 (2003)

materials [1], the field of electrocatalysis [5], and biological and chemical sensors [6,7].

Organic monolayers on well-defined metal substrates may be obtained in various ways. Typical strategies are Molecular Beam Epitaxy (MBE) [8], so-called Self-Assembled Monolayers (SAMs) [9] or Langmuir–Blodgett (LB) films [10,11]. SAMs are molecular assemblies formed by the spontaneous interaction of an active surfactant with a solid or liquid substrate. Examples are thiols on gold or silver [9,12,13] as well as alkyltrichlorosilanes on oxide surfaces [14]. The order of these two-dimensional systems is produced by a spontaneous chemical reaction at the interface followed by a slow two-dimensional ordering process as the system approaches equilibrium. Unfortunately, even when these films are made on single-crystal substrates, they appear to contain a significant number of pinholes and other defects and their mode of formation makes it unlikely that they will be close-packed [15,16,17]. Constraints exist also for monolayers based on the Langmuir–Blodgett technique. These studies are limited to the choice of molecules with a very low solubility in the subphase [10,11].

Alternatively, molecular and ionic monolayers can also be obtained on conducting surfaces in an electrochemical environment [18,19]. This approach offers the advantage that formation and structural properties of a wide variety of adlayers can be controlled by the applied electrode (substrate) potential and subsequently imaged, for instance by Scanning Tunneling Microscopy (STM). For this reason, potentiostatically or galvanostatically generated monolayers on well-defined metal electrodes have become attractive model systems and provide an important testing ground for fundamental issues in two-dimensional physics, such as phase transitions in adlayers and substrates [18,19,20,21]. Since the pioneering work of *Frumkin* [22] many "equilibrium" and "dynamic" adsorption studies at metal/electrolyte interfaces have been performed with mercury or low-melting-point electrodes of sp-metals, such as Bi, Pb, Sn or Zn, onto which organic molecules are usually weakly adsorbed [23,24]. The formation of two-dimensional condensed monolayers was first reported by *Lorenz* [25], and *Vetterl* [26] who found an unusual capacitance vs. potential hysteresis for saturated aqueous solutions of nanoic acid and the so-called "capacitance pits" for various purine and pyrimidine bases. Since then, many other examples have been reported and quantitatively analyzed (thermodynamics, kinetics; cf. the literature reviewed in [18,19]).

Despite the rather detailed phenomenological knowledge on the two-dimensional phase formation in adlayers, as obtained at the mercury/electrolyte interface, further progress has been hampered by the lack of information on the structure and molecular mechanisms of film formation. New experimental and theoretical perspectives are available when well-defined single-crystal electrodes are employed as substrate materials. This step offers the advantage of combining classical electrochemical experiments with the power of structure sensitive in situ techniques, such as scanning probe microscopy [27], surface X-ray scattering [28] and/or vibrational spectroscopies [12]. The in situ formation of condensed organic films on solid electrodes was suggested by *Batrakov* et al. in 1974 for the adsorption of camphor on Zn(0001) [29]. During the last five years a few other examples have been reported in the literature, mostly based on voltammetric and capacitance measurements: pyridine on Ag(210) [30], thymine on Cd(0001) [31], Au(100) [32], Au(111) [33] and Ag(*hkl*) [34], coumarine on Au(111) and Au(100)-(hex) [35], uridine on Au(111) [36], uracil as well as camphor on Au(*hkl*) and Ag(*hkl*)[37,38,39,40,41,42,43,44]. Quantitative thermodynamic studies have been reported for pyridine [30] and several derivatives of uracil [40]. The phenomenological kinetics of film formation have been described by models which involve a nucleation and growth mechanism [36,37,42]. The results of these "macroscopic" electrochemical experiments are still scarcely complemented by structural studies, such as IR- or Raman spectroscopies or the use of UHV techniques after emersion of the electrode [33,45,46,47]. Remarkable progress in developing a "true" atomistic/molecular picture of "clean" and adsorbate-modified single-crystal electrodes and/or the various phase formation processes taking place there was achieved with the advent of Scanning Tunneling Microscopy (STM) and Atomic Force Microscopy (AFM). Both techniques enable the imaging of bare as well as adsorbate-covered solid surfaces not only under UHV conditions, but also in air and in solution. A number of heterocyclic and sulfur-containing molecules have been studied recently by in situ scanning probe techniques. Direct evidence for the formation of two-dimensional long-range ordered structures on Au(111) has been reported for several purine and pyrimidine bases [33,37,48,49,50,51,52], phenol [53], pyridine [54], $2,2'$- and $4,4'$-bipyridine [55,56,57,58], phenanthroline [59,60], octylthiole [61], tetramethylthiourea [62,63] and cysteine [64]. *Itaya* et al. have employed iodine-modified Au(111)- and Ag(111)-electrodes to image self-organized arrays of crystal violet, methyl-pyridinium-phenylendivinylene and porphyrine derivatives [65,66,67]. The same group also reported high-resolution in situ images of aromatic molecules such as benzene, naphthalene and anthracene on Pt(111), Rh(111) and Cu(111) electrodes [68,69]. These results are complemented by a few structural studies with physisorbed films of purine bases and porphyrine derivatives at the HOPG(0001)/aqueous electrolyte interface [70,71,72,73,74,75,76]. Except for one report with uracil on Au(100) [51] all other studies have been conducted on hexagonal densely packed substrate surfaces.

Basic ordering principles of the above monolayers appear to be (i) the ability to create strong and intermolecular hydrogen bonds between adjacent molecules [33,51,72], (ii) packing constraints, (iii) the formation of interfacial stacks due to π-electron attraction, hydrophobicity and London dispersion forces [51,55] [59] as well as (iv) substrate–adsorbate co-ordination chemistry [33,61]. The structure of the adsorbed monolayers, and in particular the registry between the adlayer and the substrate, is affected by the nature and the crystallographic orientation of the metal, the adsorbate–adsorbate interactions and the corrugations in the adsorbate–substrate interaction potential [77]. The individual metal atoms of the substrate surface may already operate as potential local co-ordination centers for suitable ligands, such as rigid planar heterocyclic molecules [78,79].

In the following we shall demonstrate with a few selected examples from our laboratory the strength of in situ STM in developing a molecular-level understanding (atomistic picture) on the structure and phase formation processes of organic monolayers on electrode surfaces. The chapter is organized as follows: In Sect. 2 we will focus on steady-state properties of Au(111) and Au(100) in 0.05 M H_2SO_4. Structural transitions in adlayers of uracil and some of its derivatives will be described in Sect. 3, based on electrochemical and in situ STM experiments. In Sects. 4 and 5 we will discuss the phase behavior of $2,2'$-bipyridine and $4,4'$-bipyridine on gold single-crystal electrodes. We will conclude with a brief summary of our findings and an outlock.

2 Characterization of Selected Substrate Surfaces[1]

Macroscopic electrochemical experiments as well as structural studies with molecular adlayers were performed on single-crystal metal electrodes, such as Au(*hkl*), Ag(*hkl*) and Cu(*hkl*). In order to characterize the experimental starting conditions and the methodology used, "clean" substrate surfaces have been studied comprehensively employing cyclic voltammetry, capacitance and impedance measurements, chronocoulometry as well as in situ STM [51], FTIR [80] and surface X-ray scattering [81,82]. As examples selected results will be discussed for the two most frequently used systems Au(111)/0.05 M H_2SO_4 and Au(100)/0.05 M H_2SO_4.

2.1 Au(111) in 0.05 M H_2SO_4

The "steady-state" and dynamic properties of the Au(111)/aqueous electrolyte interface have been characterized experimentally with a variety of electrochemical and structure-sensitive methods. Examples are impedance measurements [83], electroreflectance [84], LEED [84,85], SHG [86], in situ surface X-ray scattering [87] and STM [88,89,90] experiments. In the following paragraph we will only focus on some essential results of in situ STM studies [91] which are relevant for the subsequent discussion of phase formation processes in organic adlayers.

Figure 1 illustrates the phase behavior of an Au(111)-electrode in sulfuric acid with a typical current vs. potential curve and STM images of char-

[1] All potentials in this review refer to the saturated calomel electrode (SCE).

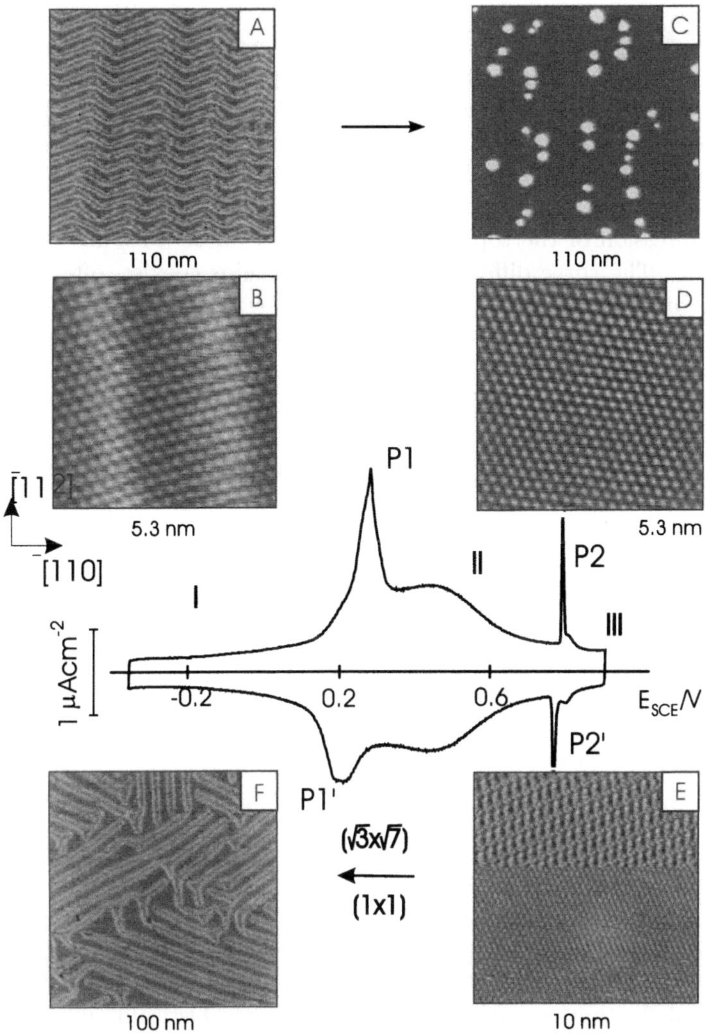

Fig. 1. Cyclic voltammogram for Au(111) in 0.05 M H_2SO_4, scan rate $10\,mV\,s^{-1}$. In situ STM images represent the substrate surface structure at various stages of the potentiodynamic experiment: (**A**) large-scale image of a thermally reconstructed Au(111)-(p$\times\sqrt{3}$) surface, $E = -0.20\,V$, (**B**) atomic resolution of (**A**), (**C**) Au(111)-(1 $\times$ 1) surface with monatomic high gold islands after lifting the surface reconstruction, $E = 0.60\,V$, (**D**) atomic resolution of (**C**), (**E**) order/disorder transition of the ($\sqrt{3} \times \sqrt{7}$) (hydrogen) sulfate overlayer, (**F**) potential-induced reconstructed Au(111)-(p$\times\sqrt{3}$) surface at $E = -0.20\,V$

acteristic "steady-state" structures. The experiment started with a flame-annealed electrode (15 min hydrogen flame) after potential-controlled immersion at −0.20 V. The thermally induced Au(111)-(p×$\sqrt{3}$) reconstruction is preserved under these conditions (negative charge densities) [101]. As an example, Fig. 1A shows a large-scale in situ STM image of a freshly prepared Au(111) electrode in 0.05 MH_2SO_4. The typical double row pattern of the reconstruction (6.3 nm periodicity, corrugation amplitude (0.20 ± 0.05) Å) due to the 4 % compression of the top layer in one of the three [110]-directions is well developed. The three different rotational domains coexist, often the transition from one domain into another one occurs by correlated bending of the corrugation lines of 120°. Entire sets of corrugation lines change their orientation in a zig-zag pattern by ±120° and thus form a periodic sequence of domain boundaries. This superperiodicity is often referred to as a "chevron" or "herringbone" structure. The reconstructed Au(111) surface in an electrochemical environment has the very same structure as its counterpart in vacuum [92,93].

Details of this uniaxially compressed structure are shown in Fig. 1B. The individual gold atoms are clearly resolved. Their positional arrangement changes periodically in the [110]-direction from face-centered cubic (fcc) through a transitional region of so-called "bridging sites" (bright corrugation lines [92]) to hexagonal close-packed (hcp). The stripe-to-stripe separation within a double row amounts to (2.6 ± 0.3) nm (gold atoms in hcp and bridge sites), while the closest distance between adjacent stripes of two different double rows (gold atoms in fcc and "bridge sites") was estimated to (3.7±0.3) nm. The transition region between fcc and hcp sites extends up to three gold atoms.

Changing the electrode potential towards positive charge densities causes the lifting of reconstruction, and the extra amount of gold atoms from the ca. 4 % more densely packed (p×$\sqrt{3}$) structure forms monatomic high gold islands on the surface (Fig. 1C). Subsequently, so-called "electrochemical annealing" takes place. Small islands disintegrate, and larger ones grow either due to statistical fluctuations at rims (two-dimensional Ostwald ripening) and/or the merging of equal size islands [94]. These gold islands are rather mobile. While many islands can be created during a potential step experiment, as for instance −0.20 V to 0.50 V, only few gold islands were found after a slow potential scan (5 mV s^{-1}) from −0.20 V to 0.50 V. In the latter case, most islands have already merged with step edges during the time of scanning the potential.

The maximum P1 in the current vs. potential curve correlates with the substrate surface transition [88]. The difference in zero charge potentials between the reconstructed (p×$\sqrt{3}$) and the unreconstructed (1×1) gold surfaces gives rise to an additional contribution due to double layer charging.

Figure 1D shows the ideal terminated (1 × 1) surface with atomic resolution. Around 0.80 V a disorder/order transition takes place within the layer of adsorbed (hydrogen-) sulfate ions [95,96,97]. The formation/dissolution of the so-called $(\sqrt{3} \times \sqrt{7})$ overlayer correlates with the current maxima P2/P2′ in the voltammogram. Figure 1E illustrates the dissolution of the $(\sqrt{3} \times \sqrt{7})$ phase after application of a cathodic potential step. The atomic/ionic resolution of the (hydrogen-) sulfate structure (top) as well as of the gold lattice (bottom) allow us to conclude that the main maxima of the overlayer pattern are located at bridge positions of the underlying Au(111)-lattice.

The phase transition between the reconstructed and the unreconstructed gold surface is reversible in aqueous electrolyte at room temperature. Stepping the potential towards negative charge densities, e.g. past the current maximum P1′ in Fig. 1, causes the reconstruction to reappear. The reconstruction lines are now no longer aligned in only one direction or "super"-domain, as in the case of the initial (thermally induced) reconstruction, but they run in a rather irregular zig-zag pattern in all three main crystallographic directions of the underlying Au(111) substrate (Fig. 1F). The rims of the monatomic gold islands formed during the transition $(p\times\sqrt{3}) \rightarrow (1 \times 1)$ serve as nucleation centres in the reformation of the $(p\times\sqrt{3})$ phase. The detailed mechanisms of these transitions are documented in [88] and [90].

2.2 Au(100) in 0.05 M H_2SO_4

The phase behavior of Au(100) in aqueous H_2SO_4 and $HClO_4$ has been studied before [98,99,100,101]. Therefore we will only focus on results essential for the understanding of some of the voltammetric and in situ STM experiments reported below.

Figure 2 shows a typical current vs. potential curve of Au(100) in 0.05 M H_2SO_4. The first cycle started with a freshly flame-annealed (e.g. thermally reconstructed) Au(100)-(hex) electrode after immersion at −0.10 V (indicated as "START", solid line). The corresponding large-scale STM image (Fig. 1A) reveals the typical one-dimensional corrugation lines superimposed by a Moire pattern, which is due to a small rotation of the (hex)-structure with respect to the quadratic bulk lattice [102]. The quadratic symmetry of the underlying (1×1) lattice allows two, 90° mutually rotated directions of the reconstructed domain. On carefully flame-annealed electrodes usually one terrace is covered by a single domain. The reconstruction lines of the top layer are either exactly aligned with the atomic rows of the quadratic lattice, or they are rotated by 0.8°. The latter case is shown in Fig. 2A. Details of the *hexagonal* arrangement of the individual gold atoms in this phase are illustrated in Fig. 1B (long-range corrugation of (1.45 ± 0.05) nm periodicity, and ~ 0.05 nm height due to the structural misfit between the reconstructed surface and the bulk).

Scanning the electrode potential towards more positive values causes the "lifting" of the thermally-induced reconstruction at potentials around the current peak P3 ($E_{P3} = 0.304$ V, $q^{m}_{P3} = 21 \mu C\,cm^{-2}$), and the Au(100)-$(1 \times 1)$

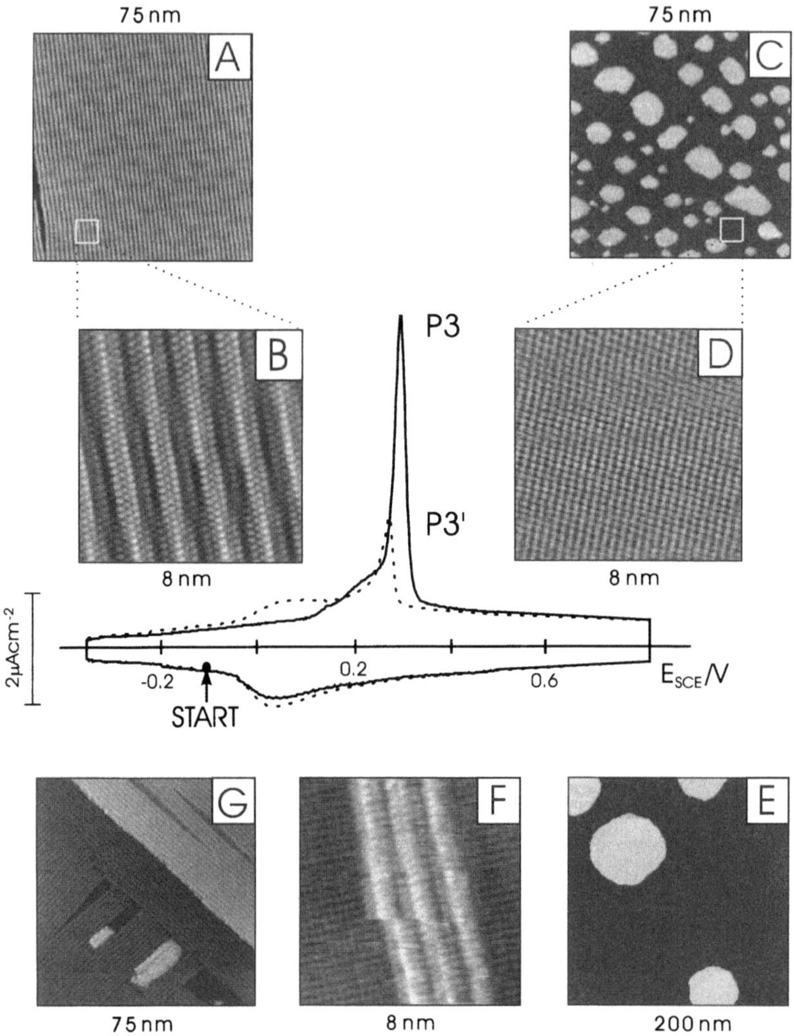

Fig. 2. Cyclic voltammograms for Au(100) in 0.05 M H_2SO_4, scan rate $10\,mV\,s^{-1}$. The first scan as obtained with a flame-annealed, reconstructed electrode after immersion at −0.10 V is plotted as a *solid line*. The steady-state voltammogram is shown as a *dotted curve*. In situ STM images represent the substrate surface structure at various stages of the potentiodynamic experiment: (**A**) large-scale image of a thermally reconstructed Au(100)-(hex) surface, $E = -0.10$ V, (**B**) atomic resolution of the reconstructed Au(100)-(hex) surface, (**C**) Au(100)-(1 × 1) surface with monatomic high gold islands after lifting the surface reconstruction, $E = 0.55$ V, (**D**) atomic resolution of the Au(100)-(1 × 1) surface structure, (**E**) "electrochemically annealed" Au(100)-(1 × 1) surface after 60 min at 0.55 V, (**F**) atomic resolution image of three newly formed hexagonal reconstruction rows on Au(100)-(1 × 1), $E = 0.0$ V, (**G**) potential-induced reconstructed Au(100)-(hex) surface at $E = -0.20$ V

surface structure is formed. The excess of gold atoms of the pre-existing (hex)-phase (24 %) is expelled onto the surface and monatomic high gold islands are created [98,99,100,101] (Fig. 1C). Similar to the case of Au(111), sufficiently long electrochemical annealing at positive potentials, for instance at 0.55 V, causes an increase of the average island size and simultaneously a significant decrease of their number (Fig. 1E). The rate of this electrochemical annealing process increases with potential and is higher in the presence of strongly specifically adsorbed anions [38]. A typical high-resolution *square* lattice arrangement of surface gold atoms under conditions such as demonstrated in Fig. 1C and Fig. 1E, is shown in Fig. 1D.

Finally, scanning the electrode potential towards more negative values causes the charge (potential)-induced surface reconstruction to appear. Morphological details of the newly formed *hexagonal* "reconstruction stripes" within the metastable gold surface top layer of *square* symmetry at $E = 0.0\,\mathrm{V}$ are depicted in Fig. 1F. The (hex)-structure is reformed to a large extent at sufficient negative potentials and long waiting times there (Fig. 1G). Typically, this phase is characterized by a significantly higher defect density, and rather small domains rotated 90° (e.g. along the two crystallographic axes) with respect to each other. Both structural features also determine the negative shift of P3 during the second and each subsequent voltammetric cycle (dotted line in Fig. 2, $E_{P3'} = 0.278\,\mathrm{V}$).

Experiments on the time-dependence of the (hex) → (1 × 1) transition demonstrate that the (1×1) phase is established clearly during a voltammetric scan with $10\,\mathrm{mV\,s^{-1}}$. The reformation of the long-range order of the (hex)-structure, after once having been lifted, requires a much longer time than is typically available within such a dynamic scanning regime (see also [98,101]). The nucleation of the charge (potential)-induced reconstruction starts at substrate defects, such as steps and remaining gold islands, and is strongly dependent on the history in the thermodynamic stability range of the (1 × 1) phase: (i) Short waiting times at positive potentials create just many small gold islands, e.g. many potential nuclei for the (hex)-phase, which forms as a rather disordered pattern after excursion towards negative charge densities. (ii) Alternatively, long rest times at positive potentials allow the formation of few large gold islands (Fig. 2E), which significantly decrease the rate of formation of the (hex)-phase during a subsequent cathodic potential step or scan. The emerging (hex)-phase is usually highly ordered (Fig. 2G).

3 Structural Transitions in Adlayers of Uracil Derivatives on Au(*hkl*)

Two papers in 1995 that presented experimental evidence for the condensation of uridine [36] and uracil [37] on different orientations of gold single-crystal electrodes have been the starting point for numerous investigations concerning the formation of organized organic monolayers of similar molecules on solid electrodes (c.f. the literature summarized in the introduction).

3.1 General Overview – Results of Classical Electrochemical Studies

The results obtained in these studies documented convincingly that the occurrence of potential-induced two-dimensional phase transitions in organic adlayers is not only restricted to the "perfect" surfaces of liquid electrode materials, such as mercury and gallium [18,19], but also may occur on gold, silver and copper as well as various other materials. By now, the most comprehensively investigated molecule on liquid and solid electrodes is uracil. The results obtained made this molecule a model substance for studies of two-dimensional phase transitions on solid electrodes.

3.1.1 Voltammetric Measurements

Below we describe some results of these electrochemical experiments with a focus on the nature of the adsorbate and the substrate material, as well as the kinetics of adlayer transitions. Complementary in situ STM and spectroscopic experiments will be reported in Sects. 3.2–3.4.

Figure 3 shows typical cyclic voltammograms for 12 mM uracil in 0.05 M H_2SO_4 (A) and 0.05 M $KClO_4$ (B) on Au(100). The first voltammetric scan, as obtained with a freshly flame-annealed (i.e. "thermally-induced") reconstructed Au(100)-(hex) electrode after immersion at −0.10 V is indicated by the solid line. For comparison, the corresponding "quasi-steady-state"

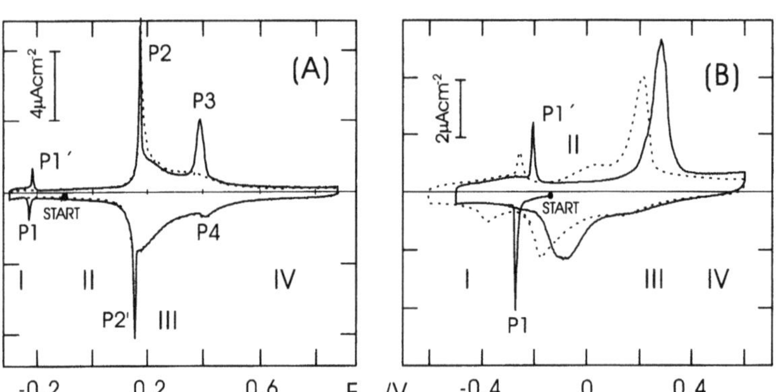

Fig. 3. Cyclic voltammograms for Au(100)/0.05 M H_2SO_4 (**A**) or 0.05 M $KClO_4$ (**B**) in the presence of 12 mM uracil, scan rate 10 mV s^{-1}. The first scan as obtained with a flame-annealed, reconstructed electrode after immersion at −0.10 V is plotted as the *solid line*. The steady-state voltammogram is shown as a *dotted curve*. The stability regions of the various adlayer phases are labeled I to IV

voltammograms as obtained after at least five subsequent complete cycles between -0.30 V and 0.80 V (A) or -0.50 V and 0.60 V (B) with a scan rate of $10\,mVs^{-1}$ are also added (dotted line). These current vs. potential curves may be understood as "quasi-diagrams of state". The phenomenological analysis of these electrochemical data allows us to distiguish four different potential regions labeled I–IV in Fig. 3 [37,40]. The following discussion will be focused on the phase behavior of uracil in 0.05 M H_2SO_4 on Au(100) (Fig. 3A) [51]. Differences to the results obtained in $KClO_4$ (Fig. 3B) [37,40] will be pointed out whenever appropriate.

In the potential range of regions I the molecules are randomly ("gaseous-like") adsorbed on the electrode. The formation of a two-dimensional condensed, physisorbed uracil film at more positive potentials is indicated by typical current spikes P1/P1′ and P2/P2′ that limit the so-called pit region II. We use the term physisorbed layer because the adsorption enthalpy of this film is in the typical energy range of physisorption $-10\,kJ\,mol^{-1}$ to $-40\,kJ\,mol^{-1}$ [40]. A marked hysteresis is observed when changing the direction of the potential scan, which is distinctly different at both edges. The physisorbed film II undergoes changes when passing the potential of the sharp current peak P2. Region III is ascribed to a complex partial charge transfer/reorientation process of *adsorbed* molecules. The total charge consumed equals about $40\mu C\,cm^{-2}$. The positive limit is marked by the broad peak P3. Comparing the properties of this feature with corresponding data of each subsequent voltammetric scan as well as with the uracil-free base electrolyte solution (Fig. 2) we hypothesize here that P3 represents the potential region, where, within the time scale of the present experiment, the substrate surface transition (hex) $\rightarrow$ (1×1) takes place, see P1 in Fig. 1. Finally, a chemisorbed phase IV is formed at sufficiently positive potentials. The latter is stable at temperatures as high as 100 °C, and significantly inhibits the onset of gold oxidation. Its dissolution at negative potentials is probably associated with the shallow feature P4, and not with a sharp peak as found previously in the case of Au(111) [37].

Comparison of the just described first voltammetric scan, obtained immediately after flame-annealing of the Au(100) electrode, with the steady-state voltammogram shows two significantly different features. For the latter the following was found: (i) The characteristic negative current spikes P1 and P1′ shift at least 0.1 V more negative, and can be observed no longer due to superposition with the bulk hydrogen evolution. (ii) The peak P3, previously found at 0.395 V, also disappears. The Au(100) electrode is not yet potential-induced reconstructed within the time-scale of the scanning experiment.

Comparing the uracil adsorption on Au(100) in 0.05 M H_2SO_4 and 0.05 M $KClO_4$, we like to emphasize that the regions I–IV exist in both electrolyte solutions. Two differences are obvious [42,51]: (i) The stability ranges of II, III and IV are shifted towards more positive potentials in acidic solutions, which is a general trend with decreasing pH for the pyrimidine bases [32,41]. (ii) The adlayer and substrate surface transition region III in $KClO_4$ solution is characterized by one broad, structureless peak. This feature is determined by kinetics. The key factor is most probably the significantly lower mobility of gold surface atoms in the presence of ClO_4^-, in comparison to SO_4^{2-}.

3.1.2 Transient Measurements on the Stability of the Physisorbed Film II

The sharp current spikes P1/P1′ as well as the hysteresis observed point to a first-order phase transition within the adlayer [19]. This hypothesis is supported by the results of potential step experiments [42]: In order to reduce the contribution of double layer charging and of uracil still adsorbed in the non-condensed state to the overall current, "primary" transients have been measured in the following double step regime (Fig. 4A): The initial potential was chosen in region I to be more negative then P1; in the experiment shown at $E_1 = -0.320\,\mathrm{V}$. After a waiting time of 5 s to equilibrate the system, the potential was then stepped to $E_2 = -0.220\,\mathrm{V}$, which is close to P1′, but still within the stability range of phase I when approaching P1′ via negative potentials. Single exponentials were obtained. After a short waiting time of $t_2 = 15\,\mathrm{ms}$, which was long enough to let the double layer charging and/or adsorption current approach zero, the potential was stepped once again, but now to final values E_3 located within the thermodynamic stability range of the 2D condensed phase II. Such transients (E_2/E_3) exhibit a clearly developed maximum, which shifts towards shorter times with increasing overvoltage, and which is only preceded by a minor contribution due to double layer charging and adsorption of additional uracil. All experimental observations, including the time constants of both features, suggest that the initial part of the transients can be approximated as being rather independent of the subsequent nucleation and growth process. The time-delay due to preadsorption in state I is taken into account as an induction time t'. With these assumptions the rising part of the transients E_2/E_3 together with the current maximum could be analyzed separately. Careful tests revealed [42] that the experimental data could be represented best by the exponential law of nucleation [103] in combination with surface-diffusion-controlled growth [104,105]:

$$i(t) = k_{1'}\left\{1 - \exp\left[-k_{2'}(t - t')\right]\right\} \exp\left[- k_{3'}\Big((t - t') - 1/k_{2'}\Big\{1 - \exp\big[- k_{2'}(t - t')\big]\Big\}\Big)\right], \quad (1)$$

where $k_{1'} = k_{2'}q_{\mathrm{m}}$, $k_{2'}$ is the nucleation rate, q_{m} corresponds to the total charge involved in the phase formation and $k_{3'}$ is a constant related to the growth process. Additional confidence for the reliability of this model is also given by the charge balance. The average value of $q_{\mathrm{m}}qk_{1'}/k_{3'}$ amounts to $2.8\,\mu\mathrm{C\,cm^{-2}}$. This value is rather close to $(3.0 \pm 0.1)\,\mu\mathrm{C\,cm^{-2}}$, the charge density obtained by integration of the respective voltammetric peak P1′.

Furthermore, the application of (1) requires (i) the random distribution of a large number of centers, which are small with respect to the total electrode area, and (ii) sufficiently large diffusion zones that linear diffusion parallel to the surface still holds. It may be assumed that both conditions are fulfilled in the present system. Finally we note that similar results have also been reported for the formation of physisorbed thymine films on Au(111) and Ag(111) [41].

In order to support the choice of the above model the growth process was studied separately by applying the following double potential step regime (Fig. 4B): The first potential step is directed from E_1 in region I to E_2, which is now located in the thermodynamic stability range of the 2D condensed phase II, e.g. past P1′. During this primary process critical nuclei of the ordered phase will be formed. After a defined waiting time at E_2 the potential will be finally stepped to E_3, located within the hysteresis region between P1′ and P1. E_3 was chosen in such a way that no additional nuclei can be formed and therefore only the secondary growth process takes place, involving nuclei formed previously, e.g. during the first potential step. A similar strategy had earlier been applied to monitor the growth of mercury nuclei on platinum electrodes [106] or the growth of isoquinoline films on mercury [107]. Typical transients are plotted in Fig. 4B. Curves 1 and 2 still represent primary transients E_1/E_2 and E_1/E_3. The first trace indicates clearly that even at 8 ms critical nuclei of the ordered phase II are formed. On the other hand, the trace E_1/E_3 (E_3 located between P1 and P1′) gives no indication of any nucleation process within the time scale of observation. These two control experiments provide the boundary conditions for the subsequent secondary growth transients with a fixed number of nuclei being formed in the primary step, and can therefore be modeled by instantaneous nucleation in combination with an appropriate law of growth. The corresponding current for a surface diffusion controlled growth process is given by:

$$i = k_{1''} \exp\left[-k_{2''}(t-t'')\right] . \tag{2}$$

Combination with the parallel contribution due to double layer charging (single exponential with the constants $k_{3''}$ and $k_{4''}$) yields

$$i = k_{1''} \exp\left[-k_{2''}(t-t'')\right] + k_{3''} \exp\left[-k_{4''}(t-t'')\right] \tag{3}$$

with $k_{1''} = k_{2''} q_s$, where $k_{2''}$ is a constant related to the rate of the secondary growth process, q_s is the charge involved and $t'' = 8\,\mathrm{ms}$ is the waiting time at E_2 used to normalize the time scale.

Indeed, the model expressed by (3) fits all secondary growth transients very well, which can be seen be comparison of the experimental data (dots in Fig. 4B) with the calculated curves (solid lines in Fig. 4B) [42].

3.1.3 Influence of Substrate and Adsorbate Structure

Two-dimensional condensed uracil films of type II were found at medium values of the surface excess $\Gamma \approx 5.0 \pm 0.5 \times 10^{-10}\,\mathrm{mol\,cm^{-1}}$, which corresponds to

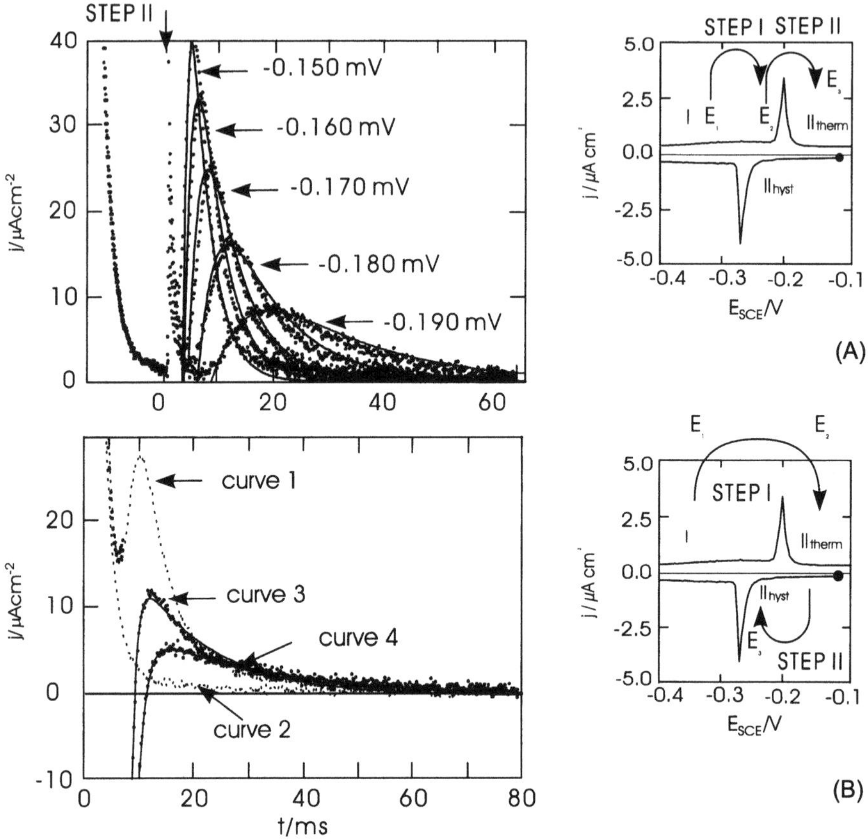

Fig. 4. (**A**) Primary current transients $i(t)$ for Au(100)-(hex)/12 mM uracil, 50 mMKClO$_4$ as obtained with a double potential step $E_1 = -0.320$ V(5 s)/$E_2 = -0.220$ V(15 ms)/E_3. The final potentials E_3 are indicated in the figure. The *solid lines* represent the theoretical curves calculated with the parameters of the numerical fit to the exponential nucleation law in combination with surface diffusion-controlled growth (2). (**B**) Secondary growth transients. The potential was stepped from $E_1 = -0.320$ V (5 s) to $E_2 = -0.170$ V (8 ms) and then to -0.215 V (curve 3) or -0.235 V (curve 4). Primary transients with $E_1 = -0.320$ V(5 s)/$E_2 = -0.170$ V (curve 1) and $E_1 = -0.320$ V(5 s)/$E_3 = -0.210$ V (curve 2) are given as reference. The *solid lines* represent the theoretical curves calculated with the parameters of the numerical fit to the growth model expressed by equation (3)

a planar orientation of the molecule, on Au(111)-(p×$\sqrt{3}$), Au(100)-(hex) [37], Au(100)-(1 × 1) [38], Ag(111) and Ag(100) [39], but not on the very open Au(110) surface. The formation of the ordered physisorbed uracil film, which mainly arises from lateral attractive interactions between adjacent molecules, is supported on smooth substrate surfaces, but sensitively inhibited by highly corrugated and/or defect-rich surfaces. Quantitative studies on the temperature dependence of the condensed phase II showed that the thermodynamic stability of the ordered phase increases in the following sequence [39,40]:

$$\mathrm{Hg} < \mathrm{Ag}(111) < \mathrm{Ag}(100) < \mathrm{Au}(111)-(\mathrm{p} \times \sqrt{3}) < \mathrm{Au}(100)-(1 \times 1) \\ \sim \mathrm{Au}(100) - (\mathrm{hex})\,.$$

This sequence does not coincide directly with the order of hydrophilicity Hg $\leq$ Au $<$ Ag, when comparing the same crystal face, or (111) $\leq$ (100) $<$ (110) when comparing the same metal, which has been derived from physisorption studies of diethylether [108,109] and aliphatic alcohols [110]. The above sequence is tentatively explained by the energetics at the Fermi level for the various substrate materials. The positions of the d-bands are located 2, 4 and 8 eV below the Fermi level for gold, silver and mercury, respectively. In consequence, the overlap of the molecular non-bonding and/or π-orbitals with the gold surface are much stronger than those for silver or mercury. It is also striking that the physisorbed uracil (thymine and 5,6-dimethyluracil) film is always more stable on the square (100)-planes than on the hexagonal densely packed (111)-planes [40]. This trend correlates with the anisotropic distribution of electron density at the more open surface, which is much less pronounced on surfaces with hexagonal symmetry. Both arguments suggest that the stability of the physisorbed film is not only determined by lateral interactions between adsorbed molecules, but also by electronic substrate–adsorbate interactions.

Furthermore, we found that chemisorbed uracil films of type IV could be formed on all low index faces of gold, silver and copper, indicating the determining role of substrate–adsorbate interactions.

Figure 5 illustrates with selected examples (12 mM adsorbate in 50 mM $KClO_4$, 10 °C) that the basic adsorption behavior of the C- and N-methyl derivatives of uracil on Au(111) is significantly different. The former group, e.g. uracil (U), 5-methyluracil (5-MU) and 5,6-dimethyluracil (5,6-DMU), exhibits four characteristic interfacial regions, labeled I–IV. Their qualitative properties have already been described for uracil on Au(100) in Sect. 3.1.1. (cf. Fig. 3). Comparing U with 5-MU and 5,6-DMU at constant concentration and temperature, we learn that the introduction of methyl-groups in the 5- and 6-position of the 2,4-dihydroxypyrimidine ring causes primarily an enlargement of the physisorbed condensed region II, which also coincides with the hydrophobicity trend of the C-methyl derivatives. All other properties are rather similar. The same conclusions can be drawn from experiments with uracil derivatives having halogen atoms as substituents (Cl, Br, I) in position

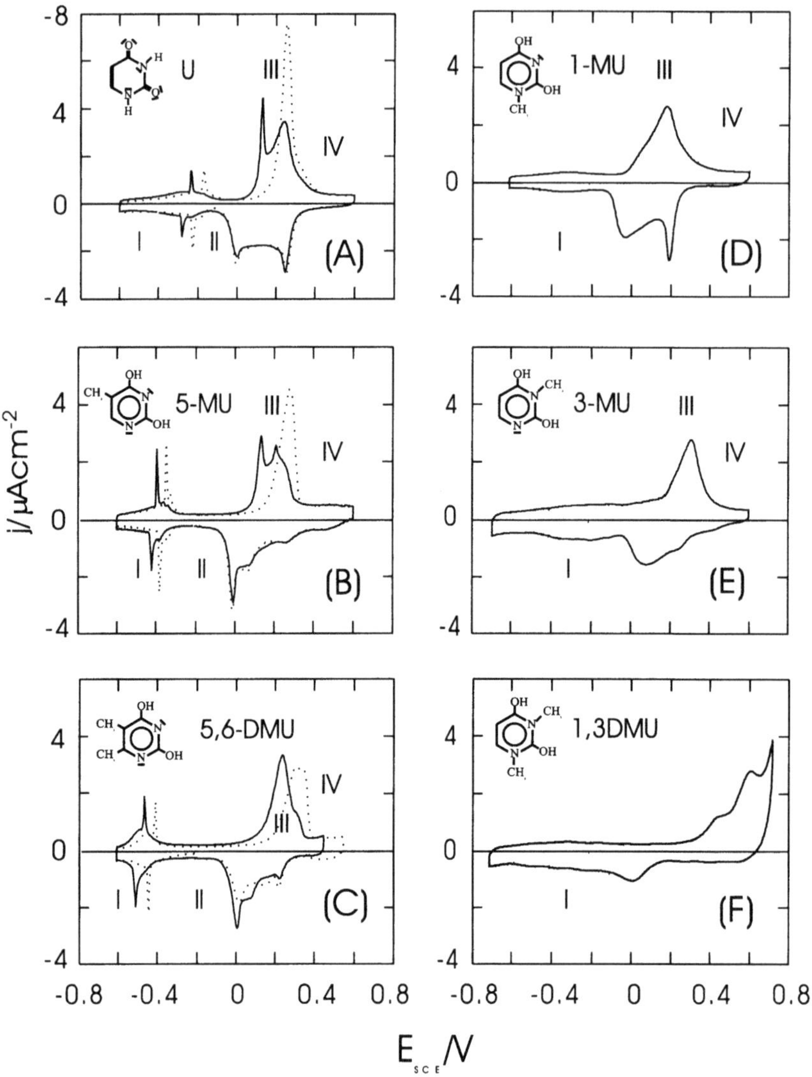

Fig. 5. Current-potential curves for Au(111)/50 mMKClO$_4$ in the presence of 12 mM organic adsorbate: uracil (**A**), 5-methyluracil (**B**), 5,6-dimethyluracil (**C**), 1-*N*-methyluracil (**D**), 3-*N*-methyluracil (**E**), 1,3-*N*, *N*-dimethyluracil (**F**). Scan rate 10 mV s^{-1}, temperature 10 °C. The *dotted line* represents the first complete cycle as obtained with a thermally reconstructed Au(111)-electrode when scanning the potential from $E_i = -0.20$ V (immersion potential) initially towards negative values. The *solid lines* show steady-state voltammograms (5th cycle). The various adsorption regions are indicated as I–IV

Table 1. Physical properties of methyl derivatives of uracil

Parameter	U	5-MU	5,6-DMU	1-MU	3-MU	1,3-DMU
μ/D [115,116]	4.16	4.12	4.66	3.89	–	3.9
pK_a*	9.7/14.2	9.9/14.0	10.07	9.77	9.85	–
$F_p/°C$*	339–341	316–317	297–300	236–238	189–191	124
A/nm^*	0.37 [113]	0.44 [114]	0.52 [115,116]	–	–	6.5

* estimated area for a densely packed arrangement of planar oriented molecules

5 or 6 of the ring [111]. In contrast, no hints of any 2D condensed physisorbed film (region II) were found for the N-methyl derivatives within the ranges of concentration (10^{-5} M – 0.1 M) and temperature (0 °C to 70 °C) studied. Region I is followed immediately by regions III and IV. The former consumes in case of 1-methyluracil (1-MU) and 3-methyluracil (3-MU) 50μC cm^{-2} and 35μC cm^{-2}, respectively. No region IV was found with 1,3-dimethyluracil (1,3-DMU). These results clearly indicate that the formation of the chemisorbed phase IV for this class of substances (2,4-dihydroxypyrimides) requires at least one ring nitrogen atom, either in the N(1) and/or N(3) position, which may actively interact with the positively charged gold surface. This argumentation is supported by Saenger's crystallographic studies of co-ordination compounds between Au(III) and uracil derivatives, which yield the following stability sequence of co-ordination sites: $N(1) > N(2) \gg O(1), O(4)$ [112]. Crystallographic studies also help address the lack of a condensed physisorbed film in case of the N-methyl derivatives of uracil. No 2D-layered structures could be found for the latter group of substances. These systems are not capable of forming highly ordered planar networks of hydrogen-bonded molecules, which contrasts with the behavior reported for U [113], 5-MU [114] and 5,6-DMU [117]. The probable correlation between the packing in a 3D solid and a 2D condensed film is also obvious when referring to the melting temperatures F_p of the derivatives studied (Table 1). They are significantly higher for U, 5-MU and 5,6-DMU than for 1-MU, 3-MU and 1,3-DMU.

These qualitative structural considerations served as the starting point of a systematic chronocoulometric study [40]. The analysis of these experiments yielded comparative data on energetic parameters such as film pressure, surface excess Γ, electrosorption valency, surface dipole, and Gibbs energy of adsorption for regions I and II. The Gibbs energies of adsorption range for the derivatives studied between $-10\,\mathrm{kJ\,mol^{-1}}$ and $-40\,\mathrm{kJ\,mol^{-1}}$. For the condensed film II, values of the maximum surface excess Γ_{IIm} of U, 5-MU and 5,6-DMU have been estimated ranging from $3.4 \times 10^{-10}\,\mathrm{mol\,cm^{-1}}$ up to $4.3 \times 10^{-10}\,\mathrm{mol\,cm^{-1}}$, which corresponds to an area per adsorbed molecules of 39–49 Å^2. These results point to a planar orientation of the pyrimidine bases in state II. The area per molecule in the gaseous-like state I was estimated to be 70–80 Å^2 per molecule. Referring to the role of ring substitution and the layered structure of the two-dimensional crystallographic lattices of U, 5-MU and 5,6-MU we hypothesize that condensed films of type II are the result of the replacement of water molecules of the Helmholtz layer followed by the formation of a network of planar oriented, most probably hydrogen-bonded molecules.

Because of the partial oxidation of these pyrimidine bases at $E > 0.20\,\mathrm{V}$ no reliable thermodynamic parameters of the adsorption states III and IV were obtained. Nevertheless, the charge balance points to a significant electron transfer of 0.5–0.7 electrons per molecule with the respective substrates. These results qualitatively support the importance of N(1)- and/or N(3)-substrate co-ordination in the formation of the chemisorbed phase IV.

The decisive prerequisite for the reliability of the above thermodynamic experiments was the knowledge, from in situ SXS studies, that the pyrimidine bases stabilize the thermally induced ($p\times\sqrt{3}$)-reconstruction of Au(111) [28]. With these results the conditions of the thermodynamic measurements could be chosen in such a way that the substrate surface was always defined and stable.

3.2 In-Situ STM Experiments with Uracil Adsorbed on Au(*hkl*)

3.2.1 General Overview

In order to address structural aspects of the above reported phase behavior of uracil derivatives on Au(*hkl*) combined structural studies have been performed employing in situ STM [50,51], SXS [81,82] as well as sum frequency generation experiments [118] at various stages of the electrochemical experiment. Some results for uracil on Au(100) will be reported below (Fig. 6).

A well-prepared flame-annealed Au(100)-electrode, immersed in region II into a solution containing 0.05 M H_2SO_4 and uracil, not only shows the characteristic one-dimensional corrugation of the (hex)-reconstructed substrate (cf. Fig. 2A); in addition a new STM contrast pattern was found in the potential region marked by P1/P1′ and P2/P2′, which provides clear evidence for the existence of an ordered overlayer (Fig. 6A). This pattern represents the two-dimensional condensed physisorbed film II, which was previously proposed just based on macroscopic electrochemical experiments [37,40]. Passing the voltammetric peak P2 towards positive potentials "blurrs" out the contrast pattern of the organic phase within a narrow potential interval ($0.15\,\mathrm{V} < E < 0.20\,\mathrm{V}$) (Fig. 6B). Despite these changes within the organic monolayer (most probably the partial oxidation), individual molecules can still be resolved in $0.20\,\mathrm{V} < E < E_{P3}$, which indicates their rather low mobility in region III. At potentials more positive than P3 the gold reconstruction is lifted and monatomic high, nearly square-shaped gold islands appear. The rather broad size distribution, which was, for instance, not observed on Au(111) in the presence of uracil [50], is remarkable. Simultaneously, the

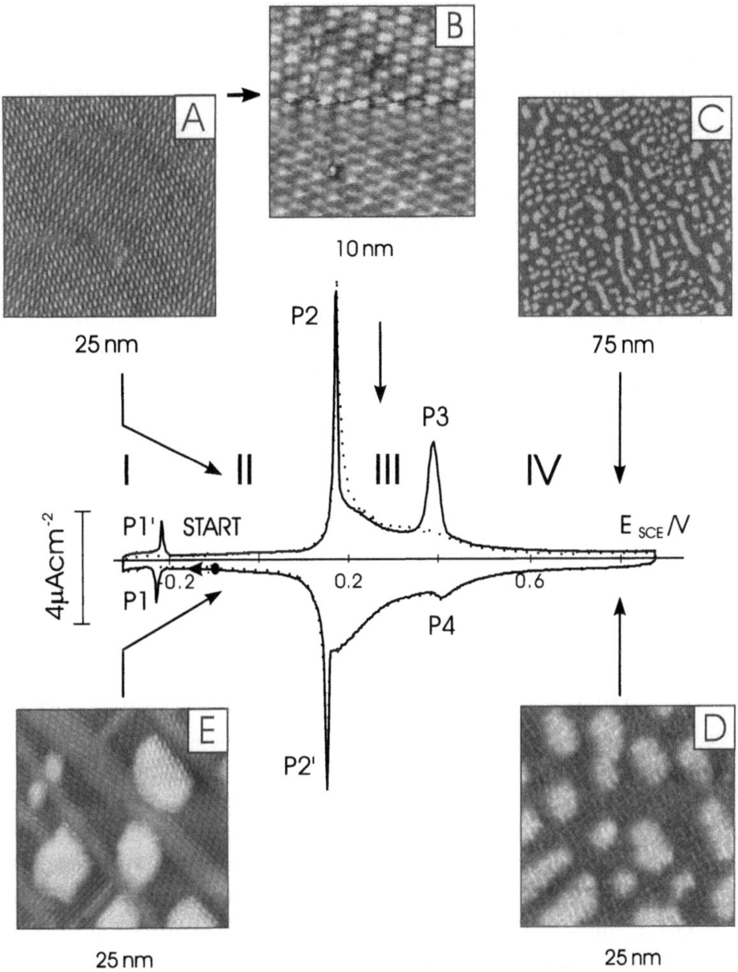

Fig. 6. Cyclic voltammograms for Au(100)/0.05 MH_2SO_4 in the presence of 12 mM uracil, scan rate 10 mV s^{-1}. The first scan as obtained with a flame-annealed, reconstructed electrode after immersion at −0.10 V is plotted as *solid line*. The steady-state voltammogram is shown as *dotted curve*. The stability regions of the various adlayer phases are labelled I–IV, and they are illustrated with typical in situ STM images: (**A**) physisorbed uracil film on a thermally reconstructed Au(100)-(hex) surface, $E = 0.0$ V, (**B**) contrast pattern for the transition from −0.05 V (region II) to 0.30 V (region III), (**C**) "snapshot" of the unreconstructed Au(100)-(1 × 1) surface in the presence of the chemisorbed uracil film, $E = 0.55$ V, (**D**) high-resolution image of (**C**), (**E**) partially potential-induced reconstructed Au(100) surface covered by the physisorbed uracil film II, $E = -0.05$ V

contrast pattern of the uracil film III has disappeared, and a completely different overlayer structure is found. A *snapshot* of this phase transition is shown in Fig. 6C. Zooming into smaller spots in region IV, as for instance in Fig. 6D, reveals that the entire electrode surface, including all islands, is covered with a new, more or less ordered (chemisorbed) uracil film. This overlayer inhibits the onset of gold oxidation up to $-1.25\,\mathrm{V}$, and dissolves slowly after passing the weak voltammetric peak P4 towards more negative values of the electrode potential. Closing the potential cycle causes the reformation of the physisorbed film II, but now on an island-rich and only partially reconstructed gold surface (Fig. 6E). The potential-induced (hex) top-layer structure of the gold surface starts to develop. The completion of this process requires, for instance at $E < E_{P1}$, waiting times longer than one hour. This time scale also supports the notion that during cycling with the potential with a scan rate of $10\,\mathrm{mV\,s^{-1}}$ in $-0.30\,\mathrm{V} < E < 0.90\,\mathrm{V}$ only the first scan represents features of a hexagonal reconstructed Au(100)-surface, and there only at $E < E_{P3'}$.

Before addressing these dynamic changes in Sect. 3.3 we will first focus on some structural details of the two ordered uracil adlayers.

3.2.2 Physisorbed Uracil Film in Region II

On thermally reconstructed Au(100)-(hex) electrodes uracil forms a highly ordered physisorbed adlayer in the potential region between P1/P1′ and P2/P2′. High-resolution images such as those shown in Fig. 7, reveal a two-dimensional pattern of regularly shaped objects, which we hereafter assign to individual molecules (c.f. discussion in [50]). The image contrast varies little with tunneling voltage or current. Two average values of characteristic center-to-center "molecular" distances could be estimated: $a = (6.4 \pm 0.3)\,\mathrm{Å}$ and $b = (8.7 \pm 0.4)\,\mathrm{Å}$, and the corresponding angle of the "primitive unit cell" amounts to $\alpha = (68 \pm 3)^\circ$ (Fig. 7A). These data give rise to $A_{\mathrm{exp}} = (52 \pm 5)\,\mathrm{Å}^2$ as the approximate area per molecule.

The physisorbed uracil film II exhibits characteristic rotated domains. The domain boundaries are usually sharp, with an angle of $(138 \pm 7)^\circ$, which is approximately twice the angle α between three adjacent uracil molecules and is in the typical range of intermolecular hydrogen bonds [112], e.g. it points to a certain way of "chemical communication" and *not* to a structure-determining role of the symmetry of the underlying substrate (Fig. 7B). Rotated angles significantly different from 138° or translational domain boundaries were seldom found. Nevertheless we like to emphasize that the substrate geometry is not completely negligible. The reconstruction of the surface seems to act as a weak template for the organic adlayer. In many but not all images analyzed, the short vector $\boldsymbol{a}$ of the "primitive unit cell" is practically aligned with the [011]-direction of the substrate corrugation lines. The average rotation angle was estimated to be $\gamma = (2 \pm 2)^\circ$ (Fig. 7C). A second rotation angle $\gamma = (42 \pm 2)^\circ$ was also found. The latter could be rationalized

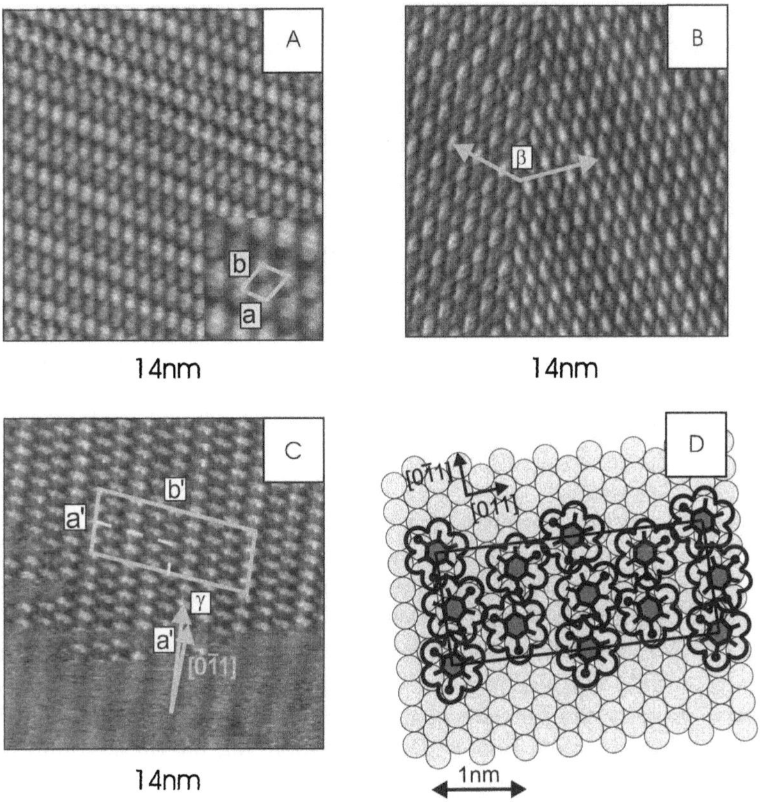

Fig. 7. Unfiltered in situ STM images of the physisorbed uracil film on Au(100)-(hex)/0.05 M H_2SO_4 + 3 mM uracil: (**A**) high-resolution image at $E = -0.05$ V, $i_T = 2$ nA, $v_T = +0.01$ V. The primitive unit cell is indicated, (**B**) typical pattern of a rotational domain boundary, (**C**) "high order" coincidence cell and orientation with respect to the reconstruction rows of Au(100)-(hex) after tip-induced removal of the adlayer in the lower part of the image. (**D**) Proposed packing model. One quarter of the adsorbate mesh that is coincident with the underlying substrate lattice is shown

by $(180° - 2 \times 68°) = 44°$, where 68° represents the angle of the "primitive unit" cell.

On the basis of these experimental data and the previously reported observations for uracil on Au(111)-(p×$\sqrt{3}$) [50] the following model is proposed for the molecular packing arrangement on Au(100)-(hex) (Fig. 7D): The covalent bonding lengths between atoms in each uracil molecule and the van der Waals radii for H, C, N and O were taken from [112] and [113]. Based on the molecular area it can be concluded that uracil molecules are lying flat on the gold surface, rather similar to their hydrogen-bonded two-dimensional structure in the crystalline state [113]. Taking into account the dimensions and the "majority" rotation angle of the "primitive unit cell" with respect to the [011]-direction of the substrate reconstruction lines, a "higher order" coincidence mesh with $a' = 26\,\text{Å}$, $b' = 65\,\text{Å}$, $\alpha' = 90°$, $\gamma'' = 0°$ is suggested (Fig. 7C,D). Each uracil molecule is hydrogen-bonded to three of its neighbors with three hydrogen bonds. The model cell contains 32 molecules, and γ'' is the "theoretical" rotation angle with respect to the [011]-substrate direction (marked by the reconstruction lines in Fig. 7C). The "theoretical area" per uracil molecule is $A_{\text{th}} = 53\,\text{Å}^2$. Both A_{exp} and A_{th} are rather close to our previous STM ($A_{S,U} = (47 \pm 6)\,\text{Å}^2$ [50]) and chronocoulometric ($A_{C,U} = (42 \pm 5)\,\text{Å}^2$ [40]) results for uracil on Au(111)-(p$\times\sqrt{3}$), and therefore point out the structure determining role of attractive adsorbate–adsorbate interactions due to directional hydrogen bonds.

The aspect of commensurability between the physisorbed uracil adlayer and the underlying gold substrate was complemented by additional measurements with "island-free" Au(100)-(1 × 1) electrodes [38], which exhibit a *square* substrate surface geometry. The experiments demonstrate convincingly that the STM contrast pattern of the physisorbed organic film II is rather similar to previous results on reconstructed Au(100)-(hex) and Au(111)-(p$\times\sqrt{3}$) electrodes [50,51], which exhibit a *hexagonal* surface symmetry (Fig. 8). We conclude that the structure of physisorbed uracil films on the low-index gold surfaces investigated is primarily determined by lateral attractive interactions between adjacent molecules due to their ability to form directional hydrogen bonds. The corresponding interaction energy between neighboring uracil molecules was recently estimated in a quantum chemical study to be $-44.5\,\text{kJ}\,\text{mol}^{-1}$ [119]. The substrate surface geometry

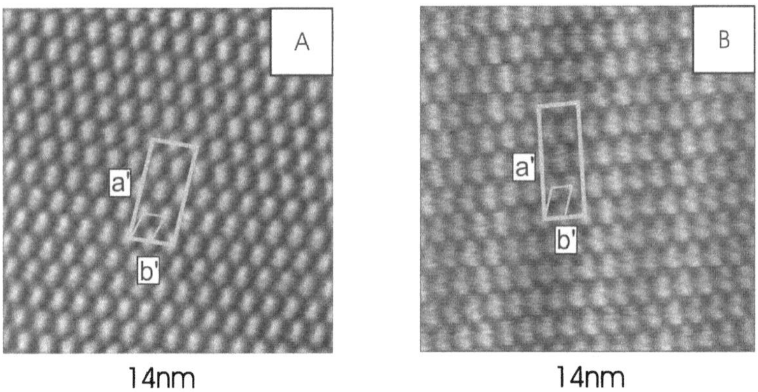

Fig. 8. STM image of physisorbed uracil on Au(100)-(1 × 1), $E = -0.05\,\text{V}$, $i_T = 1\,\text{nA}$, $v_T = -0.01\,\text{V}$ (**A**) and Au(111)-(p$\times\sqrt{3}$), $E = -0.05\,\text{V}$, $i_T = 0.5\,\text{nA}$, $v_T = 0.03\,\text{V}$ (**B**). The suggested coincidence adsorbate mesh (a', b', α', dashed parallelograms) as well as the basis vectors $\boldsymbol{a}$ and $\boldsymbol{b}$ of the primitive unit cell are indicated

seems to be of somewhat minor importance for the arrangement of the individual molecules within the adlayer as long as sufficiently large terraces with a rather low defect density are available.

3.2.3 Uracil Adlayer in Region III

The STM contrast pattern of the organic film changes when passing the sharp current peak P2 (Fig. 3) towards positive potentials. High-resolution images become more noisy. This observation is attributed to (i) the break-off of the hydrogen-bonded network and (ii) to a different electronic state of the molecule as the result of a partial charge transfer process. It is quite remarkable that the positional arrangement of individual molecules in region III appears to be rather similar to that in region II. Figure 9 illustrates these features. The estimated dimensions of the oblique primitive unit cell amount to $a = (6.4 \pm 0.5)\,\text{Å}$, $b = (8.6 \pm 0.6)\,\text{Å}$, $\alpha = (68 \pm 2)°$. Imaging of molecular structures becomes impossible when approaching potentials close to the broad voltammetric peak P3 ($E \sim 0.39\,\text{V}$) in $0.05\,\text{M}\,H_2SO_4 + 3\,\text{mM}$ uracil.

The substitution of sulfate by the less specifically adsorbed perchlorate ions reveals that the perchlorate ions do not influence the structure III of the organic adlayer. Their major effect is (i) on the potential range of the adjacent adlayer phases II and IV, as it is also reflected in the morphology of the corresponding voltammogram (Fig. 3), and (ii) on the stability of the (hex) and of the (1×1) substrate surface.

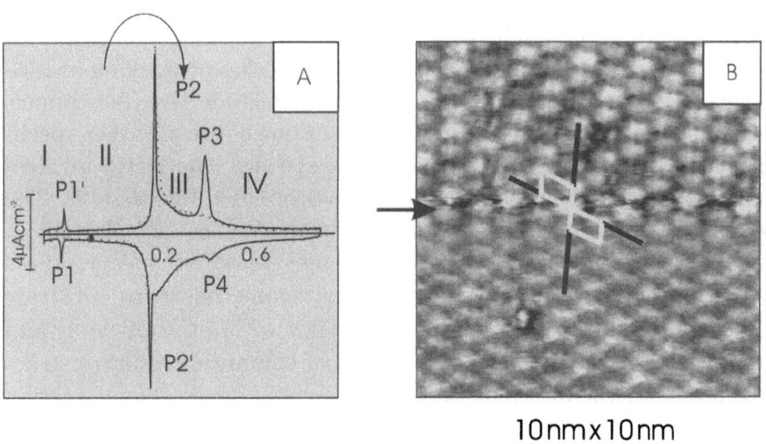

Fig. 9A,B. Change of the image contrast when scanning the electrode potential out of the stability range of the physisorbed film II into region III corresponding to $-0.05\,\text{V} \rightarrow 0.30\,\text{V}$ ($50\,\text{mV}\,\text{s}^{-1}$). The beginning of the scan is indicated with an arrow. $i_T = 2\,\text{nA}$, Au(100)-(hex)/$0.05\,\text{M}\,H_2SO_4 + 3\,\text{mM}$ uracil

3.2.4 Chemisorbed Uracil Film in Region IV

At sufficiently positive electrode potentials on all three low index gold faces, voltammetric and capacitance measurements indicate that uracil forms a second ordered adlayer, which (i) occurs rather independently of adsorbate concentration, (ii) is stable at temperatures as high as 100 °C, and (iii) significantly inhibits the onset of the gold oxidation. Figure 10 summarizes typical steady-state images of the chemisorbed uracil film on a quadratic Au(100)-(1 × 1) electrode (Fig. 10A–C) as well as on the hexagonal surface of Au(111)-(1 × 1) (Fig. 10D–F).

In the case of the island-free Au(100)-(1 × 1) electrodes a regular stick-like pattern was obtained, after equilibration of the entire system. One stick always consists of four blobs, which have been assigned tentatively to individual molecules facing the gold surface either with the N(1)- or with the N(3)-ring-nitrogen atom. Two characteristic directions of the molecular stacks, indicated in Fig. 10B as a' and b' were found. The corresponding angle amounts to 107°. The four-molecule stacks formed along a' appear to be somewhat longer, (12.6 ± 0.2) Å, than those along b', which are (11.4 ± 0.4) Å. The distance of neighboring uracil molecules within one stack ranges between 0.36 Å and 0.40 Å. This repeat pattern suggests a "primitive" unit cell that contains sixteen molecules, with $a = b = (20.3 \pm 0.8)$ Å and $\alpha = (108 \pm 2)°$. The corresponding experimental area per molecule is estimated as $A_{ex} = (23 \pm 3)$ Å^2, which clearly points to a perpendicular orientation of the molecules in state IV.

The STM contrast pattern, as described above, points to characteristic stacking between adjacent uracil molecules. Vertical stacking was found for many heterocyclic species [112,120]. Experimental studies of stacking in various bulk solutions showed that the formation of stacks is additive (not cooperative!), reversible, and stabilized by weak interactions between rather specifically arranged molecular segments of adjacent species. The latter involves electron correlation, London dispersion and hydrophobic forces [112]. The interaction energy between uracil stacks was calculated as $-27.3\,\mathrm{kJ\,mol^{-1}}$, which is weaker than the total hydrogen-bonding interaction energy. These short-distance interactions are superimposed on a strong adsorbate–substrate coupling. The conformation of the individual chains within the assembly and their packing with respect to each other (the final equilibrium adlayer structure) is determined by the interplay between interchain forces, the interaction with the surface as well as entropic effects.

On the basis of the above data the molecular packing arrangement of uracil on Au(100), as depicted in Fig. 10C, was proposed. The model reproduces all experimental observations and dimensions, such as the distances and angles between adjacent molecules and stacks, respectively [51]. In analogy to the results of quantum chemical calculations for pyridine on Ag(hkl) [121] and alkylthioles on Au(hkl) and Ag(hkl) [9,122], and referring to the dimensions of the possible molecular and substrate co-ordination sites (N-N-

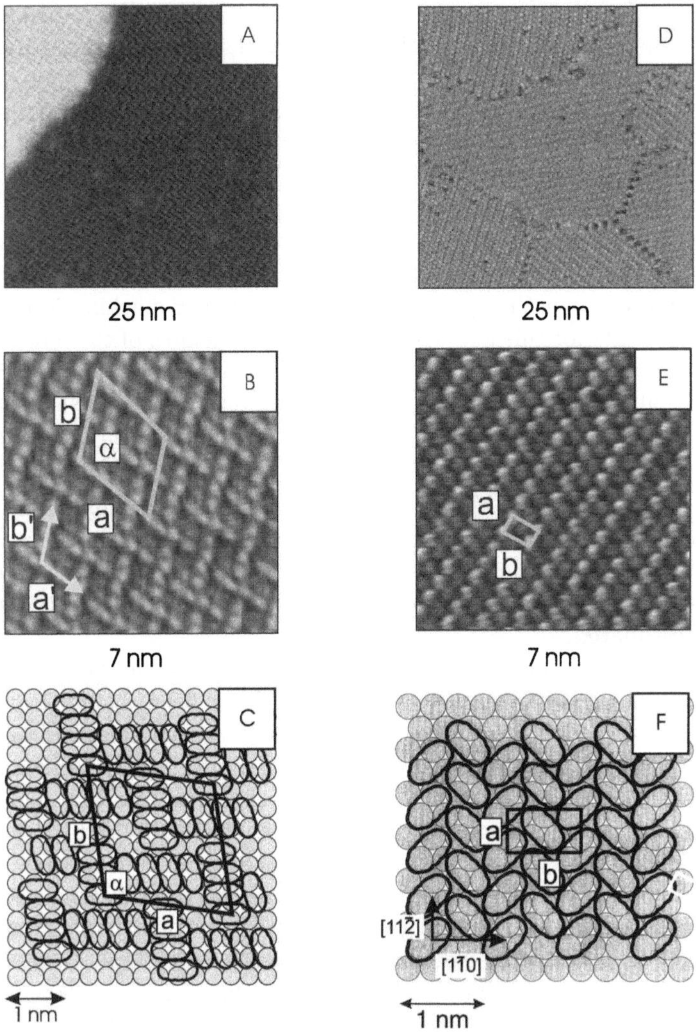

Fig. 10. STM images and packing models of chemisorbed uracils on Au(100)-(1 × 1) (**A–C**); E = 0.60 V, i_T = 1.8 nA, v_T = −0.52 V) and Au(111)-(1 × 1) (**D–F**); E = 0.60 V, i_T = 2 nA, v_T = −0.65 V) in 0.05 M H_2SO_4. Characteristic dimensions of the adsorbate mesh and the respective elementary cells are indicated in the high-resolution patterns (**B,E**) as well as in the proposed packing models depicted in (**C,F**), respectively

distance, a_{Au}) the above adlayer model was constructed with the constraints of a maximum number of equivalent high-order co-ordination sites (four-fold hollow sites). The second assumption was the introduction of a tilt angle of 20° between the axis of the molecule and the normal to the respective stick-direction, rather typical for π-stacking systems [112,120]. The experimentally observed preferential occurrence of four-molecule stacks can be rationalized as follows. Due to their molecular dimensions, only the two middle uracil molecules of each stack can occupy the high co-ordination four-fold hollow sites, while adjacent molecules shift towards energetically less favorable bridge positions. Eventually, the enthalpy gain due to attractive π-stacking is not sufficient to compensate for this loss when increasing the length of the molecular chains to units longer than four.

The geometrical arrangement of the stacks allows us to derive a "theoretical" primitive unit cell with $a = b = 20.5$ Å, and $\alpha = 106°$, which contains sixteen individual molecules (indicated in Fig. 10C). The area per molecule amounts to $A_{th} = 25.2\,\text{Å}^2$ ($\Gamma_m = 6.6 \times 10^{-10}\,\text{mol}\,\text{cm}^{-2}$), which is close to the experimental result reported above. With $q^m = 44\,\mu\text{C}\,\text{cm}^{-2}$ [51], the charge obtained for the formation of a complete chemisorbed uracil monolayer on Au(100) from voltammetric data, a charge transfer of 0.60 to 0.66 electrons per molecule is estimated.

The structure-determining role of the substrate geometry for the chemisorbed uracil film is also evident in experiments with an Au(111)-electrode of hexagonal symmetry [50] (Fig. 10D,E). At positive charge densities the entire Au(111)-electrode surface is covered with a highly ordered overlayer. Rotational as well as translational domains were found. The rotation angle between adjacent domains lies almost exclusively within the range $(120 \pm 5)°$, thus exactly following the symmetry lattice of the underlying gold substrate (Fig. 10D). The size of the individual domains varies between $10\,\text{nm}^2$ and $50\,\text{nm}^2$ and increases with observation time (Ostwald ripening). Analyzing the STM contrast pattern of single domains, we found a hexagonal arrangement of slightly asymmetrical "blobs", which are assigned to individual molecules. The corresponding lattice constant was estimated as (4.9 ± 0.2) Å. This value suggests a $(\sqrt{3} \times \sqrt{3})$R30° overlayer, but closer inspection of the experimental molecular-resolution images shows an additional striped modulation (Fig. 10E). Every other row of molecules appears brighter. Based on the above observations, the dimensions of the primitive unit cell were determined as $a = (4.9 \pm 0.2)$ Å, $b = (8.7 \pm 0.4)$ Å, $\alpha = (90 \pm 2)°$ (Fig. 10F). This cell is commensurate with the underlying substrate, $(\sqrt{3} \times 3)$ in terms of Wood's nomenclature, and contains two molecules. The uracil molecules are assumed to be located in the three-fold hollow sites of the hexagonal gold lattice. The corresponding area per molecule amounts to $(21 \pm 4)\,\text{Å}^2$, which translates into the surface concentration of $7.9 \times 10^{-10}\,\text{mol}\,\text{cm}^{-2}$. These data are no longer compatible with a flat interfacial orientation of uracil. With the charge per chemisorbed uracil monolayer on Au(111) $q^m = 60\,\mu\text{C}\,\text{cm}^{-2}$, as determined by chronocoulometry [40], a partial charge transfer of 0.8 electrons per adsorbed molecule is calculated The charge transfer on Au(111) is somewhat larger than on Au(100), which correlates with the higher percentage of occupied hollow sites in the former case. And it is in line with theoretical calculations for pyridine on Ag(*hkl*) by *Rodriguez* [121], who concluded that the charge transfer during adsorption at hollow sites is larger than on bridge- or atop-positions.

Electrochemical studies of the pH-dependence revealed that the formation of the chemisorbed uracil films on Au(*hkl*) involves the active participation of one proton [32,64]. Combining this observation with the results of the in situ STM experiments reported above and with the chronocoulometric studies, the following mechanism is suggested for the formation of phase IV (Fig. 11): Uracil molecules approach the positively charged Au(111)-(1 × 1) surface, and via the abstraction of one proton either in N(1)- or, eventually, in N(3)-position a surface co-ordination complex is formed.

This model is supported by three arguments: (i) Voltammetric experiments revealed that the adsorption of 1,3-dimethyluracil, where both ring nitrogen atoms (N(1), N(3)) are chemically blocked with methyl groups, do not show any evidence of the chemisorbed phase [40]. (ii) Crystallographic studies of co-ordination complexes of ions like Au^{3+} and Ag^{+} with uracil or 5-methyluracil as ligands demonstrate that the preferred binding site for metal ion co-ordination is the N(1) ring nitrogen (N(1) $\gg$ N(2) $>$ O(1) $>$ O(4)) [112,123]. Similar properties of these molecules at a *positively charged gold surface*, as is the case within the thermodynamic stability range of the chemisorbed uracil film, seem to be feasible. (iii) The energetics of the molecular orbitals of uracil and thymine (HOMO) [124,125] and the position of the

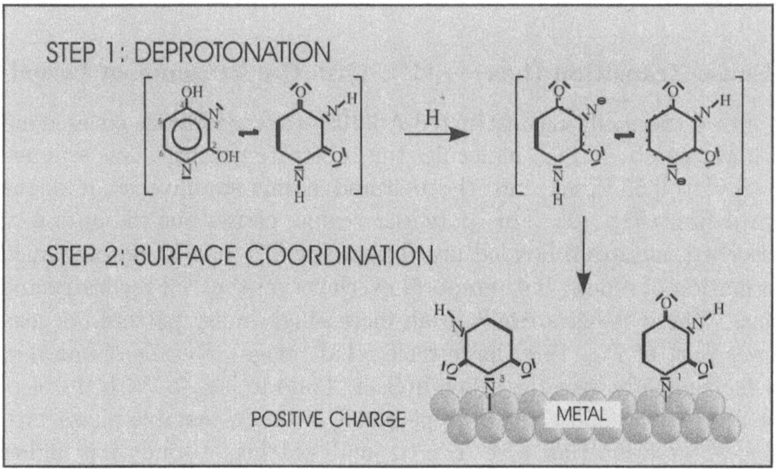

Fig. 11. Mechanistic model of the formation of the chemisorbed uracil film on Au(111)

d-bands of gold (shifted with respect to the corrected work function [126,127]) compare favorably on a common scale. Unfortunately, the d-bands of the metal are completely occupied and therefore this coupling does not support any binding interaction between the gold electrode and the electronically "undisturbed" uracil molecule [121]. This situation changes for the mono-anions. All orbitals shift towards lower binding energies, which may lead to the superposition of the Fermi-level of the gold electrode and the HOMO of the uracil anion [112,124,125]. In consequence, chemisorption according to the mechanism proposed in Fig. 11 is possible. Some evidence for the preferential co-ordination of the N(3) ring nitrogen atom, instead of N(1), was recently given in an in situ FTIR study of 5-MU on Au(111) [46].

3.3 Dynamic Processes on the Au(100) Surface in the Presence of Uracil

The formation of the various "steady-state" organic phases on single-crystal gold and platinum electrodes is often accompanied by complex changes of the substrate surface structures as well as of the respective adlayers. The mechanisms of these processes have been studied in detail for several systems employing electrochemical measurements under well-defined conditions in combination with structural techniques, such as electroreflectance [128], in situ surface X-ray scattering [28,129] and in situ scanning tunneling microscopy [49,50,51,55,56,57,64,73,74].

As an example, in the next section we will describe dynamic changes of the uracil/Au(100) system, as they occur in a typical voltammetric scanning experiment, on a structural basis. The results presented below are rather typical for the entire group of pyrimidine and purine bases on Au(100) as well as Au(111).

3.3.1 Phase Transition (hex)→ (1 × 1) in the Presence of Uracil

Starting with a thermally reconstructed Au(100)-(hex) electrode covered entirely with physisorbed uracil molecules the substrate potential was scanned from −0.05 V to 0.55 V, e.g. into the thermodynamic stability range of the chemisorbed film (Fig. 12). This potential regime causes the disruption of the physisorbed, hydrogen-bonded uracil phase at $E > 0.20\,\mathrm{V}$, accompanied by the formation of a new, but temporal overlayer on the still reconstructed gold surface. These changes result in an increasingly noisy pattern, as seen in the lower part of Fig. 12A (large-scale STM image). Details of the new structure have already been described in Sect. 4 and in Fig. 9. With the electrode potential set at 0.55 V the disappearance of the metastable phase with progressing observation time gives rise to small patches of somewhat higher contrast, of which examples are shown in Fig. 12B by white arrows. On some of these areas we also found evidence of the first monatomic high gold islands on newly formed Au(100)-(1 × 1) spots (Fig. 12B,C), which indicate the local

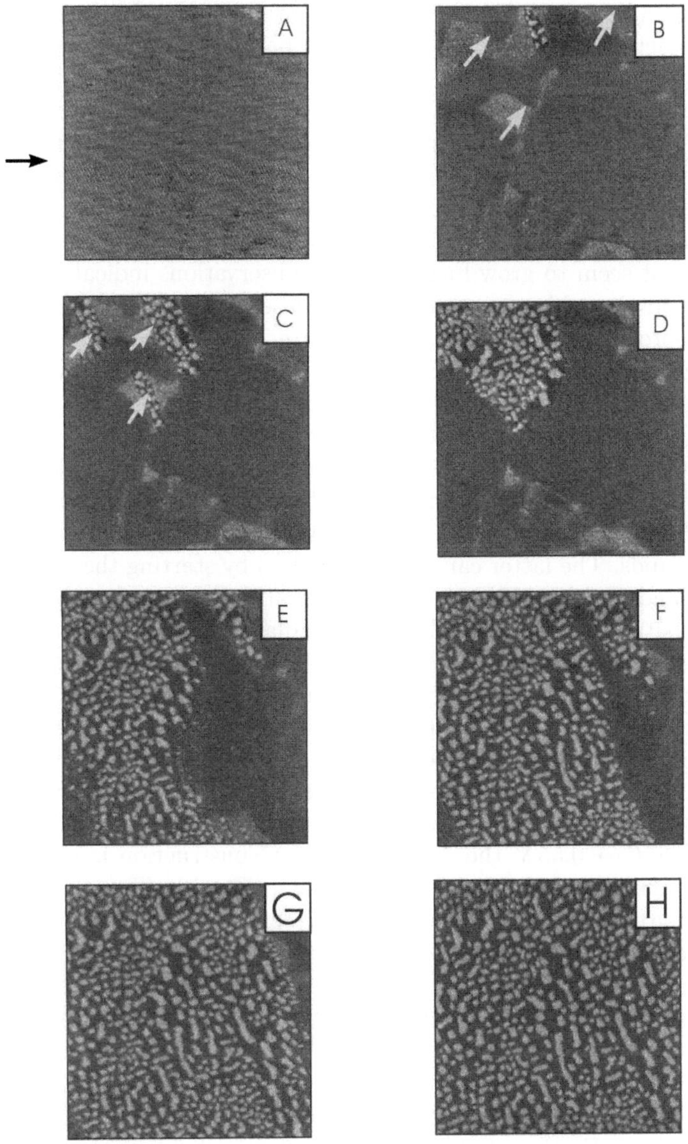

Fig. 12. Sequence of STM images (120 nm × 120 nm) for an Au(100)-electrode in 0.05 M H_2SO_4 + 3 mM uracil after a single potential scan from −0.05 V to 0.55 V(50 mV s^{-1}): (**A**) scan to 0.55 V as indicated by the arrow, (**B**) t = 6 min, E = 0.55 V, (**C**) t = 15 min, (**D**) t = 22 min, (**E**) t = 28 min, (**F**) t = 33.5 min, (**G**) t = 39.5 min, (**H**) t = 48 min, i_T = 2 nA

lifting of the Au(100)-(hex) reconstruction. The unreconstructed areas grow anisotropically. Their growth rate, in the direction parallel to the reconstruction rows, is significantly larger than the rate perpendicular to these rows. As indicated in Fig. 12D–H new Au(100)-(1 × 1) centers nucleate, individual growing areas merge and finally the surface is completely unreconstructed. The overall process reaches in this experiment a "steady-state" after 48 min The rather small Au(100)-(1 × 1) islands are variable in size and have an anisotropic shape (not round, but often elongated or angular). Once formed these islands do not seem to grow further. These observations indicate that the islands just contain gold atoms, which are released from adjacent lattice positions of the previously reconstructed substrate, and immediately after this step, their further mobility is "frozen in". (Rather similar observations have been recently reported for uracil on Au(111) [50,81,82].) The blocking is most probably caused by the chemisorbed uracil film in region IV. Initially, the newly formed adlayer in between islands is rather disordered, but the degree of order increases significantly with time at sufficiently positive potentials [51]. Large-scale long-range order is prevented by the high defect density of gold islands. The latter can be decreased (i) by starting the experiment with an island-free Au(100)-(1 × 1) electrode at positive potentials [51] and/or (ii) by in situ thermally annealing of the adlayer [130].

The above structural changes of the Au(100)-(hex) surface are much slower in the presence of uracil than in adsorbate-free 0.05 M H_2SO_4. For comparison, the same potential scan regime has been chosen in both sets of experiments. The individual scan experiments were repeated several times, and image sequences such as those shown in Fig. 12 and Fig. 13 represent typical and rather intrinsic properties of both systems. Following the potential scan −0.05 V → 0.55 V the Au(100)-(hex) reconstruction is almost completely lifted after 20 min in the *absence* of uracil (Fig. 13). These results compare with those obtained after 48 min in the *presence* of uracil. The lifting of the reconstruction proceeds in the former case by nucleation at substrate defects, such as step edges (Fig. 13C) and growth of the unreconstructed areas along the direction parallel to the reconstruction rows. The growth rate in perpendicular direction is much slower. The monatomic gold islands in the absence of the organic molecule are significantly larger than in the presence of uracil, and their size distribution steadily changes due to surface diffusion and Ostwald ripening. Further details on the mechanisms involved in this process have been extensively described in the literature [98,99,100,101].

3.3.2 Phase Transition (1 × 1) → (hex) in the Presence of Uracil

Returning to the Au(100)/uracil system, the first voltammetric cycle is closed and the electrode potential is scanned back from 0.55 V (region IV) to −0.05 V (region II) (Fig. 14). Passing the voltammetric peak P4 (c.f. Fig. 6) causes an increase in the size of the gold islands (arrow in Fig. 14). This process is triggered by the dissolution of the chemisorbed uracil film, which previously

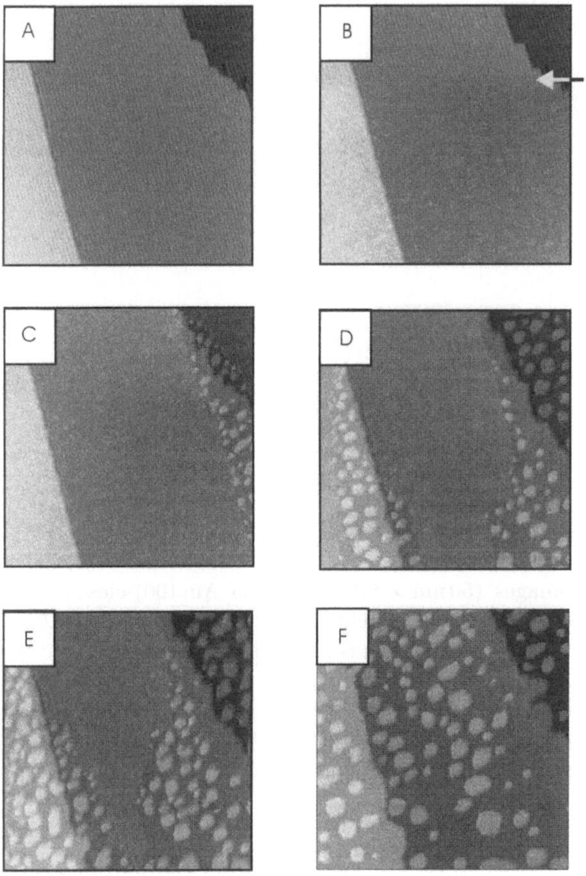

Fig. 13. Sequence of STM images (100 nm × 100 nm) for Au(111)/0.05 MH_2SO_4 after a single potential scan from −0.05 V to 0.55. (**A**) $E = -0.05$ V, (**B**) $E_i = -0.05\,V \rightarrow E_f = 0.55$ V (50 mV s^{-1}),arrow indicates the start of the potential scan, (**C**) $t = 1.5$ min, $E = 0.55$ V, (**D**) $t = 10$ min, (**E**) $t = 11.5$ min; (**F**) $t = 18$ min

inhibited the long-range mobility of these gold islands (Fig. 12). When passing P4 the islands try to reach their "equilibrium" shape. This trend does not last, because the unreconstructed surface is no longer thermodynamically stable at more negative potentials (or charge densities), and the transformation into the reconstructed top-layer structure begins (Fig. 14B,C). The newly formed reconstruction rows nucleate at substrate defects, such as steps and islands, and their growth is mainly one-dimensional. The expected two domains of mutually perpendicular, reconstructed areas were found, rather similar as in organic-free solution (c.f. [98,99]). The entire process is very slow in the presence of uracil, and is still far from completion after 45 min. The physisorbed organic film is formed once again, in-between as well as on top of the gold islands. The time scale of the entire substrate surface transition

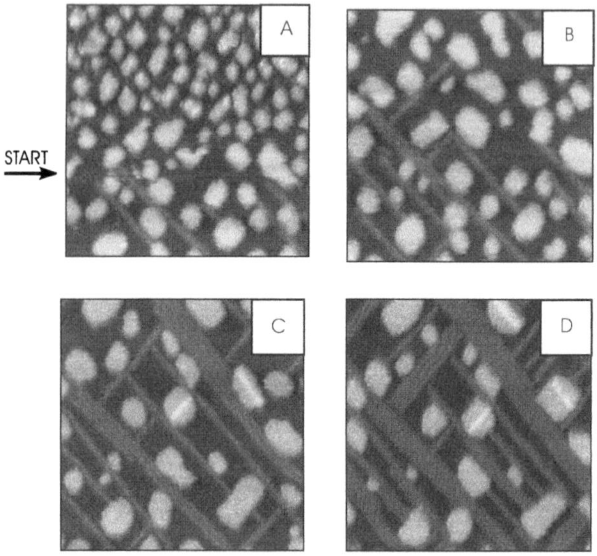

Fig. 14. Sequence of STM images (50 nm × 50 nm) for an Au(100)-electrode in 0.05 M H_2SO_4 + 3 mM uracil after a single potential scan from 0.55 V to −0.05 V (50 mV s^{-1}): (**A**) The start of the potential scan is indicated by an arrow. Note the remarkable change in island distribution and size! (**B**) $E = -0.05$ V, $t = 2.5$ min, (**C**) $E = -0.05$ V, $t = 14.5$ min, (**D**) $E = -0.05$ V, $t = 45$ min. The physisorbed film covers islands as well as reconstructed and unreconstructed patches of the electrode surface. The substrate transformation is not yet completed

clearly demonstrates that not only the lifting but also the potential-induced formation of the reconstructed Au(100)-(hex) surface is significantly inhibited by the presence of uracil. The detailed quantification of these processes and mechanisms is, unfortunately, hampered by the fact that repeated scanning of an area disturbs the reformation of the physisorbed film due to proximity effects of the STM tip. Similar artifacts were found on the Au(111)-surface [50]. Despite this limitation, the above results demonstrate convincingly that each voltammetric cycle following the first one will start in regions I or II with an Au(100)-surface, which is partly reconstructed and still contains gold islands. This substrate structure is more open than the surface of a thermally reconstructed Au(100)-(hex) electrode, as obtained just after flame annealing (first voltammetric scans in Fig. 2, Fig. 5 and Fig. 6). *Lecoeur* et al. found that the potential of zero charge shifts towards negative values with increasing defect density, e.g. steps and islands [131]. This fact explains the shift of the current spikes P1/P1′, which represent the dissolution and formation of the physisorbed uracil film, in the negative direction by more than 0.1 V on such a non-ideally reconstructed gold surface (Fig. 6). It is important to note that the onset of hydrogen reduction prevents their observation at $E < -0.30$ V during the second and each subsequent scan. An additional effect is expected from islands and domain boundaries of adjacent reconstruction elements, which could act as active centers for the formation/dissolution of the ordered physisorbed film. The latter seems to influence the transition between the organic adlayers in regions II and III. The positions of the current spikes P2/P2′ in Fig. 6 do not change despite the shift in zero charge potential. The temporal structure III is then formed on-top and in-between the still existing and only partially reconstructed gold islands. The charge (current) contribution due to the complete lifting of the substrate surface reconstruction in the presence of uracil at $E > 0.20\,\mathrm{V}$ during the second cycle is rather small, and does not give rise to a distinct voltammetric feature, such as P3 in the first scan with an island-free reconstructed surface. The final contrast pattern in region IV and structural changes during the subsequent cycle to negative potentials are similar to results already documented in Fig. 12 and Fig. 14. The described structural behavior is repeated during multiple scanning. In conclusion, the changes of the voltammetric behavior between the first complete and each subsequent potential scan could be traced back primarily to kinetically controlled changes of the substrate surface structure, as caused by the simultaneous action of electrode potential and adsorbed organic material.

3.4 Thymine (5-Methyluracil, 5-MU) on Au(111)

In order to generalize the results obtained for uracil on Au(*hkl*) several other derivatives were studied on Au(*hkl*), Ag(*hkl*) and Cu(*hkl*) employing structure-sensitive techniques, such as in situ STM [33,48,49,132,133], FTIR [46,130] and rather recently sum and/or difference frequency generation (SFG, DFG) [118,134]. As an example some results for 5-MU on Au(111)/ $0.05\,\mathrm{M}\,KClO_4$ shall be described next (Fig. 15).

Analogous to uracil, 5-MU forms, as a function of potential, a "dilute" phase I of randomly adsorbed molecules, a two-dimensional condensed physisorbed film II of planar oriented and via hydrogen bonds interconnected molecules, a partial charge transfer/reorientation region III and a chemisorbed film IV [33,40]. In situ STM studies revealed that II represents a highly ordered monolayer of planar oriented molecules, arranged in a pattern of one-dimensional hydrogen-bonded ribbons (Fig. 15B and E), which correlates with the layered structure of the 3D crystals [114]. In the chemisorbed phase IV 5-MU is perpendicularly oriented and, most probably, bound via the N(3) ring nitrogen towards the positively charged electrode surface [33,40,46]. At first glance, in situ STM images at room temperature, such as shown in Fig. 15C, reveal a locally rather disordered structure. This pattern is determined by kinetics due to the low mobility of thymine in the chemisorbed state. Extended in situ thermal annealing increases the long-range order of this phase dramatically, and finally gives rise to a "pseudo-commensurate" phase $c(\sqrt{3} \times 4)$ [130].

The vibrational (chemical) properties of 5-MU on Au(111) have been explored by FTIR [130] and in situ DFG as well as SFG experiments [118,134].

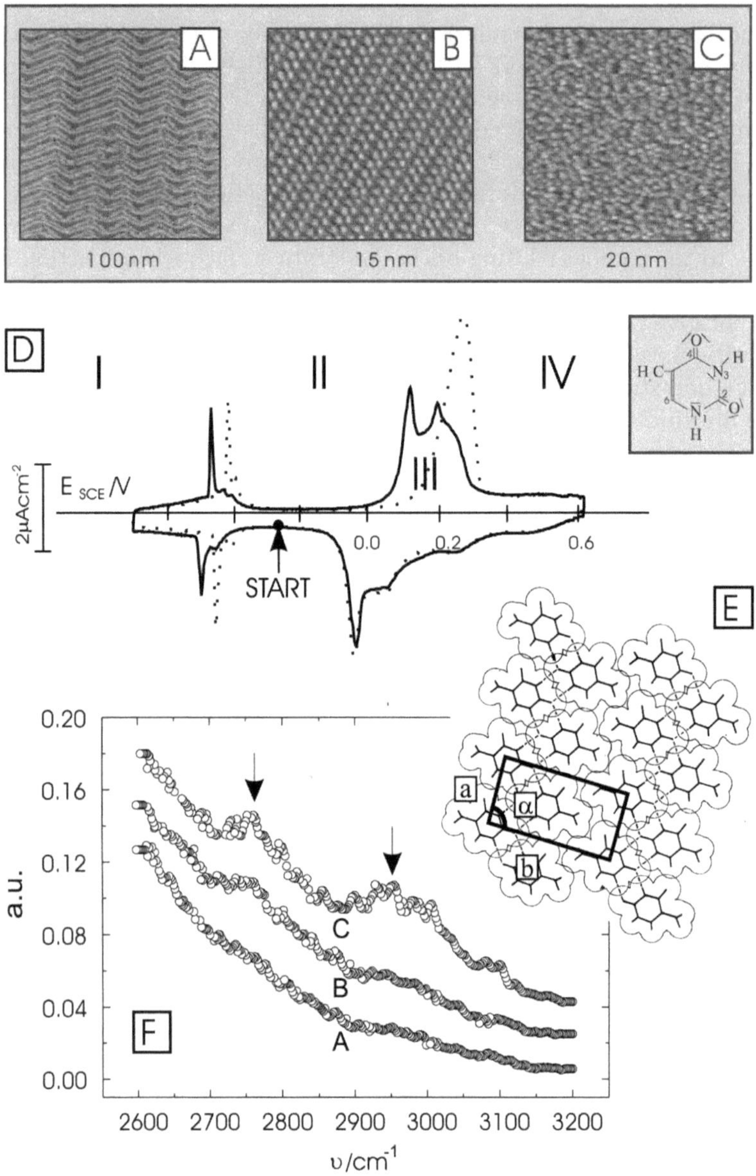

Fig. 15. Phase behavior of 5-methyluracil on Au(111)/0.05 MKClO$_4$: (**A**) disordered adsorption at −0.60 V, only the reconstructed Au(111)-(p×$\sqrt{3}$) surface pattern is seen; (**B**) physisorbed film at 0.0 V; (**C**) chemisorbed film at 0.60 V; (**D**) cyclic voltammogram, first *(dotted line)* and "steady-state" *(solid line)* cycles; (**E**) model of the physisorbed thymine film in region II; (**F**) DFG-spectra for 12 mM thymine in regions I (−0.60 V), II (0.0 V) and IV (0.60 V)

In the study described in [118] an IR-tunable laser, which uses a picosecond optical parametric oscillator, developed around an $AgGaS_2$ single-crystal, was combined with a YAG-laser (11 ps duration, bandwidth $2\,cm^{-1}$). The infrared pulses were mixed with 10 ps long pulses of the pump YAG laser beam. The IR and visible beam impinged on the sample surface with incident angles of 65°and 55°, respectively. The excitation based on the difference frequency generation enabled us to decrease considerably the non-resonant contribution of the substrate to the overall signal. (The interference with the interband transition threshold of the gold could be reduced. Further details of the method are described in [135,136].) The frequency range was optimized to the CH-stretching region. With this experimental configuration two potential-dependent vibrational modes were found for 5-MU on Au(111) (Fig. 15F). These results correlate nicely with the electrochemical and STM data: No resonance occurs in region I (curve A in the DFG-spectrum), were 5-MU is planar, but disorderly adsorbed. A weak C–H mode develops around $2750\,cm^{-1}$ (curve B in the DFG spectrum) as soon as the highly ordered 2D condensed physisorbed 5-MU film II is formed. The z-component of the transition dipole, or of the polarizability of the $-CH_3$ group is still rather small. Changing the orientation of 5-MU from planar to (most probably) perpendicular (II $\rightarrow$ III $\rightarrow$ region IV, curve C in the DFG spectrum) gives rise to a resonant feature in 2700–$2800\,cm^{-1}$, and a second band arises in 2900–$3000\,cm^{-1}$. The comparison with previous ex situ SFG studies of LB films [137] allows a tentative assignment of the observed bands to the symmetric and antisymmetric C–H stretching modes of the CH_3-group of 5-MU in the 5-position, respectively. The most important result of this work is the discovery of two potential dependent resonance modes of an ordered organic film (5-MU) on Au(111) under electrochemical conditions. This observation opens the door to a fascinating new application of SFG/DFG: The study of highly interface-sensitive vibrational properties of organic monolayers per se, or during electrode-potential-induced 2D phase transitions on single-crystal electrodes. These data could complement other results from electrochemical, STM and SXS studies, and may also (especially within the framework of a wealth of additional information from these other techniques) provide new insight into the nature of the SFG/DFG process. Substantial work on this topic is in progress.

4 2,2-Bipyridine Adlayers on Au(111) and Au(100)

The search for new strategies to create, characterize and manipulate "functional" monolayers on solid surfaces offers challenging analogies with "supramolecular chemistry" [138]. Strategies, based on ion pairing, or on molecular interactions such as hydrophobic or hydrophilic, hydrogen-bonding, host-guest, π-stacking, or donor–acceptor interactions [139] may also be utilized at electrode surfaces. A recent trend has been to use donor-acceptor interactions, e.g. specific molecular recognition features between metal ions and ligands, to control the assembly and properties of novel molecular architectures [140]. Here, aromatic nitrogen-containing heterocycles represent an important class of ligands, which eventually might form highly specific host lattices on defined surfaces [141]. Monodentate ligands, such as pyridine and chelating ligands, for instance 2, 2′-bipyridine[2] as well as higher oligopyridines, readily form stable mono- and/or multinuclear complexes with various transition metal ions [142]. On the other hand, the individual metal atoms (ions) of the substrate surface may already operate as potential local co-ordination centers for a suitable ligand, in the absence or in the presence of additional transition metal ions supplied with the bulk solution. To approach an understanding of these processes at a molecular level in situ scanning tunneling microscopy was employed. Experiments with defined individual components may show the way towards more complicated molecular architectures at electrochemical interfaces with rather new properties.

4.1 General Overview – 2,2'-Bipyridine on Au(111)

In this section, structural properties of the "apparently" rather simple system 2, 2′-bipyridine/Au(111) or Au(100) will be described. 2, 2′-bipyridine is a classical didentate ligand in co-ordination chemistry. Two planar pyridine rings are connected via a C–C bond (10 % double-bond character [143]). In the ground state in crystalline form [143] or in solution [144] 2, 2′-bipyridine assumes a coplanar *s*-trans conformation. Electrochemical studies revealed that *n*, *n*-bipyridine [145,146] and several of its Co^{2+} and Ni^{2+} complexes [147] form two-dimensional condensed films at the atomically "smooth" mercury/ electrolyte interface. The adsorption of 2, 2′-bipyridine on Ag(poly) and Au(111) electrodes was studied by Raman scattering [148], chronocoulometry and second harmonic generation (SHG) [149], surface X-ray scattering [28] and scanning tunneling microscopy [55,56,57,58]. *Lipkowski* et al. [149] found, based on electrochemical data, that 2, 2′-bipyridine exhibits multistate adsorption on Au(111).

Typical capacitance vs. potential curves for 3 mM 2, 2′-BP in 0.05 M H_2SO_4 on Au(111) are plotted in Fig. 16A. The experiment started with a thermally reconstructed Au(111)-(p$\times\sqrt{3}$) electrode after immersion at open circuit potential and subsequent equilibration (40 min) at 0.50 V. The first (positive as well as negative going) scan is indicated by solid lines. A set of the corresponding first and "quasi-steady-state" voltammograms is shown in Fig. 16B,C. Characteristic current peaks and "peak-like" features are indicated as P1–P5 (negative going scan) or P1′–P5′ (positive going scan). The dotted lines represent the base electrolyte system 0.05 M H_2SO_4/Au(111). Five characteristic potential regions could be distinguished when analyzing these data. They are labeled in Fig. 16 as I–V [57]. Region I, at $E > 0.40$ V

[2] 2, 2′-bipyridine will be abbreviated 2, 2′-BP

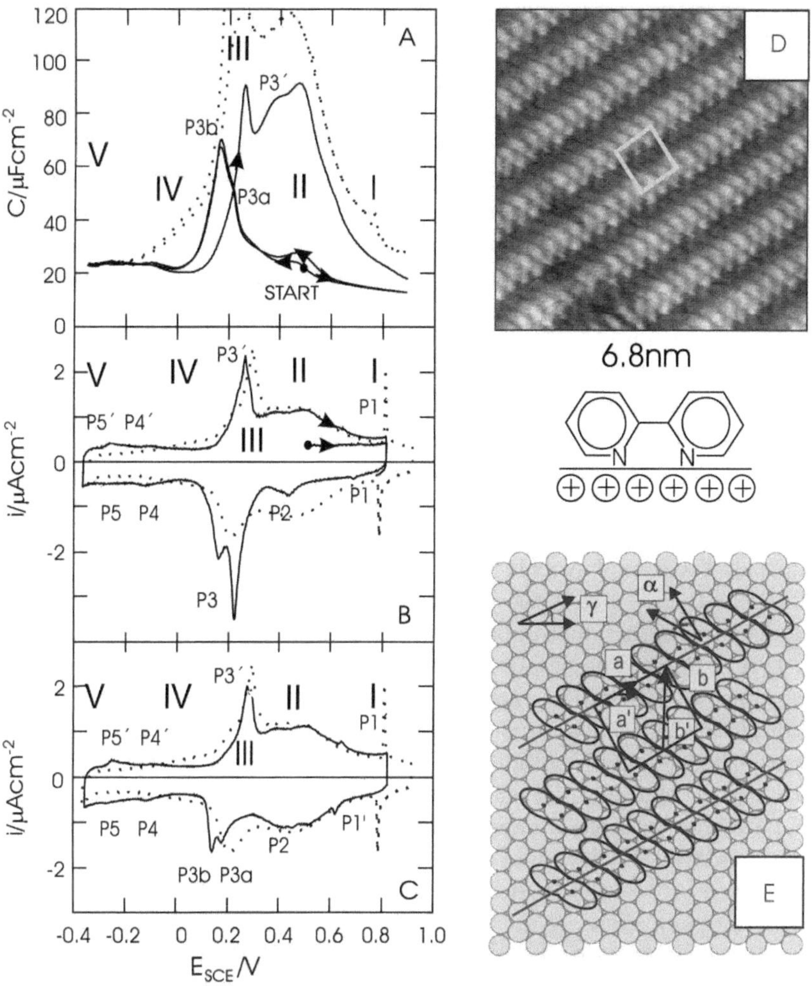

Fig. 16. Capacitance vs. potential (**A**) and cyclic current vs. potential curves (**B, C**) for 3 mM 2, 2′-bipyridine on Au(111) in 0.05 MH_2SO_4, scan rate 10 mV s^{-1}. The first scans as obtained with a flame-annealed electrode after immersion at OCP and equilibration at 0.50 V are indicated as *solid lines* in (**A**) and (**B**). (**C**) represents the "quasi steady-state" voltammogram recorded after five cycles between −0.35 V and 0.80 V. The *dotted curves* represent the adsorbate-free supporting electrolyte 0.05 MH_2SO_4/Au(111). (**D**) Unfiltered high-resolution image of the 2, 2′-BP stacking structure at $E = 0.50$ V, $i_T = 2$ nA, $v_T = 10$ mV. The suggested commensurate unit cell is indicated. (**E**) Proposed packing model for 2, 2′-BP. The numerical data are summarized in Table 2. The molecular dimensions were taken from crystallographic data [143], the positions of the nitrogen atoms are indicated by dots

(e.g. $E >$ P2/P2′) exhibits a low capacitance and pronounced hysteresis. The saturation capacitance of about $12\,\mu F\,cm^{-2}$ is established only after rather long "annealing" times at sufficiently positive potentials [57]. Between P2/P2′ and P3/P3′ (region II) the capacitance increases markedly, and a "double peak" region III develops in the voltammogram around 0.20 V. The sharp feature P3a appears as a discontinuity in the capacitance curve, while P3b is represented by a rather broad maximum, both significantly smaller than the corresponding capacitance of the base electrolyte. At $E <$ P3 a broad capacitance minimum IV develops. At $E < -0.10$ V two small voltammetric peaks P4/P4′ and P5/P5′ appear, which mark region V. The latter is delimited by region IV on one side and the onset of hydrogen evolution at more negative potentials on the other side, respectively. Region V is not a characteristic desorption feature. Closing the first voltammetric cycle by scanning the potential back from −0.35 V to 0.80 V leads to the appearance of two voltammetric features, P5′ and P4′, followed by a prominent charging current peak P3′ in region III. Remarkably, the positions of this peak and its cathodic counterpart P3 coincide with the characteristic transitions Au(111)-(p$\times\sqrt{3}$) → Au(111)-(1 × 1) and Au(111)-(1 × 1) → Au(111)-(p$\times\sqrt{3}$) in the absence of 2, 2′-BP. At $E >$ 0.25 V the charging current as well as the capacitance are significantly higher, as compared with the response after equilibration at 0.50 V (Fig. 16A,B). The first cycle can be reproduced only after an extended waiting time ($t \geq 40$ min) at this potential.

Multiple cycling between 0.80 V and −0.35 V with $10\,mV\,s^{-1}$ reveals the "quasi-steady-state" voltammogram as plotted in Fig. 16C. In addition to higher charging currents and its superposition with the 0.05 M H_2SO_4/Au(111) response a more pronounced pair of peaks P1/P1′ was observed around 0.60 V, which may indicate the metastable formation of patches of the ordered ($\sqrt{3} \times \sqrt{7}$) (hydrogen) sulfate structure [95] between the still disordered organic phase. (Note that 2, 2′-BP adsorption shifts the potential of zero charge significantly towards more negative values as compared with the "clean" base electrolyte [149].)

In order to have a first clue about the role of the substrate reconstruction on the phase formation of 2, 2′-BP in H_2SO_4 the above experiments were repeated with a "defect-free" unreconstructed Au(111)-(1 × 1) electrode [38]. Surprisingly, the same voltammetric results were obtained as in the above experiments, which started with a thermally reconstructed Au(111)-(p$\times\sqrt{3}$) electrode. These observations are independent of the choice of the starting potential in the range from −0.35 V to 0.80 V, and therefore support the notion that the P3/P3′ features are not only adsorbate-controlled, but associated with the reconstruction/deconstruction of the substrate surface in *acidic solutions*. The presence of 2, 2′-BP seems to support the formation of Au(111)-(p$\times\sqrt{3}$) at $E <$ 0.20 V as well as the lifting of this substrate surface reconstruction at $E >$ 0.25 V under these conditions, which implies that the gold surface is involved actively in the adlayer phase formation process. This behavior appears to be unique, and was previously observed neither in neutral solutions [81,82,129,149] nor for organic monolayers of pyrimidine and purine derivatives on Au(111) (Sect. 3).

In order to address directly the structural aspects of this complicated phase behavior for 2,2′-BP on Au(111) "steady-state" in situ STM studies at various stages of the electrochemical experiment were performed [57].

The potentiostatic deposition of 2,2′-BP on Au(111)-(1 × 1) at positive potentials (charge densities), e.g. in region I, causes the formation of a highly ordered monolayer of parallel molecular stacks (Fig. 16D). The distance between adjacent rows amounts to (9.6 ± 0.5) Å. At well-prepared surfaces defect-free domains of 25 up to 40 nm^2 can be found. The angles of 60° and 120° between adjacent ordered domains reflect the characteristic symmetry elements of the substrate. A rotation angle $\gamma = 30°$ between the stacking rows and the [112]-direction of the Au(111)-(p×$\sqrt{3}$) reconstruction lines was obtained (cf. Sect. 4.3.2). This superposition implies that the distance between adjacent stacking rows in the [110]-direction amounts to $b' = (11.1 \pm 0.3)$ Å, which is close to $4a_{\mathrm{Au}} = 11.54$ Å ($a_{\mathrm{Au}} = 2.885$ Å). The perpendicular distance between neighboring stacks was estimated to be b = (9.6 ± 0.3) Å.

High-resolution experiments, such as shown in Fig. 16D, reveal within each stripe the internal structure of the individual molecules. They appear as rather spherical "bright blobs" at low tunneling currents ($i_{\mathrm{T}} < 2$ nA) and moderate tunneling voltages v_{T}. The bright spots are better resolved as somewhat more asymmetrical features at tunneling currents higher than 2 nA. The estimated intermolecular distance amounts to $a = (3.80 \pm 0.20)$ Å. Furthermore, molecules are tilted from the normal to the axis of the chain by an angle $\alpha = (28 \pm 2)°$, and the relative positions of molecules in every other parallel chain seem to be identical. The characteristic experimental parameters of the 2,2′-BP adlayer on Au(111) are summarized in Table 2. Similar results were also reported by *Tao* et al. in neutral 0.1 M$NaClO_4$ [55,56].

Referring to a previous thermodynamic and SHG-study of 2,2′-BP [149] as well as to related work on the phase behavior of pyridine and pyrazine on group IB metals in the high-coverage range [121,150,151,152,153] it is assumed that in region I 2,2′-BP forms a stable adsorbate–substrate surface co-ordination complex with vertically oriented molecules in the sterically stressed *s*-cis state, e.g. both nitrogens pointing towards the gold surface. Crystallographic studies showed that the two nitrogen atoms are 2.8 Å apart in this configuration [143]. This distance is rather close to $a_{\mathrm{Au}} = 2.885$ Å. Based on all experimental data, the geometry of 2,2′-BP and of the hexagonal Au(111) lattice, a commensurate (4 × 2$\sqrt{3}$) adlayer unit cell, which contains three molecules, is proposed. The area per molecule is estimated as 38 Å^2, which is close to Lipkowski's chronocoulometric data [149] and to previous STM data of *Tao* et al. [55]. The corresponding molecular packing arrangement is drawn in Fig. 16E. All characteristic dimensions are summarized in Table 2. The model represents nearly exactly all experimental observations

Table 2. Dimensions of the 2, 2′-bipyridine stacking phase I

System	a/Å	b/Å	$\alpha/°$	$\beta/°$	$\gamma/°$	A/Å^2
H_2SO_4	4.0	9.6	28	58	30	38.4
Na_2SO_4	3.6	9.6	29	60	-	34.6
$KClO_4$	3.8	10.1	27	60	-	38.4
model	3.85	10.0	30	60	30	38.5
$NaClO_4$	3.7	8.8	30	60	-	33.3

and dimensions, such as the distance between adjacent molecules and stacks a (a'), b (b'), their relative shift as well as the angles α and γ, and does not violate packing constraints. In analogy to the results of a quantum chemical calculation of *Rodriguez* for pyridine on Ag(111) [121] and referring to the dimensions of the possible molecular and substrate co-ordination sites (N-N-distance, a_{Au}) the unit cell was constructed with the constraints of a maximum number of equivalent high order co-ordination sites. This configuration was accomplished by placing the "corner molecules" of the unit cell with their nitrogen atoms (indicated by "+" in Fig. 16E) in two-fold bridge positions. This choice also ensures that all other 2, 2′-BP molecules occupy substrate co-ordination sites with nitrogen atoms, which are only slightly off bridge positions. This configuration also minimizes the "adlayer corrugation", and therefore, ensures an optimal arrangement for lateral π-stacking between the parallel aromatic rings [112]. The existence of a commensurate 2, 2′-BP overlayer at the positively charged Au(111)-electrode is also supported by the strong influence of the substrate surface crystallography as will be demonstrated for Au(100)-(1 $\times$ 1) in Sect. 4.3 [58].

The above experimental studies reveal that the packing arrangement of 2, 2′-BP on Au(111) at positive charge densities (phase I) is identical in neutral ($KClO_4$, Na_2SO_4) and acidic solutions ($HClO_4$, H_2SO_4), despite the pK_a-values of -0.20 and 4 [37,154]. The authors in ref. [57] propose that the 2, 2′-BP-molecules deprotonate in acidic solutions at positive charge densities, and subsequently co-ordinate with the gold surface in the *unprotonated s-cis conformation.*

Despite an extensive search, no other "steady state" ordered structure of 2, 2′-BP was found in (neutral) aqueous Na_2SO_4 ($KClO_4$) solutions. The situation is different in H_2SO_4($HClO_4$). The ordered adlayer of type I "dissolves" at $E < 0.20\,\mathrm{V}$. But surprisingly, at $E < -0.10\,\mathrm{V}$, where the electrode surface is negatively charged, a new overlayer emerges (Fig. 17). This region was labeled V in Fig. 16.

The following structural details were found: (i) V is stable up to the onset of proton reduction. (ii) Nearly defect-free molecular rows decorate the reconstruction lines of the underlying Au(111)-(p$\times\sqrt{3}$) substrate surface. $\gamma = 86°$ is estimated as the characteristic rotation angle. (iii) The angle $\beta = 60°$ between adjacent rotational domains of V reflects the symmetry

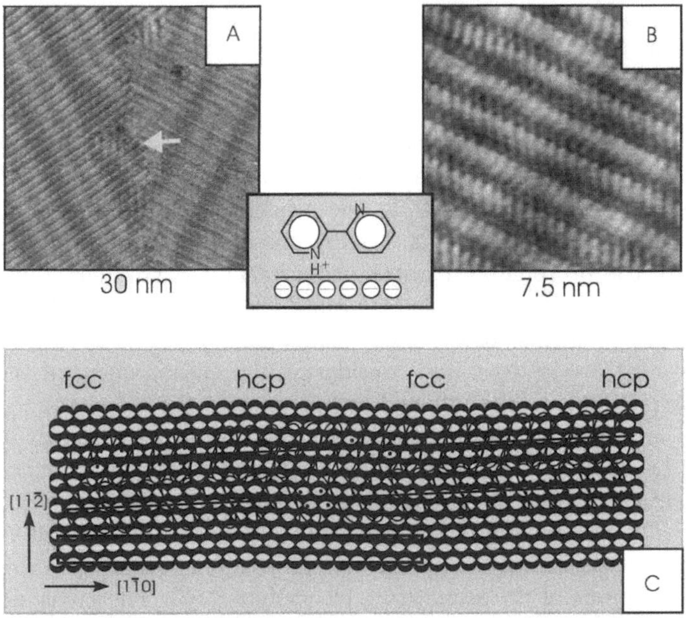

Fig. 17. STM images of the 2, 2′-BPH$^+$ layer V in 0.05 M H_2SO_4/Au(111)-(p×$\sqrt{3}$) at −0.20 V. The arrow in **A** indicates a characteristic domain boundary as governed by the direction of the substrate surface reconstruction lines. (**B**) High-resolution image of V. Note the alternate tilting angles of the stacking rows and their equidistant arrangement. (**C**) Proposed packing model for 2, 2′-BPH$^+$ (c.f. Table 3)

Table 3. Characteristic dimensions of the cathodic stacking phase V of 2, 2′-BPH$^+$

System	a/Å	b/Å	α/°	β/°	γ/°
H_2SO_4	3.6	9.9	+23 or −16	60 or 120	86
model	3.6	10.0	-	-	87.5

of the hexagonal substrate surface. (iv) High-resolution experiments show that adjacent parallel stripes are separated by (9.9 ± 0.2) Å. This distance represents approximately $(2\sqrt{3})a_{\mathrm{Au}} = 10$ Å. Each stripe consists of regularly shaped objects, which were assigned to individual molecules. A repeat distance of (3.6±0.3) Å was estimated. The area per molecule therefore amounts to (36 ± 3) Å^2. The image shows further that each molecule is tilted from the axis of the chain by either +23° or −16°, respectively. Identical tilting angles repeat every other row.

Based on these experimental observations the following understanding for the molecular arrangement of 2, 2′-bipyridinium ions (2, 2′-BPH$^+$) in H_2SO_4 was developed: Because of packing constraints a possible planar interfacial orientation of the adsorbate, which could be stabilized by $d - \pi^*$ backbonding [141], is not compatible with the observed high-resolution STM contrast pattern (Fig. 17B) and the corresponding molecular dimensions [143]. The same argument excludes a torsional configuration (one pyridyl ring planar to the surface, and the other rotated by 80°) as suggested by *Hubbard* et al. on Pt(111) [155]. All experimental data point to rather similar adlayer structures in regions V and I. The only two differences seem to be the registry with the Au(111)-(p$\times\sqrt{3}$) reconstruction lines and the tilt angle α. There is also no reason that 2,2-BPH$^+$ will deprotonate at the negatively charged electrode surface. Therefore it is proposed that 2,2′-BPH$^+$ co-ordinates in an s-trans configuration with the protonated nitrogen atom facing the gold electrode, and the other one directed towards the solution. Placing the molecules in "close to higher order co-ordination sites" (bridge and/or threefold hollow positions of the substrate lattice), and considering the misalignment of the gold atoms along [110] due to the uniaxial compression in the reconstructed phase (ca. 2.5°), an incommensurate stacking structure, such as shown in Fig. 17C, could be imagined. The model does not present strong evidence for the experimentally observed characteristic tilt angles of +23° and −16° between 2,2′-BPH$^+$ and the axis of the individual stacks, or for their alternating appearance. Speculations might involve the role of coadsorbed water molecules [156] or anions of the supporting electrolyte [157]. The latter do not seem to be specific because structure V was observed in aqueous H_2SO_4 as well as in $HClO_4$ solution.

4.2 Dynamic Processes

4.2.1 Potential- and Temperature-Induced Dissolution of the Stacking Phase I

In these sections dynamic changes of the 2,2′-BP/Au(111)/0.05 M H_2SO_4 system as they occur in typical "cathodic" potential scanning/step experiments will be described.

The first experiment started with a highly ordered 2,2′-BP stacking layer I, equilibrated at $E_i = 0.50$ V on Au(111), and then the potential was scanned slowly (10 mV s^{-1}) towards negative values (Fig. 18). The range was restricted in order to ensure that the gold electrode still bears a positive charge [149]. At $E = 0.24$ V first point dislocations occur (Fig. 18A). The active centers are formed not only at defects of the adsorbate or substrate lattices, such as point defects or domain boundaries, steps or kink positions, but instead are distributed evenly across entire terraces. The nucleated holes grow anisotropically with time and more negative electrode potential, until the ordered adlayer is dissolved completely. Dissolution proceeds preferentially along the direction of the commensurate molecular stacking rows. During this transition one can identify two phases clearly: solid-like patches exhibiting the characteristic long-range order of 2,2′-BP in region I (paragraph 1.3.1.) and a fluid medium disordered on the atomic scale (Fig. 18C,D). The observed order/disorder transition is first order.

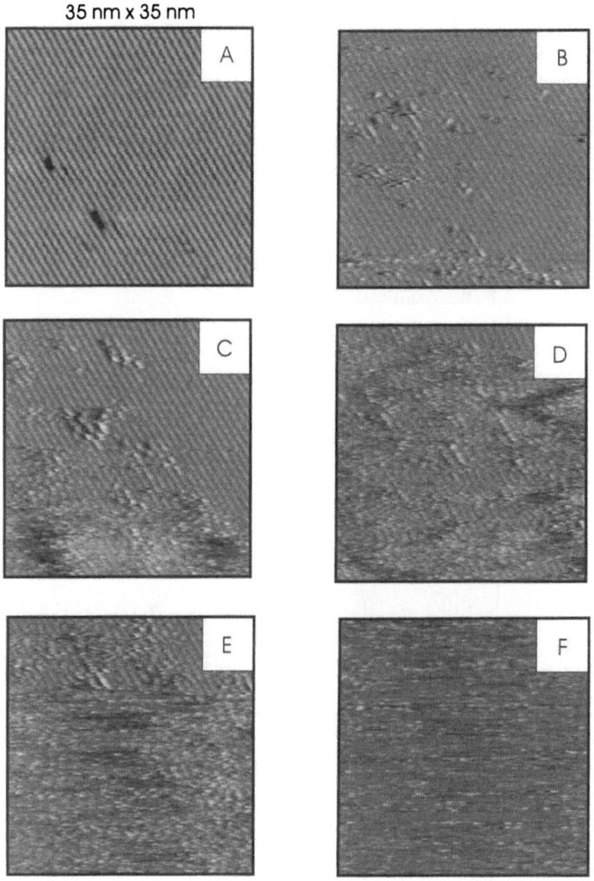

Fig. 18. Sequence of STM images for 3 mM 2, 2′-BP on Au(111)/0.05 MH_2SO_4 after a single potential scan ($10\,mV\,s^{-1}$) from 0.50 V to 0.24 V (potential-induced dissolution of I): (**A**) $E = 0.50\,V$, (**B**) $E = 0.24\,V$, $t = 1\,min$ at 0.24 V, (**C**) $E = 0.24\,V$, $t = 3\,min$, (**D**) $E = 0.24\,V$, $t = 3.75\,min$, (**E**) $E = 0.17\,V$, $t = 4.5\,min$, (**F**) $E = 0.17\,V$, $t = 8.0\,min$; $i_T = 2\,nA$

The above interpretation is supported by in situ STM experiments on the temperature-induced dissolution of the ordered 2, 2′-BP phase I [158]. The system was stabilized at $E_i = 0.20\,V$ and $T = 29.7\,°C$ (Fig. 19A), and then the temperature was ramped slowly at $0.02\,°C\,min^{-1}$. First point and line dislocations ("one-dimensional holes") are shown in Fig. 19B. They grow anisotropically by successive stripping of molecular stacks, fast along the directions of the stacking rows, and rather slowly perpendicular to it. Comparison of Fig. 19D and E also reveals the appearance of positional and directional fluctuations within the ordered as well as the disordered phases. Under the present experimental conditions the ordered 2, 2′-BP stacking layer dissolved completely at 33.6 °C. Closer inspection of the disordered regions

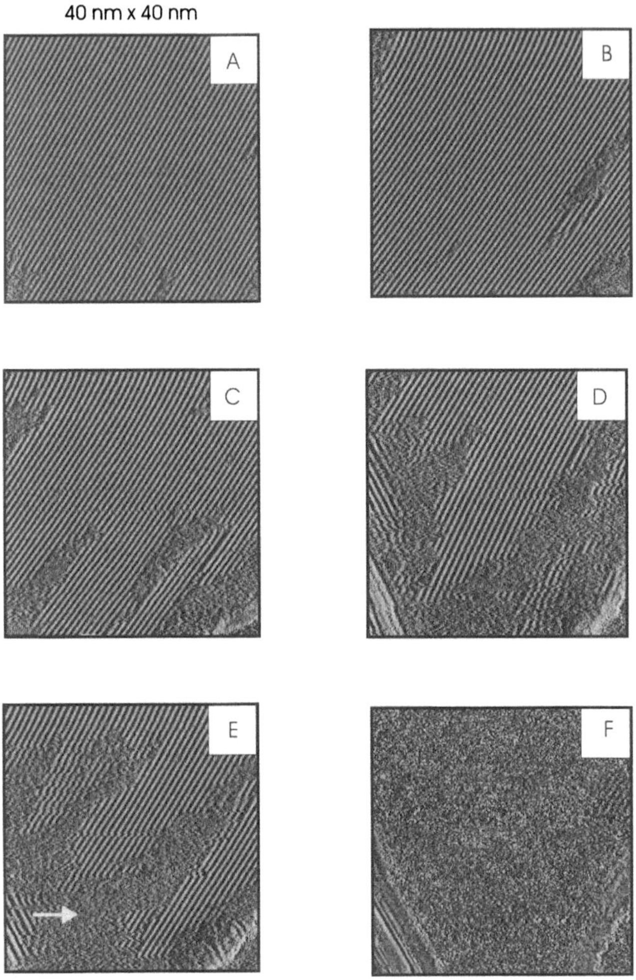

Fig. 19. Temperature-induced dissolution of the 2, 2′-BP stacking phase I for 3 mM 2, 2′-BP in 0.05 M H_2SO_4/Au(111). (**A**) The system was equilibrated at $E = 0.20$ V and 29.7 °C for 20 min. Then the temperature was ramped with 0.02 °C min^{-1}. (**B**) 31.95 °C, (**C**) 33.14 °C, (**D**) 33.35 °C, (**E**) 33.51 °C, (**F**) 33.60 °C. All images were obtained in constant current mode ($i_T = 0.25$ nA) at the same surface area

reveals a granular structure on a length scale exceeding the pixel resolution of the image in x–y direction. This means that a noticeable part of the modulation in the disordered area is not a time effect caused by the fast motion of adsorbed species, but rather reflects, at least in part (cf. arrow in Fig. 19E) the static positions of particles in the disordered phase. This suggests the existence of adsorbate–substrate co-ordination complexes. Similar observations were reported recently for the two-dimensional fluid–solid equilibrium of cesium and oxygen coadsorbed on Ru(0001) under UHV conditions [159].

In the present case, the adsorbed species obviously "feel" the effect of the substrate corrugation potential. This situation is different from purely two-dimensional systems, like guanine or adenine on HOPG [72] or the physisorbed films of uracil or thymine on Au(*hkl*) [50,51], where the adsorbate–substrate interactions are very weak. There the phase transition is governed entirely by the applied electrical field and by directional, lateral attractive interactions via hydrogen bonds.

Dramatic changes of the interfacial structure occur after application of a single potential scan/step from E_i in region I to final potentials corresponding to a negatively charged gold electrode, e.g. $E_f < -0.00$ V. The ordered 2, 2′-BP adlayer rapidly dissolves, and the entire substrate surface appears very mobile (Fig. 20B). Fractal, monoatomically high gold structures, such as shown in Fig. 20C, grow at step edges. The entire gold surface is immediately reconstructed. Zooming into smaller areas after approaching the final potential initially reveals holes and local defects (Fig. 20D). Both heal very

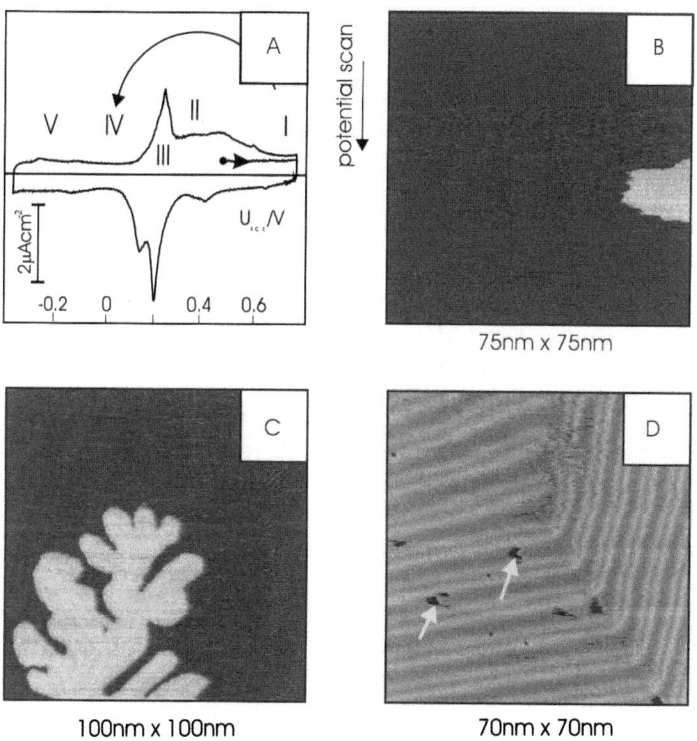

Fig. 20. Structural features of the Au(111) surface after a single potential scan ((**A**), 10 mV s^{-1}) from 0.50 V to −0.05 V (negatively charged electrode) for 3 mM 2, 2′-BP/0.05 M H_2SO_4. (**B**) "growing step"; (**C**) monatomic high gold "structure" decorating a former step. (**D**) High-resolution image of the Au(111)-(p×$\sqrt{3}$) surface immediately after arriving at −0.05 V; local defects are indicated by white arrows

quickly, and soon afterwards a new STM contrast pattern develops, which represents the $2,2'$-BPH$^+$ phase V previously introduced in Sect. 4.1.

Starting with an analogy to the arguments of Sondag and Huethorst for thiols on gold [160] the following interpretation of the above observation is suggested: Neutral (deprotonated) $2,2'$-BP forms stable and highly ordered substrate (gold)-adsorbate complexes via N-co-ordination with the positively charged electrode surface. The chemisorbed $2,2'$-BP molecules weaken the bonds between gold atoms of the first and the second substrate layer, implying that the Au-[Au($2,2'$-BP)] bonding might be somewhat weaker than the bonding in Au-($2,2'$-BP). The excursion towards negative electrode charges destabilizes the organic adlayer as well as its components. Due to the rather weak π-stacking interactions [119] relatively mobile Au-($2,2'$-BP) complexes may first be expelled from the formerly ordered organic patches, followed by dissociative desorption. The former causes the generation of local substrate defects. The apparent high mobility of gold atoms generated in this process (the initial coverage of $2,2'$-BP is approximately $4.6 \times 10^{-10}\,\mathrm{mol\,cm^{-2}}$ [149]) facilitates the annealing and quick healing of these defects as well as the deposition of additional material on steps of the substrate surface (Fig. 20B,C). Simultaneously, the created and initially free (neutral) $2,2'$-BP molecules protonate (solution pH $\sim$ 1.5), most probably still within the double layer region or via a fast exchange with solution species. Subsequently they form the newly ordered organic adlayer already know as V, which is stabilized by electrostatic interactions of the positively charged $2,2'$-BPH$^+$ cation with the negatively charged gold surface as well as lateral stacking forces between adjacent pyridyl rings.

4.2.2 Potential-Induced Formation of the $2,2'$-Bipyridine Stacking Phase I

The potential and/or temperature-controlled dissolution of the above order/disorder transition is completely reversible. A representative time sequence, after application of a potential scan from $E_\mathrm{i} = 0.10\,\mathrm{V}$ (still positive charge densities) to $E_\mathrm{f} = 0.50\,\mathrm{V}$, is plotted in Fig. 21. Within a few seconds at E_f, the $2,2'$-BP molecules start to order. No preferential nucleation sites, which might correlate with the defect structures of the substrate surface, where found (Fig. 21A). First stacks are formed, their directions correlate but exhibit a "missing row" arrangement. The distance between adjacent molecular rows amounts to 18.5 Å (Fig. 21B). With increasing observation time this metastable phase is displaced by a densely packed stacking domain (Fig. 21C,D) with dimensions typical for phase I (Sect. 2). The latter grows anisotropically in two directions: (i) parallel along the main stacking direction. Individual molecules and "small" stacks are still in search of energetically favorable adlattice positions at the "end of the growing front" (arrow in Fig. 21E). (ii) The second growth mechanism proceeds via incorporation of entire molecular rows perpendicular to the main stack direction.

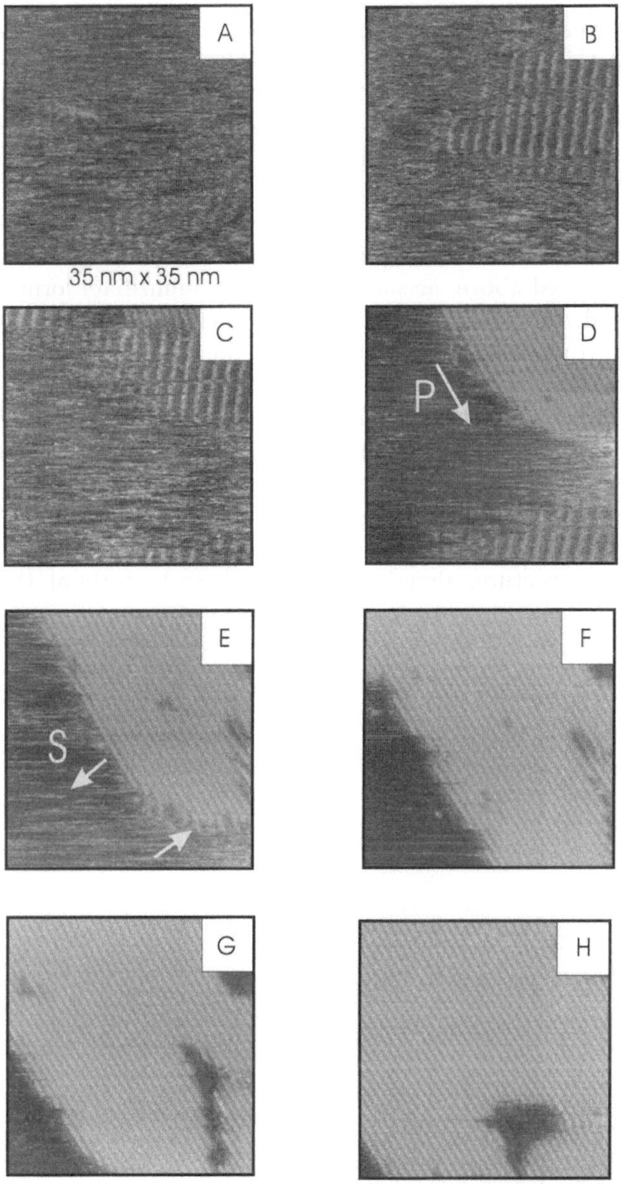

Fig. 21. Potential induced formation of the $2,2'$-BP stacking structure I on Au(111) after a potential scan from region II ($10\,\mathrm{mV\,s^{-1}}$): (**A**) $E = 0.10\,\mathrm{V} \rightarrow E_\mathrm{f} = 0.50\,\mathrm{V}$, (**B**) $E = 0.50\,\mathrm{V}$, $t = 1\,\mathrm{min}$, (**C**) $E = 0.50\,\mathrm{V}$, $t = 2\,\mathrm{min}$, (**D**) $E = 0.50\,\mathrm{V}$, $t = 5\,\mathrm{min}$, (**E**) $E = 0.50\,\mathrm{V}$, $t = 8\,\mathrm{min}$, (**F**) $E = 0.50\,\mathrm{V}$, $t = 11\,\mathrm{min}$, (**G**) $E = 0.50$, $t = 19\,\mathrm{min}$, (**H**) $E = 0.50\,\mathrm{V}$, $t = 42\,\mathrm{min}$; $i_\mathrm{T} = 2\,\mathrm{nA}$. Solution composition 3 mM $2,2'$-BP/0.05 M H_2SO_4

Defects within the adlayer change their shape or anneal with increasing observation time. The detailed mechanism most probably involves place exchange processes and surface diffusion of 2, 2′-BP-gold complexes. A similar mechanism is proposed for the "aging" of self-assembled monolayers formed by alkanthioles on gold surfaces, which follows the initial fast adsorption step [161,162,163,164].

In a second series of experiments the 2, 2′-BP stacking phase I was dissolved by potential excursion towards *negative* charge densities (region V), which is, as already reported above, accompanied by the immediate formation of the reconstructed Au(111)-(p$\times\sqrt{3}$) surface. Subsequently, the potential was stepped back to more positive values, for instance to $E_f = 0.40\,V$ (Fig. 22). It was discovered that the rims of the Au(111)-(p$\times\sqrt{3}$) reconstruction lines, which mark the [112]-direction, act as nucleation centers of an "angular" one-dimensional stacking structure (Fig. 22B). This phase is characterized by a tilt angle $\alpha = 60°$ between neighboring stacks. The distance between adjacent molecular stacks amounts to 13.5 Å, e.g. it is somewhat larger than in regions I. Subsequently, a two-dimensional growth of the densely packed stacking structure, already known as phase I, starts at the

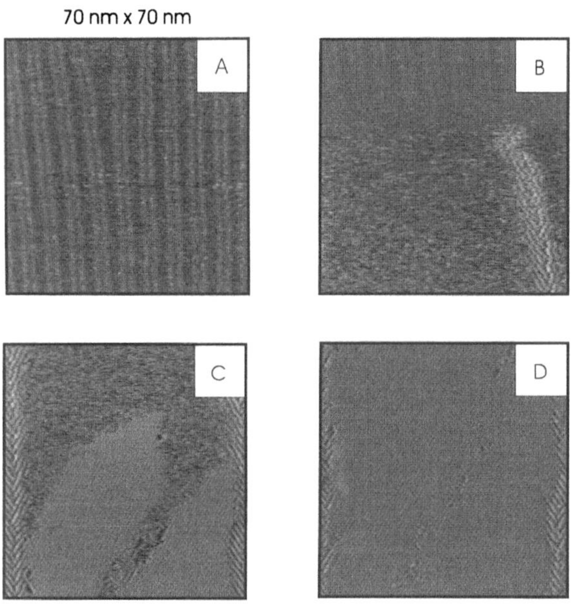

Fig. 22. Potential-induced reformation of the 2, 2′-BP stacking structure when stepping the potential from −0.20 V to 0.40 V. (**A**) $E = -0.20\,V$, (**B**) $E = -0.20\,V \rightarrow E = 0.40\,V$. Note the decoration of the reconstruction lines with the herringbone structure, (**C**) $E = 0.40\,V$, $t = 1.00$ min. Growth of two long-range ordered stacking domains out of the one-dimensional herringbone phase at an angle of 30°. (**D**) $E = 0.40\,V$, $t = 2$ min

edges of the metastable "angular" phase, smaller domains merge, and line defects heal (Fig. 22C and 22D). The process ends in acidic solutions with a quasi-perfectly ordered 2, 2′-BP stacking adlayer. This experiment demonstrates also that the molecular stacking rows in phase I are 30° rotated with respect to the reconstruction lines of the gold surface.

The active role of the Au(111)-(p×$\sqrt{3}$) reconstruction elements as preferred nucleation sites is not unique to the present system. Poirier [165] and Knoll [166] reported on the active role of the herringbone elbows in the formation process of mercaptohexanol and 4-mercaptopyridine monolayers, respectively. Heteroepitactic experiments revealed that these surface-confined dislocations act as loci of nucleation in the deposition of nickel, cobalt and iron on Au(111) under UHV [167,168,169,170] as well as under electrochemical conditions [171].

Finally we would like to comment that the long-range order of 2, 2′-BP adlayers on Au(111) at $E < 0.60$ V is significantly more pronounced in acidic solutions than in experiments under comparable conditions in neutral Na_2SO_4 or $KClO_4$ solutions. The surface mobility of gold atoms and substrate–adsorbate complexes appears to be much lower in the latter systems, which often causes defect-rich, "apparently frozen" adlayer patterns [172].

4.3 2, 2′-Bipyridine on Au(100)

4.3.1 General Overview and "Steady-State" STM

A set of capacitance vs. potential curves for 3 mM 2, 2′-BP in 0.05 MH_2SO_4 is plotted in Fig. 23. The first potential scan started with an Au(100)-(1 × 1) surface at 0.60 V (solid line). The capacitance curve of the supporting electrolyte is shown as the dotted trace. The phenomenological analysis of these electrochemical data allows us to distinguish four different potential regions, labeled I to IV. After equilibration of the entire system at $E = 0.60$ V a saturation capacitance of $\sim$ 8 μF cm^{-2} was reached in $0.40\,\mathrm{V} \leq E \leq 0.90\,\mathrm{V}$ (region I). At $E < 0.40$ V the capacitance increases gradually (region II) until a characteristic double-peaked feature is reached in $-0.25\,\mathrm{V} < E < 0.0\,\mathrm{V}$ (region III). The latter may indicate multistate desorption or restructuring of the adlayer and/or substrate surface. Finally, at $E < -0.25$ V all capacitance curves merge with the trace measured in the absence of 2, 2′-BP. This observation suggests the rather complete desorption of the organic molecules at the negatively charged electrode. Changing the direction of the potential scan gives rise to a marked hysteresis, up to $E < 0.20$ V. The capacitance curve of the base electrolyte is retraced. Obviously, the adsorption of 2, 2′-BP proceeds rather slowly, and does not immediately give rise to a homogeneous phase, which is indicated by the position of peak P3.

In situ STM measurements revealed an ordered monolayer of parallel molecular stacks with irregular kink dislocations in region I, e.g. at the positively charged Au(100)-(1 × 1) electrode (Fig. 23A,B). Scanning the elec-

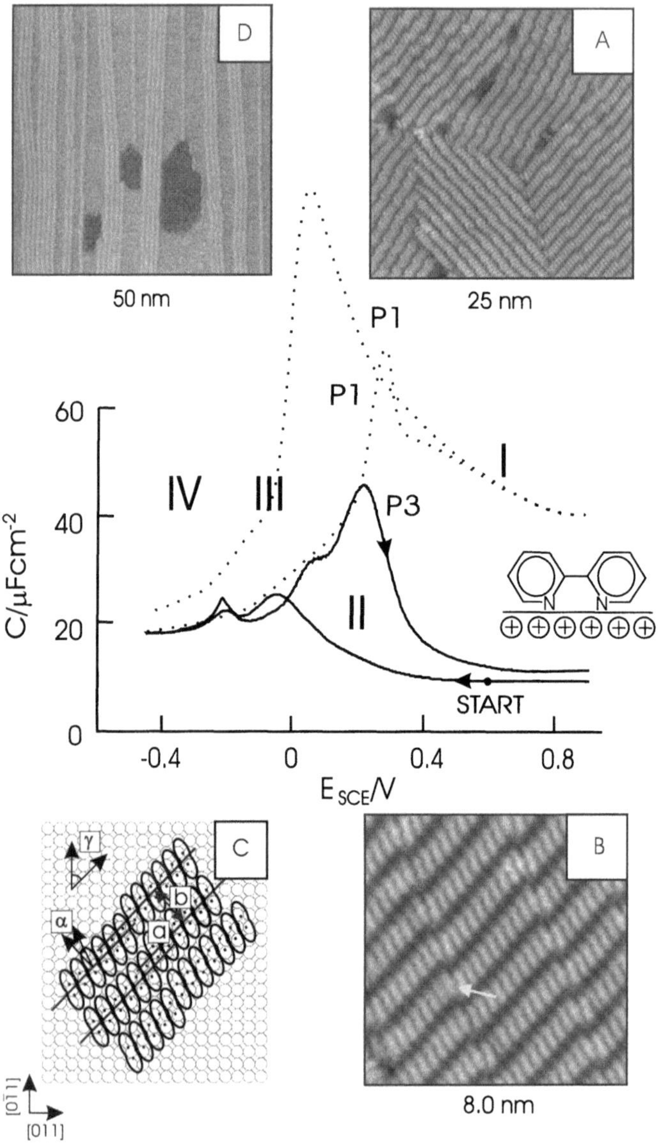

Fig. 23. Capacitance vs. potential curve for Au(100)/0.05 M H_2SO_4 in the absence *(dotted line)* and in the presence of 3 mM 2, 2′-BP, scan rate $10\,\mathrm{mV\,s^{-1}}$. The first scan as obtained with an electrochemically annealed Au(100)-(1 × 1) electrode started after equilibration at 0.60 V, and is indicated as the *solid line*. The stability regions of various adlayer phases are labeled I to IV. (**A**) equilibrium structure of 2, 2′-BP on Au(100)-(1 × 1) at $E = 0.60$ V; (**B**) unfiltered high-resolution images of the 2, 2′-BP stacking structure at 0.50 V, tunneling current $i_T = 2$ nA or $i_T = 4$ nA, respectively; (**C**) Proposed packing model. Characteristic distances and angles are indicated. (**D**) non-equilibrated Au(100)-surface after dissolution of the 2, 2′-BP stacking phase I at −0.25 V

trode potential towards more negative values, e.g. passing region II, gradually dissolves the ordered 2, 2′-BP phase. Dissolution starts preferentially in regions of domain boundaries or high defect ("hole") density, respectively. At $E \leq -0.15\,\mathrm{V}$ (region III) the first potential-induced reconstruction lines of the hexagonal substrate surface structure Au(100)-(hex) appear (Fig. 23D). The density of molecular stacks decreases drastically, and they seem to occupy exclusively the still remaining quadratic Au(100)-(1×1) areas of the electrode surface.

Complete dissolution (desorption) of the 2, 2′-BP stacking layer was found at $E < -0.25\,\mathrm{V}$. The electrode is still partially unreconstructed. Careful test experiments revealed that no additional ordered adsorbate phase exists at more negative electrode potentials (or charge densities). This is different from our results on Au(111), where we found at $E < -0.05\,\mathrm{V}$ an incommensurate stacking pattern of singly protonated 2, 2′-BP species [57]. Closing the potential cycle causes the reformation of the 2, 2′-BP stacking phase at $E > 0.20\,\mathrm{V}$. Extended electrochemical annealing in this region I (one hour or longer) gives rise to an improved molecular order, comparable with that depicted in Fig. 23A.

After equilibration in $0.40 \leq E \leq 0.80\,\mathrm{V}$ high-resolution images of the ordered 2, 2′-BP phase such as shown in Fig. 23B were measured. Individual molecules appear as elongated blobs, arranged in parallel stripes. The distance between adjacent rows amounts to $b = (10.2 \pm 0.4)\,\text{Å}$. The intermolecular distance was estimated to be $a = (3.7 \pm 0.2)\,\text{Å}$. Furthermore, molecules are tilted from the normal to the axis of the chain by an angle $\alpha = (18 \pm 2)^\circ$. The most characteristic difference with the 2, 2′-BP stacking phase on Au(111) is the existence of kinks. Rather irregularly, a sudden displacement of a section of a chain (four up to twelve molecules) in the direction perpendicular to its main axis by $\sim 2.1\,\text{Å}$ was observed. A typical example is marked by the arrow in Fig. 23B. The tilt angle α of molecules at kink positions increases somewhat from 18° to 25°. Furthermore, these kink positions propagate among parallel stacking rows of one domain. Their existence reflects a misfit in substrate–adsorbate co-ordination geometry. Similar kinks have also been observed with phenanthroline monolayers on Au(111) [59].

An analysis of domain properties revealed that individual domains are seldom larger then $15\,\mathrm{nm} \times 15\,\mathrm{nm}$ (Fig. 23A). The 2, 2′-BP adlayer exhibits two types of domain boundaries: (1) rotated domains with characteristic angles $\beta = (90 \pm 10)^\circ$ reflecting the symmetry of the underlying substrate and (2) rotated domains with $\beta' = (10 \pm 5)^\circ$. The latter are most probably created upon merging stacking rows of only slightly different displacement (due to the existence of non-matching kink-positions). Polygonal "holes" filled with disorderly adsorbed 2, 2′-BP were also found in these cases.

Analyzing those images, which were recorded on partially reconstructed Au(100) electrodes in the presence of 2, 2′-BP (Fig. 24), we found that the majority of molecular stripes is tilted by an angle $\gamma = \pm 45^\circ$ with respect to

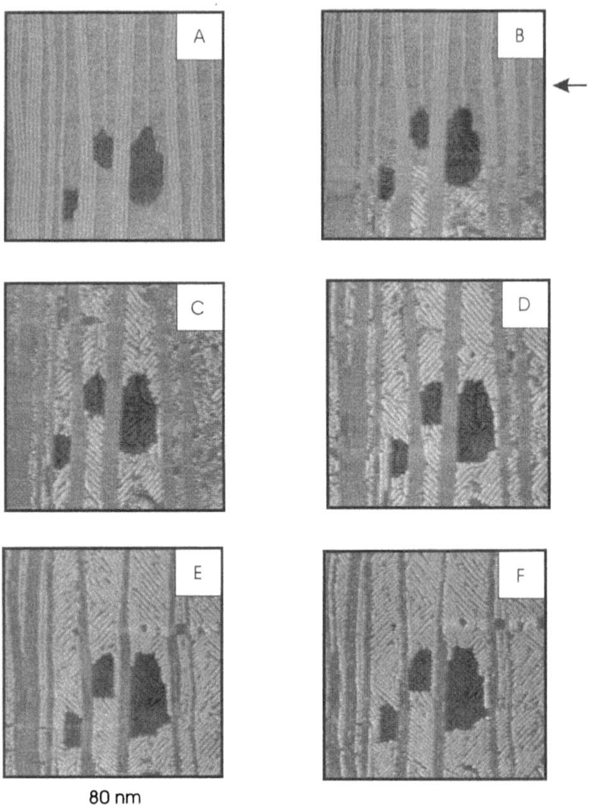

Fig. 24. Selective nanodecoration of Au(100) with 2, 2′-bipyridine. (**A**) Au(100) with reconstructed (hex-rows) and unreconstructed (1 × 1) patches as obtained at −0.17 V (5 min) after a potential scan from 0.50 V for 3 mM 2, 2′-BP in 0.05 M H_2SO_4, (**B**) potential scan $E_i = -0.17\,V \rightarrow E_f = 0.0\,V$ (arrow indicates the start, $10\,mV\,s^{-1}$); (**C**) $E = 0.0$ V, 4 min after (**B**); (**D**) $E = 0.10$ V, 6 min after (**B**); (**E**) $E = 0.15$ V, 20 min after (**B**); (**F**) $E = 0.20$ V, 30 min after (**B**); (**G**) $E = 0.35$ V, 35 min after (**B**); (**H**) $E = 0.50$ V, 40 min after (**B**)

the Au(100)-(hex) reconstruction rows. The latter are aligned to the [011]- or the [011]-directions of the square gold lattice. This observation reveals that the preferential orientation of the 2, 2′-BP stacking rows with the Au(100)-(1 × 1) surface assumes an angle $\gamma = \pm 45°$. We also note that the distance between adjacent 2, 2′-BP rows in the [011]-direction amounts to $b' = 14.4$ Å, which is practically identical to $5 \times a_{Au} = 14.42$ Å ($a_{Au} = 2.885$ Å).

The experimental results lead to an incommensurate molecular packing arrangement of 2, 2′-BP on Au(100)-(1 × 1), see Fig. 23C [58]. The model represents well all experimentally observed dimensions, such as the distance between adjacent molecules and stacks, a and b ($b' = 5 \times a_{Au}$), the tilt angle α as well as the registry between the square surface and the majority direction of molecular stacks ($\gamma = 45°$), and does not violate packing principles. In analogy to the results of a quantum chemical calculation of *Rodriguez* for pyridine on Ag(*hkl*) [121], and referring to the dimensions of the possible molecular and substrate co-ordination sites (*N*-*N*-distance, a_{Au}), the above adlayer model was constructed with the constraints of a maximum number of equivalent high-order co-ordination sites. This concept could only be accomplished by placing $2,2'$-BP with its nitrogen atoms (indicated by dots in Fig. 23D) in "close to" twofold bridge positions. This configuration also minimizes the "adlayer corrugation", and therefore ensures an optimal arrangement for lateral π-stacking between parallel aromatic rings [112]. Here we would like to stress that the experimentally observed tilt angle $\alpha = 18°$ and the size of $2,2'$-BP [143] do not allow a perfect alignment of the organic molecule with bridge positions of the substrate surface. Aligned molecules within one row approach with increasing number of species energetically unfavorable a-top positions. The emerging misfit to favorable adsorbate and substrate co-ordination sites causes the displacement of small sections of the chain (between four and twelve molecules) by $1/2\sqrt{2}a_{\mathrm{Au}} = 2.05\,\text{Å}$ into the nearest, slightly more stable bridge-type co-ordination sites. This parallel shift propagates among rows of one domain, therefore still minimizes the steric repulsion between chains (*Ulman*'s concept of "commensurability of intra-assembly planes" [173]), and does not violate the ability of lateral attractive π-stacking.

The observed adlayer patterns of $2,2'$-BP on Au(100)-(1×1) and Au(111) [57] clearly illustrate the structure-determining role of substrate surface geometry and of the strong mixing of the non-bonding orbitals of the two nitrogen heteroatoms with the electronic states in the metal, e.g. strong adsorbate–substrate interactions. Secondly, the "short-distance" alignment of molecules within one row is stabilized by lateral π-stacking. The larger geometrical misfit between potential adsorbate and substrate co-ordination sites seems to be responsible for the higher degree of disorder of the $2,2'$-BP adlayer on Au(100)-(1×1), in comparison to Au(111) [57]. Similar results were also obtained for uracil on Au(*hkl*) electrodes [51] and thymine on Cu(*hkl*) [174]. The extension of the stability range of the $2,2'$-BP stacking phase I on Au(100) towards more negative potentials (approximately 200 mV) reflects the difference in zero charge potentials for both substrate surface geometries [101].

4.3.2 Selective Decoration of an Au(100)-Electrode

The above experiments revealed that an ideal Au(100)-(1×1) surface, prepared by electrochemical annealing at 0.60 V (30 minutes) and subsequent deposition of $2,2'$-BP reconstructs partially at -0.17 V, simultaneously with the dissolution of the ordered $2,2'$-BP stacking phase I. A snapshot of this process is presented in Fig. 24A. The surface appears structurally inhomogeneous: Parallel *hexagonal* Au(100)-(hex) patches separate regions of *quadratic* symmetry. In subsequent experiments a surface spot with three monoatomic deep holes to mark the observation area was chosen. Scanning the electrode potential slowly ($10\,\mathrm{mV\,s^{-1}}$, start is indicated by the arrow in Fig. 24B) to 0.0 V, e.g. already within the thermodynamic stability range of the chemisorbed stacking phase I, demonstrates that $2,2'$-BP adsorbs only on Au(100)-(1 × 1) patches. Most of the molecular stacks are rotated by ±45° with respect to the [011] direction of the neighboring (hex) reconstruction rows. Figure 24B and C also reveal that the stacks not only nucleate at the boundary between (hex) and (1 × 1) substrate surface patches, but also in the middle of the (1 × 1) terraces. The long-range order increases with observation time. The selective decoration of the (1 × 1) regions reflects the local positive surface excess charge, which results from the more negative position of the pzc of this surface [101] (note that the adsorption of $2,2'$-BP on Au(*hkl*) shifts the pzc even more negative as compared with the adsorbate-free base electrolyte [149]). This observation also supports the notion that $2,2'$-BP forms stable adsorbate–substrate co-ordination compounds at positively charged gold electrodes.

Raising the potential towards more positive values (Fig. 24D–G) shows that the reconstruction elements are still not covered with ordered $2,2'$-BP stacks. They disintegrate slowly, e.g. transform into the (1 × 1) structure. Remarkably, no gold islands, which one actually would expect for the substrate surface transition (hex) → (1 × 1), were found. Only after extended waiting time (> 40 min) did a few islands appear at $E = 0.50\,\mathrm{V}$. The $2,2'$-BP stacks on the previously reconstructed surface spots are short and their mutual orientation is rather disordered. Both observations indicate that a non-negligible percentage of extra gold atoms (ions) of the former (hex) phase is not released but consumed during the formation of the $2,2'$-BP substrate co-ordination complex. The selective nanodecoration of Au(100) with $2,2'$-BP is reversible (usually in $-0.25\,\mathrm{V} < E < 0.35\,\mathrm{V}$). The presence of gold islands disturbs this process. The latter, once formed, act immediately as nucleation centers for new reconstruction elements in the [011]- and [011]-directions during a subsequent potential excursion towards negative values, and reduce rapidly the metastability of the Au(100)-(1 × 1) patches.

The above observation complements three other regimes of nanostructuring of gold surfaces, of which recently several examples were published: (1) decoration of entire Au(111)-($\mathrm{p}\times\sqrt{3}$) reconstruction lines with alkylthioles [165,166,175] or C_{60} molecules [176] at low coverages; (2) decoration of discommensurations at the elbows of the Au(111)-($\mathrm{p}\times\sqrt{3}$) reconstruction elements with small nickel clusters (in situ [171] and ex situ [167]), or (ex situ) islands of iron [170], cobalt [169], palladium [177] as well as rhodium [178]; (3) decoration of steps with UPD-deposited Pb and Tl atoms on Au(111) and Au(100) [179].

Better understanding and consequently controlling of these nanodecoration processes on model surfaces, such as Au(*hkl*), may allow us to derive general principles of template-structure manufacturing and of the formation of functional low-dimensional adlayer structures.

5 4, 4′-Bipyridine on Au(111)

The phase behavior of n, n'-BP on single-crystal electrodes is not only dependent on the properties of the substrate surface, but also on the arrangement of potential molecular co-ordination centres. 2, 2′-BP and 4, 4′-BP are instructive examples. In the latter case the nitrogen atoms are located at opposite ends of the two rigid pyridyl rings. This structure offers a fascinating bifunctionality. 4, 4′-BP might be bound via one nitrogen on appropriately prepared electrode surfaces, such as Au(hkl), Ag(hkl) or Cu(hkl), while the other nitrogen might offer a reactive position towards the solution side to build up vertical architectures. Indeed, recent work with 4, 4′-BP and cytochrome C revealed that 4, 4′-BP immobilzes cytochrome C on electrodes, most probably close to its native conformation, and therefore promotes efficiently long-range electron transfer reactions [180,181]. The structural understanding of this catalytic process as well as the interfacial behavior of the bridging ligand 4, 4′-BP are still rather limited.

Based on voltammetric and electroreflectance experiments *Hill* et al. [180,181] and *Niki* [182,183,184] suggested that 4, 4′-BP is oriented perpendicularly on gold and silver electrodes. Co-ordination is accomplished via one of the two nitrogen atoms. This hypothesis is supported by recent SERS experiments [185,186] and the STM work of *Cunha* et al. on Au(111) [56].

Figure 25 summarizes some results of combined electrochemical and in situ STM experiments on the phase behavior of 4, 4′-BP on "island-free" Au(111) electrodes in 0.05 M $KClO_4$. Analysing a typical current vs. potential curve allows us to distinguish four different potential regions labelled I–IV in Fig. 25: In the potential range IV the molecules are randomly adsorbed on the electrode. The transitions IV $\leftrightarrow$ III, III $\leftrightarrow$ II and II $\leftrightarrow$ I are represented by rather sharp and asymmetrical maxima of the charging current, which, in addition, exhibit a characteristic hysteresis. The corresponding current vs. time transients point to rather complicated nucleation and growth processes involved in these transitions. Furthermore, capacitance vs. potential curves (not shown) exhibit typical "pit-like" features in regions III, II and I, respectively. All these phenomenological observations indicate that 4, 4′-BP forms on Au(111)-electrodes at least three distinct ordered structures, as a function of the applied potential.

High-resolution in situ STM experiments showed in region I (positive charge densities) a periodic contrast pattern composited of main "dots" and intervening "ribbons" each consisting of four small dots (Fig. 25A). The repeat pattern of the main dots does not change with the tunneling conditions and was therefore chosen to define the unit cell with $a = (11.1 \pm 0.6)$ Å, $b = (12.8 \pm 0.7)$ Å and $\alpha = (81 \pm 8)°$ [56,187]. The experimental STM con-

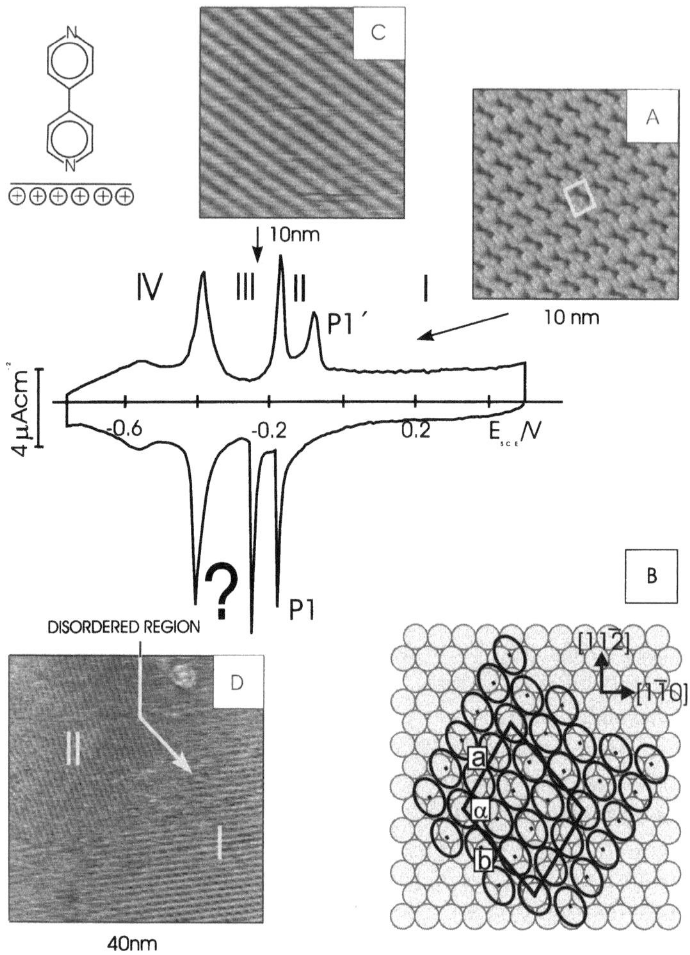

Fig. 25. Steady-state cyclic current vs. potential curve for 3 mM 4,4′-bipyridine on Au(111) in 0.05 M$KClO_4$, scan rate 10 mV s^{-1}. The stability regions of the various adlayer phases are labeled I to IV. (**A**) Unfiltered high resolution image of the 4,4′-BP stacking structure I at $E = 0.25$ V, $i_T = 0.3$ nA, $v_T = 10$ mV. The suggested commensurate unit cell is indicated. (**B**) Proposed packing model for 4,4′-BP. (**C**) Unfiltered image of the "striped" 4,4′-BP structure II at $E = -0.18$ V, $i_T = 0.11$ nA. (**D**) Snapshot of the transition I → II after a potential step −0.05 V → −0.17 V

trast features were interpreted by assuming a densely packed ordered stacking structure of mutually parallel oriented 4,4′-BP molecules at positive charge densities. Each dot is assigned to an individual molecule. This assumption implies also the occupancy of different substrate co-ordination sites. The suggested "theoretical" unit cell (Fig. 25B), with $a' = 11.6$ Å, $b' = 13.3$ Å and $\alpha' - 77.5°$ contains nine 4,4′-BP molecules: This result gives rise to an area per molecule of 16.7 Å^2, which points to a perpendicular orientation of 4, 4′-BP on Au(111) at positive charge densities. Additional and complementary details of the molecular adlayer I might be available from high-resolution AFM experiments, such as indicated in [56].

Changing the electrode potential towards more negative values in region II gives rise to a very fragile, stripe-like overlayer, which could only be resolved at low tunneling currents. No molecular resolution was achieved. The transition I $\rightarrow$ II proceeds isotropically according to an order/order mechanism, where both ordered phases temporarily coexist, separated locally by a small disordered region (Fig. 25D).

The suggested ordered adlayer in region III, based on voltammetric data, could not yet be resolved by in situ STM/AFM studies.

The interplay of molecular structure, substrate geometry and electrode potential, as demonstrated for 4, 4′-BP and in Sect. 4 for 2, 2′-BP, offers fascinating opportunities to functionalize metal surfaces by substrate–adsorbate co-ordination and stacking, and may lead the way into new dimensions of molecular electrochemistry.

6 Concluding Remarks

Combined electrochemical, in situ STM and spectroscopic measurements revealed a detailed molecular-level understanding of structural transitions in adlayers of pyrimidine derivatives and n, n'-bipyridines on Au(hkl) as well as of the respective substrate surfaces. The results presented offer the potential to be generalized towards a host of N- and S-heterocyclic compounds and various substrate surfaces, such as Au(hkl), Ag(hkl), Cu(hkl) and Pt(hkl). The ordering within the two-dimensional assembly is the result of the interplay of molecular structure, substrate geometry, electrode potential and temperature. It involves specific adsorbate–adsorbate and adsorbate–substrate interactions, such as electrostatic forces, π-stacking, van-der-Waals forces, intermolecular hydrogen bonding and partial charge transfer. The understanding of these complex co-operative mechanisms offers exciting possibilities to modify metal surfaces via substrate–adsorbate and/or ion–adsorbate co-ordination, molecular recognition ("key–lock" principle and/or "host–guest" chemistry and physics) and self-organization employing the *electrode potential* as an active *tool* and/or *detector*. Structural, electrical and mechanical properties and functions of single molecules and ions in template structures seem to be approachable. The various scanning probe microscopies offer a powerful group of methods to study these phenomena at a local level in situ with unprecedented resolution in real space and, under some circumstances, in real time. The combination with other structure-sensitive techniques may soon lead to fascinating in situ reactivity studies at solid/liquid interfaces at the molecular/atomistic level.

Finally we would like to stress that defect structures and surface reorganization contribute significantly to equilibrium structures and reaction pathways of two-dimensional assemblies. The exact control and monitoring of the substrate surface properties may allow one to generate templates for the defined modification of horizontal as well as vertical adlayer architectures with a great variety of functions.

Acknowledgements

The present work was supported by the Volkswagen Foundation (I/73 025) and the Deutsche Forschungsgemeinschaft through Wa 879/3-2, and a Heisenberg–Fellowship for one of us (T.W.). It is a pleasure for the authors to thank Prof. D. M. Kolb for continuous support in these studies. We also like to thank Prof. R. de Levie, Prof. N. J. Tao and Prof. A. Tadjeddine for critical comments and many helpful discussions. Finally, we express our gratitude to Molecular Imaging for the generous loan of a Pico-SPM with temperature stage.

References

1. J. D. Swalen, D. L. Allara, J. D. Andrade, E. A. Chandross, S. Garoff, J. Israelachvilli, T. J. Carthy, R. Murray, R. F. Rease, J. F. W. Rabold, K. J. Wynne, H. Hu: Langmuir **3**, 932 (1987)
2. A. Ullman: *Introduction to Thin Solid Films* (Academic Press, Boston 1991)
3. R. N. Parkins, in: J. O. M. Bockris, B. E. Conway, E. Yeager, R. E. White (Eds.): *Comprehensive Treatise of Electrochemistry* Vol. 4 (Plenum Press: New York 1980) p.307
4. W. Plieth: Electrochim. Acta **37**, 2115 (1992)
5. M. M. Baizer: *Organic Electrochemistry* (Plenum Press: New York 1991)
6. R. W. Murray, in Murray, R. W. (Ed.): *Molecular Design of Electrode Surfaces* (Wiley: New York 1992) p. 1
7. J. Janata, M. Josovic, D. M. DeVaney: Anal. Chem., 207R (1994)
8. E. Altman, R. Colton: Surf. Sci. **295**, 13 (1993)
9. A. Ullman: Chem. Rev. **96**, 1533 (1996)
10. R. E. Pagano, R. Miller: J. Colloid. Interface Sci. **45**, 126 (1973)
11. A. Nelson, N. Auffret: J. Electroanal. Chem. **244**, 99 (1988)
12. M. Porter, T. B. Bright, D. Allara, C. E. D. Chidsey: J. Am. Chem. Soc. **109**, 3559 (1987)
13. L. Strong, G. M. Whitsides: Langmuir **4**, 546 (1988)
14. J. Sagiv: J. Am. Chem. Soc. **102**, 92 (1980)
15. H. Finklea, D. A. Snider, J. Fedyk: Langmuir **9**, 3360 **(1993)**
16. P. Fenter, P. Eisenberger, K. S. Liang: Phys. Rev. Let. **70**, 2447 (1993)
17. O. M. Magnussen, B. M. Ocko, M. Deutsch, M. J. Reagan, P. S. Pershan, D. Abernathy, G. Grübel, J. F. Legrand: Nature **384**, 150 (1996)
18. R. deLevie: Chem. Rev. **88**, 599 (1988)
19. C. Buess-Herman: Progr. Surf. Sci. **46**, 335 (1994)

20. L. Stolberg, J. Lipkowski: in: J. Lipkowski, P. N. Ross (Eds.) *Adsorption of Organic Molecules at Metal Electrodes* (VCH, New York 1992)
21. A. A. Kornyshev, L. Vilfan: Electrochim. Acta **40**, 109 (1995)
22. A. N. Frumkin: Z. Phys. Chem. **116**, 466 (1925); Z. Phys. **35**, 972 (1926)
23. B. Damaskin, O. A. Petrii, V. V. Batrakov: *Adsorption of Organic Compounds on Electrodes* (Plenum Press, New York 1972)
24. R. Parsons: Chem. Rev. **90**, 813 (1990)
25. W. Lorenz: Z. Electrochem. **62**, 192 (1958)
26. V.Vetterl: Collect. Czech. Chem. Commun. **31**, 2105 (1966)
27. D. A. Bonell: *Scanning Tunneling Microscopy and Spectroscopy* (VCH, New York 1993)
28. Th. Wandlowski, B. M. Ocko, O. M. Magnussen, S. Wu, J. Lipkowski: J. Electroanal. Chem. **409**, 155 (1996)
29. V. V. Batrakov, B. B. Damaskin, Y. B. Ipatov: Elektrokhimiya **10**, 216 (1974)
30. A. Hamelin, S. Morin, J. Richer, J. Lipkowski: J. Electroanal. Chem. **304**, 195 (1991)
31. A. Popov, R. Naneva, K. Dimitrov, T. Vitanov, V. Bostanov, R. de Levie: Electrochim. Acta **37**, 2369 (1992)
32. B. Roelfs, H. Baumgärtel: Ber. Bunsenges. Phys. Chem. **99**, 677 (1994)
33. B. Roelfs, E. Bunge, C. Schröter, T. Solomun, H. Meyer, R. Nichols, H. Baumgärtel: J. Phys. Chem. B **101**, 754 (1997)
34. M. H. Hölzle, D. Krznaric, D. M. Kolb: J. Electroanal. Chem. **386**, 235 (1995)
35. M. H. Hölzle, D. M. Kolb: Ber. Bunsenges. Phys. Chem. **98**, 330 (1994)
36. M. Scharfe, A. Hamelin, C. Buess-Herman: Electrochim. Acta **40**, 61 (1995)
37. M. H. Hölzle, Th. Wandlowski, D. M. Kolb: Surf. Sci. **335**, 281 (1995)
38. M. H. Hölzle, Th. Wandlowski, D. M. Kolb: J. Electroanal. Chem. **394**, 271 (1995)
39. Th. Wandlowski: J.Electroanal. Chem. **395**, 8 (1995)
40. Th. Wandlowski, M. H. Hölzle: Langmuir **12**, 6597, 6604 (1996)
41. D. Krznaric, B. Cosovic, M. H. Hölzle, D. M. Kolb: Ber. Bunsenges. Phys. Chem. **100**, 1779 (1996)
42. Th. Wandlowski, Th. Dretschkow: J. Electroanal. Chem. **427**, 105 (1996)
43. R. Guidelli, M. L. Foretsi, M. Innocenti: J. Phys. Chem. **100**, 18491 (1996)
44. H. Striegler, M. H. Hölzle, Th. Wandlowski, D. M. Kolb: D. M. Poster 1a-41, presented at the 47th Annual Meeting of the ISE, Budapest (1996)
45. Th. Boland, B. Ratner: Langmuir **10**, 3845 (1994)
46. W. Haiss, B. Roelfs, S. N. Port, E. Bunge, H. Baumgärtel, R. J. Nichols: J. Electroanal. Chem. **454**, 107 (1998)
47. D. M. Kolb: Z. Phys. Chem. NF **154**, 179 (1987)
48. N. J. Tao, J. A. DeRose, S. M. Lindsay: J. Phys. Chem. **97**, 205 (1993)
49. Th. Wandlowski, D. M. Lampner, S. M. Lindsay: J. Electroanal. Chem. **404**, 215 (1996)
50. Th. Dretschkow, A. S. Dakkouri, Th. Wandlowski: Langmuir **13**, 2845 (1997).
51. Th. Dretschkow, Th. Wandlowski: Electrochim. Acta **43**, 2991 (1998)
52. S. Sowerby, W. M. Heckel: Origin of Life and Evolution of the Biosphere, J. Electroanal. Chem. **28**, 283 (1998)
53. K. M. Richard, A. A. Gewirth: J. Phys. Chem. **99**, 12288 (1995)
54. G. Andreasen, M. E. Vela, R. C. Salvarezza, A. J. Arvia: Langmuir **13**, 6814 (1997)

55. F. Cunha, N. J. Tao: Phys. Rev. Lett. 75, 2376 (1995)
56. F. Cunha, N. J. Tao, X. W. Wang, Q. Jiang, B. Duong, J. D'Agnese: Langmuir **12**, 6410 (1996)
57. Th. Dretschkow, D. Lampner, Th. Wandlowski: J. Electroanal. Chem. **458**, 121 (1998)
58. Th. Dretschkow, Th. Wandlowski: J. Electroanal. Chem. **467**, 207 (1999)
59. F. Cunha, Q. Jing, N. J. Tao: Surf. Sci. **389**, 19 (1997)
60. O. Dominguez, L. Echegoyen, F. Cunha, N. Tao: Langmuir **14**, 821 (1998)
61. J. Pan, S. M. Lindsay, N. J. Tao: Langmuir **9**, 1556 (1993)
62. E. Bunge, R. J. Nichols, H. Baumgärtel, H. Meyer: Ber. Bunsenges. Phys. Chem. **99**, 1243 (1995)
63. E. Bunge, R. J. Nichols, B. Roelfs, H. Meyer, H. Baumgärtel: Langmuir **12**, 3066 (1996)
64. A. S. Dakkouri, D. M. Kolb, R. Edelstein-Shima, D. Mandler: Langmuir **12**, 2849 (1996)
65. N. Batina, M. Kunikate, K. Itaya: J. Electroanal. Chem. **405**, 245 (1995)
66. M. Kunikate, N. Batina, K. Itaya: Langmuir **11**, 2337 (1995)
67. K. Ogaki, N. Batina, M. Kunikate, K. Itaya: J. Phys. Chem. **100**, 7185 (1996)
68. S. L. Yau, Y. G. Kim, K. Itaya: J. Am. Chem. Soc. **118**, 7795 (1996)
69. L. J. Wan, K. Itaya: Langmuir **13**, 7173 (1997)
70. R. Srinivasan, J. C. Murphy, R. Fainchtein, N. Pattabiraman: J. Electroanal. Chem. **312**, 293 (1994)
71. R. Srinivasan, R. Gopalan: J. Phys. Chem. **98**, 8770 (1993)
72. N. J. Tao, Z. Shi: J. Phys. Chem. **98**, 1464 (1994)
73. N. J. Tao, Z. Shi: Surf. Sci. **321**, L 149 (1994)
74. N. J. Tao, Z. Shi: J. Phys. Chem. **98**, 7422 (1994)
75. N. J. Tao: Phys. Rev. Lett. **76**, 4066 (1996)
76. S. W. Sowerby, G. B. Petersen: J. Electroanal. Chem. **433**, 85 (1997)
77. B. M. Ocko, O. M. Magnussen, J. X. Wang, R. R. Adzic, Th. Wandlowski: Physica B **221**, 238 (1996)
78. E. Constable: Progr. Inorg. Chem. **42**, 67 (1994)
79. P. Steel: Coord. Chem. Rev. **106**, 227 (1990)
80. Th. Wandlowski, Th. Dretschkow, T. Iwasita: Langmuir, in preparation, 2002
81. Th. Wandlowski, B. M. Ocko, O. M. Magnussen, S. Wu, J. Lipkowski: J. Electroanal. Chem. **409**, 155 (1996)
82. S. Wu, J. Lipkowski, O. M. Magnussen, B. M. Ocko, Th. Wandlowski: J. Electroanal. Chem. **446**, 67 (1998)
83. A. Hamelin in: A. G. Gewirth, H. Siegenthaler (Eds.) *Nanoscale Probes of the Solid/Liquid Interface* (Kluwer, Dordrecht 1995) 285
84. D. M. Kolb, J. Schneider: Electrochim. Acta **31**, 929 (1986)
85. M. S. Zei, G. Lehmpfuhl, D. M. Kolb: Surf. Sci. **221**, 23 (1985)
86. A. Friedrich, B. Pettinger, D. M. Kolb, G. Lüpke, R. Steinhoff, G. Marowsky: Chem. Phys. Lett. **163**, 123 (1989)
87. J. Wang, B. M. Ocko, A. J. Davenport, H. S. Isaacs: Phys. Rev. **B46**, 10321 (1992)
88. O. M. Magnussen: PhD Thesis, University of Ulm (1993)
89. X. Gao, A. Hamelin, M. J. Weaver: J. Chem. Phys. **95**, 6993 (1991)
90. N. J. Tao, S. M. Lindsay: Surf. Sci. **274**, L546 (1992)
91. Th. Dretschkow, Th. Wandlowski: Electrochim. Acta **45**, 731 (1999)

92. D. D. Chambliss, R. J. Wilson: J. Vac. Sci. Technol. B **9**, 928 (1991)
93. J. V. Barth, H. Brune, G. Ertl, R. J. Behm: Phys. Rev. **42**, 9307 (1990)
94. R. Randler: Diploma Thesis, University of Ulm (1995)
95. O. M. Magnussen, J. Hageböck, J. Hotlos, R. J. Behm: Farad. Disc. Chem. Soc. **94**, 329 (1992)
96. G. J. Eden, X. Gao, M. J. Weaver: J. Electroanal. Chem. **375**, 357 (1994)
97. Th. Dretschkow, Th. Wandlowski: Ber. Bunsenges. Phys. Chem. **101**, 749 (1997)
98. O. M. Magnussen, J. Hotlos, R. J. Behm, N. Batina, D. M. Kolb: Surf. Sci. **296**, 310 (1993)
99. X. Gao, G. J. Edens, A. Hamelin, M. J. Weaver: Surf. Sci. **296**, 333 (1993)
100. N. Batina, A. S. Dakkouri, D. M. Kolb: J. Electroanal. Chem. **370**, 87 (1994)
101. D. M. Kolb: Surf. Sci. **51**, 109 (1996)
102. B. M. Ocko, J. Wang, A. Davenport, H. Isaacs: Phys. Rev. Lett. **65**, 1466 (1990)
103. M. Y. Abyaneh, M. Fleischman: Electrochim. Acta **27** (1982) 1573
104. F. C. Franck: Proc. R. Soc. (London) A **201** (1950) 586
105. R. Philipp, J. Dittrich, U. Retter, E. Müller: J. Electroanal. Chem. **250** (1988) 159
106. R. Kaishew, B. Mutatschiew: Electrochim. Acta **20** (1965) 643
107. G. Quarin, Cl. Buess-Herman, L. Gierst: J. Electroanal. Chem. **123** (1981) 35
108. C. Hinnen, C. N. van Huong, J. P. Dalbera: J. Chim. Phys. (Paris) **79**, 38 (1982)
109. J. Lipkowski, C. N. Huong, C. Hinnen, R. Parsons, J. Chevalet: J. Electroanal. Chem. 143 (1983) 375
110. R. Guidelli in *Abstracts of the IUVSTA-Workshop on Surface Science and Electrochemistry*, San Benedetto del Tronto (1994), p. 46
111. Th. Wandlowski, unpublished
112. W Saenger: *Principles of Nucleic Acid Structure* (Springer-Verlag, Berlin, Heidelberg 1984)
113. G. S. Parry: Acta Crystallogr. **7** (1954) 313
114. K. Ozeki, N. Sakabe, J. Tanaka: Acta Crystallogr. B **25**, 1038 (1969)
115. A. Pokorilla, J. Jawarski: Biochim. Biophys. Acta, **331**, 1 (1973)
116. I. Kulakowska, M. Geller, B. Lesyng, K. Wierzhowski: Biochim. Biophys. Acta **361**, 119 (1974)
117. J. G. Baker, S. D. Christian, M. H. Kim, G. Dryhurst: Biophys. Chem. **9**, 355 (1979)
118. Th. Wandlowski, Th. Dretschkow, A. Tadjeddine, W. Q. Zheng: LURE Highlights 1997, p. 20
119. J. Sponer, J. Leszynsky, P. Hobza: J. Phys. Chem. **100**, 5590 (1996)
120. N. J. Tao: In: J. Lipkowski, P N. Ross (Eds.): Imaging of Surfaces and Interfaces (Wiley-VCH 1999), p. 211
121. J. A. Rodriguez: Surf. Sci. **226**, 101 (1990)
122. H. Sellers, A. Ulman, Y. Shnidman: J. Am. Chem. Soc. **115**, 9389 (1997)
123. I. Mutikainen: Ann. Acad. Sci. Fenn. Ser. A **217**, 1 (1988)
124. C. Nagata, A. Imamura, H. Fujita, in: M. Ktani (Ed.): *Advances in Biophysics* (Univ. of Tokyo Press, Tokyo 1973) Vol. 4 p. 1
125. L. Baralda, M. C. Bruni, P. M. Costi, P. Pecorari: Photochem. Photobiol. **52**, 361 (1990)

126. H. G. Günther, A. Goldman, R. Courths: Surf. Sci. **176**, 115 (1986)
127. P. N. Ross, A. T. D'Agostino: Electrochim. Acta **37**, 615 (1992)
128. F. Henglein, D. M. Kolb, L. Stolberg, J. Lipkowski: Surf. Sci. **317**, 325 (1993)
129. S. Wu, J. Lipkowski, O. M. Magnussen, B. M. Ocko, Th. Wandlowski: J. Electroanal. Chem. **446** (1998) 67
130. W. Li, W. Haiss, S. Floate, R. Nichols: Langmuir **15**, 4875 (1999)
131. J. Lecoeur, J. Andro, R. Parsons: Surf. Sci. **114**, 320 (1982)
132. M. Cavallini, G. Aloisi, M. Bracali, R. Guidelli: J. Electroanal. Chem. **444**, 75 (1998)
133. Th. Dretschkow, Th. Wandlowski: Langmuir, in preparation (2002)
134. Th. Wandlowski, Th. Dretschkow, A. John-Annacker, M. D. Martinez, A. Tadjeddine, W. Q. Zheng, A. Bittner: Surf. Sci., in preparation (2002)
135. A. de Rille, A. Tadjeddine, W. Q. Zheng, A. Peremans: Chem. Phys. Lett. **271**, 95 (1997)
136. A. Tadjeddine, A. Peremans: in: R. J. H. Clark, R. E. Hester (Eds.): *Spectroscopy for Surface Science* (Wiley, New York 1998) p.159
137. C. D. Bain: J. Chem. Soc. Faraday Trans. **91**, 1281 (1995)
138. J. M. Lehn: Angew. Chem. **100**, 91 (1988)
139. J. M. Lehn: *Supramolecular Chemsistry* (VCH, Weinheim 1995)
140. E. Constable: Progr. Inorg. Chem. **42**, 67 (1994)
141. P. Steel: Coordn. Chem. Rev. **106**, 227 (1990)
142. E. B. Constable; Adv. Inorg. Chem. **34**, 1 (1989)
143. L. L. Merritt, E. Schroeder: Acta Crystallogr. **195**, 801 (1956)
144. J. W. Ernsley, J. G. Garnett, M. A. Long, L. Lunazzi, G. Spunka, C. A. Keraclini, A. Zanadel: J. Chem. Soc. Perkin II 853 (1979)
145. E. A. Mambetkaziev, A. M. Shaldybaeva, V. N. Statsyuk, S. I. Zhdanov: Elektrochimija **11**, 1750 (1975)
146. N. K. Akhmetov, R. I. Kaganovich, E. A. Mambetkaziev, B. B. Damaskin: Electrochimija **13**, 280 (1977); Elektrochimija **14**, 1761 (1978)
147. L. Pospisil, J. Kuta: J. Electroanal. Chem. **101**, 391 (1979)
148. M. Kim, K. Itoh: J. Electroanal. Chem. **188**, 137 (1985); J. Phys. Chem. **91**, 126 (1987)
149. D. Yang, D. Bizotto, J. Lipkowski, B. Pettinger, S. Mirwald: J. Phys. Chem. **98**, 7083 (1994)
150. J. Lipkowski, L. Stolberg: in: J. Lipkowski, P. N. Ross (Eds.) *Adsorption of Molecules at Metal Electrodes* (VCH, New York 1992) p. 171
151. J. E. Demuth, K. Christmann, P. Sanda: Chem. Phys. Lett. **76**, 201 (1980)
152. P. Avouris, J. E. Demuth: J. Chem. Phys. **75**, 4783 (1981)
153. U. W. Hamm, V. Lazarescu, D. M. Kolb: J. Chem. Soc. Faraday Trans. **92** 3785 (1996)
154. H. H. Perkampus, H. Koehler: Z. Elektrochem. **64** , 365 (1960)
155. S. A. Chaffins, J. Y. Gui, B. E. Khan, C. H. Liv, F. Lu, G. Salaita, D. A. Stern, D. C. Zapiev, A. T. Hubbard: Langmuir **6**, 957 (1990)
156. A. M. Funtikow, U. Linke, U. Stimming, R. Vogel: Surf. Sci. **324**, L343 (1995)
157. B. E. Conway, R. G. Barradas: Electrochim. Acta **5**, 319, 348 (1961)
158. D. Lampner, Th. Dretschkow, Th. Wandlowski; MI-Application Note, in preparation (2002)
159. J. Trost, J. Wintterlin, G. Ertl: Surf. Sci. **329**, 583 (1995)

160. J. A. Sondag-Huethorst, C. Schoenenberger, L. G. J. Fokkink : J. Phys. Chem. **98**, 6826 (1994)
161. G. Hahner, Ch. Wöll, M. Buck, M. Grunze: Langmuir **9**, 1955 (1993)
162. R. L. McCarly, D. J. Dunaway, R. J. Willicut: Langmuir **9**, 2775 (1993)
163. G. E. Poirier, M. Tarlov: Langmuir **10**, 2854 (1994)
164. G. E. Poirier: Chem. Rev. **97**, 1117 (1997)
165. G. E. Poirier: Langmuir **13**, 2019 (1997)
166. M. Hara, H. Sasabe, W. Knoll: Thin Solid Films **273**, 66 (1996)
167. D. D. Chambliss, R. J. Wilson, S. Chiang: J. Vac. Sci. Technol. B **9**, 933 (1991)
168. B. Voigtländer, G. Meyer, N. M. Armer: Surf. Sci. Lett. **255**, L529 (1991)
169. B. Voigtländer, G. Meyer, N. M. Armer: Phys. Rev. B **44**, 10354 (1991)
170. J. A. Stroscio, D. T. Pierce, R. A. Dragoset, P. N. First: J. Vac. Sci. Technol. A **10**, 1971 (1992)
171. F. A. Möller, O. M. Magnussen, R. J. Behm: Phys. Rev. Lett. **77**, 5249 (1996)
172. Th. Dretschkow: Diss. Univ. Ulm (1999)
173. A. Ulman, R. P. Scaringe: Langmuir **8**, 894 (1992)
174. Th. Dretschkow, Th.Wandlowski: unpublished results
175. M. H. Disher, J. C. Hemminger, F. J. Feher: Langmuir **13**, 2318 (1997)
176. E. Altman, R. J. Colton: Surf. Sci. **279**, 49 (1992)
177. A. W. Stephenson, C. J. Baddeley, M. S. Tihkov, R. M. Lambert: Surf. Sci. **398**, 172 (1998)
178. A. I. Altman, R. Colton: Surf. Sci. **304**, L400 (1994)
179. E. Budevski, G. Staikov, W. J. Lorenz: *Electrochemical Phase Formation and Growth* (VCH, Weinheim 1996)
180. M. J. Eddowness, H. A. O. Hill: J. Chem. Soc. Chem. Commun. 771 (1977)
181. F. A. Armstrong, H. A. O. Hill, N. J. Walton: Acc. Chem. Res. **21**, 407 (1988)
182. T. Sagara, K. Niwa, A. Stone, C. Hinnen, K. Niki: Langmuir **6**, 254 (1990)
183. T. Sagara, H. Murakami, S. Igarashi, H. Sato, K. Niki: Langmuir **7**, 3190 (1991)
184. A. Czerwinski, S. Zamponi, J. Sobkowski, R. Marassi: Electrochim. Acta **35**, 591 (1991)
185. T. M. Cotton, M. Varga: Chem. Phys. Lett. **106**, 491 (1984)
186. T. Lu, T. Coton, R. L. Birke, J. R. Lombardi: Langmuir **5**, 406 (1989)
187. Th.Wandlowski, Th.Dretschkow: unpublished results

Assembly of Au-Cluster Superstructures by Steering the Phase Transitions of Electrochemical Adsorbate Structures

Xinghua H. Xia[1], Lorraine C. Nagle[2], and Ralf Schuster[3]

[1] Debye Instituut afdeling Natuurkunde, Gecondenseerde Materie
Princetonplein 1, 3584 CC Utrecht, The Netherlands
X.Xia@phys.uu.nl

[2] Nanochemistry group, Chemistry Department, University College Dublin
Belfield, Dublin 4, Ireland
Lorraine.Nagle@ucd.ie

[3] Fritz-Haber-Institut der Max-Planck-Gesellschaft
Faradayweg 4–6, 14195 Berlin, Germany
schuster@FHI-Berlin.mpg.de

Abstract. Small Au islands of nm size are arranged in 2D piles or lines on a Au(111) surface by steering the phase transitions in an underpotentially deposited Cu adlayer, while the system is observed in situ by electrochemical scanning tunneling microscopy. The phase transitions between the (1×1) and $(\sqrt{3} \times \sqrt{3})$ Cu-adlayer structures proceed via nucleation and growth processes with progressing phase boundaries. The Au clusters, which were previously formed during the lifting of the $(\sqrt{3} \times 22)$ reconstruction of the underlying Au(111) surface, tend to cling to these phase boundaries during the phase formation process. Their arrangement on the surface is therefore determined by the morphology and kinetics of the phase transitions in the Cu UPD layer, which can be controlled by the starting electrochemical potential and the width of the potential step.

The 2D and 3D arrangement of nanometer particles has attracted a lot of attention, both theoretically and experimentally [1,2,3,4]. Potential applications of such aggregations of colloidal particles range from optoelectronic devices [5] to new materials with extraordinary mechanical properties [6]. The size and arrangement of the particles determines the electronic and mechanical characteristics. However, many of these studies pertain to extended 3D and 2D close packed structures or to approximately statistical ensembles of such particles. Only recently were successful attempts made to intentionally structure such close-packed structures of nanoparticles: 2D annular rings of close-packed nanoparticles were created during the evaporation of a thin wetting film of a suspension of thiol-passivated Ag clusters [7]. The arrangement of the particles represents a fingerprint of the process of evaporation of the solvent and the rupture of the film, whereas the local order is still determined by the particle interactions. Similar to this "wires" of functionalized Ag nanocrystals were found to form spontaneously at an air/water interface [8].

K. Wandelt, S. Thurgate (Eds.): Solid–Liquid Interfaces, Topics Appl. Phys. **85**, 323–331 (2003)

In this chapter we demonstrate how small Au islands can be arranged in densely packed 2D piles or lines on a Au(111) surface by controlling the phase transitions of the "solvent" in which they are immersed. This "solvent" is a 2D adsorbate film of copper and SO_4^{2-} ions on the Au(111) surface, whose phases are controlled by the electrochemical potential of the surface in a $CuSO_4$+H_2SO_4 solution. The Au islands or clusters are formed on this surface upon lifting of the $\sqrt{3} \times 22$ reconstruction of the Au(111) surface by suitably adjusting the electrochemical parameters. Upon the lifting of the reconstruction, Au atoms of the topmost surface layer are expelled onto the surface and agglomerate into small Au islands. These islands are partly mobile on the surface and tend to cling to the phase boundaries of the underpotentially deposited (UPD) Cu structures, which are formed on the surface, depending on the electrochemical potential. Hence, the arrangement of the Au clusters is determined by the morphology and kinetics of the formation of the Cu UPD structures, which can be controlled by the starting potential and the width of the potential changes initiating the phase transitions.

250 nm thick Au films evaporated on Cr covered (2 nm) glass were used as samples after multiple rinsing with triply destilled water and subsequent annealing with a butane burner. This procedure achieved large (111) oriented terraces of up to 100 nm diameter [9]. The experiments were conducted in a STM equipped with an electrochemical minicell open to air. The potential of the sample was potentiostatically controlled and was referenced to a Cu wire immersed in the $CuSO_4$ electrolyte. Details of the experimental setup can be found in [10]. The $CuSO_4$ was of 99.999 % purity, supplied by Aldrich. H_2SO_4 was of suprapure quality supplied by Merck. Triply destilled water was used throughout the experiment.

Figure 1 shows the cyclic voltammogram (CV) of the Au(111) surface in 1 M $CuSO_4$, which was recorded in the STM cell. It exhibits two prominent pairs of peaks characteristic of the formation of the different underpotentially deposited Cu phases. The right pair, at about 280 mV, stems from the formation/dissolution of the $\sqrt{3} \times \sqrt{3}$ Cu phase with 2/3 ML Cu coverage from a disordered lattice gas. The other pair at $\approx$ 120 mV is characteristic of the phase transition between a 1×1 and a $\sqrt{3} \times \sqrt{3}$ Cu structure. At potentials below 0 mV, Cu bulk deposition starts on the surface. The CV is in qualitative agreement with those published in the literature, although the peaks, characteristic of the $1 \times 1 \leftrightarrow \sqrt{3} \times \sqrt{3}$ transition, are usually larger than those of the $\sqrt{3} \times \sqrt{3}$ formation from a disordered lattice gas [11]. We attribute these slight differences to the high ion concentration employed here, the cell which is open to air, and the use of Au films which exhibit a lot more defects than single-crystal surfaces.

Figure 2 shows an STM image of a Au(111) terrace after the potential was lowered from the lattice gas region at $\phi = 400$ mV to the $\sqrt{3} \times \sqrt{3}$ Cu UPD region at 90 mV with a scan speed of about 50 mV/s. To avoid kinetic limitations and shielding effects of the tip during the structure formation,

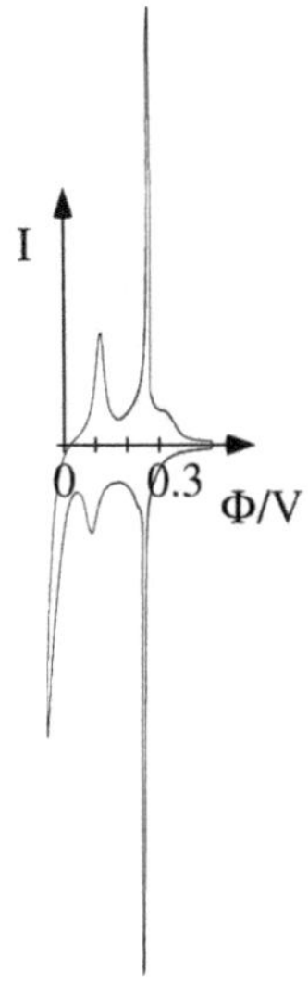

Fig. 1. Cyclic voltammogram of a Au(111) surface in 1 M $CuSO_4$. The stability ranges of the Cu-UPD phases are indicated

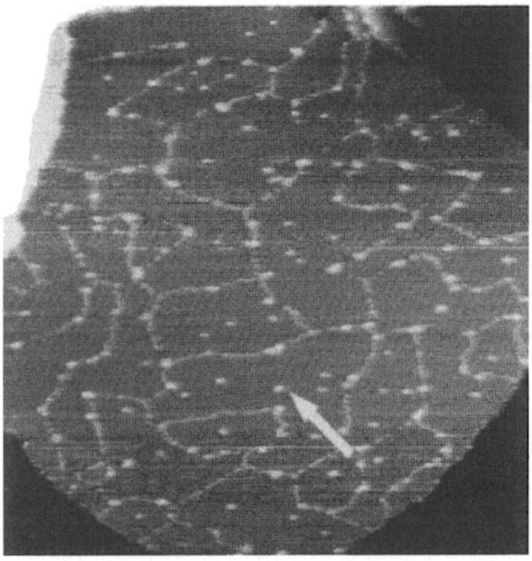

Fig. 2. STM image of the Au(111) surface at $\phi = 90\,mV$ in 1 M $CuSO_4$ + 0.5 M H_2SO_4, where it is covered with the $\sqrt{3} \times \sqrt{3}$ underpotentially deposited Cu adlayer. The *gray lines* represent the domain walls between the three translational domains of that structure. The *bright dots*, mostly lined up in the domain walls, are monoatomically high Au islands, stemming from the lifting of the substrate reconstruction. ($100\,nm \times 100\,nm$; $I_t = 1\,nA$; $U_t = 200\,mV_{Cu|Cu^{2+}}$)

we used a highly concentrated solution of 1 M $CuSO_4$ + 0.5 M H_2SO_4 as electrolyte [10]. In this solution, due to the presence of the additional SO_4^{2-} ions, the potential of the $1 \times 1 \leftrightarrow \sqrt{3} \times \sqrt{3}$ phase transition is negatively shifted by about 70 mV, relative to that in the pure 1 M $CuSO_4$ electrolyte of Fig. 1. Translational domains of the $\sqrt{3} \times \sqrt{3}$ structure (which are atomically resolved in high-resolution images not shown here) are separated by domain walls which are imaged as faint gray lines.

Mostly amongst these domain walls, bright white dots of nanometer dimensions are visible, which form relatively straight rows. As discussed in more detail in [12,13] these dots represent islands of Au atoms, which were expelled from the topmost Au layer upon lifting of the $\sqrt{3} \times 22$ reconstruction of the Au(111) surface. In the absence of Cu^{2+} the Au(111) surface is known to exhibit a $\sqrt{3} \times 22$ reconstruction at low enough potentials [14,15,16,17]. In this reconstruction the topmost surface layer of the Au(111) surface is uniaxially contracted [12,18]. This leads to areas with alternating fcc and hcp stacking, separated by dislocation lines. Domains of that reconstruction are arranged in a herringbone-like pattern with stripe widths of about 12 nm. Upon increasing the potential positive to about $350\,mV_{Cu|Cu}2+$ in perchloric acid, this reconstruction is lifted due to the adsorption of anions from the solution [17]. However, *Magnussen* found that in $0.1\,MH_2SO_4$ the lifting of the reconstruction starts at $\approx 260\,mV$, but it proceeds in a two-step mechanism and is completed only at potentials between $0.4\ V_{Cu|Cu}2+$ and $0.5\,V_{Cu|Cu}2+$ [12]. The surplus Au is expelled onto the surface, where it orders into small islands [12,16]. Lowering the potential below about $100\,mV_{Cu|Cu}2+$ leads to complete restoration of the reconstruction with somewhat degraded long-range order [12,16]. Although this hysteresis points to the kinetic limitations of the formation and dissolution, due to the large mass transport involved, both the lifting and reformation of the reconstruction take place within several tens of seconds at appropriate potentials. In SO_4^{2-} containing solutions the potentials for the formation and dissolution potentials of the $\sqrt{3} \times 22$ are slightly shifted towards more negative values due to the stronger specific adsorption of the SO_4^{2-} ions. Nevertheless, Magnussen found in $0.1\,MH_2SO_4$ that the rearrangement of the Au(111) surface (indicated by the dissolution of the Au islands) already started at about $350\,mV_{Cu|Cu}2+$, which is more positive than the formation of the $\sqrt{3} \times \sqrt{3}$ Cu UPD structure in Cu^{2+} containing electrolytes [12]. It is therefore conceivable that this reconstruction was at least partly formed during the potential transient into the $\sqrt{3} \times \sqrt{3}$ phase when passing the potential for the surface reconstruction. However, ordering into bigger domains, which would be easily visualized by STM, is anticipated by the formation of the $\sqrt{3} \times \sqrt{3}$ Cu UPD layer causing the instantaneous lifting of the reconstruction [19]. The expelled Au atoms then form small islands. Such a counterbalance between reconstruction of the surface and its lifting by the formation of the $\sqrt{3} \times \sqrt{3}$ structure also explains that the number of Au islands is dependent on the starting potential, the time spent in the respective phase region, the potential scan rate and the history of the sample, i.e. previous potential cycling.

But what causes the arrangement of the Au clusters along the straight domain walls? There are in principle three possibilities: (i) The islands are expelled at the position of the domain walls of the $\sqrt{3} \times \sqrt{3}$ Cu structure, after the Cu structure was formed. This is at once ruled out by the observation that in some cases lines of Au clusters do not coincide with the domain boundaries, as marked by the arrow in Fig. 2. (ii) Secondly, it is possible that the network of Au islands was already formed before the completion of the $\sqrt{3} \times \sqrt{3}$ structure and afterwards pins the domain walls upon the growth of the $\sqrt{3} \times \sqrt{3}$ Cu phase. However, upon reconstruction, the Au(111) surface is uniaxially contracted forming domains with fcc and hcp stacking of the atoms [18]. But the formation of a hexagonal domain pattern would require the existence of at least three domains to form the vertices where three domain walls meet. During very slow lifting of the reconstruction in 0.1 MH_2SO_4 the expelled Au atoms were found to gather in islands along the hcp stripes of the herringbone reconstruction [12]. However, as mentioned above, hcp domains cannot form a hexagonal pattern.

Therefore, everything points towards the third possibility (iii) that the Au islands are assembled upon formation of the Cu UPD structure. The $\sqrt{3} \times \sqrt{3}$ structure nucleates practically instantaneously upon exceeding the potential of its formation. *Hölzle* et al. found in an investigation of the kinetics of the Cu UPD phase formation on Au(111) that the $\sqrt{3} \times \sqrt{3}$ was formed within 0.1 s after the potential step [11]. In our previous STM investigations at high Cu^{2+} concentrations its formation was also too fast to be followed by STM [13]. However, the domain pattern was indicative of 2D nucleation and growth of $\sqrt{3} \times \sqrt{3}$ domains upon the potential change. This is in accordance with the observation that at a very low Cu^{2+} concentration of 10^{-6} M, where significant depletion of the bulk electrolyte takes place upon the $\sqrt{3} \times \sqrt{3}$ formation, islands of that phase become stable on the surface [12]. It is conceivable that the surplus Au atoms of the topmost surface layer are expelled during the very beginning of the $\sqrt{3} \times \sqrt{3}$ formation most probably at uncovered surface areas. These Au atoms cannot easily be embedded into the $\sqrt{3} \times \sqrt{3}$ phase and are therefore washed away by the proceeding phase border. This process stops when phase boundaries proceeding from different nuclei converge and form a domain wall.

At first glance this explanation seems somewhat speculative, but it was possible to follow the sweeping process of the Au islands in situ during a further phase transition of the $\sqrt{3} \times \sqrt{3}$ structure. As already mentioned above, the $\sqrt{3} \times \sqrt{3}$ Cu UPD structure undergoes a reversible transition into a 1×1 Cu phase at more negative potentials. This $\sqrt{3} \times \sqrt{3} \rightarrow 1 \times 1$ transition proceeds via heterogeneous nucleation at the domain walls, whereas the reverse $1 \times 1 \rightarrow \sqrt{3} \times \sqrt{3}$ transition can emanate either from homogeneous or heterogeneous nucleation [13]. The detailed behavior depends on the width of the potential change and the starting potential.

Figure 3a shows the Au(111) surface at a potential of 9 mV, where the surface is completely covered with the 1×1 Cu UPD layer. Since this structure was formed by lowering the potential from 400 mV, Au islands are homogeneously spread over the surface. Upon changing the potential to 40 mV,

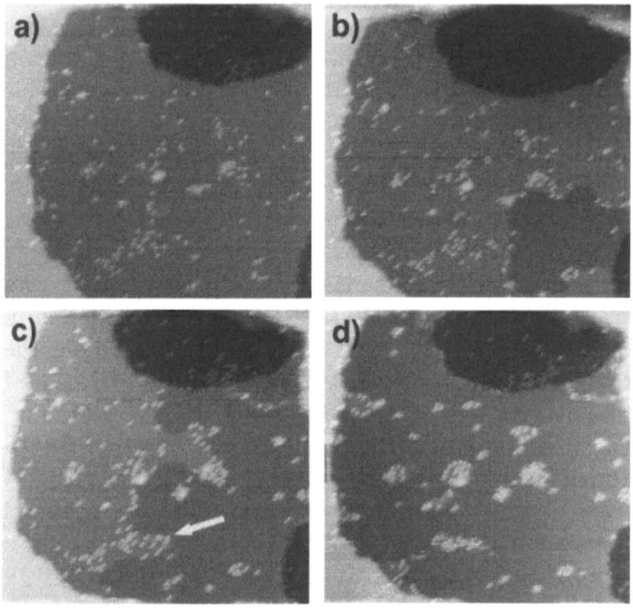

Fig. 3. (**a**) Small Au islands embedded in a 1×1 Cu UPD phase on a Au(111) surface in 1 M $CuSO_4$ + 0.5 M H_2SO_4 ($\phi = 9\,mV$). (**b**) 1 minute after increasing the potential to 40 mV the $\sqrt{3} \times \sqrt{3}$ structure nucleated at the step edge and shifts the Au particles in front of the phase boundary. (**c**) After 3 minutes part of the Au islands are piled up. (**d**) The $\sqrt{3} \times \sqrt{3}$ formation is completed after 9 minutes, leaving piles of Au clusters without tendency to coalesce. (60 nm × 50 nm, $I_t = 1\,nA$; $U_t = 200\,mV_{Cu|Cu^{2+}}$)

which is just inside the stability range of the $\sqrt{3} \times \sqrt{3}$ structure in the 1 M $CuSO_4$ + 0.5 M H_2SO_4 electrolyte, the new phase nucleates at the right step edge (Fig. 3b). The phase boundary is slowly proceeding over the terrace, dragging the Au islands with it. Small piles of clusters gradually form at the boundary; eventually the frictional force between the clusters and the surface becomes too high and they do not move along any further with the proceeding phase boundary. The phase boundary bends around these obstacles and eventually ruptures, leaving the piles in the $\sqrt{3} \times \sqrt{3}$ area (arrow in Fig. 3c). This process is reminiscent of the behavior of small particles of dust or droplets of oil on a water surface, where the surface tension or the high interface energy prevents the transition into the bulk water. Obviously the 1D "surface tension" of the $\sqrt{3} \times \sqrt{3}$ phase boundary prevents the Au clusters from being incorporated into the $\sqrt{3} \times \sqrt{3}$ phase. Hence, they are pushed over the surface similar to dust which is washed off by a water wave. More strictly speaking, the interface between the Au clusters and the $\sqrt{3} \times \sqrt{3}$ Cu UPD phase is energetically unfavorable and the length of the same is minimized by forming compact piles of Au clusters.

After the phase boundary has crossed the whole terrace, the clusters are either moved into small piles or they are aggregated at the step edges, where the phase boundary naturally stops. This piling up of clusters is reversible to a great extent. Figure 4 shows the surface after partial reformation of the 1×1 three minutes after lowering the potential to 9 mV. Inside the 1×1 area, which was mostly expanding from the terrace edges, the piles of Au islands (visible in Fig. 3) partly dissolved and the clusters spread amongst the 1×1 phase. This indicates perfect wetting behavior, i.e. the Au clusters maximize their interface length with the 1×1 phase. After completely reforming the 1×1 phase the Au clusters are more or less homogeneously redistributed over the surface. Reversing the transition back into the $\sqrt{3} \times \sqrt{3}$ structure again leads to piles of Au clusters.

If the potential step is large enough, the formation of the $\sqrt{3} \times \sqrt{3}$ phase from a 1×1 phase proceeds via homogeneous nucleation where several nuclei of the new phase are formed on one terrace. These nuclei grow independently until the proceeding phase boundaries intercept each other, similar to the formation of the domain wall pattern shown in Fig. 2. Employing this kinetics leads to lines of Au clusters on the surface, as shown in Fig. 5. This image was recorded after a potential step from 9 mV to 100 mV, which is well inside the $\sqrt{3} \times \sqrt{3}$ stability region. The domain walls of the $\sqrt{3} \times \sqrt{3}$ are visible as faint brighter lines on the terrace. Similar to Fig. 2 which was formed upon the homogenous nucleation of the $\sqrt{3} \times \sqrt{3}$ from a lattice gas, many of the Au islands are now arranged in rather straight domain walls. The quickly proceeding phase boundary leaves no time for the Au clusters to eventually form small immobile piles on the surface, which would lead to rupture of the boundary. Now the Au is dragged with the phase borders until their movement stops upon the formation of the domain wall. This directly supports the above tentative explanation for the lining up of the Au islands

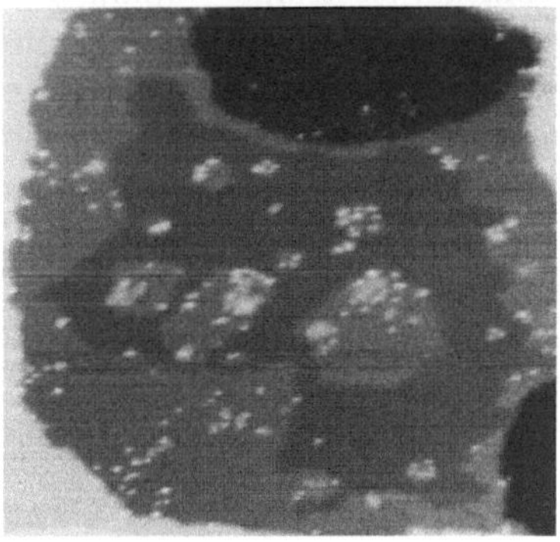

Fig. 4. The piling up of Au islands is partly reversible: upon formation of the 1×1 phase in the $\sqrt{3} \times \sqrt{3}$ structure the Au clusters increase their mutual distance and partly spread over the surface. (60 nm × 50 nm, $I_t = 1\,\mathrm{nA}$; $U_t = 200\,\mathrm{mV}_{\mathrm{Cu|Cu^{2+}}}$)

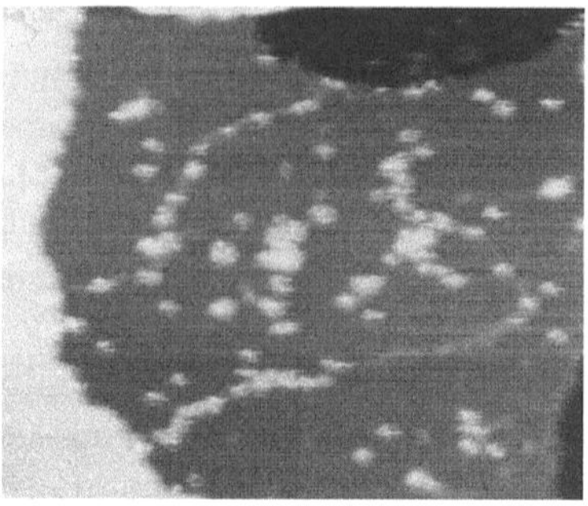

Fig. 5. A wide potential step from the stability region of the 1×1 phase into the $\sqrt{3} \times \sqrt{3}$ phase region led to domain walls in the $\sqrt{3} \times \sqrt{3}$ structure, where the Au islands are concentrated. ($60\,\mathrm{nm} \times 50\,\mathrm{nm}$, $I_t = 1\,\mathrm{nA}$; $U_t = 200\,\mathrm{mV}_{\mathrm{Cu}|\mathrm{Cu}^{2+}}$)

in the domain walls upon the formation of the $\sqrt{3} \times \sqrt{3}$ from the disordered lattice gas.

It is noteworthy at this point that in many experiments the Au clusters exhibited a tendency towards forming bigger islands. These islands are then in general immobile and are not influenced by the retracting phase boundaries. We do not know the particular reason that prevents the clusters in the experiment shown in Figs. 3–5 from coalescing. Their particular behavior might be partly attributed to contamination, which passivates the clusters. However, this does not alter the conclusion that the kinetics of the phase transitions influences the morphology and arrangement of nanoparticles on a surface. In fact, the passivation of clusters is commonly used to stabilize colloidal suspensions of such clusters [20]. In particular it was demonstrated that such organic molecules can be used as spacers between the particles arranged in a 2D structure. The width of the spacer is adjustable, e.g. by the 2D pressure in such films and determines the mutual electronic coupling and characteristics of the material [21,22]. Future experiments could concentrate on assemblies of such properly passivated clusters, deposited on the surface for example from a suspension of colloidal particles. As demonstrated in the present paper, the large-scale morphology of such "nanocrystalline materials" can than be steered by the kinetics of the structural phase transitions of a 2D surface layer in which these particles are embedded. Electrochemistry offers an ideal playground for achieving such control over the phases and their transitions, mainly due to the easy access of the main control parameter, the chemical potential of the respective system.

References

1. J. R. Heath: Science **270**, 1315 (1995)
2. C. B. Murray, C. R. Kagan, M. G. Bawendi: Science **270**, 1335 (1995)
3. W. D. Luedtke, U. Landman: J. Phys. Chem. **100**, 13323 (1996)

4. M. Antonietti, C. Göltner: Angew. Chem. Intern. Ed. **36**, 910 (1997)
5. L. V. Colvin, M. C. Schlamp, A. P. Allvisatos: Nature **370**, 354 (1994)
6. S. Förster, M. Antonietti: Adv. Mater. **10**, 195 (1998)
7. P. C. Ohara, J. R. Heath, W. M. Gelbart: Angew. Chem. Intern. Ed. **36**, 1078 (1997)
8. S.-W. Chung, G. Markovich, J. R. Heath: J. Phys. Chem. B **102**, 6685 (1998)
9. W. Haiss, D. Lackey, J. K. Sass, K. H. Besocke: J. Chem. Phys. **95**, 2193 (1991)
10. X. H. Xia, R. Schuster, V. Kirchner, G. Ertl: J. Electroanal. Chem. **461**, 102 (1999)
11. M. H. Hölzle, U. Retter, D. M. Kolb: J. Electroanal. Chem. **371**, 101 (1994)
12. O. M. Magnussen: Ph.D. thesis, Universität Ulm (1993)
13. X. H. Xia, L. Nagle, R. Schuster, G. Ertl, O. Magnussen, R. J. Behm: Phys. Chem. Chem. Phys. **2**, 4387 (2000)
14. D. M. Kolb, J. Schneider: Electrochim. Acta **31**, 929 (1986)
15. B. Pettinger, J. Lipowski, S. Mirwald: Electrochim. Acta **40**, 133 (1995)
16. D. M. Kolb: Prog. Surf. Sci. **51**, 109 (1996)
17. F. Silva, A. Martins: Electrochim. Acta **44**, 919 (1998)
18. J. V. Barth, H. Brune, G. Ertl, R. J. Behm: Phys. Rev. B **42**, 9307 (1990)
19. Y. Nakai, M. S. Zei, D. M. Kolb, G. Lehmpfuhl: Ber. Bunsenges. Phys. Chem. **88**, 340 (1984)
20. L. Bronstein, M. Antonietti, P. Valetsky: In J. H. Fendler (Ed.): *Nanoparticles and Nanostructured Films* (Wiley-VCH, Weinheim 1998)
21. C. P. Collier, R. J. Saykally, J. J. Shiang, S. E. Henrichs, J. R. Heath: Science **277**, 1978 (1997)
22. G. Markovich, C. P. Collier, J. R. Heath: Phys. Rev. Lett. **80**, 3807 (1998)

Part IV

Interface Reactions

Imaging Localized Reactivities of Surfaces by Scanning Electrochemical Microscopy

Gunther Wittstock

Department of Chemistry and Institute of Chemistry and Biology of the Marine Environment, Carl von Ossietzky Universität Oldenburg
PF 2503, 26111 Oldenburg, Germany
gunter.wittstock@uni-oldenburg.de

Abstract. Scanning ElectroChemical Microscopy (SECM) is a scanning-probe technique that relies on the measurement of faradaic currents at a microdisk electrode surrounded by an insulating shielding. Typical electrode radii are in the micrometer range. The microelectrode is imersed in an electrolyte solution and scanned in a distance of some electrode radii over the specimen surface. The microelectrode currents orginate from the electrochemical conversion of dissolved species at the microdisk electrode. As the species detected at the microelectrode are formed at the sample surface, the currents at the local probe provide information about the local reactivity of the specimen region directly beneath the microelectrode. The technique can be operated in the generation-collection mode and the feedback mode. Beside mapping local heterogenous reactivities the setup has also been used to induce local surface modifications by local reagent generation at the microelectrode, local pertubation of an equilibrium at the sample surface, or exploitation of the distribution of the electric field strength in the interelectrode space. The explanation of the working principles is illustrated by representative applications in corrosion research, in the study of biochemically active samples, and mass transport phenomena through pores. A concluding section on experimental requirements leads to an outview of possible future developments.

1 Introduction

In the early 1980s UltraMicroElectrodes (UME) gained popularity among electrochemists because of the new possibilities offered by micrometer-sized electrodes for the study of fast electrochemical reactions or for the detection of substances in restricted space [1], e.g. the detection of neurotransmitters in living organisms [2]. In 1986 *Engstrom* et al. [3] expanded this concept by using a UME to measure local concentrations of reactants and products converted or generated at a macroscopic specimen electrode – the first experiment with scanning electrochemical microscopy. At about the same time the group of *Bard* reported experiments with a ElectroChemical Scanning Tunneling Microscope (ECSTM) in which currents were observed at unusual large sample–tip distances. The signals were considered to be faradaic currents at the tip modulated by the presence of the sample [4]. By 1989 *Bard* et al. had worked out a quantitative description of the faradaic current at a microdisk

K. Wandelt, S. Thurgate (Eds.): Solid–Liquid Interfaces, Topics Appl. Phys. **85**, 335–364 (2003)

electrode as a function of microelectrode-specimen distance and the chemical nature of the specimen for the simplest cases [5,6]. The term Scanning Electrochemical Microscopy (SECM) was coined to name a new scanning probe technique that relied entirely on the measurement of faradaic currents. As with other scanning probe techniques, the acronym is used for both the method and the instrument. Very soon it became clear that SECM is not just suitable to measure local solute concentrations but also, and more importantly, represents a tool to map local (electro)chemical reactivities, to induce localized electrochemical surface modifications, or to investigate heterogeneous and homogeneous kinetics.

Besides the aforementioned groups, several researchers have contributed to the refinement of the technique and have demonstrated its applicability in such diverse areas ranging from semiconductor characterization to biomedical research. The theoretical description and the accumulated experience has been documented in a number of reviews which appeared between 1989 and 1999 [7,8,9,10,11,12,13,14,15,16,17,18,19,20,21]. As a SECM instrument is finally commercially available[1], it is expected that the possibilities of the new method will be used increasingly often for problem-solving in industry and by academic research groups that do not consider fundamental electrochemistry as their scientific center of gravity. This review aims to provide both an explanation of the operation principles of SECM as well as commented examples on how this method may contribute to an improved understanding of macroscopic phenomena by allowing a microscopic view on reactivities at solid–liquid interfaces. Mathematical artwork is kept to a minimum but references are provided that will guide the interested reader to the original literature on the theoretical foundation and quantitative description of the described effects.

2 Working Principle

The term "ultramicroelectrode" refers to electrodes with at least one dimension being smaller than the diffusion distance for dissolved molecules within the time scale of the experiment. This leads to new response characteristics of UMEs compared to conventional electrodes (Fig. 1).

2.1 Properties of Amperometric Ultramicroelectrodes in Bulk Solution

For voltammetric experiments performed in a time scale of seconds, UME characteristics will prevail when at least one dimension of the electrode is in the micrometer range. Although such electrodes are still large compared to electrodes used by electrophysiologists, the term "ultramicroelectrode" has

[1] CH Instruments, Inc., 3700 Tennison Hill Drive, Austin, TX 78733, USA

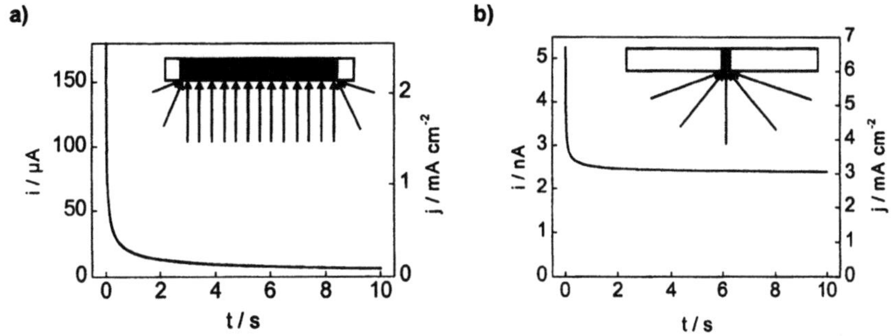

Fig. 1. Chronoamperometric response of Pt disk electrodes in an aqueous solution of 2 mM ferrocene methanol +100 mM Na_2SO_4. Electrode diameters 3 mm (**a**), and 10 µm (**b**). The insets symbolize the diffusional flux at the end of the experiment. *Left scale* shows current, *right scale* shows current density

found wide acceptance probably because "macroelectrode" is appropriate for electrodes used for large-scale bulk electrolysis and "microelectrodes" had already been established for analytical electrodes with the smallest dimension as large as about one centimeter.

UMEs with disk geometry are most important for SECM and the discussion will be restricted to this geometry. Important deviations like conical UMEs are treated briefly in Sect. 4.2. Among the special features of UMEs [22,23,24], one is of particular importance for SECM: Under diffusion-controlled conditions hemispherical diffusion causes strongly non-linear concentration profiles with very high diffusional flux in the vicinity of the electrode. This limits the expansion of the diffusion layer into the solution with important consequences:

A steady-state current is established and maintained even under prolonged electrolysis.

The steady-state current is relatively immune against forced convection in solution because forced convection can add very little to the already very high diffusional flux at the electrode surface.

The steady-state current at a disk-shaped UME in a plane with an infinite large insulator in the solution bulk was derived by *Saito* [25] and is

$$i_{T\infty} = 4nFDc^* r_T \,, \tag{1}$$

where i is the current, n is the number of electrons transferred per molecule, F is the Faraday constant, D the diffusion coefficient, c^* is the concentration of the redox-active molecule in the solution bulk, and r_T is the radius of the microelectrode. In a SECM configuration the UME represents the local probe and in analogy to other scanning probe techniques it is frequently called the "tip" from which the subscript "$_T$" was taken to specify quantities of the UME. Correspondingly, the subscript "$_S$" is used for quantities of the specimen electrode (synonyms are sample, substrate). Please note that

Table 1. Times to reach a steady state for UMEs with radius r_T after a potential step large enough to drive the local concentration of the redox-active species at the electrode surface to zero; $D = 5 \times 10^{-6}\,\mathrm{cm^2\,s^{-1}}$

$r_T/\,\mu\mathrm{m}$	0.25	0.5	2.5	5	12.5	25
$t_{ss}/\,\mathrm{s}$	0.0024	0.010	0.245	0.98	6.12	24.5

the current at a UME is proportional to the radius, in contrast to larger electrodes where it is proportional to the electrode area. As a rule of thumb one may remember that a 2 mM solution of a redox-active species ($D = 5 \times 10^{-6}\,\mathrm{cm^2\,s^{-1}}$) exchanging one electron per molecule with the electrode will yield a current of 2 nA at a UME of 10 µm diameter.

The time t_{ss} necessary to establish a steady state after a potential step sufficiently large to drive the diffusion-controlled electrochemical reaction of a dissolved species can be approximated by comparing the time-dependent current described by the Cottrell equation ($i(t) = nFAc^* D^{1/2} \pi^{-1/2} t^{-1/2}$) with the values obtained from (1). Table 1 lists the times after which the Cottrellian contribution is less than 10 % of the steady state current of (1).

From these data it becomes clear that smaller UMEs are not just advantageous for obtaining a higher lateral resolution but also allow higher mechanical translation rates and hence faster imaging.

2.2 Basic Setup

In order to perform a spatially resolved measurement, the UME and the specimen surface are mounted in a setup that enables a relative movement of the UME versus the specimen in three orthogonal directions (Fig. 2). The UME is connected as the working electrode to a potentiostat. A reference electrode and an auxiliary electrode complete the electrochemical cell. In some situations the specimen itself is connected as a second working electrode to a bipotentiostat. Movement and data aquisition are controlled by a PC equipped with interfacing boards. A more detailed consideration on desirable instrumental characteristics is given in Sect. 4.1.2. For obtaining an image the UME is scanned horizontally across the sample (xy-plane). Current values are recorded at cross-sections of an orthogonal grid and transformed to a 3D representation as usual in other scanning probe techniques. Another basic SECM experiment consists in the observation of i_T when the UME approaches the specimen in the vertical (z) direction. Such a measurement is called an approach curve.

2.3 Feedback Mode

2.3.1 Qualitative Understanding

For measurements in the feedback mode the working solution contains one redox form of a quasi-reversible redox couple (Red $=$ Ox $+\, ne^-$). For this

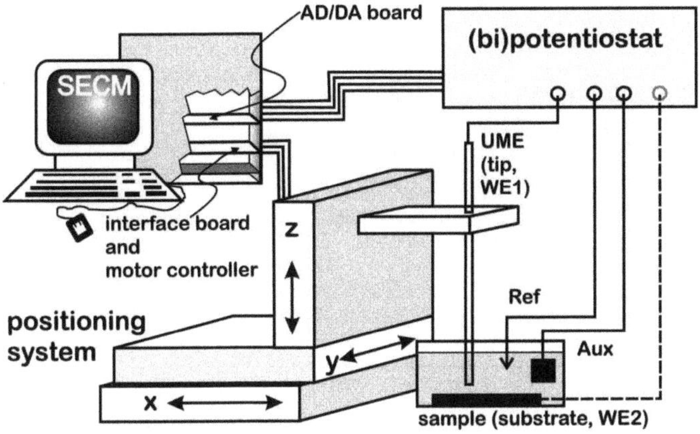

Fig. 2. Schematic setup of a basic SECM apparatus. The UME, a reference electrode (Ref) and an auxiliary electrode (Aux) are connected to a potentiostat. If needed, the sample might be connected as a second working electrode to a bipotentiostat

discussion it is assumed that initially only the reduced form Red is present. This compound serves as electron mediator and is added typically in millimolar concentrations to an excess of an inert electrolyte. Of course, analogous experiments may be carried out if the oxidized form Ox is provided. The UME is poised at a potential sufficiently large to cause the diffusion controlled oxidation of Red. Far away from the sample a steady-state current $i_{T\infty}$ is measured as described in Sect. 2.1. The subscript "$_\infty$" indicates that the UME has a quasi-infinite distance to the surface. Practically $i_{T\infty}$ is observed if the UME–sample distance d is larger than 20 times the UME radius r_T. Figure 3 shows how the steady-state current i_T is modulated if the UME approaches a specimen surface. For a unified description the UME current is normalized to the $i_{T\infty}$ and the UME–sample distance d is normalized to r_T.

If the specimen is an inert, insulating material, it simply blocks the access of Red to the microelectrode and i_T falls below $i_{T\infty}$. In order to reach the active area of the UME, Red has to diffuse though the gap formed by the specimen surface and the insulating shielding of the UME. The thicker the shielding the higher will be the mass transfer resistance. Consequently, i_T is decreased for any given d if electrodes with thicker shielding are used. While many schematics in this contribution as well as elsewhere highlight the different reactions occurring at the UME and at the sample, Fig. 4 gives a realistic picture of the relative size of the active electrode area, the UME–sample gap and the insulating shielding. The shape of the UME is described by the RG ratio, i.e. the ratio of the radius of the insulating shielding r_glass to the radius of the active electrode area r_T.

If the UME approaches a conducting surface, an electrochemical recycling process can occur. While Red undergoes electrooxidation to Ox at the UME,

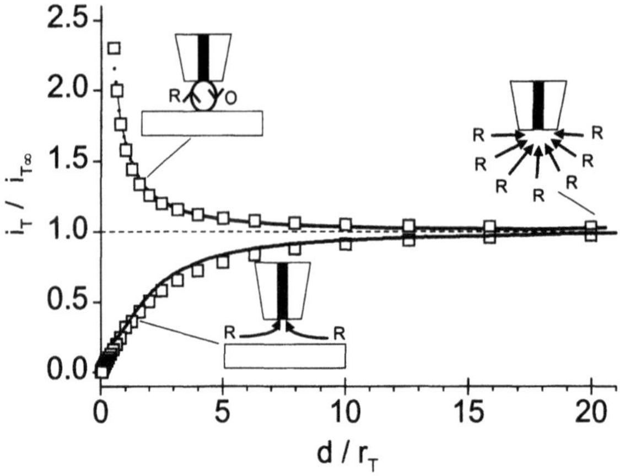

Fig. 3. Normalized current as a function of normalized UME–sample distance and schematic representation of the diffusion at the UME; (◇) theoretical values from [6], (•) experimental points

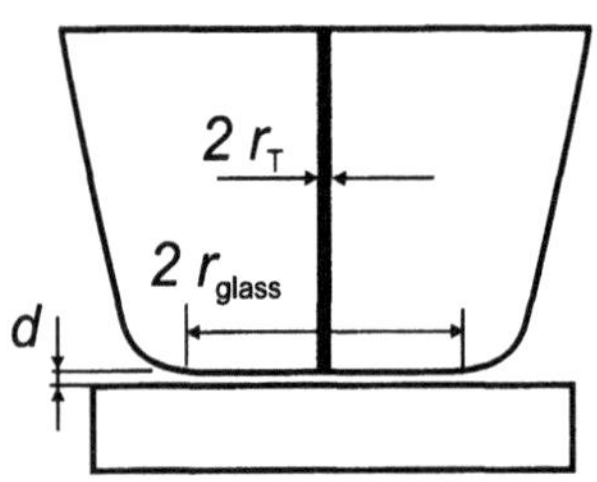

Fig. 4. Geometrical arrangement for a conventional UME ($RG = r_{\text{glass}}/r_{\text{T}} = 10$) in a working distance d above a planar sample

Ox diffuses to the sample and will be reduced back to Red. Thus the sample represents an additional source of Red for the reaction at the UME. As d is decreased the mass transport between UME and the sample becomes faster and i_{T} exceeds $i_{T\infty}$. The term "positive feedback" was coined for the communication between UME and sample by a diffusing redox mediator [5]. Correspondingly, the term "negative feedback" gained general popularity to describe the decrease of i_{T} caused by an insulating sample surface [8]. In the context of SECM the term "feedback" does *not* refer to an electronic circuit and actuator maintaining a constant probe–sample interaction as is common in other scanning probe microscopies. In positive-feedback experiments the size of the insulating shielding does not significantly influence i_{T} if $RG > 10$. Only at very small RG does the total UME geometry become important [26].

The mediator regeneration at the conducting sample can be caused by connection to a potentiostat. However, in many circumstances an external contact is neither possible nor necessary (Fig. 5). In this case the electrochemical open circuit potential (OCP) of the macroscopic substrate is de-

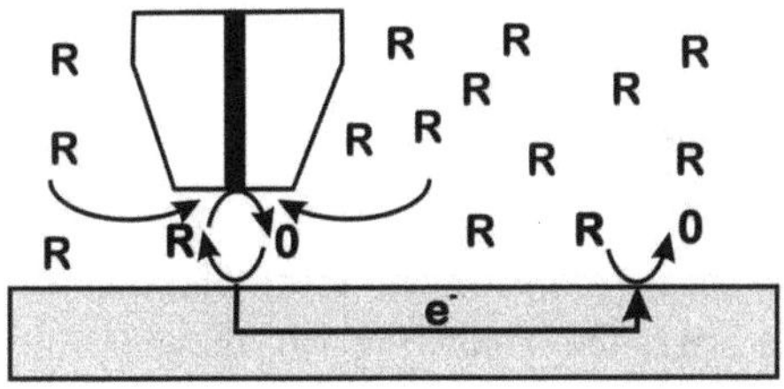

Fig. 5. Positive feedback at a sample which is at open circuit potential

termined by the activity ratio of the oxidized and reduced mediator form at the macroscopic sample and can be calculated from the known formal potential $E^{0'}$ of the mediator according to the Nernst equation $E_S = E^{0'} + (RT/nF)\,ln(c_{Ox}/c_{Red})$. The electrochemical conversion at the UME causes a locally enhanced concentration of Ox directly underneath the UME while everywhere else in the electrochemical cell Red is present. Therefore the OCP of the sample will be negative with respect to the formal potential and causes the reduction of Ox underneath the UME under nearly diffusion-controlled conditions. An equivalent amount of Red is oxidized at the sample surface far away from the UME. The net effect corresponds to an accelerated dilution: Instead of diffusion of Ox into the bulk solution electrons are transported in the conducting sample. The OCP of the sample can be used to cause a positive feedback only if the imaged region is much larger than the UME or if it is electronically connected to a larger conducting region in contact with the mediator solution. The situation is completely analogous to a concentration cell which is employed among others in microstructured electrochemical detectors [27].

2.3.2 Quantitative Treatment

As early as 1989 *Kwak* and *Bard* obtained a quantitative description of the feedback effect by digital simulation [6] resulting in tabulated data of the normalized current $I = i_T/i_{T\infty}$ as a function of $L = d/r_T$ for (i) infinite large conducting or infinite large insulating samples, (ii) diffusion-controlled electrochemical conversions at UME and sample, and (iii) *RG* ratios of 1000, 100, and 10. Besides the equality of the diffusion coefficients of Ox and Red, it was assumed that the local concentration at the edge of the insulating shielding is equal to the bulk concentration of Red. Analytical approximations [28] have the following mathematical structure for insulating samples

$$\frac{i_T}{i_{T\infty}} = \frac{1}{k_{ins,1} + \frac{k_{ins,2}}{L} + k_{ins,3}\exp\left(\frac{k_{ins,4}}{L}\right)} \qquad (2)$$

and for conducting samples

$$\frac{i_T}{i_{T\infty}} = k_{\mathrm{con},1} + \frac{k_{\mathrm{con},2}}{L} + k_{\mathrm{con},3}\exp\left(\frac{k_{\mathrm{con},4}}{L}\right). \quad (3)$$

These expressions have been widely used to obtain the UME–sample distance d from experimental i_T values with the following set of constants derived for $RG = 10$: $\mathrm{k_{ins,1}} = 0.292$, $\mathrm{k_{ins,2}} = 1.151$, $\mathrm{k_{ins,3}} = 0.6553$, $\mathrm{k_{ins,4}} = -2.4035$, $\mathrm{k_{con,1}} = 0.68$, $\mathrm{k_{con,2}} = 0.78377$, $\mathrm{k_{con,3}} = 0.3315$, $\mathrm{k_{con,4}} = -1.0672$ [10]. The analytical approximation has been derived from data that used the assumption $i_{T\infty} = 4nFDc^*r_\mathrm{T}$ ((1), steady-state current of microdisk UME in infinite insulating plane). This is a reasonable assumption only for $RG \geq 10$. For smaller RG the current at the UME far away from any surface is considerably larger then predicted by (1) because diffusion of Red from behind the plane of the UME contributes to the mass transport [29]. Because pipette type microelectrodes and fiber electrodes surrounded by thin insulating layers enjoy increasing popularity, several groups have published new simulations recently that avoid setting the local concentration of Red at the edge of the insulating shield equal to c^* [26,30]. The paper by *Amphlett* and *Denuault* [26] provides sets of numeric constants that, substituted into (2) and (3), yield analytical approximations with a fine graduation for RG values between 1.11 and 1002. The smooth change of the fitting parameters suggest that approach curves for UME geometries not listed in Table 1 of [26] can be interpolated from the bracketing constants. As these function also use the $i_{T\infty}$ value of a given UME geometry for normalizing the results (instead of $4nFDc^*r_\mathrm{T}$), they will be a handy tool for interpreting approach curves for UME with small RG. Please note that by substituting the experimental $i_{T\infty}$ neither the exact concentration of the mediator nor the exact diffusion coefficient need to be known. Very recently, *Martin* and *Unwin* [31] showed that unequal diffusion coefficients of Ox and Red will not change the steady state response in feedback mode used in imaging but have an influence on the transient response.

Besides the results for the limiting cases characterized by assumptions (i)–(iii) digital simulations for the most important exceptions are available [32]. A set of approach curves can be simulated if a conducting sample is not of infinite size but has radius r_S. Deviations from the curve for an infinite large sample are found if

$$r_\mathrm{S} < r_\mathrm{T} + 1.5d, \quad (4)$$

where r_S is the radius of the conducting region of the sample [32]. On the other hand it means that at small d only a region comparable in size to the UME contributes to the feedback effect facilitating the investigation of local kinetics of a macroscopic surface.

If the electrochemical reaction at the sample is not diffusion-controlled but limited by the heterogeneous electron transfer rate at the sample, a separate

I–L curve is found for each rate constant k_S of the substrate [32]. Analytical approximations for approach curves with finite kinetics at the sample are available [33]. At constant d this forms the basis of reaction rate imaging [34]. One may rationalize the effect by comparison of the SECM setup with a ferry connection across a channel (the UME–sample gap of constant width). If one of the ports is operated very fast (the reaction at the UME) and the number and speed of the ships is fixed (c and D of the mediator), the transport (i_T) depends exclusively on the loading and unloading rate of the second port (the reaction at the sample). This unique feature of SECM and the established theory that allows one to calculate i_T are mainly responsible for the popularity of SECM and compensates in many cases for the much lower lateral resolution that is usually achieved when compared to ECSTM or electrochemical scanning force microscopy (ECAFM). Many special scenarios in SECM feedback have been analyzed by digital simulation and verified on model systems. A number of reviews provides a systemic overview [10,19,20,21].

2.4 Generation–Collection Mode

At first it seems to be straightforward to use the UME to map the local concentration of species generated at the specimen electrode. Specifically such a configuration is called substrate-generation/tip-collection (SG/TC) mode (Fig. 6a).

However, there are several problems associated with the interpretation of the results: (i) if the active regions of the sample are large, no steady state situation will be reached and the local concentrations depend on the time that passed after the onset of the reaction at the sample. (ii) The diffusion layer of the specimen is disturbed by the presence of the probe (stirring, blocking

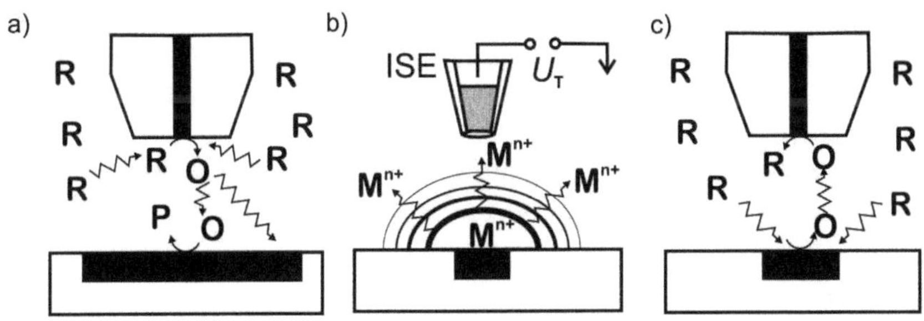

Fig. 6. SECM Generation/collection (GC) experiments: (**a**) substrate-generation/tip-collection (SG/TC) experiments, (**b**) substrate-generation and monitoring with a passive sensor (e.g. an ion-selective microelectrode, ISE), (**c**) tip-generation/substrate-collection (TG/SC) experiments: recording the sample current i_S as a function of the UME position. In contrast to the feedback mode, the reaction at the sample may lead to a product P that is not identical to Red

of reactant diffusion to the sample, overlap of the diffusion layers of the UME and the sample in case of amperometric UMEs). (iii) At the SG/TC mode there might be a current enhancement due to feedback effect if the reaction at the microelectrode is a reversible reaction and the UME–sample distance is less than about 5 r_T. These circumstances have delayed a rigorous theoretical treatment of the SG/TC mode although the development of a diffusion layer in space and time above a macroscopic specimen electrode and its influence on i_T have been approximated as early as in 1987 [35]. Furthermore, the lateral resolution of GC experiments is always inferior compared to corresponding feedback experiments. Together with the lack of a quantitative theory, rather pessimistic predictions were derived on the applicability of SG/TC experiments [10]. Meanwhile the transient currents in the SG/TC mode have been modeled and used to determine the diffusion coefficient ratio of redox couples [36] and the rate constant of a homogeneous reaction following electron transfer at the sample electrode [37].

Recently, SG/TC experiments have been reported in situations like mass transport though membranes, pores and skin, corrosion, and immobilized enzymes and antibodies. In these situations feedback experiments were either not possible or not sensitive enough (see Table 2, Sect. 3 for a detailed discussion of examples).

GC experiments can also be performed with potentiometric probes [38,39,40] (Fig. 6b). A theoretical treatment [39] of this situation is easier than in the case of the amperometric probes because a potentiometric probe is a passive sensor, i.e. a probe that does not change the local concentration of the species to be detected. It therefore does not have an own diffusion layer, and feedback effects cannot occur.

Alternatively, one can use the UME as a microscopic generator of redox-active substances that are collected by the macroscopic substrate (tip-generation/substrate-collection (TG/SC) mode, Fig. 6c). This arrangement found application as an electrochemical time-of-flight apparatus for measurements of fast homogeneous reactions occurring in the UME–sample gap [41,42]. The use of the TG/SC mode for imaging purposes has been demonstrated [43] but did not find wider application probably because the same information is obtained more easily by feedback experiments. Generation of reactive substances, e.g. Br_2, OH•, for local surface modification have been reported [17,44].

2.5 Direct Mode

The "direct mode" describes a situation where the UME and a large conducting sample form an electrochemical cell. Both electrodes are connected by an ionic conductor, typically an ion-conducting polymer. The current flow is maintained by faradaic reactions at the UME and the sample. In contrast to the GC mode or the feedback mode, the reactants are not diffusing between tip and substrate. Therefore, entirely different reactants may be consumed or generated at both electrodes. This may be used for etching and deposition processes at the sample surface or in the vicinity of the tip. The electrical field between UME and sample restricts the modification to the sample region directly underneath the UME. The classical example is the deposition of a microscopic Ag line inside a Ag^+-loaded Nafion film [45]. The experiment is carried out in air. The tip of an STM apparatus penetrates the ion-conducting film and causes the reduction $Ag^+ + e^- \rightarrow Ag$ while the tip is dragged slowly inside the film. An electronic feedback system maintains a constant current via adjustment of the penetration depth. Simultaneously, a local etching of silver occurs at the specimen surface (Fig. 7a). Other examples are the formation of fluorescence pattern [46], the deposition of conducting polymer structures in two and three dimensions [47,48,49] and the localized desorption of self-assembled monolayers (SAM) of Au alkanethiolates [50,51] (Fig. 7b). Besides those reports, several examples of nanostructuring with STM instruments in humid atmosphere are likely to be causesd by a direct mode configuration with adsorbed water serving as the ionic conductor (see for instance [52] and references cited therein). Imaging in the direct mode is feasible [53] but did not attain much popularity mainly because the sample is changed during the measurement process.

3 Application

The arrangement of a UME facing an a sample surface can be used to perform a microscopic analysis or to induce local surface modification. A more detailed systematic according to the applied working principle is attempted in Table 2. A particularly attractive feature of SECM is the possibility to induce surface reactions and check sequentially the result or monitor simultaneously the reaction rate. The latter principle is applied in a number of variations to analyze the kinetics of heterogeneous processes at solid/liquid, liquid/liquid and liquid/gas interfaces. Used as a kinetic method without x, y-positioning, SECM allows fast heterogeneous and homogenous chemical reactions to be investigated under *steady state conditions.* This avoid artifacts by finite time constants of electronic equipment, charging currents and iR drop often encountered in electrochemical relaxation experiments. A number of very elegant experimental procedures have been presented and were summarized in recent reviews [10,20,21]. The application part of this contribution focuses on the investigation of localized reactions at surfaces of technologically important materials and biochemically modified surfaces as they are used for instance in biosensorics.

3.1 Imaging of Composite Materials

From the inception of SECM, composite materials have represented a popular type of sample, because the method is uniquely suited for distinguishing

Table 2. Examples for application of scanning electrochemical microscopy

Microscopic analysis	Local surface modification
Reaction rate imaging in the feedback mode on composite samples [54,55], electrodes covered with nonconducting overlayers (passivating oxides, polymer layers, organic monolayers) [50,56,57], penetration of the UME in a loaded polymer or gel [58]	Direct mode: deposition of metals and polymers [45,46,47,48,49], etching of metals [45], localized desorption of self-assembled monolayers (SAM) [50,51]
Imaging of local generation or consumption of solution species enzyme modified surfaces in the potentiometric or amperometric GC mode [59,60,61,62], release of substances from cells and whole plants after chemical or optical stimulus [63,64], measurement of localized corrosion [65,66,67,68,69], counterion uptake and ejection during redox reactions at conducting polymers [38,70]	Local generation of reagents for surface modification (SG/TC mode) oxidizer for etching of semiconductor and metals, destruction of enzyme layers and monolayers, polymerizations [44,87,88,89,90,91], reducers for metal deposition [92], generation of metal ions at a sacrifical UME followed by deposition on the sample [93,94], initiation of pitting corrosion [67,95]
Imaging of localized mass transport in the GC mode iontophoretic transport through skin [71,72], hydrodynamic flow and diffusion though dentine [73,74] Topographic imaging in negative feedback imaging of DNA strands by a metal tip immersed in a surface water film [75] (See [76,77,78] for discussion of the mechanism)	Local perturbation of an equilibrium pH shift by electrolysis at the UME to enable a deposition process at the surface [96], reversible modulation of enzyme activity by pH shift [97], consumption of an inhibitor for a deposition process at the sample by UME-generated substances [91]
Kinetic analysis by approach curves and digital simulation (no xy movement) electrochemical time of flight setup for analysis of fast chemical reaction following electron transfer [41], exploitation of the high mass transfer to study fast electrochemical reactions at the tip [79], mass and charge transfer across liquid/liquid interfaces [80,81,82,83], mass transfer across liquid/gas interfaces [84], relaxation measurements after local perturbation of dissolution equilibria and adsorption equilibria [85,86]	

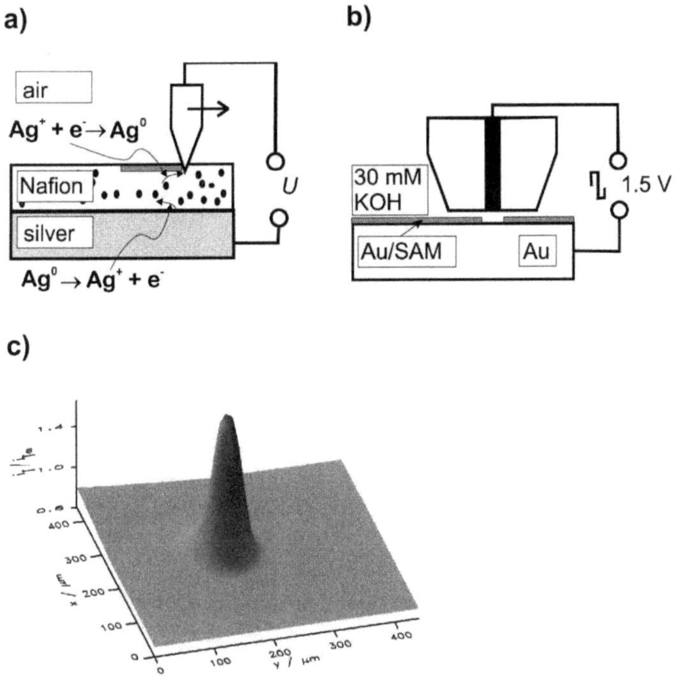

Fig. 7. Direct mode of the SECM. (**a**) Schematic representation of metal deposition inside ion-conducting films as done in [45]. Schematic representation of localized monolayer desorption, (**c**) SECM feedback image of the blank Au area (high currents) created within an alkanethiolate-covered surface (low currents)

the different reactivity of materials. Quite often the image contrast is obtained by combining microstructures of insulating and conducting materials on one surface. Such samples prepared by thin film technology continue to be very valuable for checking the performance of SECM instruments and to demonstrate the working principle [98]. Beside being of methodic interest, technologically important question can be addressed. Figure 8 shows images of a UME array, i.e. an assembly of individual UMEs. Such arrays are tested for application in hand-held sensors where the steady state response characteristics of UME are desired and several UME are operated in parallel to enhance the signal [55]. One way of manufacturing such arrangement is the use of laser ablation of an insulating material covering the conducting material. Conventional scanning electron microscopy (SEM) of such samples requires the sample to be coated with a gold or carbon film thereby irreversibly destroying the sample. Furthermore, the inspection of the SEM image (Fig. 8a) does not allow a conclusion of whether the ablation process has exposed the conducting carbon layer. In contrast, SECM images reveal

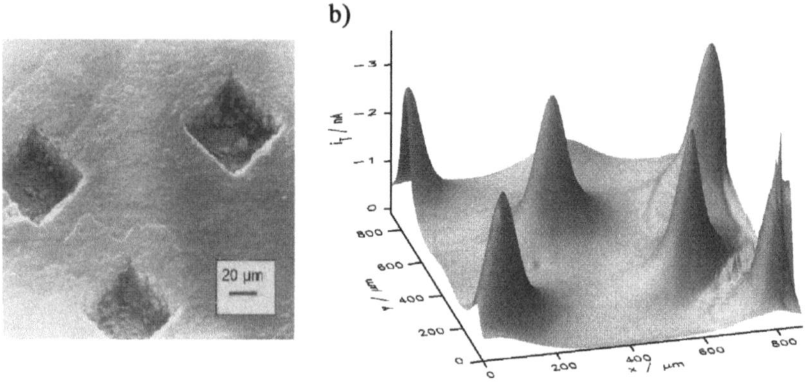

Fig. 8. UME array produced by eximer laser treatment of a 22 µm thick insulating layer to expose a buried graphite paste layer. (**a**) Scanning electron micrograph, (**b**) GC mode SECM image mediator 2 mM ferrocene monocarboxylic acid, $E_S = +400\,\text{mV}$, $E_T = -100\,\text{mV}$. Reprinted from [55] (Copyright Wiley-VCH 1998)

the sample region at which the electron transfer reaction takes place by the high currents (Fig. 8b).

The different components in blends of conducting and insulating polymers have been imaged with the feedback mode [57]. Of potential practical importance is the study of electrocatalytic materials as first demonstrated by *Wipf* and *Bard* [34] who used the different reaction rate of Fe^{2+} oxidation at Au and glassy carbon electrodes to obtain a reaction rate image of the composite surface. Methodically, partially blocked electrode surfaces represent a related problem. If the blank regions of the surface are comparable or larger then r_T the blank regions can be imaged (Fig. 7c). If the blank regions are much smaller than r_T, they cannot be resolved and the surface appears as a uniform surface with finite kinetics of the electron transfer between the mediator and the partially covered electrode. For the self-assembly of hexadecylthiolate on Au, effective rate constants k_{eff} were extracted from experimental approach curves at different stages of the self-assembly process using the theory in [32,33]. Assuming that the fraction θ of the gold surface is covered by a blocking film containing disk-shaped blank regions with the radius r_D, k_{eff} is related to the film structure by $k_{\text{eff}} = 4(1-\theta)D/(\pi r_D)$ [99]. The case of finite kinetics at the blank Au regions was treated, too.

3.2 Corrosion

Fundamental corrosion research seems to be an area where SECM will become an invaluable tool to follow localized reactions like pitting corrosion.[2]

[2] For a comparison with the Scanning Reference Electrode Technique (SRET) see [100,101]

Metallic materials are usually protected by a thin oxide layer (passive layer) from corrosion. Localized breakdown of the passive layer leads to a microscopic active area with rapid local dissolution of the metal. Once such a region has become active, extreme conditions of pH and reactant concentration may prevent re-passivation, finally leading to macroscopic destruction of the material. These processes have been followed by SECM on steel [102] and Al_3Fe [103]. While there are accepted mechanisms to explain pitting once initiated, the initiating itself is far less well understood. Consequently several studies focused on identifying precursor states for pitting corrosion on steel [104], Ti [68,105,106] and Ta [56]. The study of *Basame* and *White* [56] on Ta/Ta_2O_5 electrodes exemplifies the new information that can be gained by SECM. Before the onset of active pitting, precursor sites can be identified at which electron transfer between the metal and dissolved species occurs. These sites not only show a size distribution but they are also chemically selective for certain redox processes occurring at different potential with respect to the conduction band edge of the semiconducting metal oxide layer (Fig. 9).

This behavior constitutes a remarkable, previously unknown difference to Ti/TiO_2 electrodes, where no chemical selectivity for the precursor sites was discovered in similar experiments. Using a theory developed for the analysis of mass transport through porous membranes, it was found by an analysis of 11 sites distributed over $0.079\,cm^2$ that 69 % of the total current passed through the active sites that covered only 0.1 % of the geometric electrode area [106].

Precursor sites on stainless steel are thought to be associated with inclusions. *Williams* et al. developed a methodology to image the development of pits adjacent to such inclusions by a combination of complementary microscopic techniques including SECM [69]. Another approach tries to use the UME of the SECM to induce pitting by local generation of aggressive species like Cl^- and to monitor the flux of dissolved Fe^{2+} created by the developing pit [102,67].

Other studies followed active corrosion of diverse materials under simulated process conditions. These studies comprise such diverse materials as dental amalgams and metallic implants [65], potentiometric sensors made from AgI and exposed to cyanide solution [107], silicon in an etching bath [108], and composite electrodes used as base sensors of an extremely long-time stable amperometric biosensors with NADH-dependent alcohol dehydrogenase. SECM revealed that the electrode body acts as a reservoir for the controlled release of the soluble redox mediator [109].

3.3 Investigation of Biochemically Active Samples

Investigation of biological or biochemically relevant problems has been a main focus of SECM applications since the inception of SECM [110]. Organisms consist of cells that have extensions in the micrometer range and interact

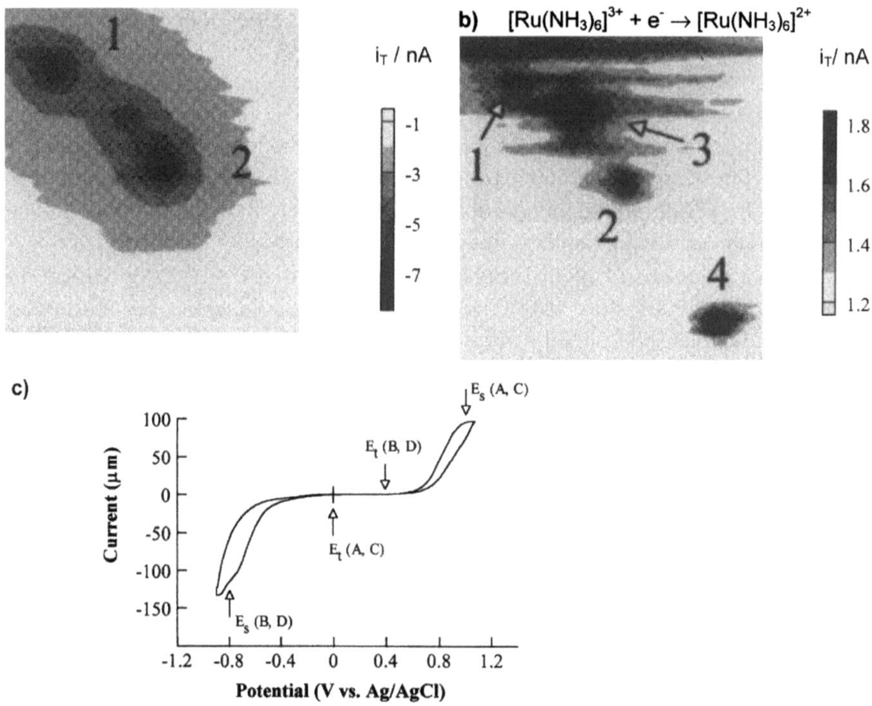

Fig. 9. Precursor sites for pitting corrosion on $200 \times 200\,\mu m^2$ Ta/Ta_2O_5 imaged in the GC mode in a solution containing both 2.5 mM $[Ru(NH_3)_6]^{3+}$ and 10 mMI^-. Oxidation of I^- (**a**) proceeds at sites 1 and 2 only, $E_S = 1.0\,V$, $E_T = 0.0\,V$. Reduction of $[Ru(NH_3)_6]^{3+}$ (**b**) is found on sites 1–4 ($E_S = -0.8\,V$, $E_T = +0.4\,V$). By changing E_S and E_T it was switched repeatedly between images such as that shown in (**a**) or in (**b**). The cyclic voltammogram of the macroscopic electrode (part **c**) shows the sum of the reaction at all sites and does not reveal the chemical selectivity of the sites; $v = 20\,mV\,s^{-1}$, ordinate label should be "Current (µA)" [G.W.] (Reprinted with permission from [56], Copyright American Chemcical Society, 1999)

with each other by release and uptake of various substances. SECM seems to be a natural choice for investigation of such problems as it is routinely operated in the micrometer range. SECM measurements can be performed in buffered solution, a preferred environment for most biological samples. The local probe does not touch the specimen therefore causing less interference then alternative scanning probe techniques like AFM. Another stimulus for SECM application comes from biotechnology where characterization methods are needed that can resolve microscopic regions of biochemical activity. Such miniaturized configurations are being developed for multianalyte assays [111], for applications in high-throughput screening [112], for advanced enzyme electrodes uniting distinct microscopic regions for enzyme immobilization and detection (microcompartmentalization) [113,114], or for clinical assays where the reduction of blood volume to be drawn from ill or injured patients has received increasing attention [115].

The specific properties of biological specimens can be exploited to generate an image in GC or feedback experiments. In the GC mode the generator is a biochemical reaction [59]. Reaction of oxidoreductases can also be monitored in feedback mode (Fig. 10). In this case the mediator couple R/O is not just a quasi-reversible redox couple but one of its forms has to be an artificial cofactor for the enzyme under study [8]. Fortunately, a number of popular redox couples can also transfer electrons to enzymes [116,117].

The enzymatic reaction has to produce a sufficient regeneration rate of Red to compete with its mass transport from the bulk solution [118]. The criterion is most easily met if the ratio of mediator concentration and immobilized enzyme activity is low. This represents an important methodic difference to the work on simple inorganic samples as shown in Fig. 8, where the mediator concentration can be increased to a level where i_T can be recorded conveniently. Despite its lower lateral resolution, GC images have often been preferred for imaging of biological samples due to their higher sensitivity to low activities of immobilized enzymes and the larger variety of enzymes that can be imaged. A detailed comparative discussion, quantitative estimations and tables of enzymes and mediators are provided elsewhere [119].

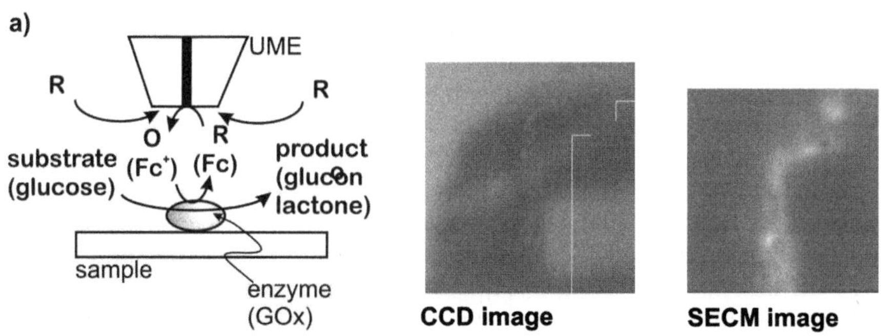

Fig. 10. Principle of enzyme-mediated feedback. (**a**) Schematic of the general principle with a possible realization given in brackets, GOx, glucose oxidase; Fc, water-soluble ferrocinium derivative; Fc^+, oxidized form of the ferrocinium derivative. (**b**) CCD image and SECM image of an agglomeration of GOx-modified magnetic beads deposited on silanized glass; $500 \times 500\,\mu m^2$, SECM working solution 1 mM ferrocene methanol and 50 mM glucose in deareated phosphate buffer, pH 7.0, $E_T = 400\,mV$ (SCE), $r_T = 5\,\mu m$. The white line in the CCD image frames the position of the beads. The dark blurred right angle is an optical landmark at the back side of the glass slide

3.3.1 Response of Intact Organs and Cultivated Cells to External Stimuli

Tsionsky et al. [63] investigated the photosynthetic oxygen production in living leaves of *Tradescantia fluminensis*. Leaves of this plant possess regions where only the guard cells contain chlorophyll that is essential for photosynthesis. In the dark (no photosynthesis) the topography was mapped in the negative feedback mode with dissolved oxygen as redox-active species. This allowed one to identify the guard cells which protrude from the leaf surface. Illumination started the photosynthetic oxygen production. A repeated image shows increased currents above the stomata due to oxygen production by guard cells. After positioning the UME above a single guard cell, the temporal response to switching the illumination on and off could be followed in vivo.

A similar approach was also successful in measuring the resorptive activity of osteoclasts. These specialized bone cells are responsible for the resorption of bone, a process in which among others Ca^{2+} is released. Bone is a dynamic system in which bone formation and bone resorption are balanced. Imbalance of both processes is found in some diseases such as osteroporosis. Conventional assays of osteoclasts activity require incubation of cells on bone slices for 18–20 h followed by examination of the resorption pits by electron microscopy. SECM simplifies the assay. A Ca^{2+}-selective potentiometric tip is used to determine Ca^{2+} released from the osteoclast against the background of the culture medium [120]. Chemical dissolution of the bone could be reduced by three orders of magnitude in a medium containing 10 mM NH_4F. Resorptive activity of osteoclast continued under these conditions although at a reduced rate. Osteoclasts respond to stimulation by parathyroid hormone (PTH) with a burst of superoxide anion generation. SECM experiments could follow this burst by using a gold microelectrode covalently modified with cytochrome c to detect the superoxide and to analyze the kinetics of its release [64]. Analysis of the kinetic response and the influence of inhibitors allow a conclusion on whether the response of the cells to PTH is mediated via other cells or results from direct action of PTH on osteoclasts [64].

3.3.2 Artificial Enzymatically Active Microstructures

SECM has been used to create artificial patterns and to map their enzymatic activity. Creation of microstructures can be performed by local destruction of an existing enzyme layer [90] or by local creation of anchor sites for subsequent covalent enzyme attachment. The latter approach was realized by deposition of a substituted polypyrrole in the direct mode [121], local desorption of a methyl-terminated SAM in the direct mode and re-adsorption of an NH_2-terminated disulfide [61], or by local generation of reactive species, e.g. hydroxyl radicals, in the TG/SC mode that attack an existing monolayer and form reactive groups required for covalent attachment of the enzyme [44]. While enzymes immobilized on insulating surfaces can be imaged either in the GC or in the feedback mode, imaging of immobilized enzymes on conducting surfaces like on the surface of enzyme electrodes is easier in the GC mode because the heterogeneous electron transfer at the metal surface interferes with the mediator regeneration by the enzymatic reaction. Figure 11a shows the local H_2O_2 production by a microscopic layer of immobilized glucose oxidase in the GC mode. The enzyme layer was produced by localized desorption of a SAM, incubation with an NH_2-terminated disulfide, followed by formation of a Schiff base with a periodate-oxidized GO_x.

In the GC image of Fig. 11a the enzyme is continuously producing H_2O_2 leading to a background current that increased from the start point of the image (front left) to the end (rear right). This situation commonly found in GC imaging can be circumvented by introducing a homogenous reaction that competes with the reaction at the UME for the species generated at the sample. This has been demonstrated in a different context with a homogenous follow-up reaction of electrochemically generated species [122]. For enzymatic reactions addition of a second enzyme that consumes the species to be detected at the UME seems to be more flexible because enzymes are available for many metabolites. Catalase enhances the rate of H_2O_2 disproportionation. The amount of catalase in the solution has to be adjusted such that the life time of H_2O_2 is long enough to diffuse across the specimen–UME gap but not long enough to accumulate in solution. Addition of minute amounts of catalase eliminates completely the rising background (Fig. 11b). The lateral resolution is moderately enhanced. The full width at half maximum (FWHM) is reduced from 350 μm in Fig. 11a to 250 μm and 220 μm in Fig. 11b and Fig. 11c, respectively. The signal height measured against the base line is decreased by 15 % only in Fig. 11b compared to Fig. 11a. Further additions of catalase quench the signal (Fig. 11c,d). It can be restored by rinsing the cell and supplying fresh, catalase-free glucose solution [61].

3.3.3 Labelling Techniques

Biomolecules that do not metabolize solution constituents, e.g. antibodies (Ab), can be imaged by saturating them with an enzyme-labeled binding partner, e.g. an enzyme-labeled antigen. This approach, first demonstrated by *Wittstock* et al. [123], has found application in miniaturized immunoassays for carcinoembryonic antigen, human chorionic gonadotropin, and human placental lactogen, and mouse IgG [124,125]. Miniaturization has been achieved either by immobilizing antigens or Ab on micromachined glass surfaces [62] or by directed deposition of paramagnetic, Ab-coated beads onto flat surfaces [125].

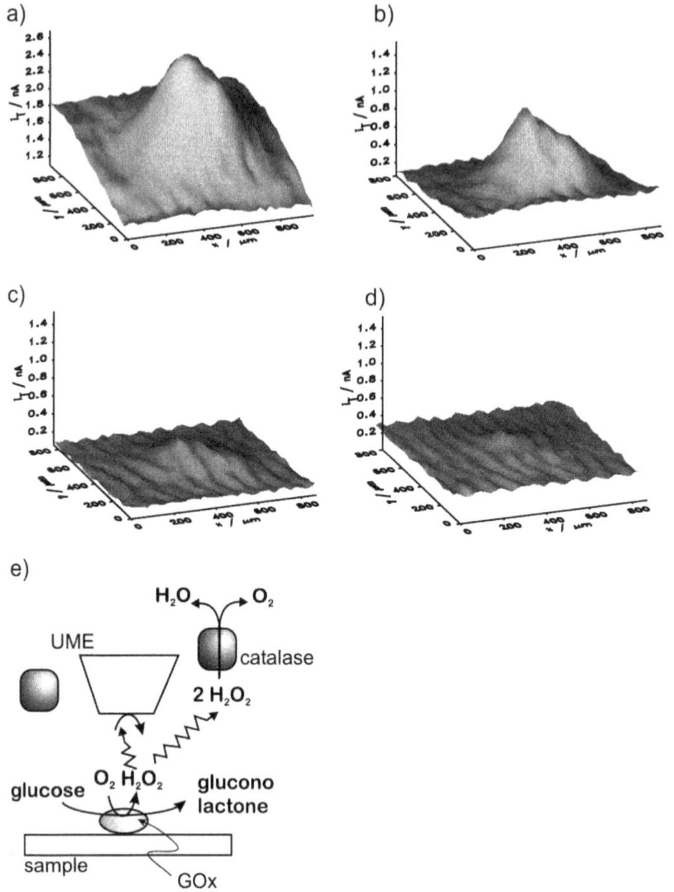

Fig. 11. H_2O_2 production by glucose oxidase (GOx) coupled to a region of NH_2-terminated monolayer. Solution composition: (**a**) 50 mM glucose oxidase, (**b**) 50 mM glucose oxidase +0.1 ng ml^{-1} catalase, (**c**) 50 mM glucose +0.6 ng ml^{-1} catalase, (**d**) 50 mM glucose +3 ng ml^{-1} catalase, (**e**) schematic representation for the reactions occurring in images (**b**)–(**f**). (Reprinted from [61], Copyright American Chemical Society 1997)

3.4 Localized Mass Transport

For the investigation of local mass transport phenomena, the sample separates two electrolyte reservoirs, a donor and a receptor compartment (Fig. 12).

The mass transport can be caused by a gradient of concentration [126], an electrical field [71,72,127,128], or hydrostatic pressure [73,129]. Iontophoretic transport through skin is explored as one mode of transdermal drug delivery. A fundamental question is to what extent the transport of ionic species is associated with appendages in the skin like hair follicles and sweat dugs. In

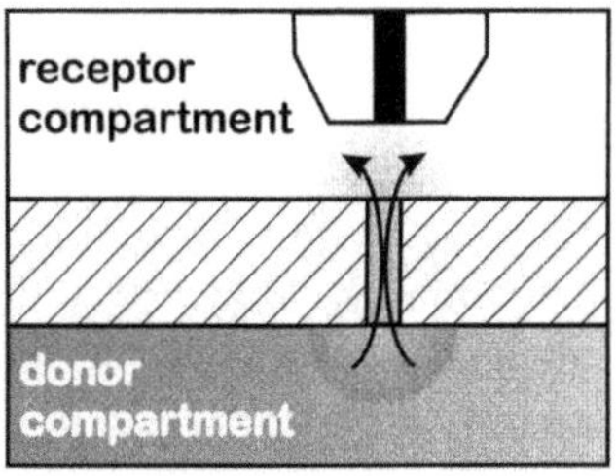

Fig. 12. Schematic setup for the investigation of localized mass transport though a porous sample

a series of papers, *Scott* et al. developed methodologies to quantify the flux associated with individual pores [127]. The method was used to measure the influence of transport-enhancing substances on the current distributions during iontophoretic transport through hairless mouse skin [72]. The transport of neutral molecules by electroosmotic flow has be investigated in a model system [71].

Macpherson et al. [129] studied the transport within tubules of dentine. Thermal, tactile or osmotic stimuli conducted along the tubules to the pulp are considered as a reason for dentinal hypersensitivity. Therefore, blocking fluid flow along the tubules is investigated as a medical treatment. Quantitative information were extracted about the action of fluid blocking agent calcium oxalate on the transport within dentinal tubules by a combination of imaging in negative feedback mode and generator collector mode. A dentine slice was used to separate the donor and receptor compartments. Both compartments contained the *same* mediator solution. When the dentine slice was imaged without an external pressure applied, the image reflects the topography of the surface due to negative feedback. Application of hydrostatic pressure across the dentine slice caused a fluid flow through the tubules that provided new reactants into the UME–sample gap that is depleted from the mediator by negative feedback imaging. Therefore, enhanced currents are recorded compared to the image without external pressure. After treatment of the dentine slice with the fluid blocking agent calcium oxalate, imaging was repeated at the same z position of the UME. Two important observations were described: No difference was found between the images recorded with and without applied external pressure demonstrating the effectiveness of the fluid blocking agent. The normalized currents were lower than those recorded before the application of the fluid blocking agent in the absence of hydrostatic pressure. This is caused by a decrease in the UME–sample distance due to the thickness of the calcium oxalate layer on top of the dentine surface. Using (3), the thickness of the oxalate layer was estimated.

4 Experimental Consideration

4.1 Hardware

Recently, a complete SECM instrument has become commercially available.[3] The majority of research groups use home-build instruments. They are built from commercial positioning systems, potentiostats and AD boards. As requirements, performance and prices of these components differ greatly, a new instruments may be tuned to meet the requirements of a particular task and to keep within budget by using simple devices for less critical components.

4.1.1 Positioning Systems

Positioning with commercial STM or AFM scanners is possible but not really convenient, because of the limited vertical translation range (around 10 µm at most) and the limited access to the software routines offered by most manufacturers. In fact, positioning has been performed mainly by Inchworm™ motors [130]. These systems were the first that combined positioning resolution in the nanometer range with travel distances in the centimeter domain. One of the disadvantages of the system is the low position repeatability and aging that require frequent re-calibration tests to obtain distorsion-free images [131]. This is particularly inconvenient if samples are scanned with widely varying scan length and grid spacings. Optical encoders eliminate the problem but increase the price substantially. The optical encoders have a lower resolution than the Inchworm motor itself, giving up one of the initial advantages of the system. Fortunately, the last few years have seen a dramatic improvement of micro- and nanopositioning instrumentation offering a wider range of alternatives than five years ago.[4] A number of groups gained excellent experience with stepper motors with polymer-coated moving parts to diminish mechanical backslash.[5] A positioning resolution of 10 nm is combined with high positioning repeatablity and travel distances in the centimeter range.

In many cases an additional piezomechanical actuator is added to the system either to increase the positioning resolution in the z direction, to provide vertical position modulation, or as part of a height adjustment mechanism. The study of heterogeneous kinetics relies on recording of approach curves. A specialized instrument may therefore use a single vertical translator only and use micrometer screws for horizontal positioning of the UME above the sample [21].

[3] CH Instruments, Inc., 3700 Tennison Hill Drive, Austin, TX 78733, USA

[4] Scientific Precision Instruments, Wormser Str. 32, 55276 Oppenheim, Germany
Newport Corp., 1791 Deere Ave., Irvine, CA 92606, USA
OWIS GmbH, Im Gaisgraben 7, 79219 Saufen i. Br., Germany
Physik Instrumente (PI) GmbH & Co, Polytec Platz 1-7, 76337 Waldbronn, Germany
Klocke & Kleindiek Nanotechnology, Soerser Weg 37, 52070 Aachen, Germany

[5] Scientific Precision Instruments, Wormser Str. 32, 55276 Oppenheim, Germany

4.1.2 Current Measurement

In most cases the potentiostat works under steady state conditions or with slow potential scan rates. Therefore, it does not need to be very fast but should have a sensitivity of at least 1 nA V. For many applications even this will not be sufficient. The potentiostat EI-400 has been reported frequently to fulfill this demand.[6] Other suitable bipotentiostats can be found among those developed for electrochemical detection in chromatography [131]. Recently, the selection was extended by dedicated bipotentiostats for SECM.[7] Often the sample does not need to be connected to a potentiostat and experiments may be performed with a monopotentiostat for which an even larger and more affordable selection is available [132]. In fact, in the author's lab the majority of experiments are carried out with an excellent monopotentiostat built according to custom specifications.[8] For the few instances where a bipotentiostat is necessary one may consider the use of two monopotentiostats [106].

For data aquisition and potential output commercial analog–digital and digital–analog converters in the form of computer boards are available from a number of manufacturers. They are used to read the signals of the potentiostat or the voltmeter in the case of a potentiometric probe.

4.2 Software

Software design for home-built instruments can be as different as the selection of the hardware components and is subjected to personal taste and preferences. Over the years we solved the programming task on two levels. Data aquisition and communication to external devices is made by low level routines mostly in the form of dynamic link libraries (DLL). This allows fast execution of time-critical steps. The libraries are called from high-level languages that manage the interface to the user (parameter input, data display). Such an approach combines the possibilities of programming close to the hardware with the convenience and speed in which entire experiments can be designed with a high-level programming language.

Off-line data processing can be done by importing SECM data into spreadsheet programs and by using their 3D data display capabilities, or by converting SECM data in the data format of a commercial STM of AFM and, if available, use that instrument to generate printouts. In the author's lab

[6] Cypruss Systems, 2449 Iowa St. Suit K, P.O. Box 3931, Lawrence, KS 66046, USA

[7] Jaissle Elektronik GmbH, Spitalhaldenweg 4, D-71336 Waiblingen, Germany
HEKA Elektronik Dr. Schulze GmbH, Wiesenstr. 71, D-67466 Lambrecht, Germany

[8] M. Schramm, H.-Heine-University Düsseldorf, Universitätsstr. 1, 40225 Düsseldorf, Germany

a specialized SECM visualization software package MIRA has been developed based on IDL.[9]

4.3 Probes

For quantitative SECM measurements probes need to have a defined geometry of the active electrode area and the insulating shielding, preferable a microdisk embedded in a insulating plane. For this reasons the most popular probes continue to be metal wires and carbon fibers of 1–25 µm diameter embedded in glass or coated with a polymer. A number of publications [10,22,23] summarize procedures to produce such UME which are also available commercially. In the case of glass-embedded UME, the insulating shielding should be shaped conically to achieve an RG of about 10. This shape is often considered a good compromise because it allows sufficient contrast when imaging insulators but reduces the risk that the insulating parts of the UME touch the sample. A procedure developed by *Kranz* et al. [47] can be recommended for shaping the insulating shielding. Outdated computer hard drives coated with grinding or lapping paper make excellent polishing wheels onto which the UME is gently pushed while spinning around its own axis.

Smaller electrodes ("nanodes") can be manufactured by etching and coating procedures or by drawing out a larger embedded Pt wire by a laser puller [10,133]. The total size is characterized by the diffusion-controlled steady state current. The shape of such electrodes can be checked by comparing approach curves towards an insulator with (3) [133]. Conically shaped tips (such as those used in ECSTM) do not show a strong current decrease when approaching an insulator. If the conical region is comparatively flat, such probes might be useful for SECM measurements. Geometric requirements can be estimated from simulation [28]. Furthermore, smaller tips also require smaller UME–sample separations thereby increasing the risk that the shielding of the UME collides with protruding regions of the sample. Ways of guiding probes above rough samples are currently under investigation.

Functionalized amperometric UMEs [59,64] are necessary for the detection of certain mediators. Their application will certainly increase if distance control mechanism become more widespread.

Potentiometric tips can be produced by aspirating an ion-selective cocktail into a micropipette equipped with an inner electrode [134] or by sealing metals into glass that make a potentiometric electrode (Sb for H^+, Ag for Ag^+, Cl^-) [38,136]. The problem of positioning such electrodes has been solved by using Sb electrodes which can act as potentiometric electrodes and as amperometric electrodes [39]. Double-barrel pipettes ("θ" cross-section) were used to combine an ion-selective pipette with a probe that could sense

[9] Interactive Data Language by Research Systems Inc., 2995 Wilderness Place, Boulder, CO 80301, USA

the distance either by amperometric feedback mode or by recording the conductivity between an inner and an outer electrode [134,135]. An all solid state potentiometric electrode for K^+ was used for SECM measurements [137]. It was made by applying a K^+-selective crown ether in plastizied poly(vinyl chloride) to a polypyrrole layer electropolymerized on an UME made from C, Au or Pt.

Use of ion transfer at amperometric liquid–liquid electrodes has been demonstrated and offers the advantage that approach in the negative feedback mode can be used to position the tip [81].

4.4 Vertical Tip Positioning

Vertically tip positioning is usually performed in the feedback mode using the distance dependence of Fig. 3. If the reactivity of the sample is unknown or when positioning the tip above an area with heterogeneous reactivity, negative feedback with an irreversible redox reaction, e.g. reduction of dissolved O_2, can be used. In this case both conductive and insulating surfaces cause negative feedback.

Another method is to record the conductivity between a pipette-type tip and a distant counterelectrode when applying small perturbation to the potential of the tip. After normalizing the conductance to the conductance far away from the surface, the distance dependence follows that of the normalized faradaic current $i_{\mathrm{T}}/i_{T\infty}$ above an insulating surface [134]. Observation of the probe approach by optical microscopy is usually less accurate than other approach mechanisms.

Finally the UME can also be brought to mechanical contact with the sample and then retracted in small increments. The working distance is reached if the sample can move freely over the sample. In a horizontal line scan, a touch between an amperometric UME and the sample can be identified by a sharp current rise (the tip snaps forward if the dragging force exceeds the friction between tip and sample) followed by a slower decay of the current.

4.5 Distance Control Mechanism

Currently several groups have worked out distance control mechanism. The earliest reports used vertical tip position modulation and lock-in amplification to detect the conducting or insulating nature of the sample and to switch between two set points [54,138]. An increase in sensitivity may also be expected from such a setup. More recent reports aim to adapt the distance control mechanism borrowed from Scanning Near-Field Optical Microscopy (SNOM) to the special requirements of SECM [139]. The UME vibrates horizontally above the sample surface. The change of resonance frequency, amplitude or phase shift is recorded if the sample approaches a surface. The detection can be done either by a laser beam [139], a combination of an excitation piezo and a detection piezo [140] or a quartz tuning fork [141]. Currently, none of these methods seem to be routine. Once they become more widespread, interesting new possibilities will emerge. Any kind of probe may be accurately positioned. The working distances can be reduced thus allowing smaller tips to be used. The information content of images will increase because sample topography and local reactivity can be deconvoluted. A tilt of the sample versus the xy plane of the positioning system is easily corrected.

Another likely field of progress is the implementation of hyphenated techniques. *Casillas* et al. [142] used a gold-coated optical fiber that could be used as a microscopic light source and as a microelectrode. The UME electrode can also be used to Electrogenerate ChemiLuminiscence (ECL) which can be used for optical imaging [143].

Acknowledgments

The author would like to thank R. Szargan for continuous support in setting up a SECM lab for characterization of solid–liquid interfaces at the Wilhelm-Ostwald-Institute. Current research on SECM in the author's lab is supported by Deutsche Forschungsgemeinschaft (Wi 1617/1-1..1-2, Wi 1617/2-1..2-4) and the Fonds der Chemischen Industrie.

References

1. M. Fleischmann, S. Pons, D. R. Rolison, P. P. Schmidt (Eds.): *Ultramicroelectrodes* (Datatech Systems Inc., Morganton 1987)
2. R. N. Adams: Prog. Neurobiol. **35**, 297 (1990)
3. R. C. Engstrom, M. Weber, D. J. Wunder, R. Burges, S. Winquist: Anal. Chem. **58**, 844 (1986)
4. H.-Y. Liu, F.-R. F. Fan, C. W. Lin, A. J. Bard: J. Am. Chem. Soc. **108**, 3838 (1986)
5. A. J. Bard, F.-R. F. Fan, J. Kwak, O. Lev: Anal. Chem. **61**, 132 (1989)
6. J. Kwak, A. J. Bard: Anal. Chem. **61**, 1221 (1989)
7. R. C. Engstrom, C. M. Pharr: Anal. Chem. **61**, 1099 (1989)
8. A. J.Bard, F.-R. F. Fan, D. T. Pierce, P. R. Unwin, D. O. Wipf, F. Zhou: Science **254**, 68 (1991)
9. M. Arca, A. J. Bard, B. R. Horrocks, T. C. Thomas, D. A. Treichel: Analyst **119**, 719 (1994)
10. A. J. Bard, F.-R. F. Fan, M. V. Mirkin: in A. J. Bard (Ed.): *Electroanal. Chem.* **18** (Marcel Dekker, New York 1994) Chap. 3
11. G. Wittstock: Chem. Lab. Biotech. (CLB) **46**, 166 (1995); Chem. Lab. Biotech. (CLB) **46**, 268 (1995)
12. A. J. Bard, F.-R. F. Fan, M. V. Mirkin: in J. Rubinstein (Ed.): *Physical Electrochemistry* (Marcel Dekker, New York 1995), Chap. 5
13. A. J. Bard, F.-R. F. Fan, M. V. Mirkin: in A. T. Hubbard (Ed.): *Handbook of Surface Imaging and Visualization* (CRC Press, Boca Rota 1995), Chap. 48
14. P. R. Unwin, J. V. Macpherson: Chem. Soc. Rev. **24**, 109 (1995)
15. M. V. Mirkin: Anal. Chem. **68**, 177A (1996)

16. C. Ricken, K. Borgwarth, J. Heinze: Nachr. Chem. Tech. Lab. **44**, 502 (1996)
17. D. Mandler, S. Meltzer, I. Shohat: Israel J. Chem. **36**, 73 (1996)
18. A. J. Bard, D. E. Cliffel, C. Demaille, F.-R. F. Fan, M. Tsionsky: Annali di Chimica **87**, 15 (1997)
19. P. R. Unwin: J. Chem. Soc. Faraday Trans. **94**, 3183 (1998)
20. M. V. Mirkin: Mikrochim. Acta **130**, 127 (1999)
21. A. L. Barker, M. Gonsalves, J. V. Macpherson, C. L. Slevin, P. R. Unwin: Anal. Chim. Acta **385**, 223 (1999)
22. R. M. Wightman, D. O. Wipf: in A.J. Bardin (Ed.): Electroanal. Chem. **15** (Marcel Dekker, New York 1989) Chap. 3
23. J. Heinze: Angew. Chem. **105**, 1327 (1993); Angew. Chem. Int. Ed. Engl. **32**, 1265 (1993)
24. A. C. Michael, R. M. Wightman: in P. T. Kissinger, W. R. Heineman (Eds.): *Laboratory Techniques in Electroanalytical Chemistry*, 2nd edn. (Marcel Dekker, New York 1996)
25. Y. Saito: Rev. Polarog. Jpn. **15**, 177 (1968)
26. J. L. Amphlett, G. Denuault: J. Phys. Chem. B **102**, 9946 (1998)
27. O. Niwa, M. Morita, H. Tabei: Anal. Chem. **62**, 447 (1990)
28. M. V. Mirkin, F.-R. F. Fan, A. J. Bard: J. Electroanal. Chem. **328**, 47 (1992)
29. D. Shoup, A. Szabo: J. Electroanal. Chem. **160**, 27 (1984)
30. Y. Shao, M.V. Mirkin: J. Phys. Chem. B **102**, 9915 (1998)
31. R. D. Martin, P. R. Unwin: J. Electroanal. Chem. **439**, 123 (1998)
32. A. J. Bard, M. V. Mirkin, P. R. Unwin, D. O. Wipf: J. Phys. Chem. **96**, 1861 (1992)
33. C. Wei, A. J. Bard, M. V. Mirkin: J. Phys. Chem. **99**, 16033 (1995)
34. D. O. Wipf, A. J. Bard: J. Electrochem. Soc. **138**, L4 (1991)
35. R. C. Engstrom, T. Meaney, R. Tople, R. M. Wightman: Anal. Chem. **59**, 2005 (1987)
36. R. D. Martin, P. R. Unwin: Anal. Chem. **70**, 276 (1998)
37. R. D. Martin, P. R. Unwin: J. Chem. Soc. Faraday Trans. **94**, 753 (1998)
38. G. Denuault, M. H.Troise Frank, L. M. Peter: Faraday Discuss. **94**, 23 (1992)
39. B. R. Horrocks, M. V. Mirkin, D. T. Pierce, A. J. Bard, G. Nagy, K. Toth: Anal. Chem. **65**, 1213 (1993)
40. P. Pohl, Y. N. Antonenko, E. Rosenfeld: Biochim. Biophys. Acta **1152**, 155 (1993)
41. F. Zhou, A. J. Bard: J. Am. Chem. Soc. **116**, 393 (1994)
42. F. Zhou, P. R. Unwin, A. J. Bard: J. Phys. Chem. **96**, 4917 (1992)
43. C. Lee, J. Kwak, F. C. Anson: Anal. Chem. **63**, 1501 (1991)
44. H. Shiku, I. Uchida, T. Matsue: Langmuir **13**, 7239 (1997)
45. O. E. Hüsser, D. H. Craston, A. J. Bard: J. Electrochem. Soc. **136**, 3222 (1989)
46. H. Sugimura, T. Uchida, N. Kitamura, N. Shimo, H. Matsuhara: J. Electroanal. Chem. **361**, 57 (1993)
47. C. Kranz, M. Ludwig, H. E. Gaub, W. Schuhmann: Adv. Mater. **7**, 38 (1995)
48. C. Kranz, M. Ludwig, H. E. Gaub, W. Schuhmann: Adv. Mater. **7**, 568 (1995)
49. C. Kranz, H. E. Gaub, W. Schuhmann: Adv. Mater. **8**, 634 (1996)
50. G. Wittstock, R. Hesse, W. Schuhmann: Electroanal. **9**, 746 (1997)
51. T. Wilhelm, G. Wittstock: Mikrochim. Acta **133**, 1 (2000)
52. H. Sugimura, T. Uchida, N. Kitamura, H. Masuhara: J. Phys. Chem. **98**, 4352 (1994)

53. H. Sugimura, N. Shimo, N. Kitamura, H. Masuhara, K. Itaya: J. Electroanal. Chem. **346**, 147 (1993)
54. D. O. Wipf, A. J. Bard, D. E. Tallman: Anal. Chem. **65**, 1373 (1993)
55. G. Wittstock, B. Gründig, B. Strehlitz, K. Zimmer: Electroanal. **10**, 526 (1998)
56. S. B. Basame, H. S. White: Langmuir **15**, 819 (1999)
57. K. Borgwarth, C. Ricken, D. G. Ebling, J. Heinze: Fresenius J. Anal. Chem. **356**, 288 (1996)
58. M. V. Mirkin, F.-R. F. Fan, A. J. Bard: Science **257**, 364, (1992)
59. B. R. Horrocks, D. Schmidtke, A. Heller, A. J. Bard: Anal. Chem. **65**, 3605 (1993)
60. B. R. Horrocks, M. V. Mirkin: J. Chem. Soc. Faraday Trans. **94**, 1115 (1998)
61. G. Wittstock, W. Schuhmann: Anal. Chem. **69**, 5059 (1997)
62. H. Shiku, Y. Hara, T. Matsue, I. Uchida, T. Yamauchi: J. Electroanal. Chem. **438**, 187 (1997)
63. M. Tsionsky, Z. G. Cardon, A. J. Bard, R. B. Jackson: Plant Physiol. **113**, 895 (1997)
64. C. E. M. Berger, B. R. Horrocks, H. K. Datta: J. Endocrinol. **158**, 311 (1998)
65. J. L. Gilbert, S. M. Smith, E. O. Lautenschlager: J. Biomed. Mater. Res. **27**, 1357 (1993)
66. S. B. Basame, H. S. White: J. Phys. Chem. **99**, 16430 (1995)
67. J. W. Still, D. O. Wipf: J. Electrochem. Soc. **144**, 2657 (1997)
68. L. E. Garfias-Mesias, M. Alodan, P. I. James, W. H. Smyrl: J. Electrochem. Soc. **145**, 2005 (1998)
69. D. E. Williams, T. F. Mohiuddin, Y. Y. Zhu: J. Electrochem. Soc. **145**, 2664 (1998)
70. J. Kwak, F. C. Anson: Anal. Chem. **64**, 250 (1992)
71. B. D. Bath, R. D. Lee, H. S. White, E. R. Scott: Anal. Chem. **70**, 1047 (1998)
72. E. R. Scott, J. B. Bradley, H. S. White: J. Invest. Dermatol. **104**, 142 (1995)
73. J. V. Macpherson, M. A. Beeston, P. R. Unwin, N. P. Hughes, D. Littlewood: J. Chem. Soc. Faraday Trans **91**, 1407 (1995)
74. S. Nugues, G. Denuault: J. Electroanal. Chem. **408**, 125 (1996)
75. F.-R. F. Fan, A. J. Bard: Science **270**, 1849 (1995)
76. R. Guckenberger, M. Heim: Science **270**, 1451 (1995)
77. M. Patel, M. C. Davis, M. Lomas, C. J. Roberts, S. J. B. Tendler, P. M. Williams: J. Phys. Chem. B **101**, 5138 (1997)
78. F. Forouzan, A. J. Bard: J. Phys. Chem. B **101**, 10879 (1997)
79. M. V. Mirkin, T. C. Richards, A. J. Bard: J. Phys. Chem. **97**, 7672 (1993)
80. Y. Selzer, D. Mandler: J. Electroanal. Chem. **409**, 15 (1996)
81. Y. Shao, M. V. Mirkin: J. Electroanal. Chem. **439**, 137 (1997)
82. M. Tsionsky, A. J. Bard, M. V. Mirkin: J. Phys. Chem. **100**, 17881 (1996)
83. A. L. Barker, J. V. Macpherson, C. J. Slevin, P. R. Unwin: J. Phys. Chem. B **102**, 1586 (1998)
84. C. J. Slevin, J. V. Macpherson, P. R. Unwin: J. Phys. Chem. B **101**, 10851 (1997)
85. P. R. Unwin, A. J. Bard: J. Phys. Chem. **96**, 5035 (1992)
86. J. V. Macpherson, P. R. Unwin: J. Phys. Chem. **100**, 19475 (1996)
87. D. Mandler, A. J. Bard: Langmuir **6**, 1489 (1990)
88. D. Mandler, A. J. Bard: J. Electrochem. Soc. **136**, 3143 (1990)
89. Y. Zu, L. Xie, B. Mao, Z. Tian: Electrochim. Acta **43**, 1683 (1998)

90. H. Shiku, T. Takeda, H. Yamada, T. Matsue, I. Uchida: Anal. Chem. **67**, 312 (1995)
91. K. Borgwarth, C. Ricken, D. G. Ebling, J. Heinze: Ber. Bunsenges. Phys. Chem. 99, 1421 (1995)
92. D. Mandler, A. J. Bard: J. Electrochem. Soc. **137**, 1079 (1990)
93. S. Meltzer, D. Mandler: J. Electrochem. Soc. **142**, L82 (1995)
94. S. Sauter, G. Wittstock: J. Solid State Electrochem. **5**, 205 (2001)
95. N. Casillas, S. J. Charlebois, W. H. Smyrl, H. S. White: J. Electrochem. Soc. **141**, 636 (1994)
96. I. Shohat, D. Mandler: J. Electrochem. Soc. **141**, 995 (1994)
97. J. C. O'Brien, J. Shumaker-Parry, R. C. Engstrom: Anal. Chem. **70**, 1307 (1998)
98. C. Lee, D. O. Wipf, A. J. Bard, K. Bartels, A. C. Bovik: Anal. Chem. **63**, 2443 (1991)
99. F. Forouzan, A. J. Bard, M. V. Mirkin: Israel J. Chem. **37**, 155 (1997)
100. H. S. Isaacs, G. Kissel: J. Electrochem. Soc. **119**, 1628 (1972)
101. H. S. Isaacs: Corros. Sci. **29**, 313 (1989)
102. D. O. Wipf: Colloid. Surf. A **93**, 251 (1994)
103. J. O. Park, C.-H. Paik, R. C. Alkire: J. Electrochem. Soc. **143**, L174 (1996)
104. Y. Y. Zhu, D. E. Williams: J. Electrochem. Soc. **144**, L43 (1997)
105. N. Casillas, S. J. Charlebois, W. H. Smyrl, H. S. White: J. Electrochem. Soc. **140**, L142 (1993)
106. S. B. Basame, H. S. White: J. Phys. Chem. B **102**, 9812 (1998)
107. K. Tóth, G. Nagy, B. R. Horrocks, A. J. Bard: Anal. Chim. Acta **282**, 239 (1993)
108. G. A. Shreve, C. D. Karp, K. E. Pomykal, N. S. Lewis: J. Phys. Chem. **99**, 5575 (1995)
109. B. Gründig, G, Wittstock, U. Rüdel, B. Strehlitz: J. Electroanal. Chem. **395**, 143 (1995)
110. J. Wang, L.-H. Wu, R. Li: J. Electroanal. Chem. **272**, 285 (1989)
111. J. W. Siltzel, B. Cercek, C. Dodson, T. Tsay, R. J. Obremski: Clin. Chem. **44**, 2036 (1998)
112. L. J. Kricka: Clin. Chem. **44**, 2008 (1998)
113. B. B. Ratcliff, J. W. Klancke, M. D. Koppang, R. C. Engstrom: Anal. Chem. **68**, 2010 (1998)
114. W. B. Nowall, D. O. Wipf, W. G. Kuhr: Anal. Chem. **70**, 2601 (1998)
115. G. Wittstock, S. H. Jenkins, H. B. Halsall, W. R. Heineman: Nanobiology **4**, 153 (1998)
116. M. L. Meckstroth, B. J. Norris, W. R. Heineman: J. Electroanal. Chem. **128**, 63 (1981)
117. M. L. Fultz, R. Durst: Anal. Chim. Acta **140**, 1 (1982)
118. D. T. Pierce, A. J. Bard: Anal. Chem. **65**, 3598 (1993)
119. B. J. Horrock, G. Wittstock: in A. J. Bard, M. V. Mirkin (Eds.): *Scanning Electrochemical Microscopy* (Marcel Dekker, New York 2001), Chap. 11
120. C. E. M. Berger, B. R. Horrocks, H. K. Datta: Electrochim. Acta **44**, 2677 (1999)
121. C. Kranz, G. Wittstock, H. Wohlschläger, W. Schuhmann: Electrochim. Acta **42**, 3105 (1997)
122. R. C. Engstrom, B. Small, L. Kattan: Anal. Chem. **64**, 241 (1992)

123. G. Wittstock, K. Yu, H. B. Halsall, T. H. Ridgway, W. R. Heineman: Anal. Chem. **67**, 3578 (1995)
124. H. Shiku, T. Matsue, I. Uchida: Anal. Chem. **68**, 1276 (1996)
125. C. A. Wijayawardhana, G. Wittstock, H. B. Halsall, W. R.Heineman: Anal. Chem. **72**, 333 (2000)
126. E. R. Scott, H. S. White, J. B. Phipps: J. Membr. Sci. **58**, 71 (1991)
127. E. R. Scott, H. S. White, J. B. Phipps: Anal. Chem. **65**, 1537 (1993)
128. E. R. Scott, A. I. Laplaza, H. S. White, J. B. Phipps: Pharm. Res. **10**, 1699 (1993)
129. J. V. Macpherson, M. A. Beeston, P. R. Unwin, N. P. Hughes, D. Littlewood: Langmuir **11**, 3959 (1995)
130. J. Kwak, A. J. Bard: Anal. Chem. **61**, 1794 (1989)
131. G. Wittstock, H. Emons, T. H. Ridgway, E. A. Blubaugh, W. R. Heineman: Anal. Chim. Acta **298**, 285 (1994)
132. A. Newman: Anal. Chem. **69**, 369A (1997)
133. Y. Shao, M. V. Mirkin, G. Fish, S. Kokotov, D. Palanker, A. Lewis: Anal. Chem. **69**, 1627 (1997)
134. C. Wei, A. J. Bard, G. Nagy, K. Tóth: Anal. Chem. **67**, 1346 (1995)
135. C. Wei, A. J. Bard, I. Kapui, G. Nagy, K. Tóth: Anal. Chem. **68**, 2651 (1996)
136. Y. N. Antonenko, A. A. Bulychev: Biochim. Biophys. Acta **1070**, 279 (1991)
137. R. E. Gyurcsányi, A.-S. Nybäck, K. Tóth, G. Nagy, A. Ivaska: Analyst **123**, 1339 (1998)
138. D. O. Wipf, A. J. Bard: Anal. Chem. **64**, 1362 (1992)
139. M. Ludwig, C. Kranz, W. Schuhmann, H. E. Gaub: Rev. Sci. Instrum. **66**, 2857 (1995)
140. R. Brunner, A. Bietsch, O. Hollrichter, O. Marti: Rev. Sci. Instrum. **68**, 1769 (1997)
141. P. I. James, L. F. Garfias-Mesias, P.-J. Moyer, W. H. Smyrl: J. Electrochem. Soc. **145**, L64 (1998)
142. N. Casillas, P. James, W. H. Smyrl: J. Electrochem. Soc. **142**, L16 (1995)
143. F.-R. F. Fan, A. J. Bard: Anal. Chem. **70**, 2941 (1998)

Surface Reactions of Hot Electrons at Metal–Liquid Interfaces

Detlef Diesing, Guido Kritzler, and Andreas Otto

Lehrstuhl für Oberflächenwissenschaft (IPkM), Heinrich-Heine-Universität Düsseldorf, 40225 Düsseldorf, Germany
diesing@rz.uni-duesseldorf.de

Abstract. Surface reactions especially with hot charge carriers (related to photoelectrochemistry) are investigated at noble metal–liquid interfaces. The hot charge carriers are produced in Metal–Insulator–Metal (MIM) junctions. The top metal is a thin ($d = 20\,\mathrm{nm}$) silver or gold film. The hot carriers are injected by a tunnel voltage into the film which simultaneously acts as a working electrode in a normal electrochemical circuit.

The light emission of the tunnel junction drastically increases when changing the electrode potential from $E = E_{\mathrm{pzc}}$ to $E = E_{\mathrm{pzc}} - 0.3\,\mathrm{V}$. A corresponding effect was found in ultra-high vacuum conditions, when the top electrode of a MIM was charged negatively by evaporation of 0.6 nm potassium. The results are discussed in terms of light emission caused by direct excitation of surface plasmon polaritons at the noble metal–liquid interface.

The injected hot electrons can change the kinetics of faradaic reactions at the surface. As an example the hydrogen evolution reaction was found to lose its overpotential at silver–neutral electrolyte interfaces by the injection of hot electrons. The results are discussed in terms of an overlap of the hot electron wavefunction with the wavefunction of the water OH-groups in the inner Helmholtz layer.

The tunnel current at constant tunnel voltage was found to increase, when surface defects are applied to a smooth silver top electrode by the deposition of submonolayers of silver at potentials $E = E_{\mathrm{pzc}} + 0.8\,\mathrm{V}$. A model is presented, which explains the tunnel process as a coherent eigenstate of the hot electron on both sides of the insulator. The tunnel current increase due to the addition of surface defects can be motivated by an enhanced dephasing of the hot electron wavefunction at the silver–electrolyte interface.

1 Introduction

Electrochemical processes are normally driven by the potential difference at the metal–electrolyte interface. This potential difference influences parameters like surface concentration, adsorption and orientation of products, educts, and intermediates. Hot electrons stimulate chemical reactions at potentials different from the redox potential, for instance in cases where faradaic reactions of the electrolyte occur. But even when the electrode–electrolyte system represents a stable interface at the desired redox potentials, electrochemical reactions are often hindered by an overpotential. It can be shown

K. Wandelt, S. Thurgate (Eds.): Solid–Liquid Interfaces, Topics Appl. Phys. **85**, 365–425 (2003)

that the overpotential of hydrogen evolution can be overcome by the use of hot electrons [1]. Thereby hot electrons may reveal reaction paths and kinetics. Further applications of hot electron techniques might be electrochemical redox sensors for the detection of oxidizable or reducable components in liquids or atmospheres. To understand electrochemical reactions by hot carriers one needs to take the wide field of electron dynamics into account.

The dynamics of electrons at metal interfaces (especially clean surfaces in Ultra-High Vacuum (UHV), adsorbate-covered surfaces in UHV, metal–liquid interfaces) comprise surface electronic excitations [2], image states [3,4], transient anionic [5] and cationic states of adsorbates [6] and bulk hot electrons and holes near the interface. These excitations or states are involved in one-photon processes, like photoemission (e.g. into the electrolyte [7]) and nonadiabatic chemical reactions (for instance Desorption Induced by Electronic Transitions (DIET) [8,9]), surface-photochemistry [10,11,12,13], also in the electrolyte [14], especially on silver [15], and in multiphoton processes like nonlinear surface optics (e.g. Second Harmonic Generation (SHG) at UHV [16] and electrolyte interfaces [17,18]), two-photon photoemission (e.g. [19] at UHV interfaces, and [20] at electrolyte interfaces), Surface Enhanced Raman Spectroscopy (SERS) at single crystal electrodes [21], Desorption Induced by Multiple Electronic Transitions (DIMET) [22], and manipulation of individual molecules by a scanning tunneling microscope tip [23].

Our original motivation was to study some of the process states above "with one photon less" by replacing one virtual intermediate hot electron or hole state (excited by a photon) by "real" hot electrons and holes injected into the surface region. These "real" hot electrons are excited by a tunneling processes in Metal–Insulator–Metal (MIM) junctions (see Sect. 2), in the following exclusively aluminum–aluminum oxide–noble metal (mainly silver) junctions (see Sect. 3).

By the injection of hot electrons into the working electrode the following correspondences are revealed:

- photoemission into the electrolyte/hot electron injection into the electrolyte (see Sect. 5);
- surface photochemistry at electrode surfaces/faradaic reactions with hot electrons (see Sect. 5.1 and [1]);
- intermediate charge transfer reaction in spontaneous SERS/optical surface absorption by charge transfer excitations which can be observed as hot electron injection into the noble metal electrode of the MIM junction [24].

MIM junctions are a long-standing instrument and object of research [25,26]; a comprehensive list of literature on light emission from MIM junctions was given in [27]. Examples of reports on the internal photoeffect in MIM junctions are references [28,29,30,31,32,33,34,35,36], on electron emission into the vacuum references [37,38,39,40,41,42,43,44,45,46], and on filamentary growth references [47,48,49].

The possibility of hot electron–surface plasmon polariton interaction at the outside surface of MIM junctions was first pointed out by *Kirtley* and collaborators [50] and clearly demonstrated in [27].

The first electrochemical reactions with hot electrons in MIM junctions have to our best knowledge been reported in [1]. At about the same time *Palmer* and collaborators [51] demonstrated that electrons emitted from a MIM device can be used to mediate simple chemical reactions by dissociative attachment. They called their device an electronic catalyst. Shortly after, *Gadzuk* pointed out the relation of resonance-assisted hot electron reactions to femtochemistry at surfaces of MIM junctions [52,53].

So far mainly Al/Al_2O_3/noble metal junctions have been employed, recently also amorphous tantalum oxide is considered for application in MIM capacitors [54]

Electrogenerated chemiluminescence by hot electrons in $Ta/Ta_2O_5/Pt$ junctions was reported in [55].

2 Creation of Hot Charge Carriers by Metal–Insulator–Metal (MIM) Junctions

MIM junctions allow electron and hole tunneling between the noble metal (in the following represented by silver) and the aluminum electrode according to the voltage $U_\mathrm{T} = \Phi(\mathrm{Al}) - \Phi(\mathrm{Ag})$, where Φ is the electrostatic potential. In the electronic ground state and at negative (positive) values of U_T electrons (holes) tunnel from aluminum into silver, where they may have maximum energy $-eU_\mathrm{T}$ above (below) the Fermi level of silver [so-called hot electrons (holes) in silver].

We will show in Sect. 3 that the lower edge of the conduction band of the aluminum oxide layer is closer to the Fermi levels of two metal electrodes than the upper edge of the valence band of the aluminum oxide (see Fig. 10).

Therefore the exponential decays of the coherent wavefunction of tunneling electrons and holes in the insulating oxide layer are approximately controlled by the energy differences between the Fermi levels and the lower edge of the conduction band of the oxide. Thus one may approximately apply models of tunneling through a vacuum gap (e.g. the Fowler–Nordheim model [57]), taking the lower edge of the conduction band as the vacuum level ("conduction band tunneling").

One may also create hot electrons or holes in the silver electrode of the MIM junction by external means, for instance by photon absorption [24] and electrochemical reactions at the silver–electrolyte interface [56]. In this case one will observe a tunnel current already at $U_\mathrm{T} = 0$. There is a possibility to measure these extra currents as function of U_T (separation from the DC current due to ground state tunneling is possible with lock-in methods) and apply the junction as an internal retarding field energy analyzer of the externally exited hot carriers. As yet some problems remain in using the MIM as an internal retarding field spectrometer, see [24], but at small excitation energies and small retardations this method works well (see [56]).

The possible hot electrons and hole states within the silver top electrode are given by the electronic band structure. In the tunneling and ballistic transport processes towards or off the silver–electrolyte interface the direction of propagation of the electronic wavepackets with respect to the surface normal depends on the orientation of the silver electrodes (e.g. (111) or (110) or (100) crystallographic planes parallel to the interface). The projected density of electronic states is given in Fig. 30.

For the (111) orientation there is a gap in the range of hot electrons between 0 and about 4 eV excitation energy. In this case, ballistic electrons have a minimum finite momentum parallel to the interface, and therefore propagate only above a finite limiting angle with respect to the surface normal. This is impressively demonstrated in real space simulations by *Gauyacq* and coworkers [58] for the time resolved emission of a hot electron at a noble metal (111) surface. This point will become important in understanding the hydrogen evolution reaction by hot electrons; see Sect. 5.2. In contrast one can expect propagation of hot electrons perpendicular to the surface ("on the shortest way between silver and aluminum") for a (110) orientation [24].

Usually a ballistic model is adopted for the hot electrons [52], separated into several independent steps:

- tunneling of the electron from the aluminum into the silver, or, in the Fowler–Nordheim case, tunneling from the aluminum in the tilted oxide conduction band and transport within this band to the oxide–silver interface;
- ballistic transport of the electron after tunneling (so-called primary hot electron) to the outer silver surface;
- reaction of the primary hot electron at the surface.

This "ballistic aspect" (see the Monte Carlo simulations by *Otto* et al. in [59]) is complemented by the unexpected aspect of the breaking of the coherence of the electronic tunneling state at the interface which allows for the variation or even control of the tunneling current between aluminum and noble metal–electrode at constant tunneling voltage by changes at the noble metal–interface, see [59] and Sect. 6.

The electron–electron interaction will create a cascade of secondary hot electrons, mainly of an energy less than half of the energy of the primary electron [59], for which the participation in hot electron chemistry is expected [53]. Below we will discuss in any case whether secondary hot electrons do contribute to the observed phenomena.

3 Preparation Techniques and their Influence on Surface and Bulk Properties of MIM Junctions

3.1 Experimental Setup

The preparation methods of tunnel junctions contain a wide range of physical and chemical methods [68]. Widely used are systems containing aluminum/aluminum oxide layers where the experiments were performed in the UHV at low temperatures [27,69]. The advantages of aluminum oxide are a considerably high specific resistance of $10^{13}\ \Omega\,\mathrm{cm}$ which is combined with an extremely small diffusion of oxygen or water (growth rate of the oxide film $< 10\,\mathrm{pm/d}$ [70]). When the tunnel junctions are transferred to atmosphere or even to an electrochemical environment, the corrosion resistance and long-term stability play an important role. The production of stable junctions is intensively discussed in [68]. Here we give only a short description of the preparation methods.

The Al film is evaporated in a high-vacuum (HV) chamber with a pressure $p = 5 \times 10^{-8}$ torr. Evaporation of Al at higher pressures leads to thin films, which have a considerably higher specific resistivity [71]. This may be explained by a "getter" effect of the Al which leads to an incorporation of oxide in the Al film.

The growth of the oxide films can be performed by:

- gas phase oxidation, further cited as GPO;
- physical vapour deposition, further cited as PVD;
- anodical oxidation in acidic buffer solutions, further cited as anodic oxide.

The gas phase oxidation is performed after evaporation of the Al film in the HV chamber by flooding the chamber with 1 bar O_2.

The "localized" anodic oxidation is performed in a droplet cell, see Fig. 1. The capillary contains the electrolyte (acetate buffer, pH = 6.0) and a gold wire, which works as a counter-electrode. The evaporated Al film is contacted with a gold wire. The localized anodization of Al films opens the possibility of programming the oxide structures. Another advantage is the reduction of stress in the Al film because only the junction area needs to be oxidized [72].

The PVD films were produced by evaporating alumina with an electron gun at 4.5 keV in the HV chamber [27].

After the oxide formation a 20 nm thick silver film is evaporated as top electrode and 30 nm thick silver stripes are evaporated for contacting the top electrode, see Fig. 2. The resulting MIM sandwich structure has a capacity of $5\,\mu\mathrm{F/cm}^2$ for an oxide thickness of 2.5 nm. This means a dielectric constant $\varepsilon = 12$ for the oxide. This value is the same as found for gas-phase oxides (GPO) and metal/insulator/electrolyte systems [84] as well. The dielectric constant ε for PVD oxides was found to be 3–4, which is caused by the porous structure of this oxide [68]. A detailed comparison of the GPO oxides and the PVD oxides can be found in [68]. In this work we present a comparison of anodic oxides and GPO oxides.

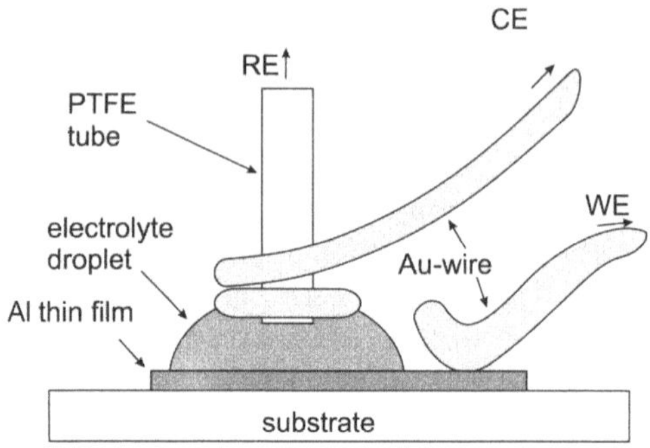

Fig. 1. Anodic oxdiation of Al thin films in a droplet cell. In the PTFE tube the electrolyte is contacted with an external reference electrode. CE = Counter Electrode is a gold wire. The Al film is contacted as the working electrode by a gold wire. The lateral extension of the droplet is about 5 mm

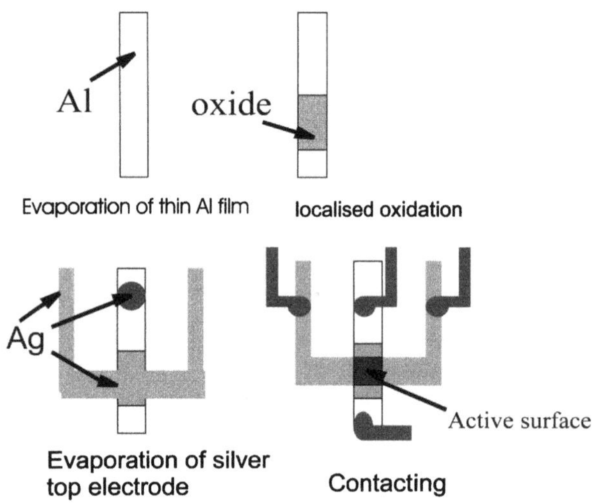

Fig. 2. Successive preparation steps of Al/Oxide/Ag junctions

3.2 Characterization of the Tunnel Barrier

For the characterization of the tunnel barrier exact information about the difference between the Fermi level of the base metal and the conduction band of the oxide, on the one hand, and the Fermi level of the top metal and the conduction band of the oxide, on the other, is needed. These parameters and the oxide thickness are important for the tunneling probability in MIM junctions [89]. For an exact determination of the energy levels in the junction we use the mechanism of Fowler–Nordheim tunneling [57]. Fowler and Nordheim mentioned that the effective length of a tunnel barrier is reduced when a certain threshold voltage is reached, so that tunneling electrons may partly use a conduction band state during tunneling. This effect is shown in

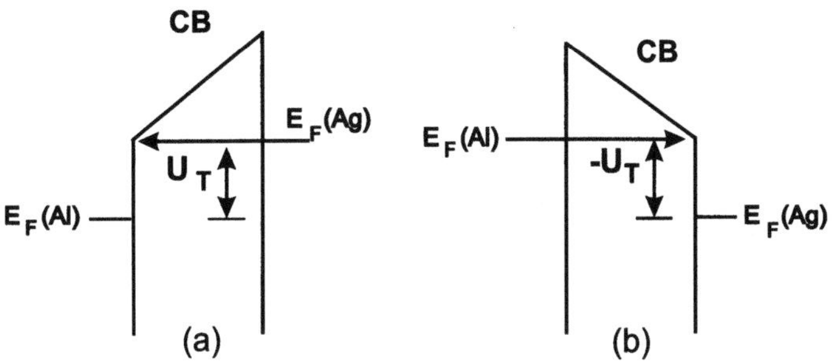

Fig. 3. Energy levels in MIM junctions in the case of starting Fowler–Nordheim tunneling, $eU_T = E_F(\text{Ag}) - E_F(\text{Al})$. At positive U_T electrons tunnel into the Al electrode (**a**), equivalent to hot hole injections into silver. Hot electron injection into the Ag electrode as shown in the right part takes place at negative tunnel voltage

Fig. 3. In the left part of Fig. 3a the Fermi level of the Ag just reaches the lower edge of the conduction band at the oxide–aluminum interface. In this case the energy eU_T of the tunneling electrons above the Fermi level of Al is equal to the difference $E_{CB}(\text{oxide}) - E_F(\text{Al})$, where $E_{CB}(\text{oxide})$ means the lower edge of the conduction band at the Al–oxide interface.

When the Fermi level of the Al is further lowered by applying a more positive tunnel voltage to the Al electrode, the tunneling electrons partly use the conduction band for tunneling through the oxide. This reduces the effective length of the tunnel barrier.

The tunnel current depends exponentially on the tunnel voltage and the effective length of the tunneling barrier. Thereby a logarithmic plot of the tunnel current versus the tunnel voltage reveals possible changes in the effective tunnel length known as the Fowler–Nordheim mechanism.

In Fig. 4 the tunnel current is plotted logarithmically and in a linear plot versus the tunnel voltage for a MIM junction with 3 nm anodic oxide. At a tunnel voltage $U_T = +2.25\,\text{V}$ the slope in the logarithmic plot increases, indicating an enhanced tunneling probability at tunnel voltages $U_T > +2.25\,\text{V}$, see Fig. 4. We assign this enhanced tunneling probability to a reduction of the effective tunnel length as shown in Fig. 3a due to the band edge $E_F(\text{Al})$/anodic oxide of 2.25 eV. By applying a negative tunnel voltage, see Fig. 5, we obtain the corresponding information on the band edge of the anodic oxide–silver interface. The slope in the logarithmic plot increases at $U_T = -3.65\,\text{V}$, indicating an $E_F(\text{Ag})$/anodic oxide band edge of 3.65 eV following the mechanism in Fig. 3b.

The oxide–metal interface is extremely sensitive to a change in the preparation conditions. In Fig. 6 the same measurement as in Fig. 4 is presented for a gas-phase oxide MIM system. The aluminum/aluminum oxide band edge slightly changes from 2.25 eV to 2.05 eV. A more pronounced change can be seen at the oxide–silver interface, see Fig. 7, where the change in the slope

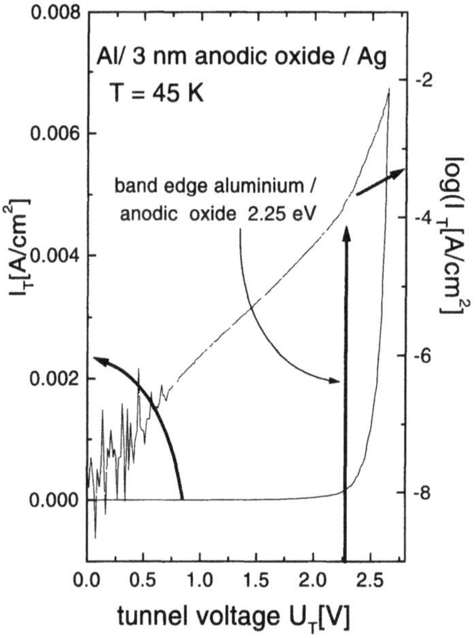

Fig. 4. Current voltage plot and Tafel plot (logarithm of current versus voltage) of a MIM junction with anodic oxide (beginning of Fowler–Nordheim tunneling indicated by a *vertical arrow*)

appears at 2.7 eV in contrast to 3.65 eV at the anodic oxide–silver interface. A more detailed investigation with GPO layers, which were formed by flooding the HV chamber with humid oxygen delivered a $E_F(Ag)$/GPO oxide band edge of 3.2 eV [72]. This leads to the assumption that an adsorbed monolayer of H_2O or OH^- plays an important role in the formation of the tunneling barriers. A table of the tunnel barriers with different preparation techniques can be found below.

The band edges of the GPO systems are used in Fig. 10.

Properties	Gas Phase Oxide	Wet Gas Phase Oxide	Anodic Oxide
Band edge: conduction band oxide/E_F(Ag)	2.7 eV	3.2 eV	3.65 eV
Band edge: E_F(Al)/conduction band oxide	2.05 eV	2.1 eV	2.25 eV

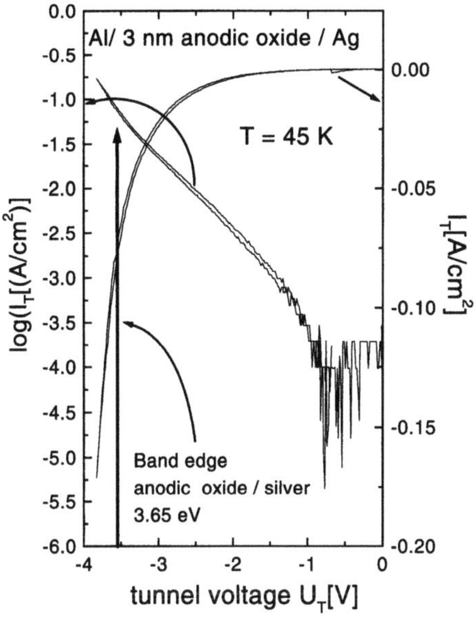

Fig. 5. Current voltage plot and Tafel plot of a MIM junction with anodic oxide, same as Fig. 4 but with negative tunnel voltage

3.3 Crystallographic Orientation of MIM Systems Grown on Glass Substrates

The crystallographic orientation of the top metal electrode can be obtained by the Bragg–Brentano method of X-ray scattering. The method delivers a signal from those lattice planes which are parallel to the macroscopic plane of the film. Of course, this method is applicable to the top electrodes of MIM junctions. The spectrum in Fig. 8 shows a MIM junction grown on glass slides. A preferentially (111) orientation with presumably azimuthal randomness can be seen with a small fraction of (100) orientation. Due to the limited penetration depth and the amorphous structure of the oxide all the discrete features in the X-ray spectrum can be assigned to the silver top electrode.

Therefore the present work is focused on hot electrons in Ag(111) films. This is especially important for the discussion of the influence of surface roughness on the transfer of hot electrons into the adsorbates in Sect. 5.2.

3.4 Electrochemical Setup

Electrons with some eV energy above the Fermi energy of silver, so-called hot electrons, are created in Al/Aluminum oxide/Ag tunnel junctions by application of a negative voltage U_{T} between the Al and the Ag electrode, see Fig. 3b and Fig. 10 below.

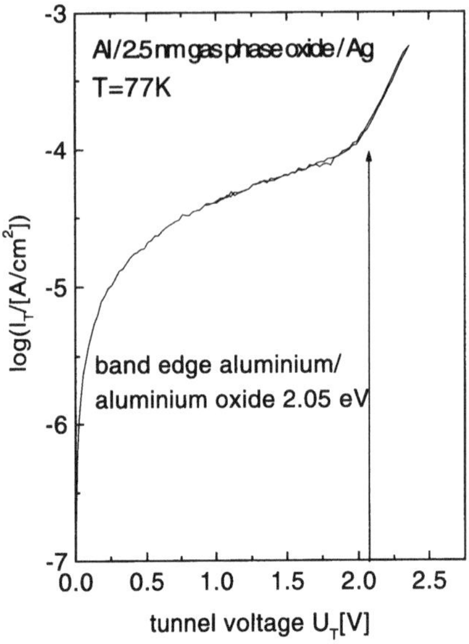

Fig. 6. Logarithmic plot of tunnel current versus tunnel voltage for MIM system with GPO oxide

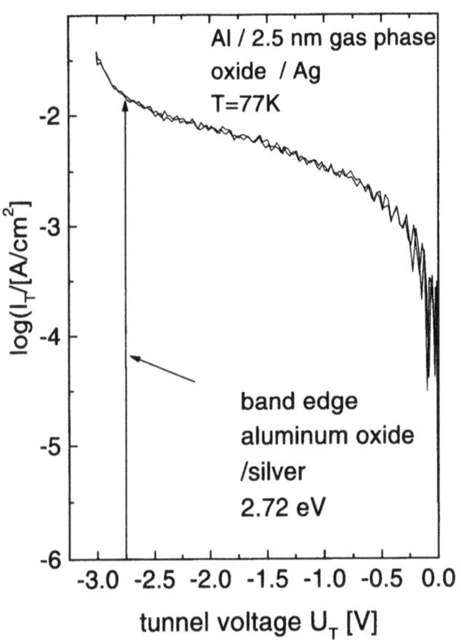

Fig. 7. Logarithmic plot of tunnel current versus tunnel voltage for MIM system with GPO oxide

In the electrochemical setup (see Fig. 9) the silver film of the MIM is made the working electrode in an electrochemical cell. Especially important for the present work is the shift of the electronic levels within the bulk electrolyte, e.g.

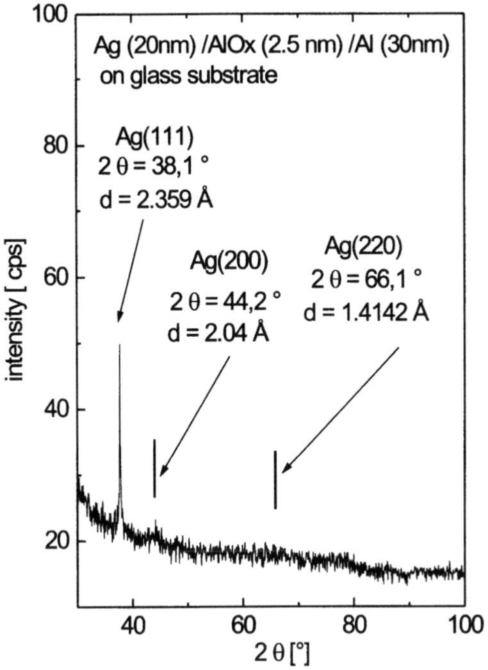

Fig. 8. X-ray diffraction versus scattering angle 2θ of a tunnel junction with Ag top electrode. The positions of the possible reflexes are indicated; d is the distance between lattice planes yielding the indicated Bragg reflections

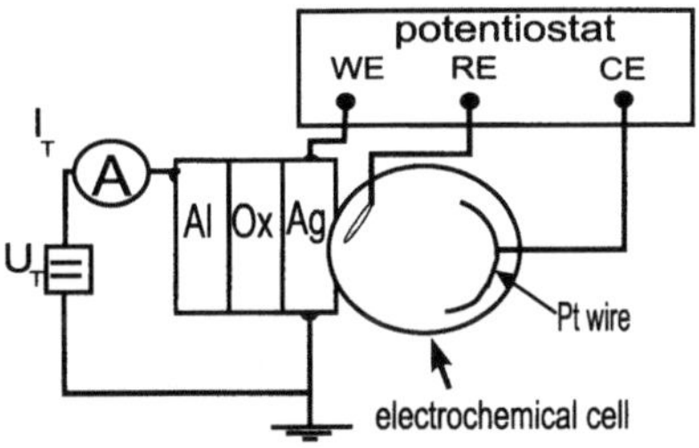

Fig. 9. Tunnel junction in electrochemical circuit. WE: working electrode, CE: counter-electrode = platinum wire, RE: reference electrode = saturated calomel electrode

the dry electron mentioned above. The experiments are carried out in highly concentrated electrolytes ($c > 5 \times 10^{-2}$M). In this way we can assume that a potential change of x eV moves the levels in the electrolyte also by x eV. To perform theses measurements two electrical circuits are necessary (see Fig. 9). One circuit contains the voltage source U_T and the ampere meter to control the current I_T through the tunnel junction. The other circuit controls the electrode potential E_{SCE} of the silver or gold top electrode of the MIM

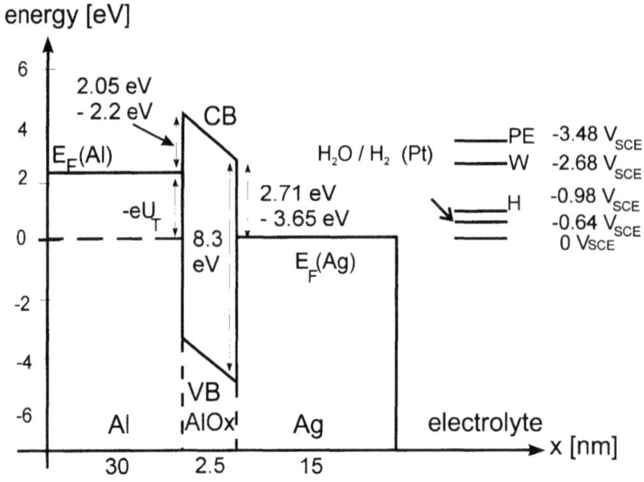

Fig. 10. Energy scheme of the tunnel junctions (in eV) in contact with the electrolyte. PE = lower edge of the electron conduction band in the electrolyte (= D state in [87], so-called dry electron), H_2O/OH^- onset potential of the hydrogen evolution at a platinum electrode both measured in 0.9 M acetate buffer(pH = 5.9) with 50% water and 50% ethylene glycol electrolyte [88] and the position of the lower edge W of the so-called wet electron state and the center H of the hydrated state of the electron in an aqueous electrolyte [87], all levels at the electrochemical potential of the top silver electrode of $0V_{SCE}$ (in this case the Fermi levels of the silver electrode and of the metallic mercury in the Saturated Calomel Electrode SCE E_F(SCE) are at the same level); 1 eV on the energetic scale corresponds to 1 V_{SCE} on the electrochemical potential scale. CB: lower edge of conduction band of Al_2O_3. VB: upper edge of valence band. U_T = tunnel voltage (the silver electrode is at ground level)

junction which is made the working electrode and acts as a common ground for both circuits. The counter-electrode is a platinum wire. The saturated calomel electrode is chosen as a reference. The active tunnel area of the samples is typically 0.12 cm^2 and in contact with the electrolyte. Outside this area the sample is covered with a protective lacquer.

3.5 Energy Levels

Now a complete energy level scheme for the MIM junction can be derived for example at $U_T = -2.2$ V, see Fig. 10, where the values for the tunnel barriers of the GPO MIM are inserted. In the electrochemical setup the silver top electrode is used as a working electrode in an electrochemical circuit. The working electrode is the common ground of the "electrochemical" and the "tunnel" circuit. The potential of the working electrode is applied by changing the potential of the counter-electrode, which is a Pt wire. Hot electrons can be injected into the working electrode by applying a negative potential to the aluminum electrode, see Fig. 10. A fraction of these hot electrons reaches ballistically the outer surface of the silver top electrode. Figure 10 shows the levels of the electrons at the silver–electrolyte interface at an electrochemical potential of $0\,\mathrm{V_{SCE}}$. This will be discussed further in Sect. 5.

4 Linear Surface Optics with MIM Systems

MIM junctions allow for new and unique studies of the optics of interfaces, especially metal–electrolyte interfaces.

4.1 The "Threshold Excitation"

Liebsch describes in his monograph [2] the theoretical results on linear electronic surface excitations obtained with the method of the time-dependent local density approximation of the density functional method. At metal surfaces of a quasi-free-electron metal (e.g. Al, alkalines) approximated by the jellium model (free electrons with bulk density n and a halfspace of a homogeneous positive background with a sharp surface) the theory predicts that the linear surface photoelectron yield $Y(\omega)$ at an incidence angle of the light of 45° is given by

$$Y(\omega) = 2\sqrt{2}\frac{\omega}{c}\left(1 - \frac{\omega^2}{\omega_\mathrm{p}^2}\right)\mathrm{Im}[d_\perp(\omega)]\,. \tag{1}$$

$d_\perp(\omega)$ is obtained from the linear screening charge distribution $n_\mathrm{linear}(z,\omega)$ (z is the distance from the sharp edge from the positive background, positive in the direction normal to the surface, towards the vacuum)

$$d_\perp(\omega) = \frac{\int z \times n_\mathrm{linear}(z,\omega)dz}{\int n_\mathrm{linear}(z,\omega)dz}\,. \tag{2}$$

At alkali metal overlayers on Ag Liebsch and co-workers predicted four types of collective modes: monopole and multipole surface plasmons, bulk like overlayer plasmons and the threshold excitation (final state near the vacuum level, photon energy $\hbar\omega \sim e\Phi$, where Φ is the workfunction).

The two parts of the yield (internal photoemission into the bulk of the silver and external photoemission into the vacuum) near the threshold excitation were calculated by reformulating the standard dipole moment expression for $\mathrm{Im}[d_\perp(\omega)]$ by an equivalent golden rule formula involving the sum over the possible final internal and external ("photoelectron") states [61], see Fig. 11 for silver with various coverages of cesium ($c = 1$ corresponds to one monolayer).

The threshold effect is completely due to internal photoemission below the onset of external photoemission, when the photon energy is smaller than the work function. But internal photoemission is still bigger than external

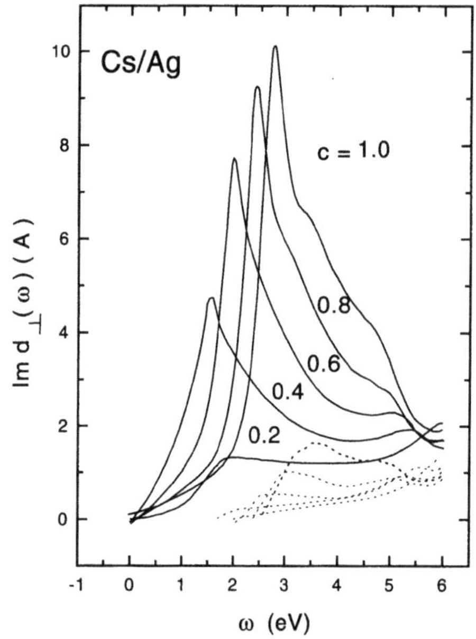

Fig. 11. Im $[d_\perp(\omega)]$ as a function of the photon energy for a jellium with the *sp*-electron density of silver, covered with increasing submonolayer coverages c of a layer of jellium with the *sp*-electron density of Cs. Separation into external *(dashed line)* and the sum *(solid line)* of internal and external photoelectron yield, setting in at the workfunction. Adopted from [61]

photoemission by a factor of 5 for photon energies 1 eV above the value of the work function. Thus internal photoemission would be the ideal technique to probe for the threshold effect, see below. Further calculations at charged surfaces concern non-linear spectroscopies: Second harmonic generation at a charged silver–jellium is calculated in agreement with the experiment at a silver electrode [60]. The recent theoretical results [62] for Sum-Frequency Generation (SFG) at a charged silver surface are displayed in Fig. 12 as a function of the tuned frequency ω_1, whereas the infrared frequency is kept constant, corresponding to 0.2 eV (about 1600 cm^{-1}). The structures in the spectra correspond to the linear response $\text{Im}[d_\perp(\omega)]$ [63]. The value n assuming one free 4 s/p electron per silver atom yields $\hbar\omega_p = 9.2\,\text{eV}$.

The polarizability of the Ag 3δ electrons has been taken into account. Therefore the threshold excitation peak at the uncharged surface at about $0.5\omega_p$ is in good agreement with the workfunction of Ag(111) of 4.5 eV [64].

Whereas at positive surface charge the threshold excitation is lost, it is shifted and increased at a charge of $-5.6\,\mu\text{C}/\text{cm}^2$. The peak at this surface charge in $\text{Im}[d_\perp(\omega)]$ should then be expected at a wavelength of light of 756 nm. Experimentally, the threshold of photoemission into aqueous elec-

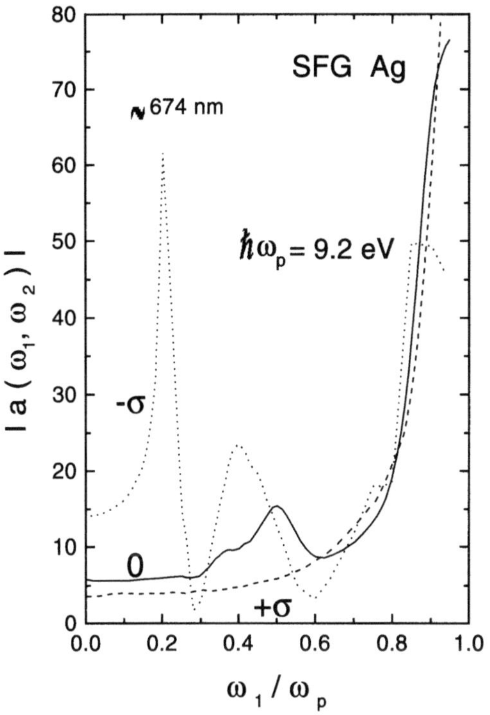

Fig. 12. Sum frequency generation spectra $|a(\omega_1, \omega_2)|$ for semi-infinite Ag (with pseudopotential $V_0 = 3\,\text{eV}$) as a function of fundamental frequency ω_1 with $\omega_2 = 0.1\,\text{eV}$. *Solid curve*: neutral surface; *dashed curves*: charged surfaces with $\sigma = \pm 0.001\text{a.u.} = \pm 5.6\,\mu\text{C/cm}^2$. The lower peaks correspond to the threshold excitation, the upper peaks to the multipole surface plasmon. Taken from [62]

trolytes is given by [7]

$$\hbar\omega = 3.23\,\text{eV} + E_{\text{NHE}}\,.$$

The threshold can be shifted to lower values at cathodic potentials of the electrode ([7,71,110] and references therein), for details see also Sect. 5.1.

4.2 Inverse Photoemission with Hot Electrons, Internal Photoemission

The threshold excitation at $\hbar\omega \sim \Phi$ can be approximately described by the photon annihilation process, i.e. by photoemission (PE):

$$e(E_{\text{F}} - \delta) + \hbar\omega \rightarrow e(E_{\text{F}} + \hbar\omega - \delta) \qquad (\delta \text{ is small})\,. \tag{3}$$

The deexcitation process of a hot electron by photon creation can be described by inverse photoemission (IPE):

$$e(E_{\text{F}} + \hbar\omega - \delta) \rightarrow e(E_{\text{F}} + \delta) + \hbar\omega', \hbar\omega' = \hbar\omega - 2\delta\,. \tag{4}$$

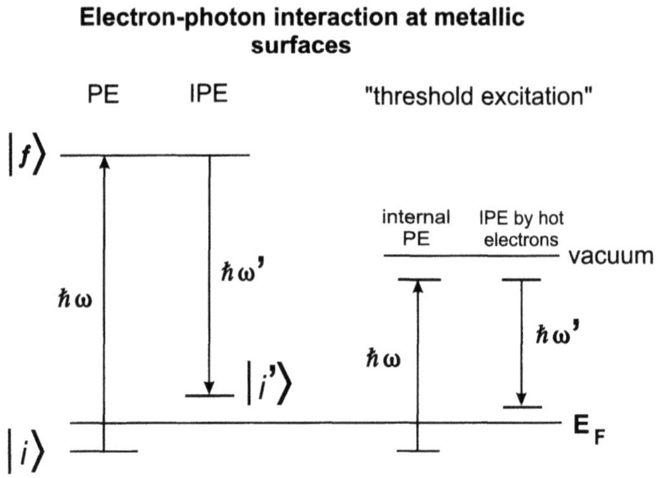

Fig. 13. Common photoemission (PE) and inverse photoemission (IPE) with state $|f>$ far above the photoemission threshold. *Right part*: threshold excitation with $|f>$ close to the photoemission threshold

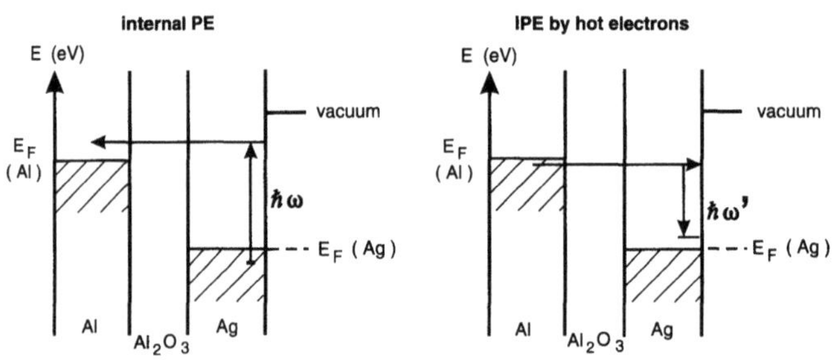

Fig. 14. Scheme of internal photoemission and inverse photoemission with hot electrons in MIM tunnel junctions. The hot electron level is below the vacuum level on the silver side

PE and IPE are time-inverted processes when one neglects δ like in the case of photoemission and inverse photoemission at photon energies exceeding Φ (see Fig. 13). Therefore it is possible to search for the threshold effect both in PE and IPE. For $\hbar\omega < \Phi$ we call the process (PE) "internal photoemission" and the process (IPE) "inverse photoemission by hot electrons". With MIM junctions both internal photoemission and inverse photoemission by hot electron spectroscopies become possible, see the schematic Fig. 14.

4.3 Light Emission from MIM Junctions via Intermediate Surface Plasmon Polaritons

Hänisch and *Otto* [27] gave the firm experimental proof that the light emission is caused by direct excitation of surface plasmon polaritons (SPP) at the

Ag–UHV interface where they are converted into propagating light modes by surface roughness. Using a so-called Weierstrass prism [65] (made of rutile, index of refraction $n = 2.8$, or ZrO_2 with $n = 2.2$) in an Attenuated Total Reflection (ATR) configuration is an alternative method to convert SPPs into light without the need to roughen the surface of the tunnel junctions. The prism enlarges the momentum $\hbar\omega/c$ of photons within the prism by the factor n. The SPP can be converted into a photon in the prism with an angle of emission α given by

$$\frac{n\omega}{c}\sin\alpha = k_{\mathrm{p}}(\omega),$$

where k_{p} is the two-dimensional wavevector of the SPP parallel to the surface.

Tunnel junctions were prepared without CaF_2 undercoating and not intentionally roughened, in Otto and Kretschmann configuration [66,67] (see Fig. 15). In Kretschmann configuration the tunnel junction is deposited on the prism. In Otto configuration the tunnel junction is deposited on a glass substrate and the surface of the silver top electrode is separated by an electrolyte gap of the order of the wavelength of the observed light from the prism surface (Fig. 15).

The advantage of the Otto configuration is the possibility to optimize the coupling between SPP and photon with distance between the top electrode and the prism, the disadvantage is the relatively slow electrochemical response due to the narrow gap. The optimal gap width d in our experiment is 200 nm. Figure 16 shows the dependence of light emission on this width.

If the gap exceeds 1 μm thickness the light emission is only caused by the scattering of the SPPs by the residual surface roughness of the silver surface of the tunnel junction. Under optimum outcoupling by ATR the integrated normalized photon yield is increased by a factor of 15.

4.4 Results on IPE by Hot Electrons

4.4.1 IPE at UHV Interfaces [73]

Figure 17 displays the changes of the light emission spectra (normalized by the tunnel currents I_{T}) of two different MIM junctions by coverage with potassium and exposure to oxygen. The upper thresholds of the inverse photoemission spectra seem to be surpassed. This is an artefact due to the crosstalk in the simple proximity image intensifiers. The absolute tunnel current (negative signum for electron current from Al to Ag) increases with potassium coverage and decreases with oxygen exposure. This is discussed in Sect. 6.1.

The upper arrow in Fig. 17 is at the wavelength of the maximum yield according to Fig. 11 at a coverage $c = 1$.

4.4.2 IPE at Electrodes [74]

Figure 18 shows a big change in the normalized intensity of a junction, which was not intentionally roughened, just by changing the potential of the silver

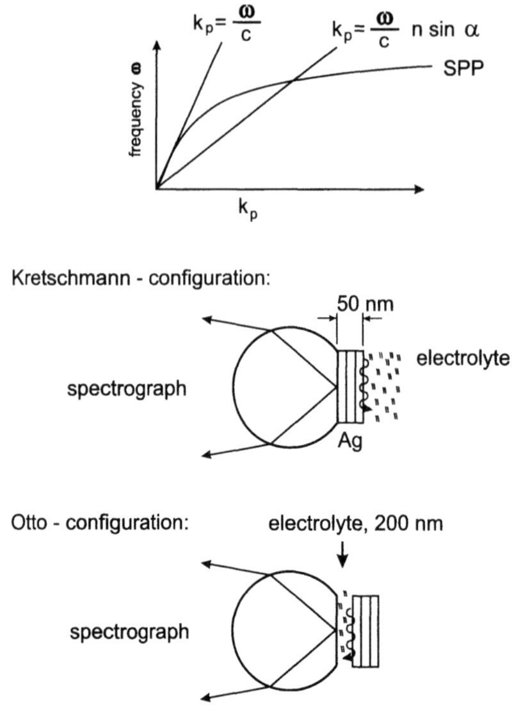

Fig. 15. *First view graph*: Dispersion of surface plasmon polaritons (SPP) and coupling to propagating light modes by attenuated total reflection (ATR). k_P is the absolute part of the two-dimensional component parallel to the interfaces of the wave vector k of the electromagnetic modes of frequency ω. The so-called light line given by $k_p = n_e\omega/c$ (n_e is the index of refraction of the electrolyte) separating propagating light modes in the electrolyte (*left* from the lightline) and evanescent waves (*right* from the lightline). The line $k_p = \frac{\omega}{cn_p \sin\alpha}$ (n_p is the index of refraction of the spheres cut to the shape of Weierstrass prisms $n_p > n_e$) gives the dispersion of the evanescent modes at the prism–electrolyte and silver–electrolyte interfaces for the light wave incident from the prism side on the prism base at an angle α. The crossing point of this line with the curved line of the qualitative dispersion of the SPP at the silver–electrolyte interface denotes the SPP resonance by ATR. *Second and third view graph*: Outcoupling from SPPs at the Ag–electrolyte interface of MIM systems in Kretschmann and Otto configuration with Weierstrass prisms

electrode by 0.1 V. Again the current increases simultaneously by a factor of 2.5, therefore the non-normalized intensity at the constant tunnel voltage U_T increases by about a factor of 25. The conversion rate at $E_{SCE} = -1.1$ V is about 10^{-4} photons/(electron in I_T).

A so-called "activation" (weak oxidation–reduction cycle with a transferred charge smaller than one monolayer) of the silver electrode of the MIM junction by a weak oxidation–reduction cycle does not change the integrated photoyield per tunnel current (Fig. 19), whereas this procedure is very im-

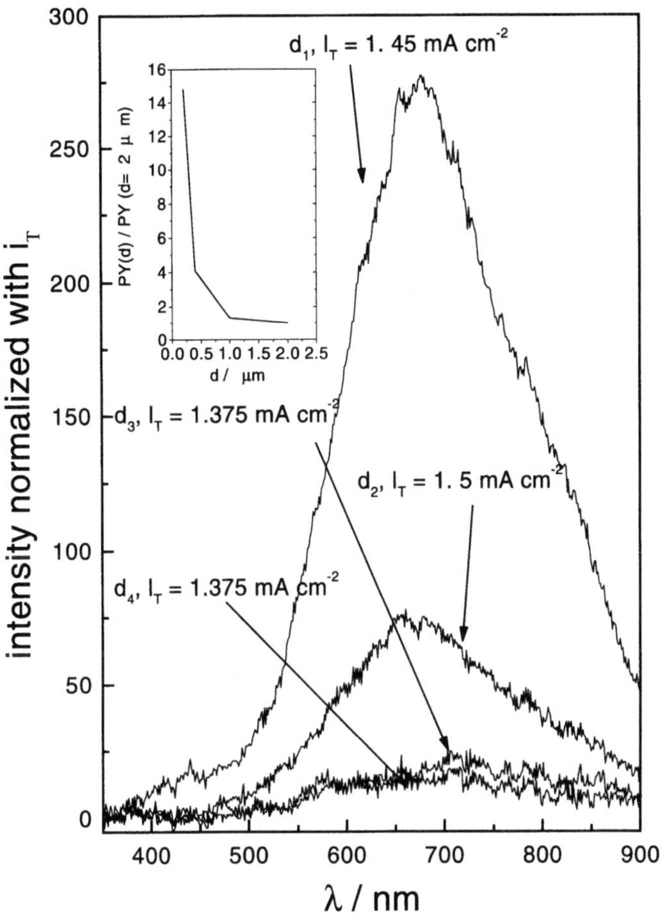

Fig. 16. normalized light emission spectra of a "smooth" tunnel junction in Otto configuration at constant tunnel voltage and electrochemical potential. The parameter d is the width of the electrolyte gap, $d_1 = 200\,\text{nm}$, $d_2 = 400\,\text{nm}$, $d_3 = 1\,\mu\text{m}$, $d_4 = 2\,\mu\text{m}$. Inset: integrated normalized emission as function of d

portant in hot electron chemistry, see Sect. 5. Therefore we are allowed to compare and discuss these results in Sect. 4.6 with the theory of the threshold excitation at smooth surfaces.

The long wavelength limit at 800 nm of the light emission spectra in Figs. 16–18 corresponds to a photon energy of 1.55 eV. This is more than half of the maximum energy of the tunneling electrons above the Fermi level of Ag. Since the number of secondary hot electrons above this limit is negligible, the light emission can be nearly exclusively assigned to primary hot electrons.

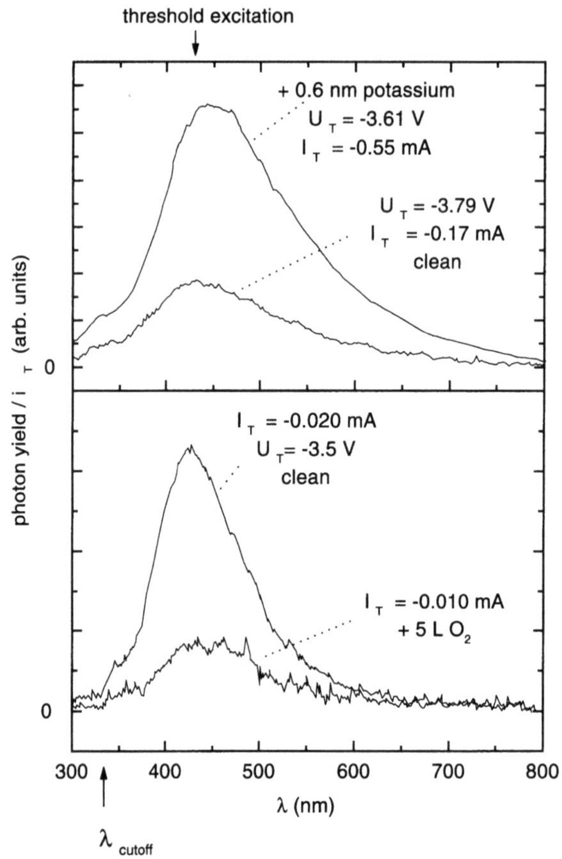

Fig. 17. Light emission spectra of rough tunnel junctions, normalized by the tunnel current I_T as a function of wavelength λ before and after the deposition of 0.6 nm K (*upper graph*) and before and after the exposure to 5Ł of O_2 (*lower graph*) at 40 K. Taken from [27]. The *upward arrows* give the expected cut-offs of the spectra, the *downward arrow* the position of the maximum of the threshold excitation in Fig. 11

4.5 Results on Internal Photoemission

MIM junctions were grown on optical gratings inscribed in the SiO_2 layer on Si wavers [75]. The grating period was 500 nm, the amplitude was 20–30 nm [76]. The size of the Ag electrode was $2 \times 2\,mm^2$. Surface plasmon polaritons were excited by a p-polarized laser beam of 632 nm wavelength, chopped with 30 Hz under the matching angle of incidence. The internal photoemission current was monitored by the current from the Al electrode to the ground with a lock-in amplifier. This allowed the discrimination of the internal photocurrent from the DC tunneling current when a voltage U_T was put on the junction. The MIM junctions were investigated in aqueous ClO_4^-

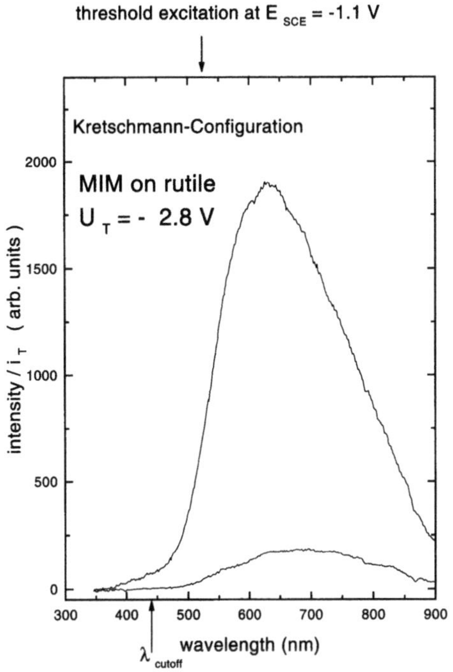

Fig. 18. Light emission spectrum of a MIM junction prepared on a rutile Weierstrass prism [65] in Kretschmann configuration in the sequence rutile /Al/AlO$_x$/Ag/electrolyte, at the potentials of the silver electrode $E_{\mathrm{SCE}} = -1.0\,\mathrm{V}$ and $E_{\mathrm{SCE}} = -1.1\,\mathrm{V}$. The upward arrow gives the expected cut-offs of the spectra at $U_{\mathrm{T}} = -2.8$ eV, the downward arrow the position of the level PE (photoemission, explained in detail in Sect. 5.1) at $E_{\mathrm{SCE}} = -1.1\,\mathrm{V}$ (2.38 eV above E_{F} of the silver electrode)

electrolytes at zero tunnel voltage U_{T}. The potential of zero charge (pzc) was evaluated by measurements of the differential capacity at about $E_{\mathrm{SCE}} = -0.8$ to $-0.9\,\mathrm{V}$. Hydrogen evolution sets in at about $E_{\mathrm{SCE}} = -1.2\,\mathrm{V}$. Both in the internal and external photocurrents there are structures at potentials near the pzc, Fig. 20.

There is a conspicuous low internal photocurrent at $-0.4V_{\mathrm{SCE}}$. By applying negative U_{T} (albeit in air) [75], the MIM junction acts as a retarding field spectrometer. The internal photocurrent was observed up to about $U_{\mathrm{T}} = -0.5\,\mathrm{V}$, which excludes a thermal origin of the internal photocurrent. However, no hot electrons above this threshold were observed.

4.6 Discussion of the Optical Results

4.6.1 Inverse Photoemission

Particularly noteworthy is the fact that for the MIM junctions in ATR configuration inverse photoemission is only observed when the junctions are biased

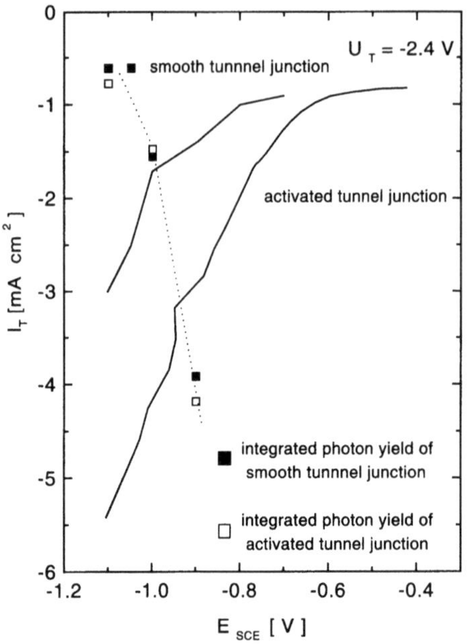

Fig. 19. Integrated photon yield of smooth and activated tunnel junctions in Kretschmann configuration as function of the electrode potential

to allow hot electrons to reach the interface. Hot holes do not deliver any light emission. The threshold excitation takes electrons from just below the Fermi level to the photoelectron emission threshold; excitations of holes from just above the vacuum level to electronic levels below the Fermi level are not involved. Thus, in the time-reversed case, inverse photoemission is only expected by primary hot electrons, in agreement with experiment. The results above are in qualitative agreement with the theory of the threshold excitation. The maxima of emission are about at the expected wavelengths. The spectral distribution $I(\omega_0)$ of the light emission spectrum depends on

- the tunnel current of the primary hot electrons as a function of their energy; this current is influenced by the dephasing at the outer silver interface (Sect. 6);
- the distribution of final states after the interacton of the primary hot electrons with the surface plasmon polaritons (Fig. 14);
- the frequency-dependent outcoupling of the SPPs at statistically rough surfaces and in the ATR configurations.

The big increase of the integrated intensity in Fig. 18 is probably not only caused by an increase of the surface charge of the silver electrode, but also by a change in the inner Helmholtz plane, e.g. by desorption of acetate anions. In this way the electronic tail, responsible for the threshold effect, can develop.

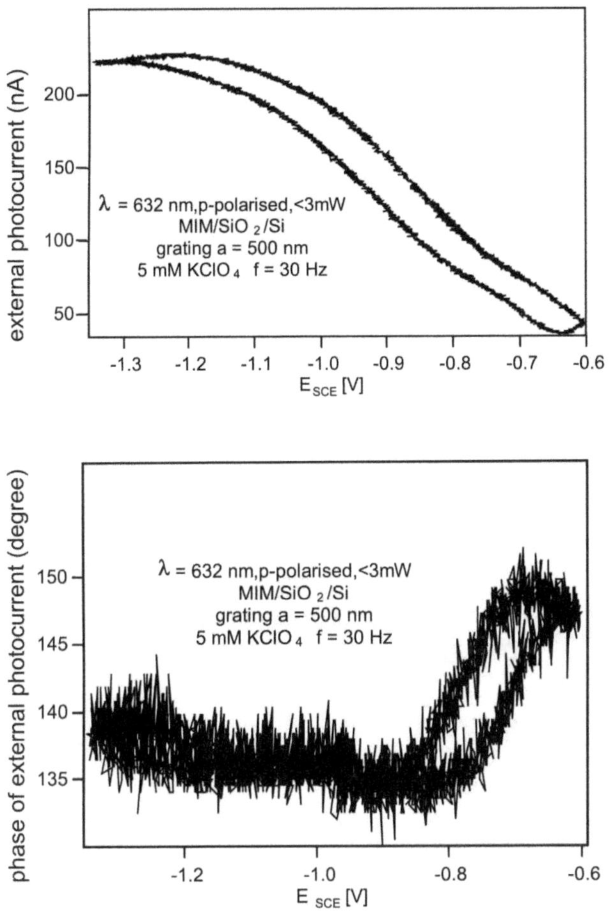

Fig. 20. Internal photo-induced current, absolute part and phase with respect to the signal from a photodiode behind the chopper for a MIM junction in perchlorate electrolyte on an optical grating

4.6.2 Internal Photoemission

The internal current in Fig. 20 is connected with a photoemission current into the electrolyte. This may be the external photocurrent connected with the threshold effect. However, *Körwer* [77] detected weak subthreshold photoemission into the electrolyte at roughened silver electrodes. We are not yet able to discriminate between the two effects. Further, there are definite retardations in the internal photoemission on the order of 1/30 s, probably indicating optical energy deposition in the Helmholtz layer within this time, eventually released by photoemission. The missing hot electrons at the energy of the incident photons in the retarding field experiment (Sect. 6) may

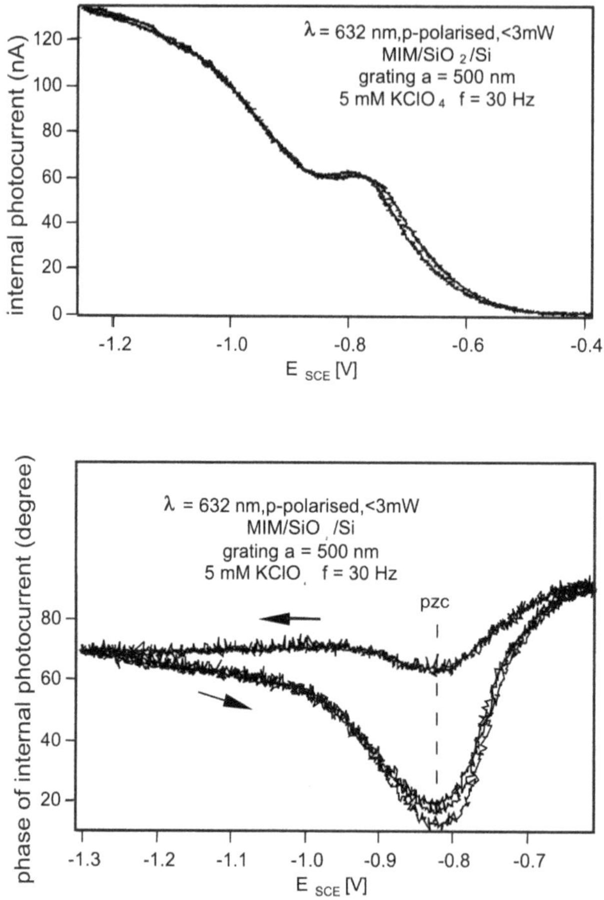

Fig. 21. Same plot as Fig. 20 for external photo current

be caused by stronger electron–electron scattering compared to inverse light emission, see below. More work is needed on this point.

4.7 Optical Surface and Bulk Absorption of Free-Electron Metals

In all experiments on inverse photoemission, interface conditions can be found (oxygen coverage, anodic potential) where the light emission is very small. This means that light emission from the bulk of silver is negligible. This is surprising because one might expect electron–photon interactions of the hot electrons in the bulk within the optical skin depth, independent of the surface or interface conditions. These hot electrons up to an energy of $-eU_{\mathrm{T}}$ above E_{F} do exist, as proven by the light emission in the visible range at other surface conditions (potassium coverage, cathodic potential). Reference [117] contains a detailed discussion of the hypothesis that the optical excitation of hot electrons at the surface or interface prevails over the bulk excitation. Here we only give a short summary.

Within the first few fs after incidence of a light pulse on the metal surface, all electrons near the surface are collectively influenced by the external field, and keep their phase relation. This leads to the collective coherent effect of specular reflection at a plane surface or to surface plasmon resonances in small clusters [78].

It was believed that hot electrons with energies several eV above E_F but below the onset of direct d to SP band transitions are excited in the bulk via phonon assisted intraband transition within the sp-band, or that before dephasing the non thermal (NT) deviation (see also Fig. 22) Δf_{NT} from the bulk Fermi–Dirac distribution $\Delta f_0(E)$ (at the sample temperature before irradiation) with photons of energy $\hbar\omega$ is given by [79,80]

$$\Delta f_{NT} = \Delta f_{NT}^0 [f_0(E - \hbar\omega)(1 - f_0(E)] - f_0(E)[1 - f_0(E + \hbar\omega)],$$

where E is the electron energy, and Δf_{NT}^o is proportional to the intensity of the irradiation (see Fig. 22).

We conjecture that hot electrons with excitation energies above 1 eV but below the onset of interband transitions are only created by the surface photo-effect. The threshold excitation, calculated by TDLDA, has a homogeneous halfwidth of about 1 eV [61] corresponding to a lifetime of about 4fs. After breaking the coherence, single photoelectrons carrying all the excitation energy of the "collective surface excitation" are emitted, as observed in external and internal photoemission. The reasons for this are further discussed in [117]. In summary, it is not easy to obtain information on the initial electronic coherent state by time-resolved spectroscopy. Though MIM junctions

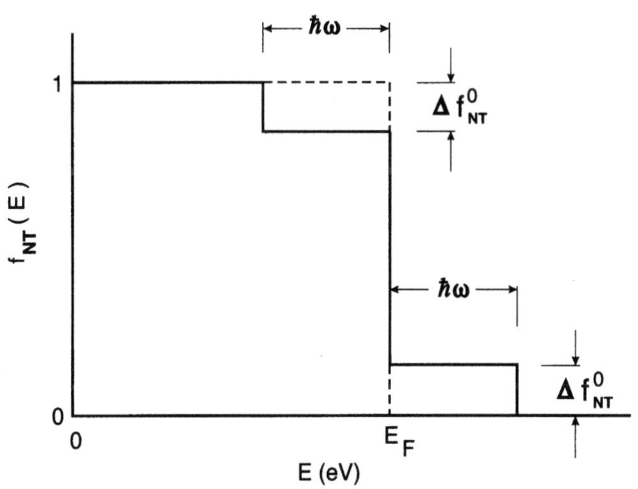

Fig. 22. Hypothetical non-thermal deviation of Fermi distribution due to excitation with photon $\hbar\omega$, according to [79,80].

are slow devices, they can deliver "ab-initio" information on the energetic distribution of the electronic excitations involved in the initial coherent state and discriminate between surface and bulk excitations.

5 Chemical Reactions with Hot Electrons

5.1 Emission of Hot Electrons into the "Dry Electron" State

Immediately after emission of the photoelectron from the metallic electrode into the electrolyte the permanent dipoles of the water molecules are not yet rotated by the photoelectron. The delocalized state is called the "dry electron state". The lowest dry electron state constitutes the lower edge of the electronic conduction band of water. This threshold for photoemission [91] is called PE in Fig. 10. The solvated electronic states cannot be reached directly by photoemission. Usually the photoemitted electron induce eventually faradaic reactions or attach to a scavenger or return to the electrode. PE can be populated by electrons in the energetic ground state at a certain electrode potential. The following values for PE can be found in the literature.

PE	$-3.4V_{\mathrm{SCE}}$	[93]
PE	$-3.3V_{\mathrm{SCE}}$	[94]
PE	$-3.26V_{\mathrm{SCE}}$	[95]
PE	$-3.23V_{\mathrm{NHE}}$	[7]

Körwer found 3.44 $\mathrm{V_{SCE}}$ for polished polycrystalline silver electrodes by photoemission [96]. A transfer of electrons to the dry electron state is also possible by hot electrons generated by tunnel junctions, which was shown first by *Diesing* [71,103] on silver electrodes.

In Fig. 23 the electrode current is plotted as a function of the tunnel voltage for different electrode potentials. One can observe for example for $E_{\mathrm{SCE}} = -0.8\,\mathrm{V}$ a sharp increase of the cathodic current at $U_{\mathrm{T}} = -2.5\,\mathrm{V}$. Comparing the different curves in Fig. 23 delivers two aspects:

- At an electrode potential of $E_{\mathrm{SCE}} = -0.2\,\mathrm{V}$ it is not possible to influence the electrode current by tunnel voltages $U_{\mathrm{T}} > -3.0\,\mathrm{V}$.
- The curves taken at $E_{\mathrm{SCE}} = -0.7\,\mathrm{V}$ and $E_{\mathrm{SCE}} = -0.8\,\mathrm{V}$ are shifted relative to each other for 110 mV tunnel voltage U_{T}.

When one plots the tunnel voltage where the cathodic current starts (Fig. 23), versus the electrode potential, one obtains the dots in Fig. 24.

For comparison a straight line fullfilling

$$eE_{\mathrm{SCE}} + U_{\mathrm{T}} = -3.3\,eV \tag{5}$$

is given. This means that the threshold for photoemission of $-3.4\,\mathrm{V_{SCE}}$ is also valid for primary hot electron emission into the dry electron state by tunnel

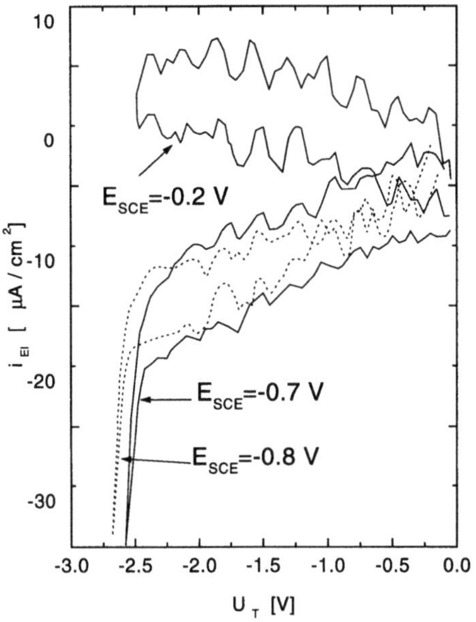

Fig. 23. Electrolyte current plotted versus tunnel voltage at fixed electrode potentials of $E_{\mathrm{SCE}} = -0.2\,\mathrm{V}$, $E_{\mathrm{SCE}} = -0.7\,\mathrm{V}$, $E_{\mathrm{SCE}} = -0.8\,\mathrm{V}$. Tunnel current $I_{\mathrm{T}} \approx 5\,\mathrm{mA/\,cm^2}$ at $U_{\mathrm{T}} = -2.6\,\mathrm{V}$

junctions. It also confirms that part of the tunneled electrons passes the top electrode elastically. The measured current densities into the PE level are several hundred times higher than that of photoemission by cw-lasers [93], when one considers the fact that a tunnel current of around $10\,\mathrm{mA/\,cm^2}$ causes primary hot electron emission into the PE of $10\,\mathrm{\mu A/\,cm^2}$. This means a transfer ratio Q of tunneled electrons from the Al-electrode into the dry electron state of around $\mathrm{Q} \approx 10^{-3}$.

5.2 Hydrogen Evolution

A fraction of the primary hot electrons reaches ballistically the outer surface of the silver top electrode. The top electrode is connected simultaneously as the working electrode within an electrolyte. Figure 10 shows the levels of the electrons at the silver–electrolyte interface at an electrochemical potential of $0\,V_{\mathrm{SCE}}$. For an applied electrochemical potential negative of $-0.64 V_{\mathrm{SCE}}$ the H_2O/H_2 level (Pt) falls below the Fermi level of Pt and hydrogen evolution sets in, see the voltammogram in Fig. 25a.

Comparison with the voltammogram of the activated silver electrode of the tunnel junction at $U_T = 0\,\mathrm{V}$ (scan b) shows the onset of the Hydrogen Evolution Reaction (HER) only at about $-1.2 V_{\mathrm{SCE}}$, corresponding to a cathodic overpotential of about $+0.56 V_{\mathrm{SCE}}$. This behavior is nearly unchanged when a negative U_{T} is applied to the junction, provided this has not been

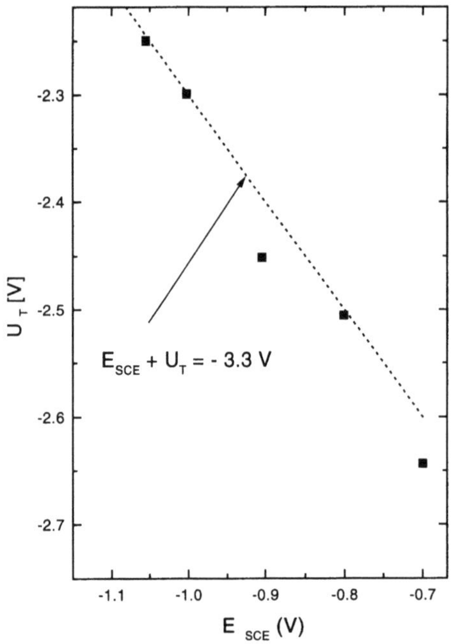

Fig. 24. Onset tunnel voltage of starting electron emission to the dry-electron level PE (see Fig. 10) plotted versus electrode potential. A *straight line* fullfilling $E_{\mathrm{SCE}} + U_{\mathrm{T}} = -3.3\,\mathrm{V}$ is given for comparison

"activated" or prepared rough intentionally. However, when one prepares the junction on a rough CaF_2 substrate [1] or "activates" the silver electrode by a weak oxidation–reduction cycle ($E_{\mathrm{SCE}} = +0.5V, t = 0.2\,\mathrm{s}$) the HER sets in at more positive potential. Moreover the onset potential of HER is shifted by 300 mV to positive potentials, when the tunnel voltage is increased from $-1.8\,\mathrm{V}$ to $-2.1\,\mathrm{V}$ (Figs. 25d,e). This is clearly anodic with respect to the redox potential $H_2O/H_2(Pt)$.

Detailed experiments under time resolution and at various values of U_{T} have demonstrated convincingly [86] that these reactions are not caused by ionic processes in the aluminum oxide film, nor by initial transition of the hot electrons into the levels of the electron in water given in Fig. 10.

In order to exclude pathways via the wet electron states W (Fig. 10), we performed analoguous experiments at tunneling voltages U_{T} between 0 and $-0.82\,\mathrm{V}$ (Fig. 26). Experiments with voltammograms at more negative U_{T} have been included in this figure.

Clearly the energy of the hot electrons at the low tunneling voltages are considerably below the wet electron states. Nevertheless about the same fraction R of the tunnel current continues as cathodic current into the electrolyte, as is evident by comparing the ratios R given in Fig. 25. The hot electrons overcome the kinetic hindrance of the HER, the hot electron reaction takes

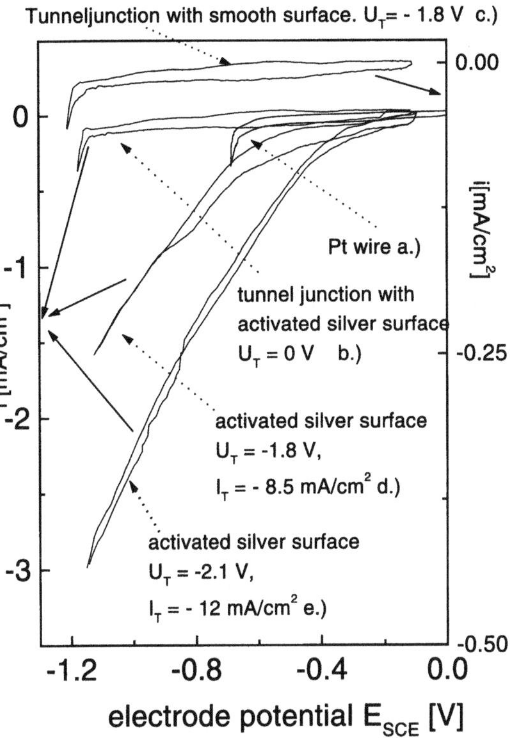

Fig. 25. Cyclovoltammograms of Pt wire and tunnel junctions in 0.9 M NaAc buffer (50% water, 50% ethylene glycole), (**a**) Pt wire, (**b**) activated top electrode $U_T = 0.0\,V$, (**c**) unactivated top electrode $U_T = -1.8\,V$, (**d**) activated top electrode $U_T = -1.8\,V$, (**e**) activated top electrode $U_T = -2.1\,V$

place in the electrochemical potential range between the potentials of the HER at platinum at about $-0.74V_{SCE}$ and at silver at about $-1.2V_{SCE}$. Weak reactions are observed even at $E_{SCE} > -0.74\,V$.

The fully hydrated *s*-like state of the electron (level H in Fig. 10) plays no role in HER because it is a stable state not known to split water.

5.3 Hydrogen Oxidation

Cathodic potential pulses from $-1.0V_{SCE}$ to $-2.0V_{SCE}$ were applied to an epitaxial Ag(111) film in neutral 0.1 M $KClO_4$ aqueous electrolyte for 0.4 seconds (Fig. 27b). During this time a faradayic cathodic charge Q^- of $4\,mC/cm^2$ due to hydrogen evolution was accumulated, which corresponds to several monolayers (the charge corresponding to one electron per Ag surface atom of Ag(111) is $222\,\mu C\,cm^2$). The integral of the anodic transient after switching back to $-1.0V_{SCE}$ is just the charge Q^+ involved in changing the voltage of the electrochemical double-layer capacity. However, if the surface and the

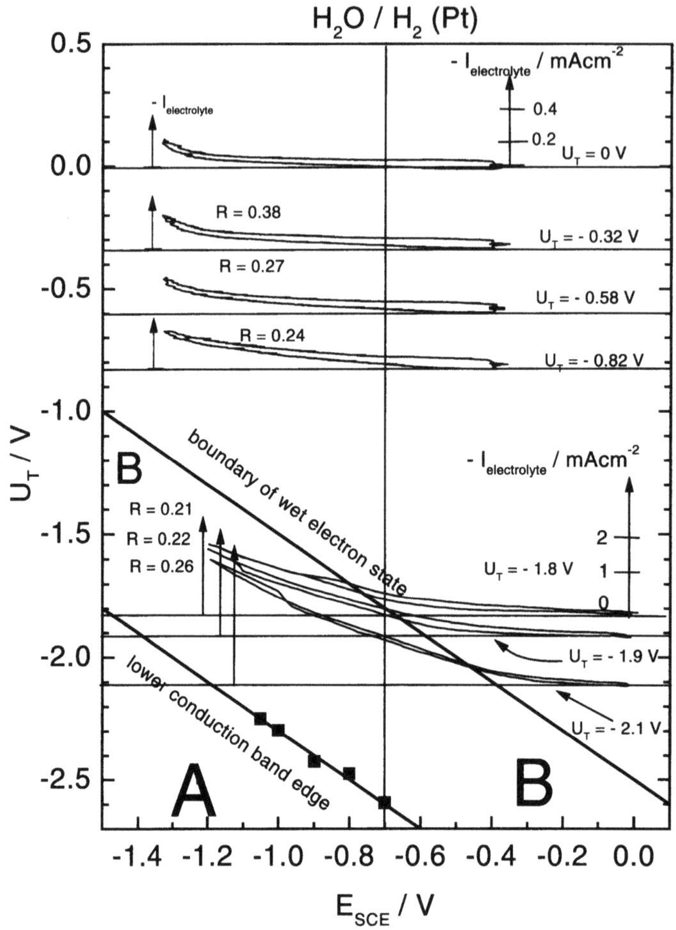

Fig. 26. *Region A:* Range of tunneling voltage U_T and of electrochemical potential E_{SCE} in which electron emission into the electronic conduction band of the electrolyte is possible; see also Fig. 24. *Region B:* Range of U_T and E_{SCE} in which electron emission into the wet state (p-type state) of the elctron in the electrolyte must be considered. Seven voltammograms (cathodic current as a function of E_{SCE} at different values of U_T. The base lines ($I_{electrolyte} = 0$) of these are positioned at the proper values of U_T. R is the ratio of $I_{electrolyte}$ induced by the tunnel current I_T and I_T at the most negative values of E_{SCE}. The vertical line H_2O/H_2(Pt) is the onset potential of the hydrogen evolution for a Pt electrode in the same electrolyte

double layer capacity of this Ag(111) film is increased by about 10% (by a short anodic pulse, leading to silver oxidation and reduction), the charge in the anodic transient corresponds to about 0.7 monolayers on top of the charge to account for the reordering of the double layer. This charge has been assigned to the reoxidation of a layer of atomically adsorbed hydrogen [90].

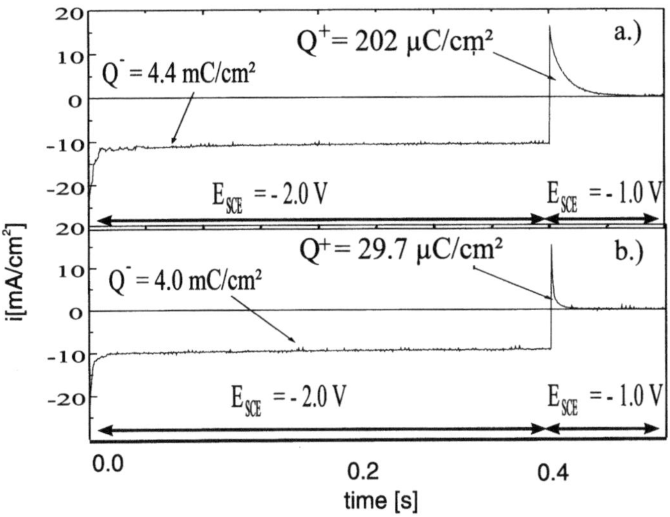

Fig. 27. Current transients on (**a**) activated Ag(111) and (**b**) smooth Ag(111) surfaces in 0.1 M $KClO_4$ after potential pulses from $E_{SCE} = -1.0$ V to $E_{SCE} = -2.0$ V and back from $E_{SCE} = -2.0$ V to $E_{SCE} = -1.0$ V

Also this reaction is related to hot electron chemistry, as demonstrated with a tunnel junction [88].

If an electrode with adsorbed hydrogen atoms is set by a potential step to an electrochemical potential of, for instance $-0.9\,V_{SCE}$, that is 0.3 V above the onset of hydrogen evolution, one may expect the injection of hot electrons with an energy of up to 0.3 eV above the Fermi level of the silver electrode. If these electrons from the reoxidation of adsorbed hydrogen tunnel into the Al electrode, one may use the junction as a retarding field energy spectrometer.

In Fig. 28 the tunnel voltage U_T is varied in steps from "accelerating" at 0.3 V to "retarding" at -0.4 V. The transients in the tunnel current I_T after switching back to $-0.9\,V_{SCE}$ are obvious. At $U_T = -0.2$ V the transient is not yet fully suppressed, demonstrating that we do not observe a thermal effect. The constant tunnel currents induced by the application of U_T is seen at about 0.1 after the end of the anodic transients.

Again we need a roughened silver top electrode of the tunnel junction to observe the anodic tunnel pulses. This and the previously mentioned results in Sect. 5.2 call for an explanation of the role of roughness in hot electron transfer.

The contribution of secondary hot electrons below 1 eV is possible, because as demonstrated in Fig. 26 primary hot electrons below 1 eV also induce HER. However the relative contribution of primary and secondary electrons to HER cannot be predicted, because primary and secondary electrons reaching the outside silver interface will have different angular distributions. In the "dephasing model" (Sect. 6), the tunnel current is partly controlled

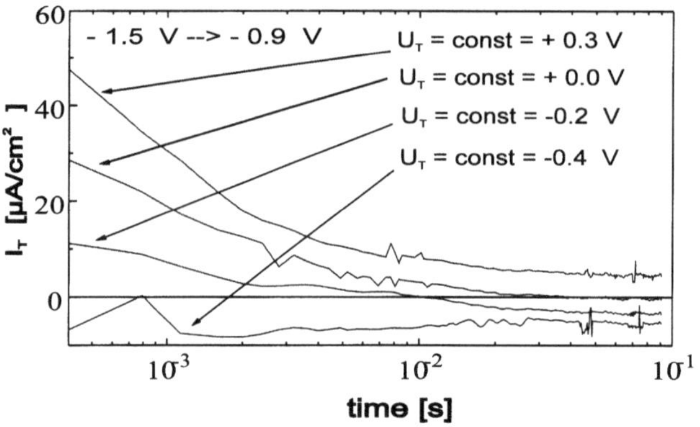

Fig. 28. Tunnel current at $E_{\mathrm{SCE}} = -0.9\,\mathrm{V}$ directly after the end ($t = 0\,\mathrm{s}$) of a cathodic pulse from $-1.5 V_{\mathrm{SCE}}$ to $-0.9 V_{\mathrm{SCE}}$ of 100 ms duration. The tunnel voltage is set from "accelerating" (+0.3 V) to retarding (−0.4 V). The time axis is scaled logarithmically

by the dephasing of the primary electrons at the outside silver interface. The incidence rate of secondary hot electrons at the interface will scale with the tunnel current, but will also depend on the energy of the secondary electrons.

5.4 Model on the Role of Roughness in Hot-Electron Transfer

5.4.1 On Normal Hydrogen Evolution

By X-ray scattering at the interface of Ag(111) with an aqueous NaF electrolyte, *Toney* et al. [97] observed a layered structure of the oxygen distribution. At a potential of $E_{\mathrm{SCE}} = -0.9\,\mathrm{V}$ the maximum of the distribution was about 0.35 nm in front of the top (111) plane of the Ag ion cores. This distance changed to about 0.28 nm at the potential of $E_{\mathrm{SCE}} = -0.15\,\mathrm{V}$. This difference was assigned to a reorientation of the water molecules in the inner Helmholtz plane, the oxygen end situated nearer to the interface at anodic potentials and the hydrogen end nearer to the interface at cathodic potentials.

Theories of hydrogen evolution in low pH electrolytes involve potential dependent reorientations of the water molecules in the first layer and hydronium ions in the second layer (the outer Helmholtz layer) and proton transfer (e.g. [101] and references therein). Hydrogen evolution in neutral to basic electrolytes probably involves tunneling of electrons from the metal into water molecules H_2O_{iHp} in the inner Helmholtz plane. This electron transfer can split the molecule and drive the Volmer reaction (for more details see [90]):

$$H_2O_{\mathrm{iHp}} + e^- \rightarrow OH^-_{\mathrm{iHp}} + H_{\mathrm{ad}} \,. \tag{6}$$

Further charge transport probably involves the steps of proton transfer between the inner and outer Helmholtz plane:

$$OH^-_{\mathrm{iHp}} + H_2O_{\mathrm{aHp}} \rightarrow H_2O_{\mathrm{iHp}} + OH^-_{\mathrm{aHp}} \tag{7}$$

and a modified Grotthus mechanism in the bulk electrolyte by proton transfer

$$OH^- + H_2O \rightarrow H_2O + OH^- . \tag{8}$$

5.4.2 On Hydrogen Evolution by Hot Electrons

Hydrogen evolution by hot electrons takes place at less cathodic potential than the usual hydrogen evolution which involves only "cold electrons" of energy at and below the Fermi surface. In this case the ordering of the water molecules is probably not the same as at the onset of the usual "cold electron hydrogen evolution". The basis of our model of hydrogen evolution from H_2O by hot electrons is as follows: The tunneling probability grows quasi-exponentially with the extension of the tail of the electronic wavefunctions into the double layer. At a certain threshold of the "penetration depth" the Volmer reaction starts. The penetration depth is a function of the electronic energy and momentum parallel to the surface and the applied electrochemical potential. The energy–momentum relation of the bulk electrons (which are involved in the observed transport processes) is given by the bulk band structure. The only information on the electron in the electrolyte we use is the energetic position PE of the lower band of the electronic conduction band, the so-called "dry electron". The penetration length of the bulk electrons of energy E into an electrolyte is calculated analogous to the penetration of free metal electrons into vacuum, with replacement of E_{vacuum} by PE. The electron wavefunction outside of the last layer of the cores of the silver surface atoms is assumed to be free-electron like (we neglect a possible increase of the mass because of interaction with the inner Helmholtz layer) with a relation between wave vector $\boldsymbol{k}$ and energy E given by

$$\frac{k^2}{2m} = E - PE, \quad k = |\underline{k}| \tag{9}$$

The wavevector $\boldsymbol{k} = (k_x, k_y, k_z)$ has the normal component k_z, which is imaginary when

$$E < PE. \qquad \text{In this case, we set} \quad k_z = \mathrm{i}\kappa . \tag{10}$$

In (9) E is the energy of the bulk electron and $k_{||} = (k_x, k_y)$ is the tangential component of the bulk part of the electronic wavefunctions, because of the continuity of the wavefunction across the metal–electrolyte interface. The decay length L of the wavefunction of electrons within the electrolyte is $1/\kappa$.

The difference PE – E_F varies with the electrochemical potential E_{SCE} according to [59,103]:

$$PE - E_F = 3.3\,\text{eV} + eE_{SCE} . \tag{11}$$

At $k_{||} = 0$, L diverges when E reaches PE, because the electron can now enter the bulk electrolyte, as observed by photoemission into the electrolyte (hence the abbreviation PE [59,103]).

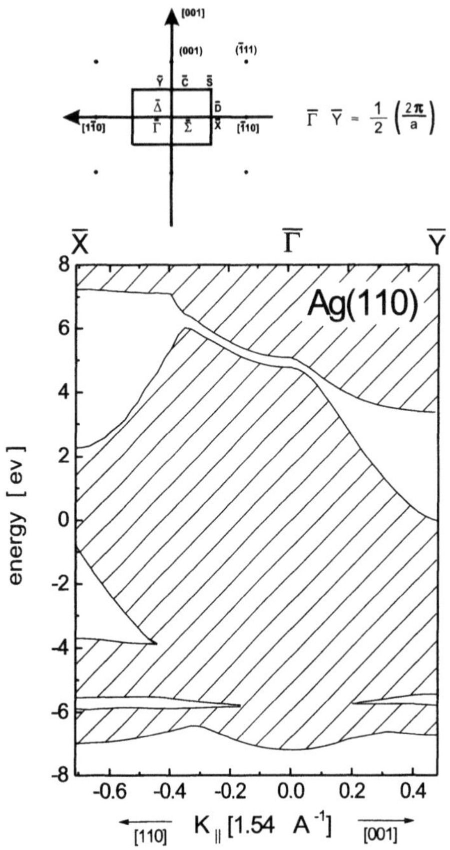

Fig. 29. Surface projected bulk electron density of states (*shaded*) of Ag(110) in the $\bar{X}-\bar{\Gamma}-\bar{Y}$ direction. The $k_{||}$ directions are indicated in the surface Brillouin zone above. The energy E is given relative to E_F. Positive energy values correspond to the energy E_{hot} of hot electrons (12). Data from a self-consistent relativistic density functional calculation [105]

5.4.3 On the Role of Roughness

Our general argument on the role of surface roughness is derived from the fact that there are gaps in the surface projected bulk electron density of states for the silver (111) and (100) surface [104].

That means that for the (111) oriented electrodes there are only propagating bulk hot electrons at $E - E_F > 2\,\text{eV}$ with components k_x and k_y of $\boldsymbol{k}$ above minimum values. Consequently hot electrons propagating normal to the surface do not exist. Related effects have been clearly observed in secondary-electron-emission and absorbed-current spectra of Ag(111), (100) and (110) [64]. The hot electrons arrive at (111) surfaces only at oblique incidence and therefore have a relativly small decay length L. But when the surface is not perfect, these electrons may be scattered by atomic scale roughness into states with larger L. Larger L means that the tunneling probability to the lowest unoccupied molecular orbital (LUMO) of adsorbed molecules, atoms, and ions grows.

We try to quantify this for the Ag(111) surface. For hot electrons in silver films of (111) orientation there is, caused by a gap in the surface projected density of bulk electronic states [104], a minimum value

$$k_m(E_{\mathrm{hot}}) \text{ of } k_{||} = (k_x^2 + k_y^2)^{\frac{1}{2}}$$

as a function of

$$E_{\mathrm{hot}} = E - E_{\mathrm{F}} = -eU_{\mathrm{T}} \tag{12}$$

(Fig. 30).

These conditions given, one easily derives:

$$L \leq L_{\mathrm{max,smooth}} = |2m/(-eU_{\mathrm{T}} - 3.3\,\mathrm{eV} - eE_{\mathrm{SCE}}) - k_m^2(E_{\mathrm{hot}})|^{-\frac{1}{2}} . \tag{13}$$

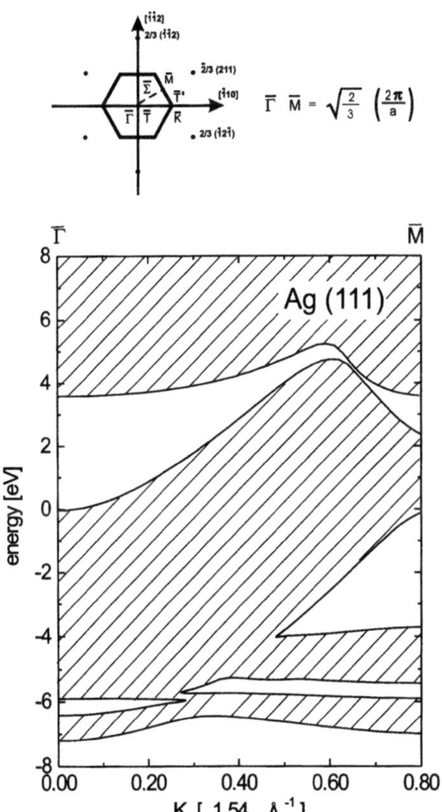

Fig. 30. Like Fig. 29, but for Ag(111) and the $\bar{\Gamma} - \bar{M}$ direction

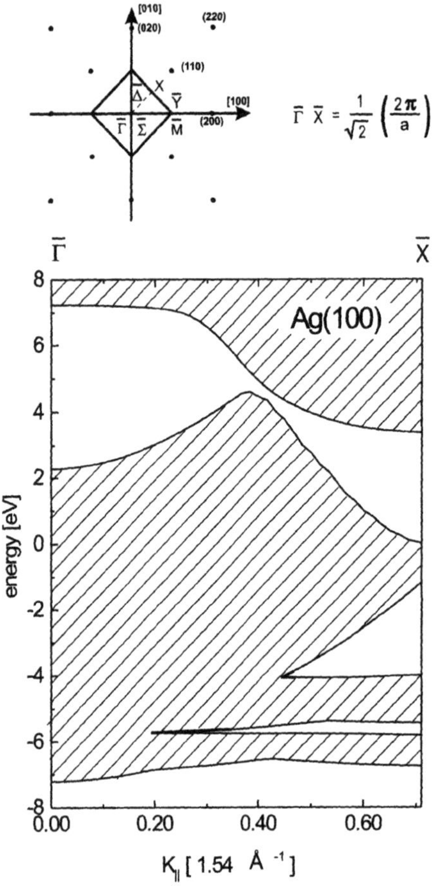

Fig. 31. Like Fig. 29, but for Ag(100) in the $\bar{\Gamma} - \bar{X}$ direction

Figure 32 shows $L_{max,smooth}$ as function of E_{SCE} for eU_T equal to 0, −1.8, and −2.1 eV, Fig. 33 at $U_T = -1.0$ V and −0.5 V

Our hypothesis is the elastic scattering of hot electrons by atomic scale surface roughness to laterally (parallel to the surface) and perpendicular (versus the interior of the crystal) damped states with the real part of the parallel wavevektor $k_{||} < k_m(E_{hot})$, so-called virtual states in the gap of the projected density of states in Fig. 30. These virtual states may comprise the Shockley surface states which are at the clean surface at $k_{||} = 0$ below the Fermi energy, but which may be pushed upward by the adsorbates. They are orthogonal to the bulk states at a perfect smooth low-index surface, but atomic scale roughness (e.g. fuzzy steps) may allow conversion of hot electrons into these virtual state electrons. The scattering into virtual states will increase the right-hand side of relation (13), reaching its maximum for $k_{||} = 0$.

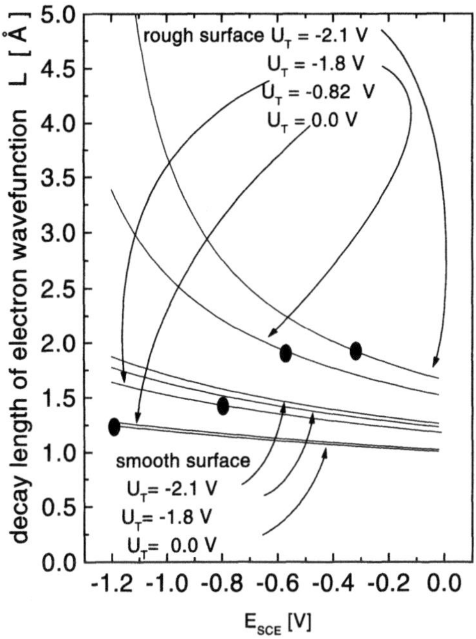

Fig. 32. Maximum decay length L of the electron wavefunction within the electrolyte as a function of electrochemical potential E_{SCE} for smooth and rough silver(111) electrodes according to (13) and (14), respectively, for the indicated tunnel voltages U_{T}. Fat points on the curves for rough surfaces, parameter U_{T}, give the potentials of the onset of the hydrogen evolution reaction, taken from Figs. 25 and 26 for parameters U_{T}

The limit is given by:

$$L \leq L_{\mathrm{max,rough}} = |2m/(-eU_{\mathrm{T}} - 3.3\,\mathrm{eV} - eE_{\mathrm{SCE}})|^{-\frac{1}{2}} \tag{14}$$

see also Fig. 32 for eU_{T} equal to 0, -0.82, -1.8, and -2.1 V. Note that $L_{\mathrm{max,rough}}$ depends only on $(U_{\mathrm{T}} + E_{\mathrm{SCE}})$.

The experimental onset of hydrogen evolution at an activated surface by hot electrons at $U_{\mathrm{T}} = -1.8\,\mathrm{V}$ is at about $E_{\mathrm{SCE}} = -0.6\,\mathrm{V}$ (Fig. 25), where $L_{\mathrm{max,rough}} = 0.195\,\mathrm{nm}$, according to Fig. 32 (fat point at $-0.6\,\mathrm{V_{SCE}}$). We obtain the same value of $L_{\mathrm{max,rough}}$ for $U_{\mathrm{T}} = -2.1\,\mathrm{V}$ at $E_{\mathrm{SCE}} = -0.3\,\mathrm{V}$ (see right fat point in Fig. 32), which corresponds with the observed onset of hydrogen evolution at $U_{\mathrm{T}} = -2.1\,\mathrm{V}$ (Fig. 25). Equation (13), valid for the smooth surface, gives $L_{\mathrm{max,smooth}} = 0.144\,\mathrm{nm}$ at $U_{\mathrm{T}} = -1.8\,\mathrm{V}$ and $E_{\mathrm{SCE}} = -0.6\,\mathrm{V}$ (Fig. 32).

The curve for a smooth surface at $U_{\mathrm{T}} = -0.82\,\mathrm{eV}$ is not included in Fig. 32. It is only about 0.05 Å above the decay length L at $U_{\mathrm{T}} = 0\,\mathrm{V}$, the difference between curves 4 and 5 $\in$ Fig. 33.

The role of the roughness can be understood when we assume that $L \approx 0.195\,\mathrm{nm}$ allows the hot electron to reach and dissociate the water molecules

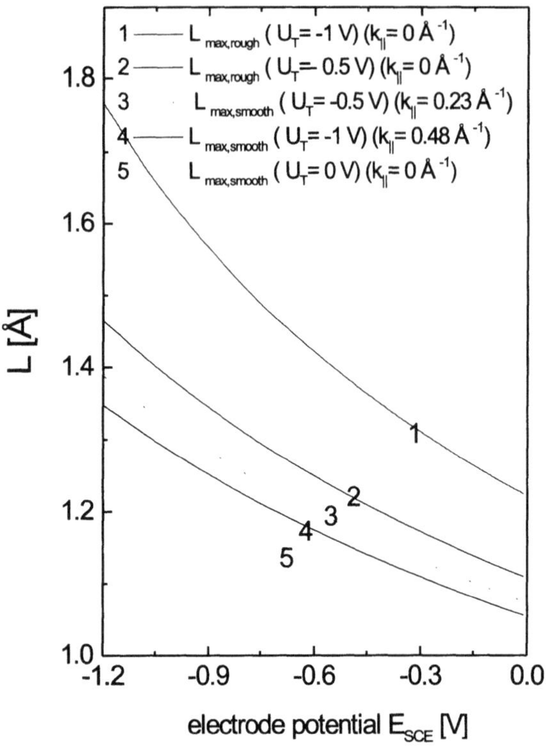

Fig. 33. Maximum decay length L (calculated in the same way as in Fig. 32 for small tunnel voltages) of the electron wavefunction within the electrolyte as a function of electrochemical potential E_{SCE} for smooth and rough silver (111) electrodes according to (13) and (14), respectively, for the indicated tunnel voltages U_T

in the inner Helmholtz layer (with the structure of the double layer at $E_{SCE} \approx -0.5\,V$, whereas the lower value of $L_{max,smooth} = 0.144\,nm$ is not sufficient for this. For $U_T = 0\,V$ "normal, cold electron hydrogen evolution" is observed at $E_{SCE} = -1.2\,V$, see Fig. 25. At this potential both the calculated values of $L_{max,rough}$ and $L_{max,smooth}$ are about 0.124 nm (see Fig. 32, left fat point). We can understand why this relatively low value of L allows hydrogen evolution, if we assume that at this rather cathodic potential the double layer is reoriented and does now allow the transfer of the "cold" electron into the inner Helmholtz layer with subsequent dissociation of water, as discussed above.

Modelling of the metal–electrolyte interface by *Pecina* and *Schmickler* [101] shows a reorientation of the water O–H groups in the inner Helmholtz layer towards the surface by the strong electric fields in the double layer at cathodic potential which may already allow the necessary electron transfer at smaller values of L.

The fat points in Fig. 32 may reflect this potential-dependent reorientation. The approximate agreement of $L_{max,rough}$ and $L_{max,smooth}$ at $E - E_F$

may explain why roughness of the silver electrode reduces the overpotential of the normal hydrogen evolution only very slightly. Our model would yield some less cathodic onset of hydrogen evolution by hot electrons at smooth silver electrodes, which is not indicated in Fig. 25. However, there may be some experimental indication for this process in [71]. To our best knowledge, this is the first model of hydrogen evolution, which not only involves orientational effects in the double layer, but also a potential-dependent penetration of the metal electrons into the inner Helmholtz layer.

5.4.4 Conjectures on Activation

Our hypothesis is based on the elastic scattering of hot electrons by atomic scale roughness of Ag(111) surfaces. This implies that two conditions must be fulfilled in order to observe the hydrogen evolution reaction by hot electrons:

(a) a sufficiently high surface concentration of special atomic scale roughness sites at a surface of prevailing (111) orientation;
(b) a sufficient overlap of bulk and virtual surface states at and near these special sites of atomic scale roughness to allow hot electron transfer.

Hydrogen evolution by hot electrons has been observed using tunnel junctions undercoated with a rough CaF_2 film of 100 nm thickness. By measuring the electrochemical capacity of these samples and a silver (111) single crystal one may define a roughness factor as the ratio of the capacities of the rough and smooth sample, both capacities normalized with respect to the apparent surface area. This roughness factor ranged between 1.2 and 1.7. The layer of adsorbed hydrogen on an epitaxial Ag(111) film was observed after an increase of the surface by 10% by a fast oxidation–reduction cycle (ORC) (for experimental conditions see Sect. 5.2). Surprisingly, the activation of the tunnel junctions on glass substrates needed only the oxidation and subsequent reduction of half a monolayer of silver. The related increase in the surface area is negligible. Therefore we think that the activation is needed to fulfil point (b) above. Maybe the activation reflects the cleaning of active sites from unknown adsorbates which prevent efficient electron transfer. However, activation was not necessary on the CaF_2 roughened samples to which these unknown adsorbates would probably stick as well. Therefore we envision the activation as transformation of already existing linear monoatomic steps into static fuzzy steps with a high density of kink sites, by the partial dissolution of terraces, starting at steps and the subsequent redeposition of the material on the terraces, from which it migrates to the steps but does not diffuse along the steps. On the other hand, a silver film contouring the rough CaF_2 substrate would have many stable kink sites, and therefore would not need activation. Checking these conjectures will need careful scanning tunneling microscopy of the non-activated and of the activated surfaces. Our model reflects the experimental observations only very approximately. This is not surprising given the very simple theoretical assumptions made for a complicated problem. Since we see no other concept yet to explain the observed influence of the activation we will resort to an experimental approach to check the concept of roughness-controlled penetration depths L. Experiments on Ag(111) electrodes with better controlled roughness are under way [116]. We tentatively explain the role of roughness in the observed reoxidation of atomic hydrogen, see Sect. 5.3 in an analogous way. Electrons in virtual bound states, which extend further out, can only hybridize with the bulk states in the presence of special roughness sites where the electron from adsorbed hydrogen can be released to Ag(111) bulk states. This process can be considered as the time-reversed reduction process discussed above. It should be noted that the critical penetration depths L necessary for the Volmer reaction and the oxidation of adsorbed atomic hydrogen are probably different. Without or with little activation, the electron transfer rate to the bulk will be low and the Tafel reaction

$$H_{ad} + H_{ad} \rightarrow H_2 \tag{15}$$

will have the higher reaction rate and will take over.

6 Dependence of the Internal Tunnel Current on the Surface Conditions of the Top Electrode

Usually, electron tunneling experiments between two stratified conductive films involve film thicknesses much larger than the mean free path of the tunneling electrons. In this case, the tunnel current density I_T is determined by the properties of the tunneling barrier and the electronic density of states within the bulk of the two films [99,100]. Outer surfaces of the two films are not involved. Changes have to be expected, when at least one film thickness is decreased and reaches the order of the mean free path of the hot tunneling carriers. Related to this is so-called ballistic electron emission spectroscopy [111,112], where electrons tunneling between a thin metal film on a semiconductor and the tip of a scanning tunneling microscope yield information on the buried interface between the metal and the semiconductor.

In the case of MIM junctions we have two rather clear-cut schemes to influence the tunnel current at constant tunneling voltage by changing conditions at the noble metal interface with an electrolyte or with UHV:

- firstly by allowing for faradaic reactions driven by hot carriers at the electrolyte interface or by emission into the electrolyte or the vacuum;
- secondly by modulating the dephasing of tunneling states, for instance at UHV interfaces by adsorption, without an emission current into the vacuum.

There are examples for dephasing without faradaic reactions both at metal–UHV and metal–electrolyte interfaces. We desribe these results first in Sect. 6.1 and give a model for dephasing at the surface, see Sect. 6.2. As yet we have not found a faradaic reaction without a strong dephasing contribution (Sect. 6.3).

6.1 Controlling the Internal Tunnel Current I_T without Faradaic Reactions

The change of tunnel current at constant tunnel voltage by adsorption of potassium from −0.17 mA to −0.55 mA and by adsorption of oxygen from −0.02 to −0.01 mA (see Fig. 17) was already reported in [27]. These strong changes were not understood at that time. They are clearly not caused by faradaic effects. To elucidate the origin of this effect we performed further experiments with $Al/AlO_x/Ag$ systems in UHV. We have chosen anodic oxide layers, because they show a very weak temperature dependence of the tunnel current from 290 K down to 45 K [106]. When evaporating a 3 nm silver film on top of a 20 nm thick silver top electrode of a MIM junction at $T = 45$ K, the tunnel current at a constant tunnel voltage of for example $U_T = -1$ V increases by a factor of 3.1. This current increase perfectly vanishes when the MIM junction is heated up to room temperature again.

This corresponds to the considerable increase of the electrical DC resistance of a smooth silver film (thickness 26 nm) from 0.69 Ω to 0.87 Ω by cold-deposition of 3 nm of silver at 59 K [113] (Fig. 34). This increase vanishes completely by annealing to room temperature (Fig. 35).

The increase of the DC resistance is caused by diffuse scattering of the electrons at the Fermi level by the unordered structure of the cold-deposited silver. In analogy, the origin of the modification of the tunnel current is a dephasing process at the top electrode surface of the MIM junction by diffuse scattering by atomic scale roughness, see below.

The increase of tunnel current I_T at constant tunnel voltage U_T by roughening at atomic scale, given by the quotient

$$Q = \frac{I_T(\text{rough}, U_T)}{I_T(\text{smooth}, U_T)}$$

depends strongly on the tunnel voltage. We found $Q = 5.7$ for $U_T = -0.4$ V and $Q = 3.1$ for $U_T = -1.0$ V. An explanation of this effect will be given below.

Recently *Kritzler* et al. observed an analogous tunnel current increase due to surface roughening of a $Al/AlO_x/Ag$ junction in a 0.05 M $KClO_4$ water/ethylene glycole electrolyte [107]. They used a small overpotential of 100 mV to bring submonolayer amounts of silver on the top electrode surface of an $Al/AlO_x/Ag$ junction (with initially flat surface). The potential was held at $E_{SCE} = +0.1$ V with tunnel voltage $U_T = \text{const.} = -0.5$ V. At

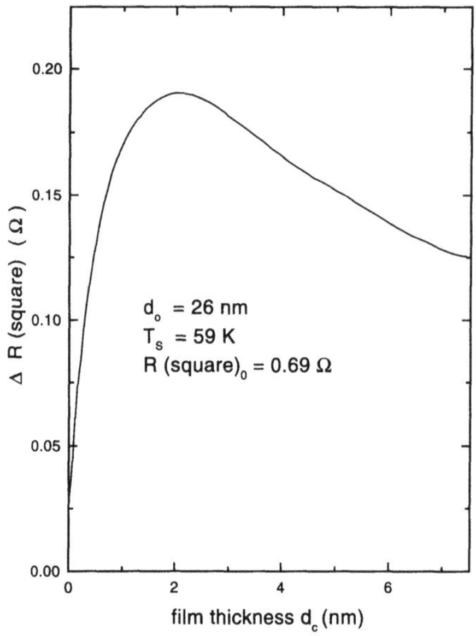

Fig. 34. Change ΔR of the square film resistance (initial value R_0) of a smooth silver film of thickness d_0 during cryocondensation at T_S of silver of additional thickness d_c. From [113]

time $t = 0\,\mathrm{s}$ $AgClO_4$ was added. The resulting concentration was 0.3 mM of $AgClO_4$.

Figure 36 shows that the tunnel current increases drastically, when the silver starts to adsorb on the flat top electrode surface. In the upper part of Fig. 36 the electrolyte current increase caused by the reduction of Ag^+ is shown, the middle part shows the integrated electrolyte current. After about 100 s just one monolayer of Ag was deposited on the surface of the tunnel junction. The lowest part of the figure shows the corresponding tunnel current. It rises within the first 3 s by about 60% although just 5% of a monolayer of Ag has been adsorbed.

At least in the case of cold-deposition of silver on a MIM junction in UHV no faradaic reactions are involved.

First we try to discuss the results described in this subsection by a ballistic model. We will assume a starting rate of J_0 electrons at at the oxide–silver interface and a ballistic elastic current J_{el} at the silver–electrolyte interface at first encounter, an elastic reflection coefficient p_{el} at the silver–electrolyte interface and a coefficient t_{el} of tunneling into Al after elastic return to the silver–oxide inteface. One easily derives:

$$J_{\mathrm{T}} = J_0\left[1 - \left(\frac{J_{\mathrm{el}}}{J_0}\right)^2 p_{\mathrm{el}} t_{\mathrm{el}}\right] .$$

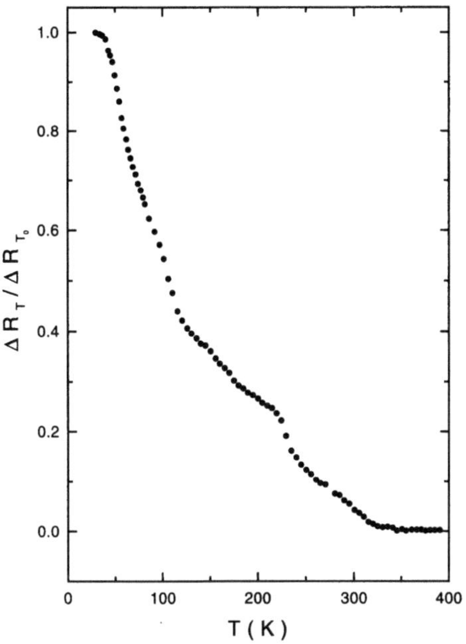

Fig. 35. Resistivity change of a 20 nm thick silver film (evaporated at $T = 293\,\mathrm{K}$), which was covered with 0.02 monolayers of silver at $T = 43\,\mathrm{K}$. Resistivity was monitored during subsequent heating from $T = 43\,\mathrm{K}$ to $T = 250\,\mathrm{K}$. From [114]

If one wishes to explain the observed change of J_T at $U_\mathrm{T} = -2.5\,\mathrm{V}$ (see for instance Fig. 42) by several 100% in terms of a variation of p_el from 1 to 0, one needs to assume $\mathrm{J_{el}/J_O} > 0.9$, even when one assumes $t_\mathrm{el} = 1$. The ratio $\mathrm{J_{el}/J_O} > 0.9$ would correspond to a mean free path of around 150 nm. This result is very unlikely, since the mean free path of the electron–electron interaction is given by *Quinn* by the following equation [109] (see also Fig. 37)

$$\lambda_\mathrm{ee} = \frac{1.45 E_\mathrm{F}^{1.5} E}{\sqrt{\frac{m^*}{m}} \arctan(\beta^{\frac{-1}{2}}) + \frac{\beta^{\frac{1}{2}}}{\beta+1}} \frac{1}{(E - E_\mathrm{F})^2} \tag{16}$$

with $m^*/m = 1.1$ and $\beta = 0.166 r_s$. E and E_F are given in eV and λ_ee in nm.

The mean free path for silver ($E_\mathrm{F} = 5.5\,\mathrm{eV}$, $r_\mathrm{s} = 3.08$) is given as an example in Fig. 37.

Accordingly we have to expect a mean free path of only several nm for $U_\mathrm{T} \approx 2\,\mathrm{V}$. Only for smaller tunnel voltages $U_\mathrm{T} \leq 1\,\mathrm{V}$ do we have mean free paths of above 80 nm. We therefore propose a model of modulating the dephasing of tunneling states by surface effects.

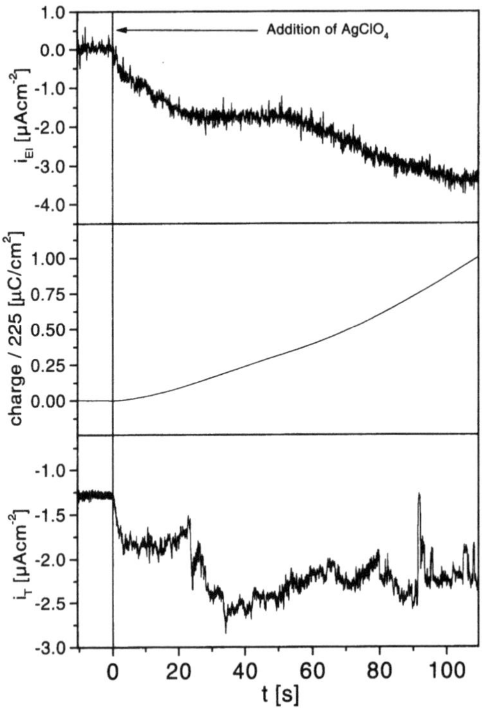

Fig. 36. Deposition of a monolayer of silver on the silver top electrode of an Al/AlOx/Ag junction. *Top graph:* Electrode current versus time at $E_{\mathrm{SCE}} = +0.1\,\mathrm{V}$ after injection of $AgClO_4$ resulting in a concentration of 0.3 mM $AgClO_4$. *Middle graph:* Deposited amount of silver (obtained by integration of curve 1) *Bottom graph:* Tunnel current at constant tunnel voltage of $U_{\mathrm{T}} = \mathrm{const} = -0.5\,\mathrm{V}$

6.2 Model of Dephasing at the Outer Interface of a MIM Junction

Our model is based on the assumption of a coherent energy eigenstate φ_{1-2} of the tunneling electron, which extends over the thin aluminum (metal 1) electrode, the oxide film and the thin silver (metal 2) top electrode. In the limit of small energies above the Fermi level of silver (corresponding to a mean free path much larger than the silver film thickness) this coherent state will have about equal probability density in both metal electrodes. This is different from the usual consideration of the tunneling process between half-spaces (metal 1 and 2) separated by a thin tunnel barrier, with usually a small transmission amplitude from metal 1 to metal 2 for an electron incident upon the tunnel barrier from metal 1. The eigenstates φ_{1-2} are a realistic approximation, when the "delocalization time" of for instance an electron initially localized in the negatively biased metal film 1 and eventually delocalized over metal films 1 and 2 is shorter than the dephasing time at the outer interface of silver (metal 2)

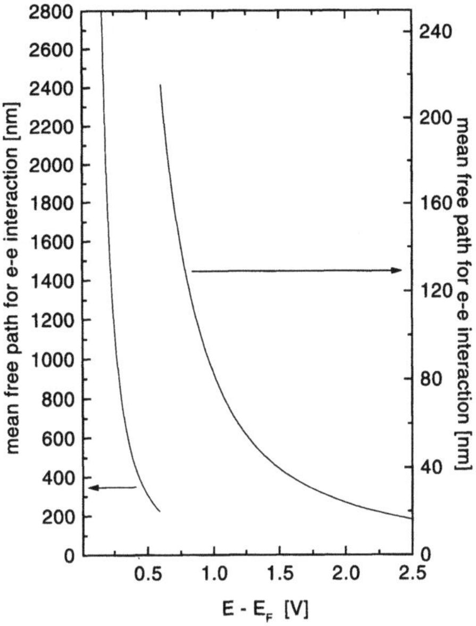

Fig. 37. Mean free path for e–e interaction [according (16)] for silver ($E_F = 5.5\,\text{eV}$, $r_s = 3.08$) plotted in nm as a function of $E - E_F$ from 0.1 eV to 2.0 eV (equivalent to tunnel voltage U_T from -0.6 V to -2.0 eV)

The "delocalization time" will decrease with decreasing film thicknesses of films 1 and 2. The dephasing time by the electron–electron interaction in the bulk will increase with decreasing film thicknesses, and the relative influence of dephasing at the surfaces increases accordingly. The electron in φ_{1-2} becomes localized as a hot electron in the silver (or at positive bias in the aluminum) electrode only by breaking the coherence of the tunneling state by dephasing through elastic or inelastic scattering in the bulk of silver (aluminum) or at the outer silver (aluminum) interface. After localization and energy relaxation a charge of one extra electron contributes to the tunnel current, which is measured in the outer circuit of the MIM contact, see Fig. 9, as a transport current involving only electrons at the Fermi levels of the circuit. At negative U_T the dephasing of electrons in φ_{1-2} with normalization volume

$A\ (d_{\text{Al}} + d_{\text{Ag}})$, where A is the area of the tunnel junction

takes place by Coulomb interaction with electrons in the Fermi sea of silver and by additional processes at the outer silver surface. The bulk dephasing is diminished (with respect to the results in Fig. 37) by a factor R given by:

$$R = \frac{|\varphi_{1-2}(\text{Ag})|^2 d_{\text{Ag}}}{|\varphi_{1-2}(\text{Ag})|^2 d_{\text{Ag}} + |\varphi_{1-2}(\text{Al})|^2 d_{\text{Al}}^2}\,, \tag{17}$$

where $|\varphi_{1-2}(\mathrm{X})|^2$ is the residence probability of the hot electron in the coherent state φ_{1-2} in film X (Al or Ag). For low energies of the state φ_{1-2}, R reaches the asymptotic value of

$$R = \frac{d_{\mathrm{Ag}}}{d_{\mathrm{Al}} + d_{\mathrm{Ag}}} \, .$$

At increasing dephasing at the outer surface R becomes smaller than this limit, because the ratio

$$\frac{|\varphi_{1-2}(Ag)|}{|\varphi_{1-2}(Al)|}$$

decreases with increasing dephasing at the outer surface.

Thus a simple approach to the tunnel current is

$$J_{\mathrm{T}} \propto J_0 \left[\left(1 - \mathrm{e}^{-\frac{R d_{\mathrm{Ag}}}{\lambda_{\mathrm{ee}}}} \right) + W_{\mathrm{S}} \mathrm{e}^{-\frac{R d_{\mathrm{Ag}}}{\lambda_{\mathrm{ee}}}} \right] , \qquad (18)$$

where J_0 is the tunnel current with an infinite thick top electrode. J_0 is given by the transmission coefficient of the tunnel barrier (exact calculations including Fowler–Nordheim tunneling are under way [72]). λ_{ee} is the elastic mean free path obtained from (16). W_{S} is the probability of dephasing at the surface.

This equation yields the usual expectation of $J_{\mathrm{T}} \propto t_{\mathrm{B}}^2$ for $d_{\mathrm{Ag}} \gg \lambda_{\mathrm{ee}}$. This is the case for high tunnel voltages, for example $\lambda_{\mathrm{ee}} = 7.4\,\mathrm{nm}$ for $U_{\mathrm{T}} = -4\,\mathrm{V}$.

For an explanation of the surface influence on the tunnel current we choose the surface factor W_{S} (18) to be 0 and 0.5. For a simplification in the present work we consider the ratio

$$\frac{J_{\mathrm{T}}(W_{\mathrm{S}} = 0.5)}{J_{\mathrm{T}}(W_{\mathrm{S}} = 0.01)} \, ,$$

for which J_0 does not need to be calculated. In Fig. 38 we have plotted this ratio versus $E - E_{\mathrm{F}}$ for $R = 0.1$ and $R = 0.5$.

Figure 38 reveals that the tunnel current becomes more and more "surface determined" at low tunnel voltages ($-eU_{\mathrm{T}} = E - E_{\mathrm{F}}$). This is reasonable, because the mean free path λ_{ee} increases [approximately proportional to $1/(E - E_{\mathrm{F}})^2$ see 16] and consequently bulk dephasing becomes less important at decreasing $(E - E_{\mathrm{F}})$. With increasing W_{s} the value of R will decrease. This adds further to the determination of J_{T} by surface dephasing (the curve for $R = 0.1$ in Fig. 38).

We assign the increase of W_{s} by deposition of unordered silver on smooth silver top electrodes, both in UHV and in the electrolyte to increasing dephasing by elastic scattering in the disordered surface layer.

It is clear that the modulation of the tunnel current I_{T} by variation of the probability W_{s} of dephasing at the surface involves only the primary hot electrons; secondary hot electron states may also dephase at the surface, but they do not contribute to I_{T}.

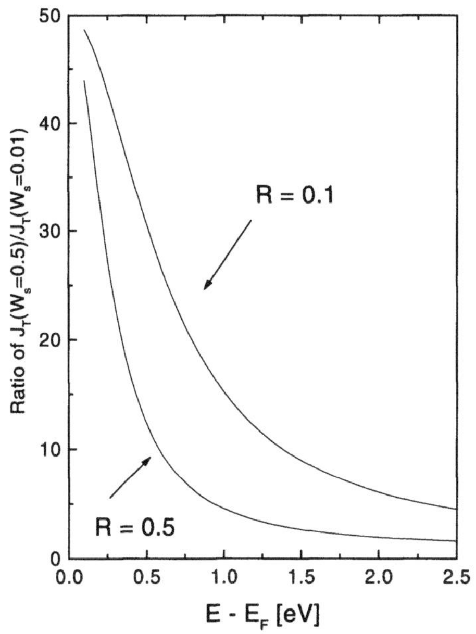

Fig. 38. Ratio of $J_T(W_S = 0.5)/J_T(W_S = 0.01)$ (tunnel current with strong dephasing at the surface/tunnel current with weak dephasing at the surface) plotted versus $E - E_F$ for 2 parameters of R.

In the case of the results described above (Sect. 6.1) the surface dephasing factor W_s is assigned to elastic scattering at the disordered silver surfaces.

6.3 Modulation of Internal Tunnel Current I_T by Faradaic Reactions at the Top Electrode

6.3.1 Electron Emission into the Electrolyte

As a test we use the lower edge of the electronic conduction band in the electrolyte, called the level of the dry electron state (here called the photoemission threshold PE, see also Sect. 5.1) as an electron acceptor.

In Fig. 39 the tunnel current of an Al/AlOx/Au system is plotted as a three-dimensional plot versus the tunnel voltage and the electrode potential.

One can see clearly that the tunnel current I_T increases drastically near $U_T = -2.5\,V$ and $E_{SCE} = -0.9\,V$, where the tunneled electrons from the Al may reach the dry electron state PE in the electrolyte situated at a potential $E_{SCE}(PE)$ given by

$$eE_{SCE}(PE) + eU_T = -3.3\,eV\,.$$

Interestingly the electron emission current into the electrolyte taken under the same conditions ($U_T = -2.7\,V$, $E_{SCE} = -1.0\,V$) increases absolutely only by about 1.5 µmA/ cm^2 (Fig. 40).

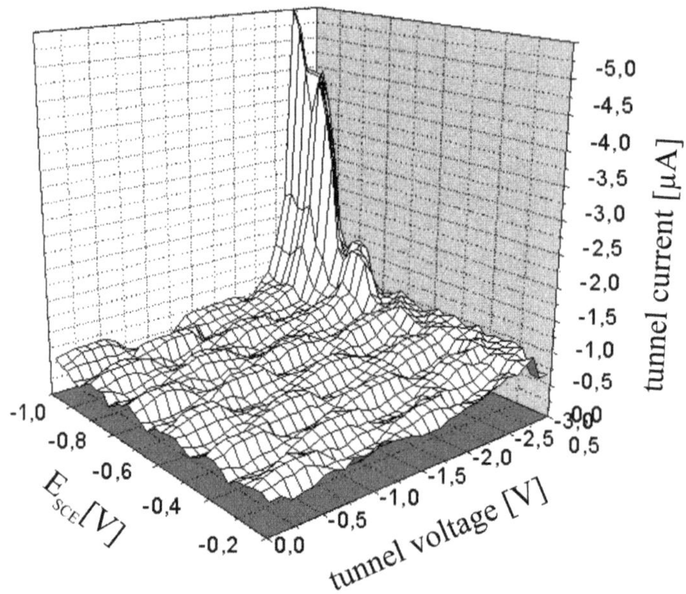

Fig. 39. Three-dimensional plot for the tunnel current of a smooth $Al/AlO_x/Au$ (anodic oxide , $d = 3.2\,\mathrm{nm}$) tunnel junction versus the tunnel voltage and the electrode potential. Sample area $A = 0.1\,\mathrm{cm}^2$

This behavior of the tunnel current and the electrode current allows two statements:

- The tunnel current (current from the Al base electrode to the Au top electrode in this case) can be clearly modulated by the electrode potential which is applied at the gold surface.
- Only a fraction (in the present case $\approx 30\%$) of the tunnel current from the Al through the oxide and to the Au top electrode continues into the dry electron states. (This "yield" is much higher than in Fig. 23. In Fig. 23 thinner oxide with 2.5 nm thickness was used causing a much higher tunnel current (several mA). In that case scavanger reactions which follow the process of hot electron injection into the PE may become determinant.)

Therefore, the modulation of I_{T} is assigned partly to dephasing by elastic scattering of the coherent state φ_{1-2} (compared to the crystal) by the disordered Helmholtz layer. When the energy of φ_{1-2} reaches the energy of level PE (see Fig. 10) this state will extend into the Helmholtz layer up to and beyond the average position of the cores of the water molecules in the relatively disorderd Helmholtz layer (Fig. 43).

Note that we do not need to involve any atomic scale roughness of the silver–electrolyte interface for our explanation.

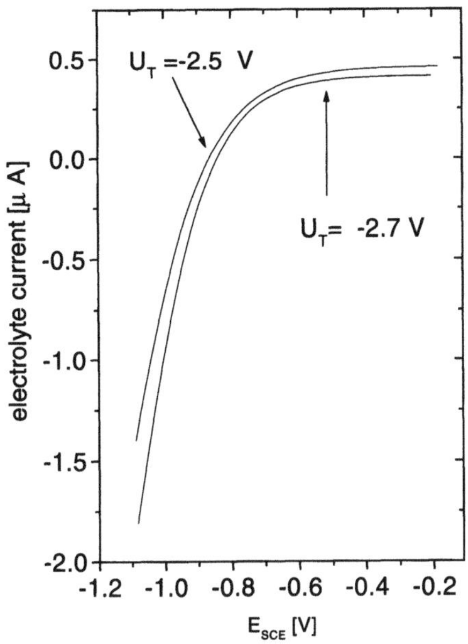

Fig. 40. Electrolyte current versus electrode potential for two different tunnel voltages, taken with the same tunnel junction as in Fig. 39. Sample area $A = 0.1\,\mathrm{cm}^2$

6.3.2 Dependence of the Tunnel Current on the Potential for Activated Surfaces

As mentioned, see Sect. 5.2, the hydrogen evolution reaction by hot electrons needs a weak activation. In this case, also the tunnel current at constant tunnel voltage increases considerably at the onset of the hydrogen evolution reaction (HER). First we plot tunnel current versus tunnel voltage of an Al/AlO_x/silver system with an activated silver top electrode in a buffered acetate electrolyte at electrode potentials $E_{SCE} = -0.8\,V$, $E_{SCE} = -1.0\,V$, $E_{SCE} = -1.2\,V$ (Fig. 41).

There is a strong dependence on the electrode potential E_{SCE}. The dependence of the tunnel current on the electrode potential at constant tunnel voltage is shown in Fig. 42.

There is almost no hysteresis. Further there is no dependence on the scan velocity from 20/ mV up to 500/ mV. (This means that the tunnel current at constant tunnel voltage is clearly influenced by the electrode potential and not by accumulated reaction products of the HER reaction.) Since the increase of the tunnel current is observed in the range of hydrogen evolution by hot eletrons (see Sect. 5), we have to discuss the assignment of the modulation of I_T to this faradaic reaction. Here, the comparison of I_T and I_{el} as function of E is very helpful.

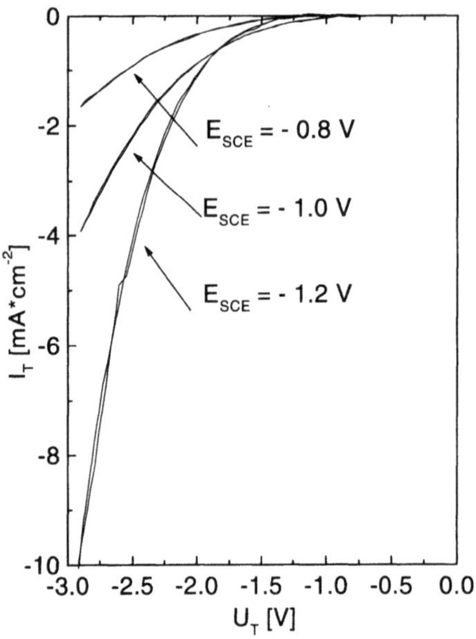

Fig. 41. Cyclic plots of tunnel current I_T versus tunnel voltage U_T of an $Al/AlO_x/Ag$ junction with activated silver top electrode in 0.9 M acetate buffer electrolyte (50% water, 50% ethylene glycole) cyclic voltammograms taken at different electrode potentials E_{SCE}

In Fig. 42 a conventional cyclovoltagramm is taken, while the tunnel voltage is set to $U_T = -2.4$ V. Additionally we plotted also the tunnel current in the same scale. The tunnel current increase due to the change of electrode potential of about 6 mA/ cm^2 exceeds the electrolyte current due to faradaic reactions of about 1.4 mA/ cm^2 by about 400% (Fig. 42).

Therefore we have to assign at least the bigger part of the modulation of I_T to a non-faradaic process. We propose like before dephasing by Atomic Scale Roughness (ASR), but in this case it is increasing with cathodic potential. This increase is caused by the lowering of the level PE and the according further penetration of the hot electron into the electrolyte with increasing cathodic potential, as schematically shown in the upper part of Fig. 43 for a smooth surface. This is analogous to the curves for a smooth Ag(111) electrode and finite negative U_T in Fig. 32. The atomic scale roughness is characterized in the lower part of Fig. 43 by a silver adatom on an Ag(111) surface. The absolute probability amplitude of the coherent state φ_{1-2} at the adatom site will increase at increasing cathodic potential and accordingly a higher probability of scattering by ASR and an increase of the factor W_S in (18) is expected.

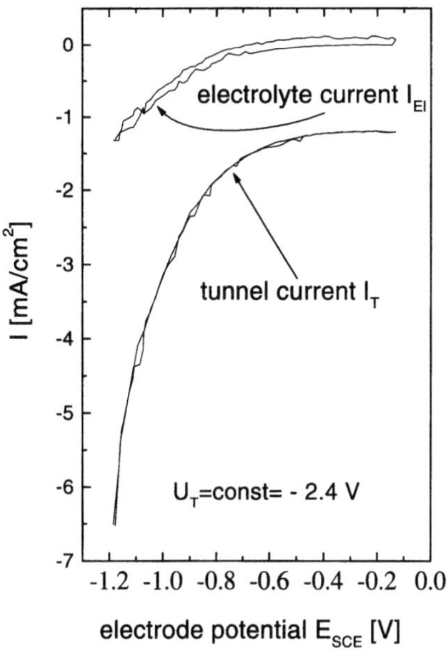

Fig. 42. Tunnel current at constant tunnel voltage of $U_{\mathrm{T}} = -2.5\,\mathrm{V}$ as a function of the electrode potential in acetate buffer electrolyte (50% water, 50% ethylene glycole) $\mathrm{d}E/\mathrm{d}t = 50\,\mathrm{mV\,s}$. Note that the scans in the negative and positive direction are very similar

6.3.3 Modulation of Tunnel Current by Surface Charging on Flat Surfaces

In Fig. 44, the differential conductivity of a MIM junction is plotted versus the electrode potential in a 0.1 M $KClO_4$ electrolyte at a constant tunnel voltage of $U_{\mathrm{T}} = -1\,\mathrm{V}$. The potential of zero charge (pzc) is determined by capacity measurements and marked in Fig. 44. The differential conductivity increases, when the silver electrode surface is charged negatively. In contrast to the measurement in Fig. 42 no faradaic reaction starts on flat surfaces by hot electrons in the perchlorate electrolyte. The voltammograms remain unchanged when the tunnel voltage up to $-2\,\mathrm{V}$ is switched on or off (see voltammogram c in Fig. 25).

To ensure that the effect shown in Fig. 44 is not caused by slow kinetics (e.g. charging effects during the poteniostatic sweep or long-term ordering processes at the electrode surface), we performed the measurement again keeping the electrode potential constant at $E = -1.2\,\mathrm{V}$ for 5 s and 50 s, respectively (Fig. 45). The increased differential conductivity due to negative potentials remains constant while keeping the electrode potential at $E = -1.2\,\mathrm{V}$.

This excludes slow kinetic processes as explanation for the increased conductivity.

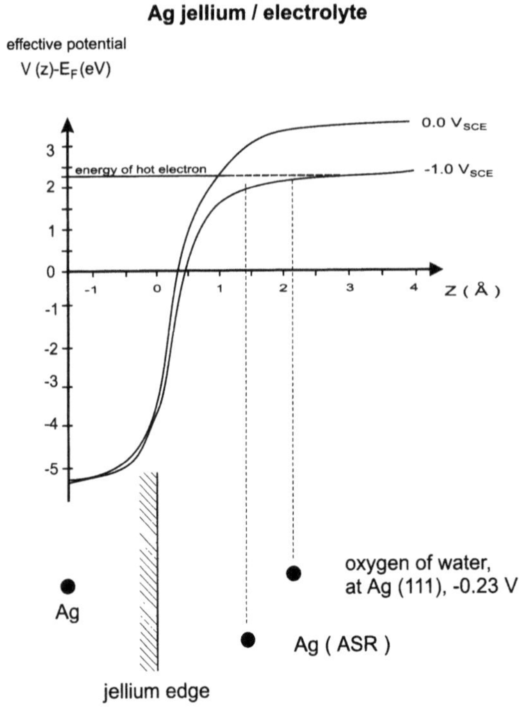

Fig. 43. *Upper*: Schematic representation of the effective potential $V(\mathrm{z})$ for electrons at the silver-jellium/electrolyte interface for $E_{\mathrm{SCE}} = 0.0\,\mathrm{V}, -1.0\,\mathrm{V}$, as a function of the distance z from the jellium edge. $V(\mathrm{z})$ has been drawn in qualitative agreement with the effective potential for jellium/UHV surfaces in [115]. $V(\mathrm{z})$ at large z corresponds to the level PE in Fig. 10. *Lower*: Position of the cores of silver atoms at the Ag(111) surface, where the absolute distance from the jellium edge is half the nearest-neighbor distance in bulk silver. The left ($z < 0$) core typifies a smooth (111) terrace, the right core ($z > 0$) an adatom on the terrace as an example of Atomic Scale Roughness (ASR). The distance between the cores of the Ag surface atoms and the oxygen cores of the inner water layer in the Helmhotz layer at cathodic potential at smooth Ag(11) has been taken from X-ray scattering experiments [97]

Positive surface charges (electrode potential $E > -0.8\,\mathrm{V} \approx \mathrm{pzc}$) do not change the differential conductivity at negative tunnel voltages (Fig. 44).

This behavior is reversed when a positive tunnel voltage is applied and hot holes are injected into the silver electrode at positive tunnel voltages U_{T} (Fig. 46). Now the differential conductivity remains unchanged when the electrode is charged negatively, whereas the conductivity increases when the electrode is charged positively.

For the small energies of the hot electrons (1 eV in Fig. 44) the decay length $L_{\mathrm{max,smooth}}$ in Fig. 33 is relatively small and dephasing by scattering from water is unlikely (Fig. 43). For hot holes (energy $E = +0.5\,\mathrm{eV}$

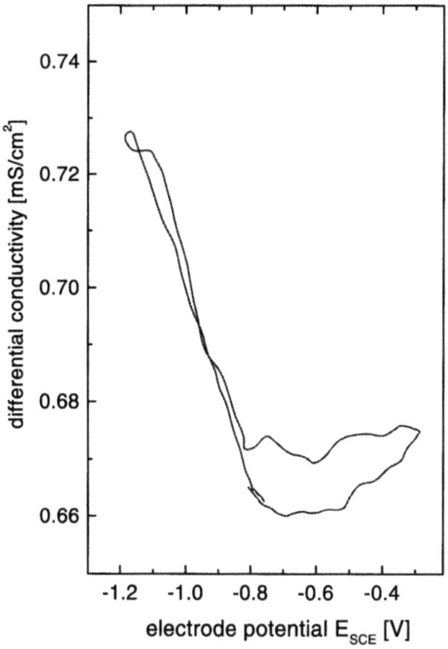

Fig. 44. Differential tunnel conductivity (electrons injected to the silver top electrode) of a MIM junction (smooth surface) as a function of electrode potential in pure aqueous 0.1 M $KClO_4$ electrolyte $U_T = \text{const.} = -1\,\text{V}$

in Fig. 46), which have the decay length $L_{\text{max,smooth}}$ of electrons with energy $E_F - E$ (see Fig. 47, right part, curve 3) this is even more unlikely. Therefore, we tentatively discuss the increased electron–electron scattering in the electron gas tail at the interface, which mainly involves electrons at and just below the Fermi energy, as an alternative mechanism to modulate W_s. Because of the low electron density in the tail there is more phase space available for scattering by the electrons. The further the tail extends (given by the value of $L_{\text{max,smooth}}$ at $E - E_F = 0$ in Fig. 33) and the further the hot electrons penetrate into this area (given by the value of $L_{\text{max,smooth}}$ at $U_T = -1\,\text{V}$ and $E = E_F + 1\,\text{eV}$, respectively), the higher is the scattering rate and W_s. Figure 47 (left part and below) demonstrates that the difference of $L_{\text{max,smooth}}$ of electrons with energies $E_F + 1\,\text{eV}$ and E_F does increase with cathodic potential, and so also will W_s. We see some relations to the change of the tunnel current in Fig. 17. The increase, respectively decrease of I_T by adsorption of potassium, repectively, oxygen is most likely caused by the increase, respectively, decrease of the extension of the electron density tail at the surface [117]. The tentative argument of increased dephasing of hot holes involves increased Auger decay in the tail of the electron distribution by the increase of phase space. The upper part of Fig. 47 demonstrates that the hot holes approach the zone of the electron tail with increasing anodic potential, thus yielding an increasing W_s. This would qualitatively explain the result in Fig. 46.

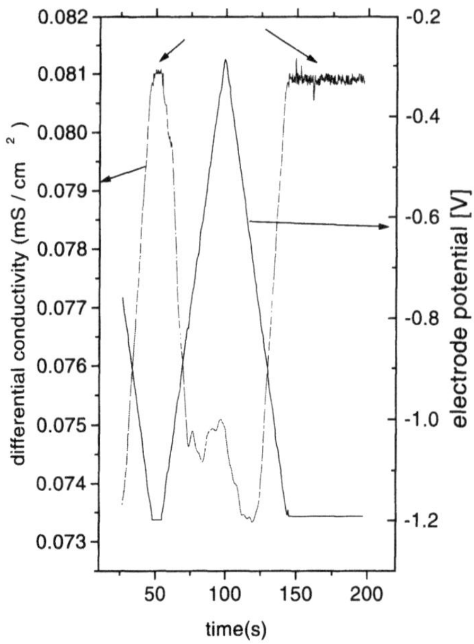

Fig. 45. Differential tunnel conductivity (electrons injected to the silver top electrode) of a MIM junction (smooth surface, anodic aluminum oxide $d = 3.2\,\mathrm{nm}$) as a function of time in pure aqueous 0.1 $MKClO_4$ electrolyte at $U_T = \text{const.} = -1\,\mathrm{V}$ during a potentiostatic sweep from $-0.2\,\mathrm{V}$ to $-1.2\,\mathrm{V}$. The electrode potential of $E = -1.2\,\mathrm{V}$ was kept constant for 5 s and 50 s, respectively, yielding also constant differential conductivity, see *upper arrows* in the figure

7 Sensorics with MIM Junctions

MIMs have the unique property of probing a metal–vacuum, –gas or –electrolyte interface "from the metal side" (from "inside"), via changes of the tunnel current in the following ways:

- The internal current of hot carriers released at the surface contributes to the tunnel current, for instance in non-adiabatic surface reactions, e.g. the oxidation of an adsorbed layer of hydrogen, see Sect. 5.3 and [56].
- The tunnel current is changed by surface disorder at the silver/UHV [106] and silver/electrolyte interfaces (see Fig. 36), by the electrode potential at silver electrolyte interfaces (Sect. 6.3.2) and maybe also by molecular adsorption, according to some preliminary results. Both points demonstrate the potential of MIM junctions for chemical and electrochemical sensing and maybe also, indirectly, for sensing of disorder caused by friction.

First promising results were obtained, when 0.1 ml methanol were injected through a capillary within about 0.5 s into 0.05 $MKClO_4$ (50% water, 50% ethylene glycole) in front of the silver electrode of a MIM junction. As demonstrated in Fig. 48, the tunnel current increases by a factor of four and regains

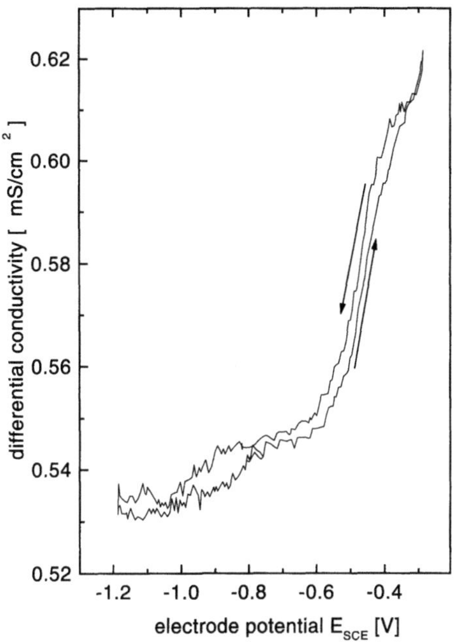

Fig. 46. Differential tunnel conductivity (holes injected to the silver top electrode) of an MIM junction (smooth surface) as a function of electrode potential in pure aqueous 0.1 M $KClO_4$ electrolyte U_T = const. = +0.5 V

the original value after 1.0 s, whereas the potential of the silver electrode at open circuit (no electrochemical current) changes only by 20 mV.

This may be due to:

(a) the electrosorption of methanol;
(b) the increased disorder of the double layer by friction caused by the liquid flow during the injection;
(c) transient adsorption and subsequent complete desorption of methanol.

To distinguish between the different possibilities the following experiments are underway:

- injection of the same electrolyte to discriminate between (b) and (a) or (c);
- injection of permanently adsorbing species to discriminate between (c) and (a) or (b);
- injection of non-adsorbing species to exclude (a) and (c).

8 Outlook

Metal–Insulator–Metal (MIM) tunneling junctions have several unique features. They allow for "surface photochemistry without photons", because

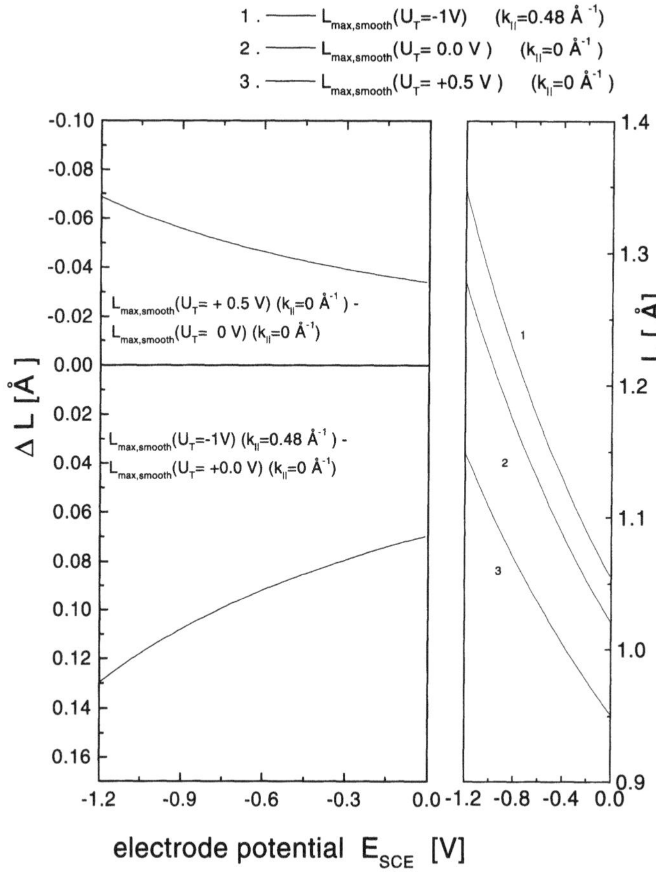

Fig. 47. *Right side:* The decay length L of the electron wavefunctions into the electrolyte as a function of the silver (111) electrode potential for electrons with energies E with $E - E_F = 1\,\text{eV}$ (curve 1), $E - E_F = 0\,\text{eV}$ (curve 2), $E - E_F = -0.5\,\text{eV}$ (curve 3) according to (13). Curve 3 is also the decay length of the wave functions of holes with energy 0.5 eV. *Left side upper part:* Difference ΔL of the decay lengths of wave functions of holes of 0.5 eV energy and of electrons at the Fermi level (note that ΔL is negative) as function of the Ag(111) electrode potential. *Left side, lower part:* ΔL for hot electrons, 1 eV above E_F and electrons at E_F.

the intermediate hot electrons in photochemistry are created by tunneling into the top electrode. Though MIM junctions are slow devices, they can deliver "ab initio" information on the energetic distribution of the electronic excitations involved in the initial coherent state and discriminate between surface and bulk excitations. Because hot electrons emitted in surface reactions can be monitored, MIM junctions can strongly contribute to the study of non-adiabatic surface reactions. This field is not easily accessible to other methods. We also see a practical aspect in sensorics, as the tunnel current

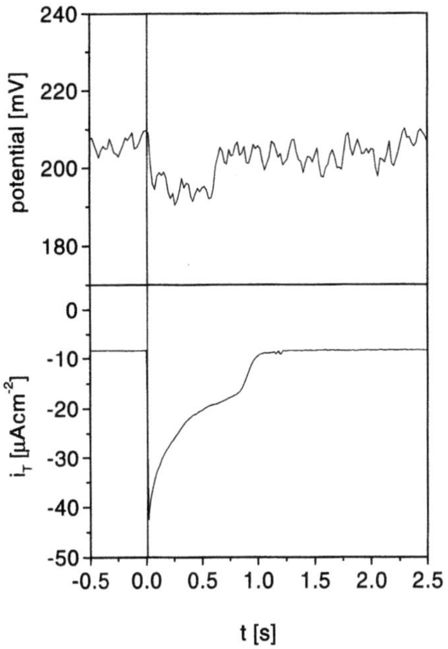

Fig. 48. Tunnel current at tunnel voltage of $U_T = -1.8\,\mathrm{V}$ and floating electrode potential (measured with open circuit) during injection of 0.1 ml methanol for 0.5 s into 0.05 M $KClO_4$ (50% water, 50% ethylene glycole)

depends on conditions at the outer surface of the top electrode of the MIM junction. They may act simultaneously both as sensor and transducer.

All these statements, for which we have given examples in this chapter, apply to interfaces with electrolytes or gases or ultrahigh vacuum. Future applications may for instance involve hot electron induced desorption, genuine optical surface absorption, non-linear optical phenomena, state and time resolved surface reactions, surface interactions of metastable species and maybe some not yet anticipated uses in sensorics.

Acknowledgements

We thank Jürgen Noffke (TU Clausthal-Zellerfeld) for the data in Figs. 29–31. The first part of this work was supported by the MWF of Northrhine Westfalia under contract number IV A5 103 00891.

References

1. D. Diesing, H. Janssen, A. Otto: Surf. Sci. **331–333**, 289 (1995)
2. A. Liebsch: *Electronic Excitations at Metal Surfaces* (Plenum, New York 1998)

3. W. Steinmann, Th. Fauster: in H. L. Dai, W. Ho (Eds.): *Laser Spectroscopy and Photochemistry on Metal Surfaces* (World Scientific, Singapore 1995), p. 764
4. R. M. Osgood, X. Wang: Solid State Phys. **51**, 2 (1998)
5. M. Bauer, S. Pawlik, M. Aeschlimann: Phys. Rev. B **55**, 10040 (1997)
6. T. Hertel, E. Knoesel, E. Hasselbrink, M. Wolf, G. Ertl: Surf. Sci. **317**, L1147 (1994)
7. T. E. Furtak, K. L. Kliewer: Comments Solid State Phys. **10**, 103 (1982)
8. F. Feulner, D. Menzel: in H.-L. Dai, W. Ho (Eds.): *Laser Spectroscopy and Photochemistry on Metal Surfaces* (World Scientific, Singapore 1995) Chap. 16
9. D. Wetzig, M. Rutkowski, W. Etterich, R. David, H. Zacharias: Surf. Sci. **402–404**, 232 (1998)
10. E. Hasselbrink: Appl. Surf. Sci. **79/80**, 34–40 (1994)
11. W. Ho: in H.-L. Dai, W. Ho (Eds.): *Laser Spectroscopy and Photochemistry on Metal Surfaces* (World Scientific, Singapore 1995) Chap. 24
12. X. L. Zhou, J. M. White: in H.-L. Dai, W. Ho (Eds.): *Laser Spectroscopy and Photochemistry on Metal Surfaces* (World Scientific, Singapore 1995) Chapt. 25
13. F. M. Zimmermann, W. Ho: in M. Anpo: *Surface Photochemistry* (Wiley, New York 1996)
14. F. Decker, B. Scrosati (Eds.): New Trends in Photoelectrochemistry, Electrochim. Acta **38**, (1993)
15. R. Kostecki, J. Augustynski: J. Appl. Electrochem. **23**, 567 (1993)
16. A. Goldmann: in E. Bertel, M. Donath (Eds.): *Electronic Surface and Interface States on Metallic Systems* (World Scientific, Singapore 1995) p. 35
17. G. L. Richmond, J. M. Robinson, V. L. Shannon: Prog. Surf. Sci. **28**, (1988)
18. B. Pettinger, C. Bilger, J. Lipkowski, W. Schmickler: in A. Wieckowski (Ed.): *Interfacial Electrochemistry* (Marcel Dekker, New York 1998) Chapt. 22
19. U. Höfer: Appl. Phys. B. **68**, 383 (1999)
20. A. M. Funtikov, S. K. Sigalaev, V. E. Kazarinov: J. Electroanal. Chem. **228**, 197 (1987)
21. A. Bruckbauer, A. Otto: J. Raman Spectrosc. **29**, 665 (1998)
22. J. A. Misewich, T. F. Heinz, D. M. Newns: Phys. Rev. Lett. **68**, 3737 (1962)
23. L. Bartels, G. Meyer, K. H. Rieder, D. Velic, E. Knoesel, A. Hotzel, M. Wolf, G. Ertl: Phys. Rev. Lett. **80**, 2004 (1998)
24. S. Schatteburg, D. Diesing, A. Otto: Appl. Phys. **B70**, 573 (2000)
25. P. K. Hansma: *Tunneling Spectroscopy* (Plenum Press, New York 1982)
26. C. B. Duke: *Tunneling in Solids* (Academic, New York 1969)
27. M. Hänisch, A. Otto: J. Phys. Cond. Matter **6**, 9659 (1994)
28. K. W. Shepard: J. Appl. Phys. **36**, 796 (1965)
29. B. Korneffel, W. Ludwig: Phys. Stat. Sol.(a) **8**, 149 (1971)
30. K. H. Gundlach, J. Kadlec: Thin Solid Films **28**, 107 (1975)
31. Z. Burshtein, J. Levinson: Phys. Rev. **B12**, 3453 (1975)
32. J. Kadlec, K. H. Gundlach: Phys. Stat. Sol.(a) **37**, 11 (1976)
33. R. K. Jain, M. G. Farrier, T. K. Gustafson: Phys. Rev. Lett. **36**, 435 (1976)
34. R. K. Jain, C. W. Slayman, M. G. Farrier, T. K. Gustafson: J. Appl. Phys. **48**, 1543 (1977)
35. J. M. P. Guedes, C. W. Slayman, T. K. Gustafson: IEEE J. Quant. Electron. **15**, 475 (1979)

36. N. Kroo, Z. Szentirmay, J. Felszerfalvi: Phys. Lett. **A101**, 235 (1984)
37. C. A. Mead: J. Appl. Phys. **32**, 646 (1961)
38. J. Cohen: J. Appl. Phys. **33**, 1999 (1962)
39. E. D. Savoye, D. E. Anderson: J. Appl. Phys. **38**, 3245 (1967)
40. R. Hrach, Czech: J. Phys. **B 18**, 1501 (1968)
41. G. Dittmer: Thin Solid Films **9**, 141 (1972)
42. Z. Hubicka, R. Hrach: Czech. J. Phys. **B 25**, 193 (1975)
43. J. Drucker, P. K. Hansma: Phys. Rev. **B 30**, 4348 (1984)
44. R. Hrach: Int. J. Electron. **73**, 841 (1992)
45. T. Kusonoki, M. Suzuki, S. Sasaki, T. Yaguchi, T. Aida: Jpn. J. Appl. Phys. **32**, L1695 (1993)
46. H. Adachi: J. Vac. Sci. Technol. B, 2093 (1996)
47. G. Dearnaley, D. V. Morgan, A. M. Stoneham: J. Non-Cryst. Solids **4**, 593 (1979)
48. A. K. Ray, C. A. Hogarth: Int. J. Electronics **57**, 1 (1984)
49. R. G. Sharpe, R. E. Palmer: J. Appl. Phys. **79**, 8565 (1996); Thin Solid Films **288**, 164 (1996)
50. J. R. Kirtley, T. N. Theis, J. C. Tsang, D. J. DiMaria: Phys. Rev. **B 27**, 4601 (1983)
51. R. G. Sharpe, S. Dixon-Warren, P. J. Durston, R. E. Palmer: Chem. Phys. Lett. **234**, 354 (1995)
52. J. W. Gadzuk: Phys. Rev. Lett. **76**, 4234 (1996)
53. J. W. Gadzuk: J. Vac. Sci. Technol. A **15**, 1520 (1997)
54. B. C. Lai, J. Y. Lee: J. Electrochem. Soc. **146**, 266 (1999)
55. Y. E. Sung, A. J. Bard: J. Phys. Chem. **B 102**, 9806 (1998)
56. G. Kritzler: PhD Thesis, University of Düsseldorf (2001)
57. R. H. Fowler, L. Nordheim: Proc. Roy. Soc. **A119**, 173 (1928)
58. A. G. Borisov, A. K. Kazansky, J. P. Gauyacq: Phys. Rev. **B 59**, 10935 (1999)
59. A. Otto, D. Diesing, H. Janssen, M. Hänisch, M. M. Lohrengel, S. Rüße, A. Schaak, S. Schatteburg, D. Körwer, G. Kritzler, H. Winkes: in M. W. Roberts (Ed.): *Interfacial Science*, A "Chemistry of the 21th Century" Monograph, (IUPAC 1997) p. 163
60. P. Guyot-Sionnest, A. Tadjeddine, A. Liebsch: Phys. Rev. Lett. **64**, 1678 (1990)
61. A. Liebsch, G. V. Benemanskaya, M. N. Lapushkin: Surf. Sci. **302**, 303 (1994)
62. A. Liebsch: Appl. Phys. **B 68**, 301 (1999)
63. A. Liebsch: private communication
64. A. Otto, B. Reihl: Phys. Rev. **B 41**, 9752 (1990)
65. W. Wittke, A. Hatta, A. Otto: Appl. Phys. **A48**, 289 (1989)
66. A. Otto: Z. Phys. **216**, 398 (1968)
67. E. Kretschmann, H. Raether: Z. Naturforsch. **23a**, 2135 (1968)
68. D. Diesing, A. W. Hassel, M. M. Lohrengel: Thin Solid Films **342**, 283 (1999)
69. M. Guedes, K. Slayman, G. Jain: IEEE J. **QE-15**, 475 (1979)
70. J. W. Diggle, T. C. Downie, C. W. Goulding: Chem. Rev. **69**, 365 (1969)
71. D. Diesing: PhD Thesis, University of Düsseldorf (1996)
72. D. Diesing, A. W. Hassel: unpublished results
73. M. Hänisch: PhD Thesis, University of Düsseldorf (1994)
74. H. Janssen: PhD Thesis, University of Düsseldorf (1997)
75. S. Schatteburg: PhD Thesis, University of Düsseldorf (1998)

76. M. Arnold, P. Bussemer, K. Hehl, H. Grabhon, A. Otto: J. Mod. Opt. **39**, 2329 (1992)
77. D. Körwer: PhD Thesis, University of Düsseldorf (1993)
78. U. Kreibig, M. Vollmer: *Optical Properties of Metal Clusters* (Springer, Berlin, Heidelberg 1995)
79. C. K. Sun, F. Vallee, L. H. Acioli, E. P. Ippen, J. G. Fujimoto: Phys. Rev. **B50**, 15337 (1994)
80. J. Hohlfeld, U. Conrad, J. G. Müller, S. S. Wellershoff, E. Mathias: in K. H. Bennemann (Ed.): *Nonlinear Optics in Metals* (Clarendon, New York 1998) p. 219
81. W. Ho: Condens. Matter Phys. **13**, 293 (1988)
82. P. J. Feibelman, M. L. Knotek: Phys. Rev. **B 18**, 531 (1978)
83. D. Pines: Phys. Rev. **92**, 626 (1953)
84. M. M. Lohrengel: Mater. Sci. Engin. **R11**, 243 (1991)
85. D. Diesing, G. Kritzler, A. Otto: unpublished results
86. D. Diesing, G. Kritzler, A. Otto: Ber. Bunsenges. Phys. Chem. **101**, 762 (1997)
87. A. Staib, D. Borgis: J. Chem. Phys. **103**, 2642 (1995)
88. D. Diesing, S. Rüße, A. Otto, M. M. Lohrengel: Ber. Bunsenges. Phys. Chem. **99**, 1402 (1995)
89. J. G. Simmons: J. Appl. Phys. **34**, 9 (1963)
90. D. Diesing, H. Winkes, A. Otto: Phys. Stat. Sol. (a) **159**, 243–254 (1997)
91. J. K. Sass, H. Gerischer: in B. Feuerbacher (Ed.): *Photoemission and the Electronic Properties of Surfaces* (Wiley, New York 1978) Chap. 16
92. Y. V. Pleskov, Z. A. Rothenberg: J. Electroanal. Chem. **20**, 1 (1969)
93. Y. V. Gurevich, Y. V. Pleskov, Z. A. Rothenberg: *Photoelectrochemistry* (Consultants Bureau 1980)
94. E. Meyer: Untersuchungen zum photoelektrochemischen Verhalten von Metallelektroden, Dissertation Technische Universität München (1973)
95. Y. V. Pleskov, Z. A. Rothenberg: J. Electroanal. Chem. **20**, 1 (1969)
96. D. Körwer: in W. Kiefer et al. (Eds.): *Photoemission and Enhanced Raman Scattering from Silver Electrodes* Int. Conf. on Raman Spectroscopy 1992 (Wiley, New York 1992)
97. M. E. Toney, J. N. Howard, J. Richter, G. L. Borges, J. G. Gordon, O. R. Melroy, D. G. Wiesler, D. Lee, L. B. Sorensen: Surf. Sci. **335**, 326 (1995)
98. D. Diesing, G. Kritzler, M. Stermann, D. Nolting, A. Otto: submitted to J. Solid State Electrochem.
99. J. Bardeen: Phys. Rev. Lett. **6**, 57 (1961)
100. B. D. Josephson: Adv. Phys. **163**, 341 (1967)
101. B. Pecina, W. Schmickler: J. Electroanal. Chem. **431**, 47 (1997)
102. P. A. Thiel, T. E. Madey: Surf. Sci. Rep. **7**, 211 (1987)
103. D. Diesing, G. Kritzler, A. Otto: in A. Kornyshev, M. Tosi, J. Ulstrup (Eds.): *Electron and Ion Transport in Condensed Media* (World Scientific, Singapore 1997) p. 257
104. H. Eckhardt, L. Fritsche, J. Noffke: J. Phys. **F 14**, 97 (1984)
105. J. Noffke: private communication
106. D. Diesing, A. Otto: unpublished results
107. G. Kritzler, D. Diesing, A. Otto: unpublished results
108. H. Busmann, S. Ewert, W. Sander, K. Seibert, P. Balk, A. Steffen: Z. Phys. B **59**, 439 (1985)

109. J. J. Quinn: Phys. Rev. **126**, 1453 (1962)
110. A. Otto, D. Diesing: J. Phys. Chem. B **103**, 2696 (1999)
111. L. Kaiser, W. Bell: Phys. Rev. Lett. **60**, 1406 (1988)
112. M. Prietsch: Phys. Rep. **253**, 163 (1995)
113. H. Grabhorn, A. Otto, D. Schumacher, B. N. J. Persson: Surf. Sci. **254**, 327 (1992)
114. A. Tonscheidt: Diploma Thesis, University of Düsseldorf (1990)
115. P. J. Jennings, R. O. Jones, M. Weinert: Phys. Rev. **B 37**, 6113 (1988)
116. R. Tölle: PhD Thesis, University of Düsseldorf (2002)
117. A. Otto, D. Diesing, S. Schatteburg, H. Janssen: Phys. Stat. Sol. A **175**, 297 (1999)

Low-Energy Negative Recoil Spectroscopy Principles and Applications

D. John O'Connor

Departement of Physics, University of Newcastle,
New South Wales 2308, Australia
John.OConnor@newcastle.edu.au

Abstract. Although low-energy ion scattering has received considerable attention in the past, less work has been focussed on the allied low energy negative recoil spectrometry. This means that there does not exist serious guidance on the optimum experimental arrangements needed to analyze materials. It is the goal of this chapter to introduce and explain the physical concepts behind this recoil technique and to provide guidance on the optimal experimental parameters.

1 Introduction

The techniques associated with low-energy ion scattering (LEIS) have been well characterized as extremely surface sensitive analysis tools [1,2,3]. Using low energy (0.1–10 keV) inert gas and alkali ions it is possible to probe the composition and structure of the outermost atomic layers of a solid. In the scattering process the projectile is scattered by surface atoms in elastic collisions best described as "billiard ball collisions". The strength of this technique is also its greatest weakness and it lies in the charge exchange processes which occur when the projectile is in the vicinity of the surface. To summarize the principal aspects of these processes, when an inert gas ion is incident upon a surface then generally most projectiles end up in a neutral charge state leaving only a small fraction (0.5–5 %) to be detected as ions (positive ions). The charge fraction is trajectory dependent which means that in measurements some particular trajectories will be favored. If alkali ions are used as projectiles then a larger fraction remains as ions and that fraction is usually independent of the trajectory of the particle. This trajectory and particle dependence of charge exchange in LEIS allows the user to tailor the projectile to the particular application under consideration.

The sensitivity of this approach is limited by the existence of a background signal which is large for alkali ion projectiles. Although it is considerably less for inert gas ion projectiles, it still sets a lower detection limit for contaminants on the surface. In a valuable variation of this technique the recoiling target atoms are detected [4] rather than scattered particles to enhance the direct structural information about the adsorption sites of species. The specificity and sensitivity of this technique can be enhanced if one uses inert gas ions as projectiles, and detects only recoiling negative species.

K. Wandelt, S. Thurgate (Eds.): Solid–Liquid Interfaces, Topics Appl. Phys. **85**, 427–433 (2003)

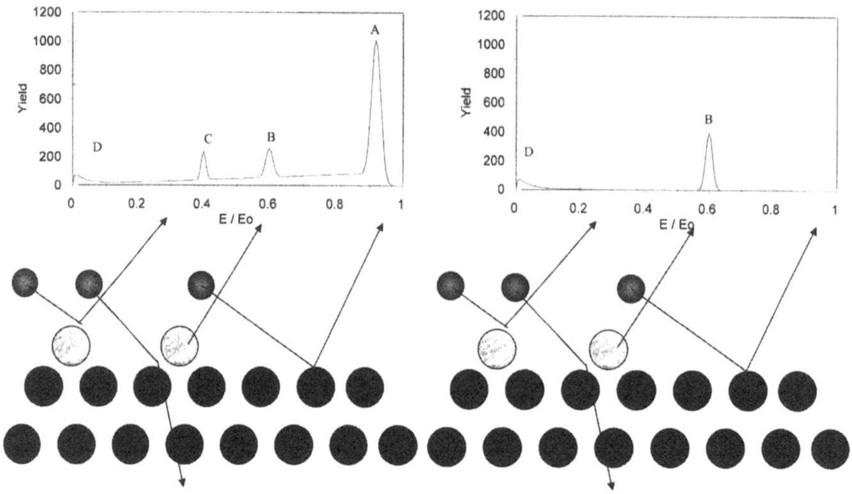

Fig. 1. Schematic comparison of the energy spectra of different ion species scattered off a surface. The energy spectra for scattered positive inert gas ions reveals a number of peaks from scattered and recoil particles as well as a sputtering peak. In (**a**) the peaks A, B and C are "signatures" for each element on the surface. B and C are on a (slightly exaggerated) background to highlight the effect of background on detection limit. In (**b**) only negative ions are detected which enhance the yield from electronegative species. Thus the signal and the background from the scattered inert gas ions is suppressed and a more prominent signal from the impurity is observed

In this way the experiment is optimized to detect the concentration and location of low levels of electronegative species on the surface without a background formed by the projectiles.

This approach has been successfully used on the following systems:

- H on Ni(110)–located adsorption site
- O on Ni(110)–located adsorption site
- O on Al(111)–proved no subsurface oxygen
- O on Mg(0001)–detected 10 % of adsorbed oxygen below the surface layer.

2 Application

Before addressing the basics of this process, consider an application which demonstrates the usefulness of this technique. The oxidation of Al(111) and Mg(0001) have been studied by a range of spectroscopic techniques. The issue to be determined is whether, during the initial oxidation phase, there is any oxygen residing between the first and second atomic layers of the surface. In both cases there are a range of possible subsurface sites which are outlined for the case of the Al(111) surface in Fig. 2. Each site has its own signature in

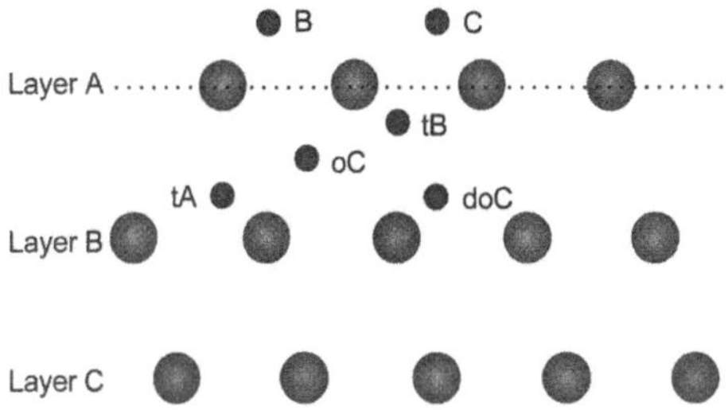

Fig. 2. The proposed oxygen adsorption sites for Al(111). There are two above surface sites (B and C) as well as four subsurface sites (tA, tB, oC and doC)

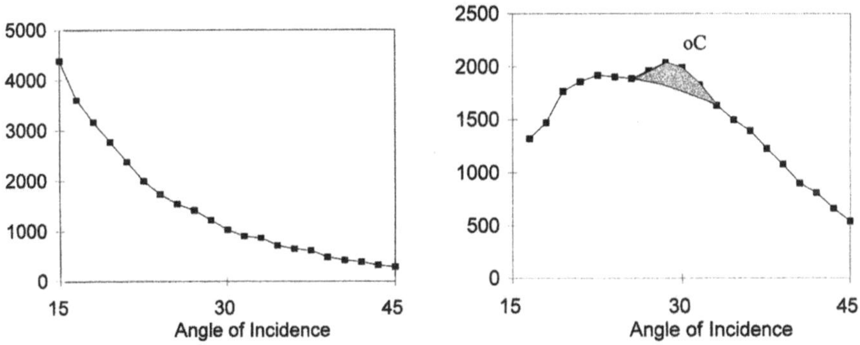

Fig. 3. The measured O^- recoil ion yield as a function of incidence angle for an oxygen exposed Al(111) (**a**) and a Mg(0001) (**b**). The lack of any fine structure in the case of Al(111) is evidence that there are no subsurface oxygen adsorption sites. The small additional peak at 30° incidence in (**b**) is indicative of oxygen sitting in the oC site in Mg(0001)

a polar scan of negative ion yield. From the identity of a peak at a particular angle it is possible to identify which sites are populated, and from the relative yields it is possible to estimate the fraction of subsurface oxygen.

Very different results have been obtained for these surfaces. In the case of Al(111) no oxygen has been observed between the first and second atomic layers in the initial phase of oxidation [5,6,7] and this is confirmed by the lack of special peaks in the polar scan (Fig. 3a). From this evidence it is possible to state that to within the detection limit no oxygen lies between the first and second atomic layer for Al(111) whereas the concentration is about 10 % for Mg(0001) [8].

3 Principles of Recoil Detection

3.1 Cross-Section

The energy of recoiling particles, E_{R}, relative to the incident energy E_0 is dependent on the recoil angle, ϕ, and the ratio, μ, of the target atom mass

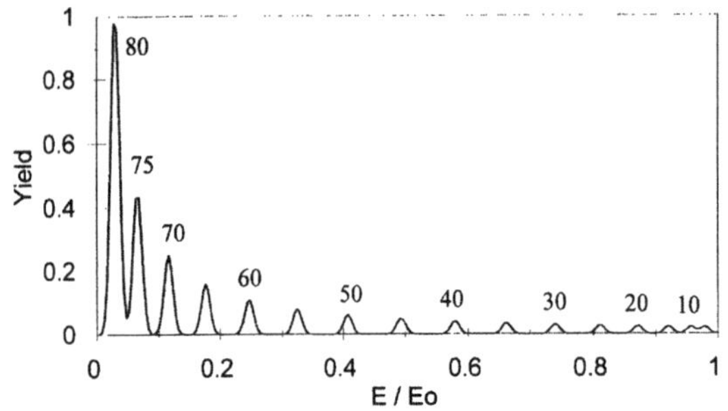

Fig. 4. The energy and yield for oxygen recoiled by 10 keV Ne. As the recoil angle increases from 5° to 80° the recoil energy decreases while the recoil yield increases

to the projectile mass:

$$E_R/E_0 = 4\mu \cos 2\phi/(1+\mu)2 .$$

It is evident from this expression that the maximum recoil energy is less than the incident energy and as the recoil angle approaches 90°, the recoil energy approaches zero.

The fundamental equation governing the recoil particle yield is

$$Y = N_i N_a \left(\frac{d\sigma}{d\Omega}\right)_{\text{recoil}} \Delta\Omega \tag{1}$$

where Y is the recoil yield, N_i is the number of incident particles, N_a is the number of target atoms per unit area, $(d\sigma/d\Omega)$ is the recoil cross-section and $\Delta\Omega$ is the collection solid angle of the detector. There is no simple expression for the recoil cross-section, however in general the cross-section will increase for:

- increasing atomic number of the target and projectile atomic numbers;
- decreasing projectile energy;
- increasing recoil angle and it diverges for $\phi = 90°$.

While it is possible to show with graphs how the recoil yield and energy depends on the experimental parameters, Fig. 5 perhaps shows it most concisely. Clearly there is a conflict here in that to maximize the recoil yield, a recoil angle close to 90° is preferable but at the same time the recoil yield will overlap with low-energy particles produced by the sputtering process. To avoid this overlap all conditions considered in this analysis will be for energies which are sufficiently high to avoid the sputtering contribution.

If the assumption is made that the measured recoil energy is set at 500 eV (well above typical sputtered particle energies), the yield as a function of incident energy and projectile (Fig. 5) reveals that to maximize yield (and hence sensitivity) one should chose a krypton projectile.

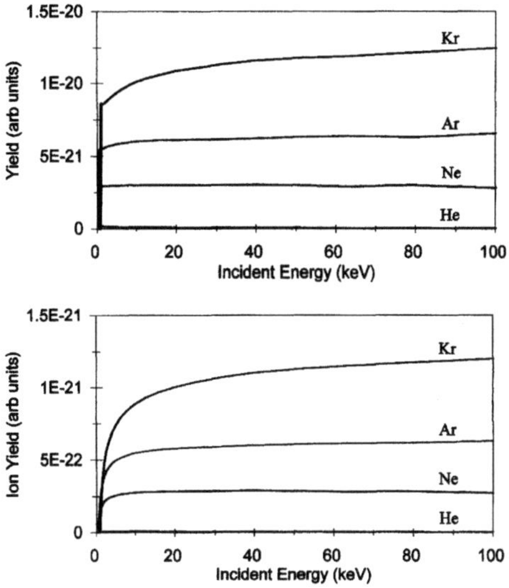

Fig. 5. (**a**) Comparison of recoil particle yield (all charge states) as a function of projectile and energy for oxygen recoil at a recoil energy of 500 eV (**b**) Comparison of the negative ion recoil yield as a function of projectile and energy for oxygen recoil at a recoil energy of 500 eV. Note the order of magnitude reduction in yield and minimal change to the shapes of curves

3.2 Charge Exchange

Consider now the role of charge exchange in the production of negative ions:

$$Y^{+0-} = N_\mathrm{i} N_\mathrm{a} \left(\frac{\mathrm{d}\sigma}{\mathrm{d}\Omega}\right)_{\mathrm{recoil}} P^{+0-} \Delta\Omega\,. \tag{2}$$

The negative affinity level of some ions overlaps strongly with the occupied states of the solid when a particle is in close vicinity of a surface. If the particle is slowly removed from the surface region then there is sufficient time for the electron to undergo a transition to unoccupied states in the substrate and no particles would escape in a charged state. As the ion's exit velocity increases, there is less time for this loss to occur so there is an increasing chance that the escaping particle will remain in the negative charged state. A complete range of measurements do not exist for the charge fraction as a function of energy and exit angle for any projectile substrate combinations. To overcome this lack of information Esaulov's charge fraction measurements [9] have been parametrized to allow the extrapolation of his findings to energies and angles not previously considered. The effect of this charge exchange to the yield dependence is not dramatic as the comparison in Fig. 5a,b demonstrates. The implication one may immediately draw from the reduction of an order of magnitude when negative ions are detected is that one is less sensitive to the presence of electronegative species. This is not the case as the use of negative

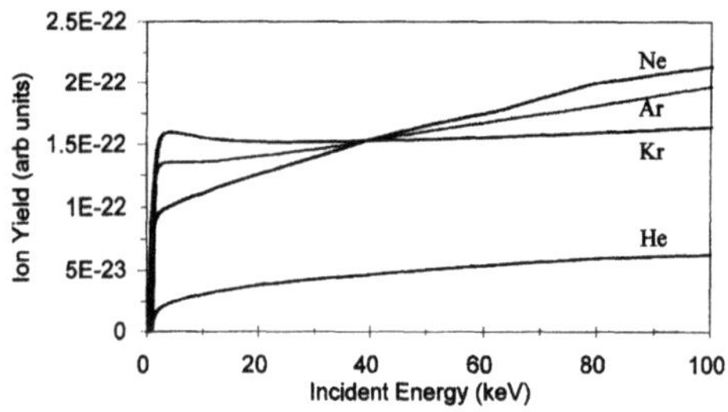

Fig. 6. The recoil ion yield per sputtered atom. The heavier projectiles also have a larger sputtering yield and hence result in more damage. This plot provides a prediction of the yield per projectile per unit damage

ions results in a dramatic reduction in background signal which overall leads to a significant improvement in detection sensitivity.

3.3 Sputtering of Sample

It is insufficient to consider ion yield in the absence of other effects that can be part of the ion bombardment of surfaces. When an energetic ion collides with a surface it distributes its energy to the atoms of the solid and they in turn pass it on to other atoms. Some atoms near or at the surface find they have sufficient energy to escape the solid interface and this process is known as sputtering. The effectiveness of sputtering is gauged by the average number of target atoms freed by this process per incident ion and this ratio, S, is referred to as the sputtering yield.

The effect of sputtering is that while a measurement is being made, the composition of the surface is changing. It is important to minimize the negative effect of sputtering so the critical figure of merit is not simply ion yield per projectile but the ion yield per projectile per sputtered atom. If we take

$$\text{Number of sputtered atoms} = SN_{\mathrm{i}}\,,$$

in this analysis the sputter coefficient for different projectiles and energies were taken from an algorithm provided by *Zalm* [10].

Hence the yield per unit damage is given by:

$$\frac{Y^{+0-}}{(SN_{\mathrm{i}})} = N_{\mathrm{a}}\left(\frac{d\sigma}{d\Omega}\right)_{\mathrm{recoil}} P^{+0-}\Delta\Omega/S\,.$$

The effect of including the sputtering yield is a great leveller. For those ions with a large scattering cross-section also have a large recoil cross-section. The outcome of this additional effect is to bring all projectiles but He together in such a way that, given the uncertainties in most terms used in this analysis, the remaining projectiles are all equally suited to the recoil analysis of surfaces using low-energy negative ions.

4 Conclusions

This is the first systematic and quantitative study of the principal terms and effects associated with low-energy negative recoil spectroscopy. While there have been numerous applications of this technique, the choice of projectile and energy have involved a limited range of effects. The choice of projectile and energy for situations where there is no charge exchange, and no consideration is given to the damage caused by the ion beam, is for the highest energy krypton as a projectile.

A more thorough analysis which incorporates the role of charge exchange and more importantly the sputtering damage caused by the ion beam, brings the conclusion to one in which the optimal projectile is less clear-cut. While helium can be confidently ruled out, the other three projectiles over a wide range of energies are very nearly equivalent.

Under such circumstances, the experimentalist is provided with a tool which can be optimized for other scattering properties while providing the optimum negative ion yield.

References

1. H. Niehus, W. Heiland, E. Taglauer: Surf. Sci. Rep. **17**, 213 (1993)
2. C. Kim, C. Hoefner, V. Bykov, J. W. Rabelais: Nucl. Instr. Methods Phys. Res. B **125**, 315 (1997)
3. D. J. O'Connor: Low Energy Ion Scattering, in D. J. O'Connor, B. Sexton, R. St. C. Smart (Eds.): *Surface Analysis Methods in Materials Science* (Springer Verlag, Berlin Heidelberg 1992)
4. P. J. Schneider, W. Eckstein, H. Verbeek: Nucl. Instr. Methods Phys. Res. B **28**, 713 (1983)
5. D. J. O'Connor, A. W. Denier van der Gon, E. R. Wouters, J. F. van der Veen, P. M. Zagwijn, J. W. N. Frenken: Surf. Sci. **296**, 131 (1993)
6. E. R. Wouters, D. J. O'Connor, J. F. van der Veen, P. M. Zagwijn, J. Vrijmoeth, W. Slijkerman, J. W. N. Frenken: Surf. Sci. **296**, 141 (1993)
7. B. C. Mitrovic, D. J. O'Connor: Surf. Sci. **405**, 261 (1998)
8. B. C. Mitrovic, D. J. O'Connor: Surf. Rev. Lett. **5**, 599 (1998)
9. M. Maazouz, L. Guillemot, T. Schlatholter, S. Ustaze, V. A. Esaulov: Nucl. Instrum. Methods Phys. Res. B **125**, 283 (1997)
10. P. C. Zalm: J. Appl Phys. **54**, 2660 (1983)

Index

Topics in Applied Physics

70 **Dye Lasers: 25 Years**
By M. Stuke (Ed.) 1992. 151 figs. XVI, 247 pages

71 **The Monte Carlo Method in Condensed Matter Physics**
By K. Binder (Ed.) 2nd edn. 1995. 83 figs. XX, 418 pages

72 **Glassy Metals III**
Amorphization Techniques, Catalysis, Electronic and Ionic Structure
By H. Beck and H.-J. Güntherodt (Eds.) 1994. 145 figs. XI, 259 pages

73 **Hydrogen in Metals III**
Properties and Applications
By H. Wipf (Ed.) 1997. 117 figs. XV, 348 pages

74 **Millimeter and Submillimeter Wave Spectroscopy of Solids**
By G. Grüner (Ed.) 1998. 173 figs. XI, 286 pages

75 **Light Scattering in Solids VII**
Christal-Field and Magnetic Excitations
By M. Cardona and G. Güntherodt (Eds.) 1999. 96 figs. X, 310 pages

76 **Light Scattering in Solids VIII**
C60, Semiconductor Surfaces, Coherent Phonons
By M. Cardona and G. Güntherodt (Eds.) 1999. 86 figs. XII, 228 pages

77 **Photomechanics**
By P. K. Rastogi (Ed.) 2000, 314 Figs. XVI, 472 pages

78 **High-Power Diode Lasers**
By R. Diehl (Ed.) 2000, 260 Figs. XIV, 416 pages

79 **Frequency Measurement and Control**
Advanced Techniques and Future Trends
By A. N. Luiten (Ed.) 2001, 169 Figs. XIV, 394 pages

80 **Carbon Nanotubes**
Synthesis, Structure, Properties, and Applications
By M. S. Dresselhaus, G. Dresselhaus, Ph. Avouris (Eds.) 2001, 235 Figs. XVI, 448 pages

81 **Near-Field Optics and Surface Plasmon Polaritons**
By S. Kawata (Ed.) 2001, 136 Figs. X, 210 pages

82 **Optical Properties of Nanostructured Random Media**
By Vladimir M. Shalaev (Ed.) 2002, 185 Figs. XIV, 450 pages

83 **Spin Dynamics in Confined Magnetic Structures I**
By B. Hillebrands and K. Ounadjela (Eds.) 2002, 166 Figs. XVI, 336 pages

84 **Imaing of Complex Media with Acoustic and Seismic Waves**
By M. Fink, W. A. Kuperman, J.-P. Montagner, A. Tourin (Eds.) 2002, 162 Figs. XII, 336 pages

85 **Solid–Liquid Interfaces**
Macroscopic Phenomena – Microscopic Understanding
By K. Wandelt, S. Thurgate (Eds.) 2003, 228 Figs. XVIII, 444 pages

86 **Infrared Holography for Optical Communications**
Techniques, Materials, and Devices
By P. Boffi, D. Piccinin, M. C. Ubaldi (Eds.) 2003, 90 Figs. XII, 182 pages

MIX
Papier aus verantwortungsvollen Quellen
Paper from responsible sources
FSC® C105338

If you have any concerns about our products,
you can contact us on
ProductSafety@springernature.com

In case Publisher is established outside the EU,
the EU authorized representative is:
Springer Nature Customer Service Center GmbH
Europaplatz 3, 69115 Heidelberg, Germany

Printed by Libri Plureos GmbH
in Hamburg, Germany